Electromechanics

PRINCIPLES, CONCEPTS, AND DEVICES

SECOND EDITION

JAMES H. HARTER
Mesa Community College

Prentice Hall

Upper Saddle River, New Jersey
Columbus, Ohio

Library of Congress Cataloging-in-Publication Data

Harter, James H.
 Electromechanics : principles, concepts, and devices / James H. Harter.—2nd ed.
  p. cm.
 Includes index.
 ISBN 0-13-097744-6
 1. Electromechanical devices. 2. Electrodynamics I. Title.

TK2000 .H34 2003
621.31'042—dc21

             2002025233

Editor in Chief: Stephen Helba
Publisher: Charles Stewart
Production Editor: Christine M. Buckendahl
Design Coordinator: Diane Ernsberger
Cover Designer: Mark Shumaker
Cover art: Mark Shumaker
Production Manager: Matt Ottenweller
Marketing Manager: Mark Marsden

This book was set in Times Ten Roman by The Clarinda Company, and was printed and bound by R. R. Donnelley & Sons Company. The cover was printed by The Lehigh Press, Inc.

Pearson Education Ltd.
Pearson Education Australia Pty. Limited
Pearson Education Singapore Pte. Ltd.
Pearson Education North Asia Ltd.
Pearson Education Canada, Ltd.
Pearson Educación de Mexico, S. A. de C. V.
Pearson Education—Japan
Pearson Education Malaysia Pte. Ltd.
Pearson Education, *Upper Saddle River, New Jersey*

10 9 8 7 6 5 4 3 2 1
ISBN: 0-13-097744-6

About the Cover

ABOUT MEMS

Imagine a machine so small that it is imperceptible to the human eye. Imagine working machines with gears no bigger than a grain of pollen being batch fabricated tens of thousands at a time, at a cost of only a few cents each. Imagine a realm where the world of design is turned upside down, and the seemingly impossible suddenly becomes easy—a place where gravity and inertia are no longer important, but the effects of atomic forces and surface science dominate. Welcome to the microdomain, a world now occupied by an explosive new technology known as MEMS (Micro-ElectroMechanical Systems) or, more simply, micromachines.

AN EXAMPLE DEVICE

The micromachine pictured on the cover of this book is that of a system used to position a hinged *silicon mirror*, reflecting laser light to activate a sensor and other circuitry when the mirror is precisely elevated. The torque delivered by the *microengine* is amplified by the gears in the transmission of the *linear rack gear reduction*

drive. The increased torque pushes the linear rack that positions the hinged silicon pop-up mirror.

The linear rack gear reduction drive, pictured here, consists of a double-reduction gear transmission driven by the microengine's 19-tooth pinion gear. The pinion gear, measuring 84 μm/.0033 in (outside diameter), is about the size of the diameter of a human hair. The gear tooth numbers, in order of torque transmission starting with the drive gear (microengine pinion gear), are 19, 57, 19, and 61, respectively, resulting in a torque amplifying gear ratio of 9.6:1. The pressure angle is a standard 20°; the module value is a microscopic 4 μm; and the tooth profile is involute. The gear train, in combination with the rack, exhibits a 2 μm/.00008 in backlash.

FABRICATION METHODS

The Sandia Ultra-planar, Multi-level MEMS Technology (SUMMiT™) fabrication process, employing conventional IC processing tools, was used to produce the hinged pop-up silicon mirror, prime mover (comb drive and microengine), and drive system (linear rack gear reduction drive). This four-level polycrystalline silicon surface micromachining process utilizes one ground plane/electrical interconnect level and three mechanical levels. Between each of these levels are sacrificial oxide layers. An additional friction-reducing layer of silicon nitride is placed between the layers to form bearing surfaces.

Devices are created using the SUMMiT™ fabrication process by alternately depositing a film, photolithographically patterning the film, and then performing chemical etching. By repeating this process with layers of sacrificial silicon dioxide and polycrystalline silicon, complex movable three-dimensional shapes can be formed. The shapes themselves result from the fabrication process in conjunction with a series of two-dimensional masks that define the patterns to be etched. At the end of the fabrication process, the silicon dioxide is chemically removed, leaving behind the mechanical structures comprised of polycrystalline silicon. These systems, batch fabricated with no piece part assembly required, are electrostatically actuated using the on-chip microengine.

The designs enabled by the four-level SUMMiT™ planarized surface micromachining process are definitely complex, however, there is further benefit in providing another layer of mechanical polysilicon. With the Sandia Ultra-planar, Multi-level MEMS Technology for Five levels (SUMMiT V™) fabrication process, a five-level polycrystalline silicon surface micromachining process incorporating one ground plane/electrical interconnect and four mechanical layers, designers are now able to create more advanced systems on moveable platforms, incorporating taller devices (up to 12 μm/.00047 in) with greater stiffness and mechanical robustness. With the additional height provided by SUMMiT V™, additional force from comb drive actuators is possible. To learn more about the SUMMiT™ and SUMMiT V™ processes and to view micromachines in action, go to www.mems.sandia.gov.

Courtesy of Sandia National Laboratories, Advanced Concepts in MEMS and Novel Si Science and Technology Dept., SUMMiT™ Technologies, www.mems.sandia.gov

Preface

Electromechanics: Principles, Concepts, and Devices, Second Edition, has been designed to give the reader an understanding of a broad segment of technology dealing with the interrelationship of electrical and mechanical machine elements and their underlying principles of operation. This book is intended for today's equipment technician, maintenance mechanic, electrician, or manufacturing technician who is responsible (in conjunction with others) for all the *know-how* beyond the initial design of a machine. This publication supports the technical staff's daily activities by providing an introduction to principles, concepts, devices, and applications related to operating, installing, troubleshooting, and servicing electromechanical systems.

The book works equally well as a self-paced study guide for employed technicians and maintenance personnel who are working independently to upgrade themselves or as a text for a course in electromechanics. The graded chapters progress from the concepts and principles of mechanics, electricity, and magnetics to the applications for electromechanical machine elements, finishing with a chapter introducing the concepts of automatic control systems.

This text has numerous figures and examples designed to help both teacher and learner. Problem-solving techniques, which are emphasized throughout the text, are presented in a conversational tone. Coupled with problem solving is the helpful technique of *dimensional analysis*, which is used to determine units for computed quantities. The use of practical design and replacement types of problems throughout the text is a strong motivator for the learner.

Electromechanics: Principles, Concepts, and Devices, Second Edition, assumes no previous mechanical training or understanding of mechanical units of measurement since it begins with a review of number notation, systems of measurement, and conversion between and within measuring systems. It continues to the principles of linear motion (Newton's laws, time, distance, speed, acceleration, etc.) and then progresses through a series of simple machines and their principles, ending with several applications dealing with lubrication and bearings. Once this series of topics is completed, the study of mechanics continues with the concepts and principles of rotary motion and its application to the transmission of power. Mechanical applications include the study of various radial and axial power-transmission machine elements (such as couplings, gearing, and belt and chain drives), oscillatory-motion mechanisms, and intermittent-motion mechanisms. The mechanical section concludes with the study of motion characteristics—constant velocity, constant acceleration, and jerk.

The remainder of the book deals with principles of electrical and magnetic circuits and devices, power, work (energy), applications of transformers, overcurrent devices, relays, contactors, starters, and solenoids. The text concludes with the study of sequential process control, ladder diagrams, and motors, both ac and dc as well as adjustable frequency ac drives, stepper motors, and an introduction to automatic control systems.

This text uses both the British Engineering System, or BES (English system), and the International System of Units, or SI (metric system), in the study of mechanics (distance, speed, acceleration, work, and power). Both systems are still studied, even today, because many parts suppliers list replacement parts, tools, and equipment in English units, whereas new equipment and products are increasingly specified in metric units. As in the past, technicians who work with mechanical equipment as well as electrical and electronic equipment must be educated in both the SI and BES systems.

The author wishes to acknowledge and thank Roger Harlow, for his steadfast support for this project from its inception through its completion; Rosalia Cahill, for her diligence in providing the solutions to the exercises and end-of-chapter questions and problems; Roger Scheunemann, for his preparation of the glossary; Tom Harter, for his creative talents in drawing the first-draft illustrations; and John Bown, for his assistance in obtaining vendor literature. Last—but certainly not least—a caring recognition of my wife, Janet, for her support during the preparation of the manuscript, for her diligence in reading the manuscript for spelling and grammar, and for her numerous suggestions for improving the readability of the manuscript.

ACKNOWLEDGMENTS

The author thanks the following reviewers, whose valued thoughts and suggestions have helped shape the book: Stephen J. Gold, South Dakota State University; Ronald Gonzales, Brigham Young University; Roger Harlow, Mesa Community College, Mesa AZ; Charles Klement, Mesa Community College; Steve Menhart, University of Arkansas at Little Rock; John Palmel, Central Arizona Community College; John A. Palmer, Jr., Central Arizona College; Richard K. Sturtevant, Springfield Technical Community College; Lloyd Temes, College of Staten Island; and Robert L. Towers, Eastern Kentucky University.

The author recognizes, with gratitude, the following companies, who supplied state-of-the-art industrial materials and photographs for inclusion in this text:

Allen Bradley Company, Inc., Milwaukee, WI

Bussmann, Cooper Industries, Inc., St. Louis, MO

Lucas Automation and Control Engineering, Inc., Vandalia, OH

Merkle-Korff Industries, Inc., Des Plaines, IL

National Fire Protection Association, Quincy, MA

Rockwell Automation, Inc., Milwaukee, WI

Square D Company, Palatine, IL

Superior Electric, Bristol, CT

TECO American Inc., Houston, TX

The L. S. Starrett Company, Athol, MA

The Torrington Company, Torrington, CT

Thomson Industries, Inc., Port Washington, NY

Westinghouse Electric Corporation, Pittsburgh, PA

The author takes this opportunity to acknowledge and thank the following individuals at Prentice Hall for their diligence in making the publication of this text possible: Charles Stewart, administrative editor; Christine Buckendahl, production editor; and Carol Mohr, copyeditor.

Contents

Introduction

The Introduction outlines the scope and structure of the text. Also, it offers assistance in selecting a calculator if you don't presently have one or in helping you match the one you have to the requirements of the text. Additionally, it gives some study hints and some general information to aid you in your preparation for the study of this text. Finally, it reminds you of the need for a constant, unfailing focus on safe working habits.

I.1 SCOPE AND STRUCTURE OF THE TEXT

Scope

The book encompasses fundamental material selected from physics, mechanics, electricity, kinematics, symbolic logic, and automatic control. The fundamental materials are coupled together to facilitate the study of *Electromechanics: Principles, Concepts, and Devices* that pertain to electromechanical mechanisms, machines, and systems. Motion, friction, lubrication, bearings, magnetic clutches, radial and axial drive systems, relays, contactors, starters, solenoids, ladder logic, sequential process-control circuits, motors, ac motor drives, and automatic control are but a few of the topics presented. The starting point for your entry into electromechanics is an understanding of elementary algebra and electrical fundamentals, including simple series and parallel circuits, a calculator, and a desire to learn.

Structure of the Text

The text is structured so that the principles and concepts of a topic are introduced first. These principles and concepts are followed by the study of an application or applications that incorporate the previously learned principle. This integration of the principles and applications gives you, the learner, the needed understanding of the technology—the *hows* and *whys*.

The chapters and the sections within the chapters are organized to encourage you to be an active participant in the process of learning. The self-educating aspects of the text are enhanced by the structure of each of the sixteen chapters. Each chapter is made up of the following parts:

- Introduction
- Performance objectives
- Sections with numbered headings
- Subsection with headings
- Chapter summary
- Selected technical terms
- End-of-chapter questions
- End-of-chapter problems

Chapters that introduce principles and concepts have one or more sets of mathematical exercises. You are encouraged to work each exercise. The *answers to the exercises* are located at the back of the book between the *Appendixes* and the *Index*.

Using the Structure of the Text

The most successful way to master the material in this text is to be an active participant in the educational process: that is, your success comes by doing, not by passively watching someone else laboring in the educational process. Begin each chapter by reading through the Introduction and the Performance Objectives for the chapter. As you read, make notes in the book's margin, underscore key ideas, and highlight important phrases and sentences.

Continue to make notes in the margin, underscore, and highlight as you read through each section of the chapter. Additionally, work through each example using pencil, paper, and calculator. Once you have an understanding of the material in a section, then test yourself by working through each problem in that section's exercises; then, using the Answers to Chapter Exercises found in the back of the book, check the answers to the problems. Rework those problems that you did incorrectly. Most problems have been solved with the full capacity of the calculator (*chain-calculated*) and then rounded to an appropriate number of significant figures to express the answer to the proper amount of precision. Because of this, you may find some small variation in your answer when compared to the book's answer.

While reading, make a list of technical terms, abbreviations, and acronyms. The definitions of many of these terms can be found in the Glossary of Selected Technical Terms, Abbreviations, and Acronyms found in the back of the book. Also, write down key ideas, rules, and formulas. Once you have a mastery of each section, then move on to the End-of-Chapter Questions and the End-of-Chapter Problems. The answers to the End-of-Chapter Questions and Problems are given in the Solutions Manual. Work with your instructor to check your work.

Once the entire chapter has been mastered, return to the performance objectives and check to see that you can meet each objective; review the chapter material for those of which you feel unsure. Finally, look at the Selected Technical Terms located near the end of the chapter after the Chapter Summary to discern if you can define each in your own words; using the book's glossary, review those with which you have difficulty.

I.2 SELECTING A CALCULATOR

To solve the examples, exercises, and end-of-chapter problems in this text, you will need a full-function calculator called a *scientific calculator*. This type of calculator is very helpful, since it has the ability to round answers to a preset number of significant figures. It also can be formatted to display answers in fixed notation (decimal notation), scientific notation, and engineering notation. If you are not familiar with these types of notation, each is explained and used in Chapter 1.

Table I-1 lists the functions and operations used in this text. If you have a calculator, you can match its capabilities to those shown in the table. Consult your

Table I.1 Calculator Functions and Operations Used in This Text

Key Symbol	Function/Operation	Comments	Alternative Key Symbols
$+$	Add		
$-$	Subtract	Simple arithmetic operations	
$\times$	Multiply		
$\div$	Divide		
CHS	Change sign		$+/-$
$1/x$	Reciprocal	Calculates the reciprocal	
$\sqrt{x}$	Square root	Calculates square root	
x^2	Square x	Squares a number	
FIX	Fix point notation		
SCI	Scientific notation	Display and rounding*	
ENG	Engineering notation		
EE	Enter exponent		EEX
e^x	Natural antilogarithm		INV LN
SIN	Sine		
COS	Cosine	Trigonometric functions	
TAN	Tangent		
SIN^{-1}	Arcsine		INV SIN
COS^{-1}	Arccosine	Inverse trig functions	INV COS
TAN^{-1}	Arctangent		INV TAN
DEG	Degree	Angular mode selection	DRG
RAD	Radian		

*Useful but not essential to the solution of the text problems.

owner's guide, your instructor, or other students for assistance if you don't know the meaning of a particular calculator keystroke.

Before you buy a calculator (assuming you need one), you should be aware that calculators are available with two different types of operating systems. Each system uses a different procedure for entering the data into the calculator. The two systems are referred to as either *reverse Polish notation* (RPN) or *algebraic entry system* (AES). The RPN system is commonly referred to as the *Polish system*, whereas the AES is simply called the *algebraic system*.

When entering data using the algebraic system, the arithmetic operators (+, −, ×, ÷) are placed between the numbers, as in 3 × 5. This, however, is not the case with the Polish system. Whether you purchase an algebraic-system or a Polish-system calculator, you might want to try out various calculators, talk to your instructor about calculators, and/or talk to other students in your department about their calculators.

Most full-function scientific calculators are programmable. If your calculator has this feature, then learn to use it. Once you can program your calculator, you will have added flexibility when solving problems with your calculator.

I.3   STRENGTHENING YOUR EDUCATIONAL EXPERIENCE

Study Hints

To educate yourself is among life's more challenging tasks. Most, if not all, of you have major commitments to your work, home, and family while attending classes to better your understanding of technology and ensure your future in a technology-related job. With limited time, how, when, and where you study will greatly influence the outcome of your endeavor in the educational process.

With the myriad commitments you have, it is important that you make time for daily study, even if it is only for a few minutes at a time. Because time is needed for material to become meaningful, a program of consistent, regular, daily study is much more beneficial to your progress than one-day-a-week study. Work with pencil, paper, and calculator at hand at all times and be an active participant in the process of educating yourself.

➤ **As a Rule** Most of your education takes place outside the classroom on your own during your study of the material.

Problem Solving

Most of us don't like word problems. However, as technicians, we are faced with interpreting word problems on a daily basis. When equipment fails to work properly, we are called in to diagnose the symptoms, prescribe and carry out the course of the repairs, and verify that the equipment is once again working properly. In the course of fixing the problem, verbal information is translated into symbolic languages (flow diagrams, schematics, control diagrams, etc.) so that a cause for the failure can be determined. Problem solving requires a thought process similar to that used in working on a failed piece of equipment.

The process of problem solving may be broken down into stages. In the first stage, the problem is read and the information is translated into symbols by *constructing and labeling* a sketch, a drawing, or a schematic. The second stage, *formulation*, involves identifying a relationship between or among the various parameters of the problem and putting this relationship into a formula. The third stage is the *evaluation* process, where value is given to each factor in the formula. *Substitution* of values for the variables takes place in the fourth stage, and *solution*, the fifth stage, completes the cycle of problem solving. In summary, the structure and flow of solving word problems encompasses the following:

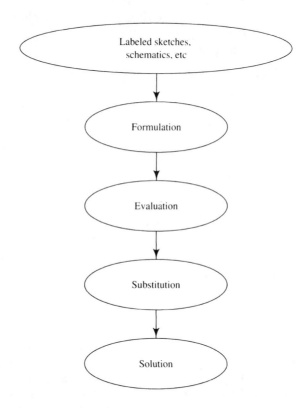

The format for solving word problems is employed in almost all of the dozens of worked examples found in each chapter. Carefully read through each example and then work through each using pencil, paper, and calculator.

General Information

- When used, $\Rightarrow$ (a right-facing arrow) is read "yields," which indicates a process has been used to arrive at the final answer. For example, 84.87 $\Rightarrow$ 84.9 indicates the answer has been rounded and expressed to three significant figures.

- The three dots arranged in a pyramid (∴) are read "therefore." This symbol will be found at the conclusion of some of the text examples.
- Selected reference tables for mechanics and electrical use are found in Appendix A.
- A table of mathematics symbols is given in Section B1 of Appendix B.
- Because the study of electromechanics uses many different symbols to formulate principles, rules, laws, and equations, the Greek alphabet is given in Section B2 of Appendix B.
- General guidelines for the maintenance, installation, and adjustment of mechanical and electrical machine elements are located in Appendix C.
- Whether reviewing for a test or looking up a concept, the index will be helpful in locating chapter topics.
- Because most exercise problems have been solved with the full capacity of the calculator (chain-calculated) and then rounded to an appropriate number of significant figures, you may find some small variation in your answers when compared to the book's *Answers to Chapter Exercises* at the back of the book.
- Use the Chapter Summary when studying for a quiz or a test covering the material in the chapter.
- Words set in **boldface type** are listed as *Selected Technical Terms* at the end of the chapter in which they are found. These terms are also found in the *Glossary* in the back of the book.
- A table with a summary of the equations used in each chapter is found near the end of each chapter. This table is a handy reference when studying for an examination or working end-of-chapter problems.

I.4 SAFETY

Safety is an attitude that you must develop if you are to work around electrical and mechanical equipment on a daily basis and not be injured. Since you can work safely only if you think safely, learn to evaluate your work habits for unsafe practices.

As a technician, you will be faced with hazards in the course of a day's work that can result in injury or death. The obvious hazards, including electric shock, moving equipment, and hand and power tools, are always present. Other less obvious hazards, such as exposure to radiation (lasers, X rays, microwaves, etc.), solvents, and chemicals, may not seem dangerous, since their threat results from repeated exposure over time.

Your understanding of technology must include a knowledge of how to operate, maintain, and use equipment, tools, and supplies in a safe manner. You must realize that safety procedures are ineffective and will not keep you from injury if your attitude and work habits are poor or careless. Freedom from on-the-job injury results from an unfailing attitude of care and confidence in yourself and your work.

I.5 CONCLUSION

I wish you well in your interactive study of this text, and I know that the field of technology will bring challenge and reward to you as it has done for so many others over the years. It is the competent technologist who truly understands the how's and why's of technology, as noted in the words of Lord Kelvin:

> I often say that when you can measure what you are speaking about, and express it in numbers, you know something about it; but when you cannot express it in numbers, your knowledge is of a meagre and unsatisfactory kind; it may be the beginning of knowledge, but you have scarcely, in your thoughts, advanced to the stage of Science, whatever the matter may be.
>
> *Lord Kelvin (1824–1907)*

Number Notation, Measurement, and Units

Chapter **1**

About the Illustration. *Measurement has played an important role throughout the ages. Here two Renaissance cannon are pictured along with aiming aids that enable them to be fired at night. In daylight, coordinates are taken with the aid of the scales (front of cannon), plumb bobs, and center line on the top of the cannon. A clinometer (see insert, lower right) is used to measure the angle of elevation of the barrel. At night when the artillery is to be fired, a cloth shield is placed in front of each cannon to occlude the light from a candle lamp while the gun is aimed using the prerecorded coordinates and elevation. Once aimed, the protective cloth is removed and the cannon is fired. (Designed by Agostino Ramelli, from* The Various and Ingenious Machines of Agostino Ramelli *(1588).)* ■

INTRODUCTION

In the study of electromechanics, you will deal with many physical quantities—some with exact numbers representing their size and others with measured numbers (approximate numbers). In order to state the precision of measured numbers correctly, you must have a knowledge of significant figures, rounding, number notation (both decimal and powers of ten), and units and unit prefixes. Also, you need a knowledge of the rules for writing answers to operations (adding, multiplying, etc.) involving physical quantities so the correct precision may be written.

CHAPTER CONTENTS

PERFORMANCE OBJECTIVES

Once you have read and studied each section; worked through each example with pencil, paper, and calculator; worked through the section and the end-of-chapter problems; and answered the end-of-chapter questions, you should be able to

- Use scientific and engineering notation.
- Apply the guidelines for significant figures to state the precision of a number.
- Round a number to the desired degree of precision.
- State the precision of an answer when adding, subtracting, multiplying, or dividing.
- Use prefixes and associated symbols with SI units of measurement.
- Convert within or between SI and BES using unit analysis.

1.1 NUMBER NOTATION

Powers-of-Ten Notation

Powers-of-ten notation, pictured in Figure 1.1, provides a procedure to express any number as a decimal number times a power of ten. This notation, in which a number is expressed as a decimal times a power of ten, enables you to work conveniently with very large and very small numbers. Figure 1.1 pictures 0.0238 expressed in powers-of-ten notation with the decimal **coefficient** and the **exponent** identified. Two valuable forms of powers-of-ten notation are *engineering notation* and *scientific notation*—each will be used in the text to express numbers to an appropriate precision.

Table 1.1 shows several examples of numbers written in powers-of-ten notation. From the examples, you can see that decimal numbers greater than one have positive powers of ten and that decimal numbers less than one have negative powers of ten. You may also have noticed that the placement of the decimal point in the powers-of-ten notation is not consistent; it may be placed anywhere in the decimal fraction. Unlike the general powers-of-ten notation, scientific notation and engineering notation have restrictions on where the decimal point is placed in the decimal fraction. The following examples will help you to understand how to write decimal numbers in powers-of-ten notation.

FIGURE 1.1

Powers-of-ten notation with each component of the notation identified.

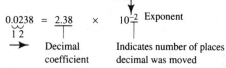

$$0.0238 = 2.38 \times 10^{-2}$$

Minus sign indicates decimal point was moved right

Exponent

Decimal coefficient

Indicates number of places decimal was moved

Example 1.1

Write .0025 in. in powers-of-ten notation. Arbitrarily place the decimal point between the 2 and the 5.

SOLUTION

Express .0025 in. using powers-of-ten notation.

Step 1

Move the decimal point:

$$.0025 \Rightarrow 2.5$$

Step 2

Count the places moved:

$$.0025$$
$$1\ 2\ 3$$

The decimal has been moved three places to the right.

Step 3

Determine the sign of the exponent. Since the decimal point was moved right, a minus sign is assigned to the exponent: (-3).

Step 4

Form the powers-of-ten notation:

$$2.5 \times 10^{-3}$$

$$\therefore \quad .0025 \Rightarrow 2.5 \times 10^{-3} \text{ in}$$

Observation

Read the symbol $\Rightarrow$ as "yields." For example, $.0025 \Rightarrow 2.5 \times 10^{-3}$ is read ".0025 yields 2.5×10^{-3}." The yield symbol implies that some operation must be carried out to get to the result.

TABLE 1.1 Powers-of-Ten Notation

Decimal Notation	Powers-of-Ten Notation
45 375	453.75×10^{2}
5382	53.82×10^{2}
0.0750	750×10^{-4}
415 800	41.5800×10^{4}
0.0015	1.5×10^{-3}
525	5.25×10^{2}

Example
1.2

Write 14,250 lb in powers-of-ten notation. Arbitrarily place the decimal point between the 2 and the 5.

SOLUTION

Express 14,250 lb using powers-of-ten notation.

Step 1

Move the decimal point:

$$14{,}250 \Rightarrow 142.50$$

Step 2

Count the places moved:

$$14{,}250$$

The decimal point was moved two places to the left.

Step 3

Determine the sign of the exponent. Since the decimal point was moved left, a plus sign is assigned to the exponent: (+2).

Observation

An implied plus sign is used; that is, the plus is left off a positive exponent.

Step 4

Form the powers-of-ten notation:

$$142.50 \times 10^2$$

$$\therefore \quad 14{,}250 \Rightarrow 142.50 \times 10^2 \text{ lb}$$

Exercise 1.1

Write each of the decimal numbers in powers-of-ten notation. The caret ($\wedge$) marks where to place the decimal point.

1. 863$\wedge$375	**2.** 0.000 55$\wedge$1	**3.** 0.009$\wedge$35
4. 18$\wedge$45	**5.** 3$\wedge$25	**6.** 0.000 000 0$\wedge$78
7. 0.52$\wedge$45	**8.** 2$\wedge$875	**9.** 0.003$\wedge$95
10. 62$\wedge$885	**11.** 1$\wedge$952.47	**12.** 0.031 68$\wedge$
13. $\wedge$2903	**14.** 0.005 180$\wedge$7	**15.** 14.90$\wedge$2

Decimal Notation

As you have learned, numbers without powers of ten are expressed in decimal notation; for example, 425 and 0.0250 are in decimal notation. When working with technological problems, you will sometimes come across a number that needs to be changed from powers-of-ten notation to decimal notation. This process is fairly simple.

■ First, form the decimal number from the decimal coefficient of the powers-of-ten number.
■ Then, place the decimal point using the size and sign of the exponent.

The following examples demonstrate the process.

Example
1.3   Write each of the following in decimal notation.
   (a) 7.58×10^3 **(b)** 4.37×10^{-2}

SOLUTION   **(a)** Form the decimal number and move the decimal point three places to the right:

$$7.58 \times 10^3 \Rightarrow 7.58 \Rightarrow 7580$$
$$\underset{1\ 2\ 3}{\longrightarrow}$$

Observation   A zero is added to the right to give enough places.

$$\therefore \quad 7.58 \times 10^3 \Rightarrow 7580$$

(b) Form the decimal number and move the decimal point two places to the left.

$$4.37 \times 10^{-2} \Rightarrow 0.0437 \Rightarrow 0.0437$$
$$\underset{2\ 1}{\longleftarrow}$$

Observation   A zero is added to the left to give enough decimal places.

$$\therefore \quad 4.37 \times 10^{-2} \Rightarrow 0.0437$$

Table 1.2 shows several more examples of numbers that have been converted from powers-of-ten notation to decimal notation.

Exercise 1.2

Write each of the powers-of-ten numbers in decimal notation.

1. 2.35×10^{-2} **2.** 2.058×10^6 **3.** 282.8×10^{-6}
4. 12.42×10^3 **5.** 6.935×10^{-3} **6.** 6015×10^2
7. 862×10^{-2} **8.** 34.86×10^{-5} **9.** 9.822×10^2
10. 0.585×10^4 **11.** 14.86×10^{-5} **12.** 0.8831×10^2
13. 0.252×10^{-1} **14.** $0.003\ 62 \times 10^5$ **15.** 14.86×10^{-4}

TABLE 1.2   Powers-of-Ten to Decimal Notation

Powers-of-Ten Notation	Movement of Decimal Point		Decimal Notation
	Direction	*No. of Places*	
5.25×10^{-3}	Left	3	0.005 25
0.318×10^5	Right	5	31 800
48.05×10^{-4}	Left	4	0.004 805
8.75×10^3	Right	3	8750

Table 1.3   Engineering Notation

Decimal Notation	Engineering Notation
6250	6.250×10^3
52 504	52.504×10^3
0.0250	25.0×10^{-3}
0.000 064	64×10^{-6}

Engineering Notation

Engineering notation is a special form of powers-of-ten notation where the exponent of ten is a multiple of three. Decimal numbers written in engineering notation are easily changed to quantities with unit prefixes forming multiples and submultiples of *metric units*. Most scientific calculators have a keystroke marked $\boxed{\textbf{ENG}}$, which automatically converts decimal notation to engineering notation. If your calculator has this feature, then use it to do Exercise 1.3 at the end of this section. The following example demonstrates the process of converting decimal numbers to engineering notation. Table 1.3 has several examples of numbers written in engineering notation.

Example 1.4

Write 0.0045 centimeters (cm) in engineering notation.

SOLUTION

Express 0.0045 cm using engineering notation.

Step 1

Form the decimal coefficient:

$$0.0045 \Rightarrow 4.5$$

Step 2

Count the places moved and determine the sign of the exponent:

$$0.0045$$
$$\underset{1\ 2\ 3}{\Longrightarrow}$$

Move the decimal point three places to the right (-3).

Step 3

Form the number:

$$4.5 \times 10^{-3}$$

$$\therefore \quad 0.0045 \Rightarrow 4.5 \times 10^{-3} \text{ cm}$$

Exercise 1.3

Write each decimal number in engineering notation.

1. 5328	**2.** 165 875	**3.** 0.000 075
4. 0.0580	**5.** 2 500 875	**6.** 0.0082
7. 0.006	**8.** 27 125	**9.** 0.000 437
10. 843.5	**11.** 1326.28	**12.** 0.006 90
13. 0.300	**14.** 902.00	**15.** 63 080.0

TABLE 1.4 Scientific Notation

Decimal Notation	Powers-of-Ten Notation	Scientific Notation
5350	53.50×10^2	5.350×10^3
0.002 87	287×10^{-5}	2.87×10^{-3}
0.0805	80.5×10^{-3}	8.05×10^{-2}

Scientific Notation

Scientific notation, like engineering notation, is a special form of powers-of-ten notation. In scientific notation the decimal point is always placed to the right of the leftmost nonzero digit. Table 1.4 shows several examples of scientific notation. You will see that in every instance the decimal point is located in the same place—to the right of the first nonzero digit. Scientific notation is commonly used with the British Engineering System (BES)—the *English system of units*—to express the precision of the numbers representing the quantities.

Most scientific calculators have a keystroke marked $\boxed{\text{SCI}}$ that automatically converts decimal notation to scientific notation. If your calculator has this feature, then use it to do Exercise 1.4. The following example will demonstrate the process of converting decimal numbers to scientific notation.

Example 1.5

Write .0752 feet (ft) in scientific notation.

SOLUTION

Express .0752 ft using scientific notation.

Step 1

Form the decimal coefficient:

$$.0752 = 7.52$$

Step 2

Count the places moved and determine the sign of the exponent:

$$.0752 \atop 12 \longrightarrow$$

Move the decimal point two places to the right (-2).

Step 3

Form the number:

$$7.52 \times 10^{-2}$$

$$\therefore \quad .0752 \Rightarrow 7.52 \times 10^{-2} \text{ ft}$$

Exercise 1.4

Write each decimal number in scientific notation.

1. 3205	**2.** 165 836	**3.** .000 075
4. .005 80	**5.** 2 506 258	**6.** 5550
7. .0683	**8.** 27 130	**9.** .000 437
10. 578 022	**11.** .008 29	**12.** .838
13. 132.8	**14.** 407.8	**15.** .013 02

1.2 MEASUREMENT

Exact Numbers

In science and technology, a few of the numbers encountered are totally free from error. These numbers are called *exact numbers* and are obtained by counting objects, by definition, or by computation with exact numbers (which are derived). Some examples of exact numbers include

> 12 eggs in a dozen (by definition)
> 9 ft^2 in a square yard (derived from 3 ft in a yard)
> 4 hubcaps on a car (by counting)

Exact numbers occur in formulas as constants. In the formula for volume of a sphere $(V = \frac{4}{3}\pi r^3)$, the constants $\frac{4}{3}$ and π are exact numbers.

Approximate Numbers

Approximate numbers result from measured quantities, and all measured quantities have some error. In the course of measuring, a technician will take a great deal of time to ensure that the amount of measurement error is reduced as much as possible.

Measurements are reported as a distance, an amount of electric current, a temperature, etc., and the **precision** of the measurement is indicated by the number of digits in the number used to report the measurement. For example, .125 in is more precise than .1 in, and 15.68 amperes (A) is more precise than 15 A. As the precision of the measuring instrument increases, more digits are reported in the measured value. The **accuracy** (how close the reading is to the true value) of the measurement is limited by the precision of the instrument, the type of instrument (mechanical, electrical, etc.), and by the technician's ability (experience and judgment) to make the reading.

Exercise 1.5

Determine whether each of the following is an exact or approximate number.

1. 3.95 minutes (min) to run 1.00 miles (mi)
2. 3 wheels on a tricycle

3. 100 cm in 1 meter (m)
4. The 2 ends of a wrench
5. A 737 airliner traveling 515 miles per hour (mi/h)
6. A temperature of 74.3°F
7. A gear with 24 teeth
8. The diameter of a mild steel rod that is 1.125 in

■

Significant Figures

In reporting a reading of a measurement, each of the stated figures in the measurement are *significant figures*. Suppose a round rod is measured with an outside **micrometer** (Figure 1.2) and the diameter is determined to be .125 in. In the measured value of .125 in, there are three significant figures. However, the last digit, 5, is uncertain because in making the reading it was estimated. It is known from this measurement that the true value of diameter is somewhere between .1245 and .1255 in, or .125 ± .0005 in. The plus or minus value indicates the **tolerance** of the reading.

When measurements are used in calculations, it is very important that the result is not overstated. That is, the act of calculating cannot improve the stated precision of a measurement. For example, suppose you measure the diameter of a pulley with a **digital caliper** (Figure 1.3) to the nearest thousandth of an inch as 2.375 in. From this measurement the circumference ($C = \pi d$) is computed and recorded as 7.461282553 in. To say that the precision of the answer is overstated is an understatement; in fact, the stated precision is an outright lie. It is absurd to claim precision to ten significant figures when the measurement of the diameter has precision only to four significant figures. The answer stated with correct precision is 7.461 in.

When stating approximate numbers (measured quantities), use the following guidelines and avoid overstating the precision of the number:

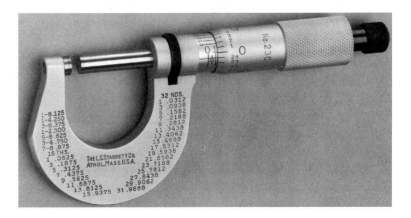

FIGURE 1.2
Outside micrometer used for precision measurement (0.0001 in). (*Courtesy of the L. S. Starrett Company*)

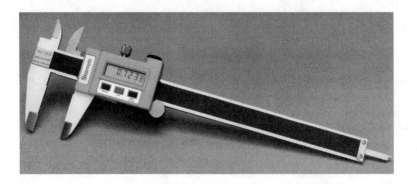

FIGURE 1.3
Digital electronic caliper used to measure inside diameter (ID), outside diameter (OD), and depth to the nearest ±.001 in (±0.0025 mm). (*Courtesy of the L. S. Starrett Company*)

1. All figures in a recorded measurement are assumed to be significant.
2. Zeros that hold decimal places are not significant. For example, the zeros in 0.**0**55 and 0.**00**25 are not significant, since they are keeping the decimal point in place. However, the zeros in 2001 and 15.50 are significant, because they indicate precision. To summarize:
   ■ Zeros at the end of a whole number are significant (as in 25**00** or 8**0**) unless indicated otherwise, as in guideline 5.
   ■ Zeros at the beginning of a decimal fraction are not significant (as in 0.**00**52 or 0.**0**95).
   ■ Zeros to the left of the decimal point in a decimal fraction are not significant (as in **0**.626 or **0**.4375).
   ■ Zeros between the first and last significant figures are significant (as in 2**00**1 or 0.05**0**2).
   ■ Zeros at the end of decimal numbers are significant (as in 25.**00** or 18.7**0**).
3. When a number lacks sufficient digits for the desired precision, as sometimes occurs in multiplication or division, then zeros are added as necessary. For example, if 0.25 is to be expressed to four significant figures, two zeros are added to 0.25 to increase the significance of the number to 0.25**00** (four significant figures).
4. All numbers (including zero) in the decimal coefficients of scientific and engineering notation are significant. Thus, 2.30×10^4 and 480×10^3 each have three significant figures.
5. A bar is placed above a zero in a whole number to indicate that it has less significance than usual (as in $25\bar{0}0$, to indicate three-place significance, not the typical four-place significance).
6. Exact numbers can be expressed to any precision with as many significant figures as may be deemed necessary. For example, in multiplying 3.28 and 5, where 5 is an exact number, 5 is thought of as being 5.00, making it compatible in precision to 3.28.
7. The precision of measured quantities or the answers to calculations should not be overstated.

Exercise 1.6

State the precision of the following measurements.
Example: 6.65 feet (ft)   **Answer:** Precise to three significant digits.

1. .345 ft   **2.** .0589 in
3. 3.005 cm   **4.** 52$\overline{0}$0 lb
5. 40 600 m   **6.** .00032 in
7. 24.008 kg   **8.** 9024.06 mi
9. 2000 lb   **10.** 0.0075 millimeters (mm)
11. 1.30×10^{-3} m   **12.** 400×10^2 kilograms (kg)

Rounding

Rounding is a process used to shorten a number to the desired amount of signifi-cance. Although there are several methods commonly used to round, the simple procedure used by most calculators will be used in this text:

- Round up when the digit after the last retained digit is five (5) or greater.
- Round off when the digit after the last retained digit is less than five (5).

Example 1.6 illustrates this procedure.

Example 1.6

Express the following decimal numbers to the indicated number of significant figures:
(a) 8515 to two (2) significant figures
(b) 0.03385 to three (3) significant figures

SOLUTION

(a) Round to two significant figures by checking the third significant digit in 8515, which is 1. Since it is less than 5, round off by replacing the digits to the right of the second digit with zeros.

$$\therefore \quad 8515 \Rightarrow 8500 = 8.5 \times 10^3, \text{to two significant figures.}$$

Observation

Zeros that are added to hold the decimal point are not significant figures.
(b) Round to three significant figures by checking the fourth significant figure in 0.03385, which is 5. Since it is 5, round up by increasing the third digit by 1 and dropping the digits to the right of the third digit.

Observation

The zero to the right of the decimal point holds the decimal point and is not significant.

$$\therefore \quad 0.03385 \Rightarrow 0.0339, \text{to three significant figures.}$$

As a final comment, note that most modern scientific calculators have both *rounding* and *display features* that may be used to set the type of display and the number of digits displayed by the calculator. These features include **FIX** (*fixed decimal*, or decimal notation), **SCI** (*scientific*, or scientific notation), **ENG** (*engineering*, or engineering notation).

Exercise 1.7

Write the following numbers so each has three (3) significant figures. In some cases, scientific or engineering notation may be used. Also, zeros may be added as needed to indicate significance. Round as needed.

1. 454.72	**2.** 88.02	**3.** 45
4. 0.005 206	**5.** 56.56	**6.** 1994
7. 0.3088	**8.** 2000	**9.** 0.0392
10. 7	**11.** 480 000	**12.** 0.07
13. 5.2×10^2	**14.** 2999	**15.** 0.08×10^{-1}
16. 5050	**17.** 0.001 889	**18.** 3 000 000

Operating with Approximate Numbers

When adding, subtracting, multiplying, dividing, extracting roots, or raising to powers, you must apply simple rules to these operations to ensure that you do not overstate the precision of the answer. Each of these operations is investigated, and examples are given to help you understand how to round the answer to the correct number of significant figures.

> **Addition and Subtraction Rule**. Round the answer (sum or difference) to the same precision as the least precise number used in the calculation.

The least precise number is the quantity that has the *least precise digit that is significant*. The place value of the least precise digit determines the precision to which the sum or difference may be stated. For example, when 125 is added to 23.2 (both are approximate numbers), the sum is written as 148, not 148.2. The reason that the sum is stated to the nearest whole unit rather than to the nearest tenth of a unit is that the 5 in the number 125 is the least precise significant digit in either of the two numbers. The 5 has a place value of 5 whole units, because it is located in the units place of the number 125. Figure 1.4 summarizes the names of the place values in decimal numbers.

Example 1.7

Add 55.2 to 62.
   (a) Determine the number with the least precise digit.
   (b) Express the answer correctly rounded.

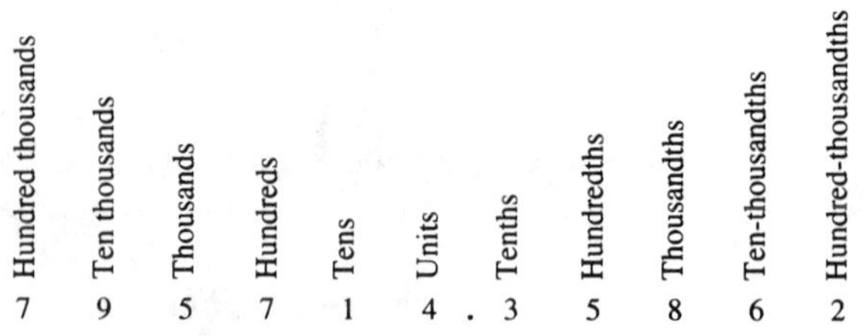

FIGURE 1.4
Place-value names for decimal numbers.

Hundred thousands	Ten thousands	Thousands	Hundreds	Tens	Units		Tenths	Hundredths	Thousandths	Ten-thousandths	Hundred-thousandths
7	9	5	7	1	4	.	3	5	8	6	2

SOLUTION **(a)** Of the two numbers, 62 is the number with the least precise digit. The 2 in the number 62 is in the units place and it is the least precise digit in the two numbers. Because 62 is precise to the nearest unit, the answer (sum) may be expressed only to the nearest unit.

Observation The number 55.2 is precise to the nearest tenth (0.1) of a unit.

(b) Add 62 and 55.2 and round the answer to the nearest unit.

$$\therefore \quad 62 + 55.2 = 117.2 \Rightarrow 117$$

Example 1.8 Subtract 1.25 from 18.4.

    (a) Determine the number with the least precise digit.

    (b) Express the answer correctly rounded.

SOLUTION **(a)** Of the two numbers, 18.4 is the least precise number and 4 is the least precise digit that is significant. Because the 4 is in the tenths place, the answer is expressed to the nearest tenth (0.1) of a unit.

Observation The number 1.25 is precise to the nearest hundredth (0.01) of a unit.

(b) Subtract 1.25 from 18.4 and round the answer to the nearest tenth (0.1) of a unit.

$$\therefore \quad 18.4 - 1.25 = 17.15 \Rightarrow 17.2$$

Example 1.9 Add 2.25×10^{-2} to 5.2×10^{-4}.

    (a) Determine the number with the least precise digit.

    (b) Express the answer correctly rounded.

Observation When quantities are expressed in a powers-of-ten notation, then each of the quantities must be written in terms of the same power of ten. This is done in order to find the number with the least precise digit that is significant.

SOLUTION **(a)** Express the quantities so that each has the same power of ten. Change 2.25×10^{-2} to 225×10^{-4}. Of the two numbers (225×10^{-4} and 5.2×10^{-4}), the least precise digit is the 5 in the units place of the number 225. The answer is expressed to the nearest whole unit times 10^{-4}.

(b) Add the two numbers and round the answer to the nearest whole unit.

$$\therefore \quad 225 \times 10^{-4} + 5.2 \times 10^{-4} = 230.2 \times 10^{-4} \Rightarrow 230 \times 10^{-4}$$

■──────────────────────────────────────

Exercise 1.8

State the precision of the following approximate numbers. Use Figure 1.4 to aid in determining the place-value name of the least precise significant digit in the number. For example, the number 1.08 is precise to the nearest hundredth unit because the 8 is located in the hundredths place of the number 1.08.

1. 520	**2.** 0.32
3. 2.725	**4.** 928 5$\overline{0}$0
5. 88	**6.** 0.0442

Solve the following and express the answer so it contains no significant figures farther to the right than occur in the least precise number. All numbers are approximate.

7. 42.750 + 2.93	**8.** 54 $\overline{0}$00 + 6875
9. 3.0505 − 2.2	**10.** 0.34 − 0.0228
11. 248.64 − 72.375	**12.** 13.4 + 8.02 + 50
13. 3.75 − 6.2 + 12	**14.** −42.05 + 3.00 + 54.25
15. $9.2 \times 10^6 + 2.5 \times 10^7$	**16.** $22.4 \times 10^3 - 1.72 \times 10^5$

➤ **Multiplication and Division Rule** Determine (by inspection) which of the quantities (numbers) has the fewest number of significant figures. The answer (product or quotient) is then rounded to the same number of significant figures as in this quantity.

Example 1.10

Determine the following:
- **(a)** The product of 56.25 and 3.8
- **(b)** The appropriate number of significant figures by rounding the answer

SOLUTION   **(a)** Find the product.

$$56.25 \times 3.8 = 213.75$$

(b) Since 3.8 has the fewest number of significant figures, the answer must be rounded to two significant figures:

$$\therefore \quad 213.75 \Rightarrow 210 = 2.1 \times 10^2, \text{ to two significant figures.}$$

Observation   The zero in 210 is not significant since it is simply a place holder. Notice that scientific notation is used to express the answer to two significant figures.

Example 1.11

Determine the following:
- **(a)** The quotient of (8.342 × 2.91)/5.620
- **(b)** The appropriate number of significant figures by rounding the answer

SOLUTION   **(a)** Find the quotient:

$$\frac{8.342 \times 2.91}{5.620} = 4.3194$$

(b) Since 2.91 has the fewest number of significant figures, the answer must be rounded to three significant figures:

$$\therefore \quad 4.3194 \Rightarrow 4.32, \text{ to three significant figures.}$$

Exercise 1.9

Simplify the following and express the answer to the appropriate number of significant figures. All numbers are approximate:

1. 4.2×74.05 **2.** 400/5.275 (400 is exact)
3. 3.14×20 (20 is exact)   **4.** 83.25/16.65
5. 0.078×3.049 **6.** 18.03×122.0
7. 23.4/(6.42 $\times$ 4.00)   **8.** $45 \times (20/12)$
9. $0.875 \times 42 \times 10^{-3}$ **10.** $(92.4 \times 10^{3})/0.0436$

➤ **Roots and Powers Rule** The root or power of a number has the same number of significant figures as the base number.

Example 1.12

In each of the following cases, determine each of the following:
    (a) The appropriate number of significant figures in the root or power
    (b) The solution
        1. $\sqrt{7.0}$ **2.** 3.31^{2}

SOLUTION   **(a) 1.** The square root of 7.0 will have two significant figures in the answer.
        2. The square of 3.31 will have three significant figures in the answer.

    (b) 1. $\sqrt{7.0} = 2.645 \Rightarrow 2.6$
        2. $3.31^{2} = 10.956 \Rightarrow 11.0$

Exercise 1.10

Solve the following and express the answer to an appropriate number of significant figures. All numbers are approximate.

1. $\sqrt{20}$ **2.** 4.5^{2} **3.** $\sqrt{7.2}$

4. 81.20^{2} **5.** 0.22^{2} **6.** $\sqrt{125.0}$

7. 30.7×5.0^{2} **8.** $8.2^{2} \times \sqrt{42.0}$ **9.** $\sqrt{286.0} \times 5.1$

1.3 UNITS AND DIMENSIONS

Units

When making measurements it is customary to record both the quantity (how much) and the unit of measure. The term *physical quantity* is used to describe the result of measuring. In technology, great importance is placed on the unit of measure. Without the unit, the measurement doesn't have much meaning. To say, for example, that a building is 20 high or that you walked 2.5 in the city leaves much to the imagination. Was the building 20 stories, 20 ft, or 20 m high? Did you walk 2.5 blocks, 2.5 mi, or 2.5 km? Without a clear statement of the unit of measure, no one knows what is meant. From these examples, it is apparent that a system of units is necessary in order to completely specify a physical quantity.

Systems of Units

Today in the United States, two systems are used to specify units. These systems are the *British Engineering System* (**BES**)—the *English System*—and the *International System of Units* (**SI**)—the *metric system*. The International System was adopted in 1960 by the General Conference of Weights and Measures, which changed and simplified the earlier metric system to form the modern metric system; it is indicated worldwide (in all languages) simply as SI. SI has since replaced BES in most applications in science and technology.

In each system of units, a set of *base units* is specified to measure the fundamental physical quantities. The base units are not related to one another, nor do they depend on each other for their definition. In each system, all additional units are derived from the base units and are specified in terms of two or more of the base units. Units derived from base units are called *derived units*.

British Engineering System (BES)

Length in feet (ft), force in pounds (lb), and time in seconds (s) are the base units in the British Engineering System (BES). All the derived units are defined in terms of these three base units. For example, the derived unit of work is expressed as foot pounds (ft-lb), and the derived unit of power is expressed as foot pounds per second (ft-lb/s). Table 1.5 summarizes the base units and several of the derived units commonly used in the study of mechanics.

BES assigns names to *dimensional quantities* in a nonsystematic way. For example, length may be expressed in inches, feet, yards, or miles. Without prior knowledge that all these units are related to length, you might think that four different dimensional quantities were being described instead of multiples and submultiples of just one, length. In BES, each time a multiple or submultiple of a unit is created, a new name is applied. The naming of the units is one of several shortcomings of BES.

Another problem associated with BES is the inconsistency in the size of the conversion factors; for instance, 12 in equals 1 ft, 3 ft equals 1 yd, and 1760 yd equals

TABLE 1.5 Selected BES Units

Physical Quantity	Unit Name	Unit Symbol
Base Units		
Length	foot	ft
Force (weight)	pound	lb
Time	second	s*
Derived Units		
Mass	slug	lb-s^2/ft
Energy (work)	foot pound	ft-lb
Power	foot pound/second	ft-lb/s
Velocity	foot per second	ft/s
Area	square foot	ft^2

*The letter s NOT sec is the unit symbol for seconds.

1 mi. Because of the inconsistencies in the conversion factors and the complexity of the names, the British Engineering System is not the standard for engineering and technology. Instead, the International System of Units is the standard system.

International System of Units (SI)

Unlike the British Engineering System, the International System of Units (SI) assigns names to dimensional quantities in a very systematic way. The unit for length is the **meter** (m). Multiple and submultiple units of length incorporate the unit name, meter, in their name. Units such as the centi*meter* (cm), milli*meter* (mm), and kilo*meter* (km) are all obvious units of length in SI.

SI uses seven base units that have been selected for their independence from one another. Included in these seven base units are the units of length—the meter (m), mass—the **kilogram** (kg), time—the second (s), and electric current—the **ampere** (A).

Like the BES, the SI has a large number of derived units that originate from two or more of the seven base units. Rather than express the derived units in terms of the base units (which can be very complex), SI derived units are given special names. Included in the derived units with special names are units of force—the **newton** (N), energy or work—the **joule** (J), and power—the **watt** (W). Table 1.6 summarizes the base units and several of the derived units used in electromechanical technology. A few guidelines are helpful when applying SI units to measurements.

Guidelines for Using the International System of Units

- When writing the symbol for the unit, do not use a period after the symbol. For example, 25 N, not 25 N., is used to express 25 newtons.

TABLE 1.6   Selected SI Units

Physical Quantity	Unit Name	Unit Symbol
Base Units		
Length	meter	m
Mass	kilogram	kg
Time	second	s
Electric current	ampere	A
Derived Units		
Force (weight)	newton	N
Energy (work)	joule	J
Power	watt	W
Velocity	meters/second	m/s
Area	square meters	m^2

- The unit symbol is not modified to include plural units. For example, 15 watts is written 15 W, not 15 Ws. The unit symbol is capitalized when the unit is named for a person. For example, the symbol for the newton is written as N, the watt is W, and the joule is J; however, the symbol for the meter is written as m, and the kilogram is kg.
- When the physical quantity is expressed as a numerical value (the usual case), then a unit symbol is used and the unit is not spelled out. For example, write 15 J, not 15 joules.
- In writing the numbers that precede the unit in the physical quantity, it is customary to set off the number in groups of three when the number is longer than four digits. For example, 2345200 is written 2 345 200. Decimal fractions are also set off in groups of three. For example, 0.002253 is written as 0.002 253. *Notice that unlike in BES quantities, commas are not used as delineators in SI quantities.*
- Decimal fractions are always written with a zero before the decimal point. For example, write 0.25, not .25 as is usually done in BES.
- The word *per* is used to indicate division. For example, the acceleration of a body due to gravity is approximately 9.81 meters per second per second. This is expressed as 9.81 m/s/s, or simply 9.81 m/s^2.

Exercise 1.11

Use the guidelines for the International System of Units to write the following correctly. It is assumed that you have no knowledge of prefixes at this time.

1. .003125 N
2. 47 watts of power
3. 12000 seconds
4. 14 thousandths of a meter
5. 234000 joules of energy
6. .525 amperes of current
7. 32 meters per second
8. 200 kilograms

Units with Prefixes

Multiple and submultiple units in SI are formed using the prefixes listed in Table 1.7. When a prefix is added to a unit, the prefix is incorporated into the unit name, as in *kilo*meter (km) and *milli*ampere (mA). In forming multiples and submultiples of SI units:

1. First, write the decimal number of the physical quantity in engineering notation (see Section 1.1), with the decimal point adjusted so the decimal coefficient is a number *from 0.1 to 1000.*
2. Next, replace the engineering notation with an appropriate prefix from Table 1.7 and form the prefixed unit using the correct prefix and unit symbols.
3. Finally, form the multiple or submultiple of the physical quantity by attaching the prefixed unit to the number of the decimal coefficient.

Example 1.13

Write 18 200 N using a prefixed unit.

SOLUTION

Express 18 200 N in engineering notation.

$$18\ 200\ N \Rightarrow 18.200 \times 10^3\ N$$

Replace the engineering notation with the appropriate prefix from Table 1.7 and form the prefixed unit.

$$10^3\ N \Rightarrow kN$$

Form the multiple by attaching the prefixed unit to the decimal coefficient.

$$18.2\ kN$$

$$\therefore\quad 18\ 200\ N \Rightarrow 18.200 \times 10^3\ N \Rightarrow 18.200\ kN$$

TABLE 1.7 Unit Prefixes for the International System of Units

Prefix	Symbol	Multiple/Submultiple	Power of Ten
tera	T	1 000 000 000 000	10^{12}
giga	G	1 000 000 000	10^{9}
mega	M	1 000 000	10^{6}
kilo	k	1 000	10^{3}
centi	c	0.01	10^{-2}
milli	m	0.001	10^{-3}
micro	μ	0.000 001	10^{-6}
nano	n	0.000 000 001	10^{-9}
pico	p	0.000 000 000 001	10^{-12}

Example
  1.14   Write 0.000 375 s using a prefixed unit.

SOLUTION   Express 0.000 375 s in engineering notation.

Observation The decimal may be moved either three places right or six places right. Either format is correct, since each meets the general guideline of expressing the decimal coefficient as a number *from 0.1 to 1000* and either has the correct amount of precision.

1. $0.000\ 375 \Rightarrow 0.375 \times 10^{-3}$ s
2. $0.000\ 375 \Rightarrow 375 \times 10^{-6}$ s

Replace the engineering notation with an appropriate prefix from Table 1.7 and form the prefixed unit.

1. 10^{-3} s $\Rightarrow$ ms
2. 10^{-6} s $\Rightarrow$ μs

Form the submultiple by attaching the prefixed unit to the decimal coefficient.

1. 0.375 ms
2. 375 μs

$$\therefore \quad 0.000\ 375 \text{ s} \Rightarrow 0.375 \text{ ms, or } 375 \text{ μs}$$

When you add or subtract prefixed physical quantities, first express each quantity to its base unit and then add or subtract. Round the answer to the least precise quantity (as in Example 1.7). The following example demonstrates this concept.

Example
  1.15   Add 23.5 kW and 120 W.

SOLUTION   Change 23.5 kW to its base unit of watts.

$$23.5 \text{ kW} \Rightarrow 23\ 500 \text{ W (zeros are place holders)}$$

Add 23 500 to 120 W.

$$23\ 500 \text{ W} + 120 \text{ W} = 23\ 620 \text{ W}$$

Round to the nearest hundred.

$$23\ 600 \text{ W} \Rightarrow 23.6 \text{ kW}$$

With units that have prefixes, all the digits (even zeros) in the prefixed number are significant, as illustrated by Example 1.16.

Example
  1.16   Add 200 kN and 10.6 kN.

SOLUTION   Express each quantity in the base unit, newtons.

$$200 \text{ kN} \Rightarrow 20\overline{0}\ 000 \text{ N}$$

$$10.6 \text{ kN} \Rightarrow 10\ 600 \text{ N}$$

Add $20\overline{0}\ 000$ N to 10 600 N:

$$20\overline{0}\ 000 \text{ N} + 10\ 600 \text{ N} = 210\ 600 \text{ N}$$

Express the answer to the nearest thousand.

$$211\ 000 \text{ N} \Rightarrow 211 \text{ kN}$$

Observation Because all the digits in the prefixed numbers are significant, the least precise digit is the rightmost 0 in the number 200 kN. Because the 0 is in the thousands place, the answer is expressed to the nearest thousand.

Example 1.17

Subtract 18 μg from 0.58 g.

SOLUTION Express each quantity in its base unit, grams.

$$0.58 \text{ mg} \Rightarrow 0.000\ 58 \text{ g}$$

$$18 \ \mu\text{g} \Rightarrow 0.000\ 018 \text{ g}$$

Subtract 0.000 018 g from 0.000 58 g:

$$0.000\ 58 \text{ g} - 0.000\ 018 \text{ g} = 0.000\ 562 \text{ g}$$

Express the answer to the nearest hundred thousandth:

$$0.000\ 56 \text{ g} \Rightarrow 0.56 \text{ mg}$$

Exercise 1.12

Write the following physical quantities in the specified multiple or submultiple unit. Express the answer with an appropriate precision.

1. 5820 J in kJ
2. 7.5×10^5
3. 0.092 A in mA
4. 3325 W in kW
5. 4.50×10^{-4} s in μs
6. 0.0025 m in mm
7. 4527 N in kN
8. 0.000 000 750 A in nA

Express the following in base units (without the unit prefix).

9. 52 cm
10. 0.182 kA
11. 335 kN
12. 2.5 mg
13. 518 μs
14. 83.2 MW

Add or subtract the following and express the answer so it contains no significant figures farther to the right than occur in the least precise number. Show all your work and, when appropriate, write the answer with a prefixed unit.

15. 50.0 μA + 0.25 mA   **16.** 845 W + 3.15 kW
17. 410 km − 0.25 Mm   **18.** 25 μs − 0.180 ms
19. 64 mg + 0.18 g   **20.** 982 mm − 37 cm

■

Conversion Within and Between SI and BES

In converting from prefixed units to base units, you may be unsure whether to multiply or divide by the conversion factor. By setting up unitless ratios, you can quickly decide the correct form for the conversion factor. This technique is called *unit analysis* and it uses the idea that any factor may be multiplied by 1 without changing its value. The following examples demonstrate the use of unit analysis. Table 1.8 lists a number of identities and conversion factors that are used in the following examples and problems. A duplicate of Table 1.8 may be found in Appendix A1, "Selected Conversion Factors for SI and BES."

Example 1.18

Convert 7500 cm to meters.

SOLUTION

From Table 1.8, select the identity 1 m = 100 cm. Multiply 7500 cm by 1 in the form of (1 m)/(100 cm).

$$7500 \ \cancel{cm} \times \frac{1 \ m}{100 \ \cancel{cm}} = 75.00 \ m$$

Observation

By setting the identity up with meters in the numerator, the centimeters will factor out, leaving the desired unit: meters.

$$\therefore \quad 7500 \ cm = 75.00 \ m$$

Example 1.19

Convert 48 000 g to kilograms (kg).

SOLUTION

From Table 1.8, select the identity 1 kg = 1000 g. Multiply 48 000 g by 1 in the form of (1 kg)/(1000 g).

$$48 \ 000 \ \cancel{g} \times \frac{1 \ kg}{1000 \ \cancel{g}} = 48 \ kg$$

$$\therefore \quad 48 \ 000 \ g = 48 \ kg$$

TABLE 1.8 Selected Identities and Conversion Factors for SI and BES*

Displacement (length)
1 m = 100 cm = 1000 mm = 3.281 ft = 39.37 in
1 km = .6214 mi = 3281 ft
1 in = 2.54 cm = 25.4 mm
1 mi = 5280 ft
1 yd = 3 ft = 36 in
1 ft = 12 in
1 revolution = 360° = 2π rad = 6.2832 rad

Area
1 m^2 = 10.76 ft^2 = 1550 in^2 **1 yd^2 = 9 ft^2**
1 cm^2 = .1550 in^2 **1 ft^2 = 144 in^2**
1 m^2 = 10 000 cm^2
1 cm^2 = 100 mm^2

Time
1 h = 60 min = 3600 s
1 min = 60 s

Force
1 N = .2248 lb   **1 lb = 16 oz**
1 lb = 4.448 N   **1 ton = 2000 lb**

Mass
1 kg = .0685 slug   1 kg = 1000 g
1 slug = 14.6 kg   1 g = 1000 mg

Velocity (speed)
1 m/s = 3.60 km/h = 2.24 mi/h = 3.28 ft/s
60 mi/h = 88 ft/s
1 ft/s = 0.3048 m/s
1 rev/min = 0.1047 rad/s   **60 rev/min = 1 cycle/s = 1 Hz**

Work (energy, torque)
1 J = .738 ft-lb   1 ft-lb = 1.36 J
1 kWh = 3.6 MJ   1 N·m = .738 lb-ft
1 Btu = 1.06 kJ = 778 ft-lb

Power
1 hp = 746 W = 550 ft-lb/s   **1 hp = 33,000 ft-lb/min**
1 W = .738 ft-lb/s = 3.412 Btu/h 1 kW = 1.34 hp

General Constants
Acceleration due to gravity (BES) = 32.2 ft/s^2
Acceleration due to gravity (SI) = 9.81 m/s^2
π = 3.1416

*__Boldface__ physical quantities are exact.

Example
1.20

Convert 0.320 h to seconds.

SOLUTION   From Table 1.8, select the identities 1 min = 60 s and 1 h = 60 min. Multiply 0.320 h by 1 in the form of (60 min)/(1 h) and in the form of (60 s)/(1 min).

$$0.320 \; \cancel{h} \times \frac{60 \; \cancel{min}}{1 \; \cancel{h}} \times \frac{60 \; s}{1 \; \cancel{min}} = 1152 \; s$$

$$\therefore \quad 0.320 \; h = 1152 \; s \Rightarrow 1.15 \times 10^3 \; s \; \text{or} \; 1.15 \; ks$$

Example
1.21

Convert 50.5 lb of weight (force) to newtons.

SOLUTION   From Table 1.8, select the conversion factor:

$$1 \; lb = 4.448 \; N$$

Multiply 50.5 lb by 1 in the form of (4.448 N)/(1 lb).

$$50.5 \; \cancel{lb} \times \frac{4.448 \; N}{1 \; \cancel{lb}} = 225 \; N$$

$$\therefore \quad 50.5 \; lb = 225 \; N$$

Example
1.22

Convert 100 mi/h to meters per second.

SOLUTION   From Table 1.8, select the conversion factors:

$$60 \; m/h = 88 \; ft/s \quad \text{and} \quad 1 \; m/s = 3.28 \; ft/s$$

Multiply 100 mi/h by 1 in the form of (88 ft/s)/(60 mi/h) and in the form of (1 m/s)/(3.28 ft/s).

$$100 \; \cancel{mi/h} \times \frac{88 \; \cancel{ft/s}}{60 \; \cancel{mi/h}} \times \frac{1 \; m/s}{3.28 \; \cancel{ft/s}} = 44.7 \; m/s$$

$$\therefore \quad 100 \; mi/h = 44.7 \; m/s$$

■────────────────────────────────────

Exercise 1.13

Solve the following problems using the identities and conversion factors in Table 1.8. Use rounding to express the final answer with an appropriate precision. When appropriate, write the answer with a prefixed unit.

1. Convert 3.50 yd to inches.
2. Change 54.2 oz to pounds.
3. Express 25 mi/h in feet per second.
4. Convert 5.0 ft^2 to square yards (yd^2).
5. Express 1250 ft-lb/s as horsepower (hp).
6. Change 71.9 cm to millimeters.
7. Convert 18.2 MJ to kilowatthours.
8. Express 850 mg as grams.
9. Express 465 m/min as meters per second.
10. Change 2.80 m^2 to square centimeters (cm^2).
11. Convert 3682 in to meters.
12. Convert 12.5 N to pounds.
13. Convert 61.2 ft-lb to joules.
14. Convert 70.0 m/s to miles per hour.
15. Convert .375 in to millimeters.

CHAPTER SUMMARY

- Any number can be expressed as a decimal number times a power of ten.
- Engineering notation is used with unit prefixes to form prefixed SI units.
- Exact numbers do not represent measured quantities and are free from error.
- Approximate numbers come from measured quantities, which contain error.
- The precision of a number is indicated by the number of significant figures used in writing the number.
- All figures in a recorded measurement are significant unless indicated otherwise, as noted in the guidelines of Section 1.2.
- When representing measured quantities, all numbers, including zero, in the decimal coefficient of a prefixed number or in a powers-of-ten number are significant.
- When representing measured quantities, all numbers, including zero, in the number of a prefixed unit are significant.
- Rounding is used to write a number with the desired amount of significance.
- The result of adding or subtracting is rounded to the same precision as the least precise number used in the calculation.
- The result of multiplying or dividing is rounded to the same precision as the least precise number in the calculation.
- The two systems of units in use in the United States are SI (metric) and BES (English).
- The base units in BES are the foot (length), the pound (force), and the second (time).
- The base units of SI are the meter (length), the kilogram (mass), and the second (time).
- A system of prefixes is used with SI to form prefixed units.
- Conversion factors are used to convert within or between SI and BES.

SELECTED TECHNICAL TERMS

The following technical terms, abbreviations, and acronyms are defined in the glossary located after Chapter 16. You are encouraged to use the glossary to aid your understanding and to test your knowledge of these important terms.

accuracy	meter
addition and subtraction rule	micrometer
ampere	multiplication and division rule
BES	newton
coefficient	precision
digital caliper	roots and powers rule
engineering notation	scientific notation
exponent	SI
joule	tolerance
kilogram	watt

END-OF-CHAPTER QUESTIONS

Write T if the statement is true and F if the statement is false.

1. In powers-of-ten notation, decimal numbers greater than 1 are written with a negative power of ten.

2. The terms *precision* and *accuracy* mean the same thing and may be used interchangeably.

3. Engineering notation is a form of powers-of-ten notation.

4. The definition of a gross (144 items) is an exact number.

5. The number 0.05020 is precise to three significant figures.

6. In this book, when a bar is placed above a digit in a decimal number this indicates the number has been rounded.

7. When two numbers are added, the sum is rounded to the precision of the most precise quantity in the process.

8. The unit of force in BES is the pound.

9. In SI, the kilogram is the unit of force.

10. When using prefixed units, it is conventional to express the decimal coefficient as a number from 0.8 to 900.

In the following, select the word or phrase that makes the statement true.

11. An exact number is totally free from (tolerance, error, precision).

12. Approximate numbers originate from (calculated, physical, measured) quantities.

13. The number 0.02500 is expressed to a precision of (two, three, four) significant figures.

14. Three of the seven base units in SI are (time, force, mass; energy, mass, time; length, mass, time).

15. The powers-of-ten equivalent to the prefix *kilo* is ($10^6, 10^3, 10^{-3}$).

Answer each of the following questions with a short answer in the form of a complete sentence. Include a restatement of the question in your answer.

16. What is wrong with the way 0.028×10^{-2} mm is written? Explain, using the guidelines for the use of the International System of Units and the guidelines for forming multiples and submultiples of SI units (pages 17–19).

17. How should 0.0028 A be written?

18. What physical quantity is measured in units of slugs? Use Table 1.5 to find your answer.

19. Why shouldn't the answer to the expression $(\pi 2.375^2)/4$ be written as 4.430136515? Note that π and 4 are exact numbers.

20. When using your calculator, why is it convenient to have the answer expressed in engineering notation (rather than decimal notation) on the screen of your calculator?

END-OF-CHAPTER PROBLEMS

Determine the value of the indicated physical quantity and express the precision of the answer to the appropriate number of significant figures. Use a unit symbol.

1. $F = ma$; determine force (F) in pounds given $m = 15.0$ slugs and $a = 10.0$ ft/s^2.

2. $v = s/t$; determine the average velocity (v) in meters per second given $s = 50.2$ m and $t = 8.2$ s.

3. $v = (v_i + v_f)/2$; determine the average velocity (v) in feet per second given $v_i = 32$ ft/s and $v_f = 88$ ft/s.

4. $s = v_i t + \frac{1}{2}at^2$; determine the distance (s) in meters given $v_i = 0.26$ m/s, $t = 3.0$ s, and $a = 5.12$ m/s^2.

5. $a = (v_f^2 - v_i^2)/(2s)$; determine the average acceleration (a) in feet per second per second given $v_f = 24$ ft/s, $v_i = 3.0$ ft/s, and $s = 822$ ft.

Using the identities in Table 1.8, convert the following to the specified unit. Express the answer to an appropriate number of significant figures and, where appropriate, use prefixed units with SI or use scientific notation with BES. Clearly show how the identities are used to solve the problems by structuring your work like that shown in Example 1.18, i.e.,

$$7500 \text{ cm} \times \frac{1 \text{ m}}{1000 \text{ cm}} = 75.00 \text{ m}.$$

6. Convert 2.25 km of distance to yards.

7. Convert .028 lb of force to newtons.

8. Convert 2754 g of mass to slugs.

9. Convert 2 450 MJ of energy to foot-pounds.

10. Convert 72.0 ft/s of velocity to meters per second.

11. Convert 62.0 mi/h of speed to meters per second.

12. Convert 4 350 ft-lb/s of power to watts.

Chapter **2**

Linear Motion

About the Illustration   *This Renaissance machine uses a horse-powered rope tow to move carts loaded with excavated dirt. A number of simple machines are compounded to make this machine. Included are the wheel (lever to which the horse is attached) and axle (drum—V), and pulleys (A, E). The circular motion of the horse is translated into linear motion of the carts through the rope wound around the drum (V). By wrapping the rope around the drum several times and with the empty cart attached, sufficient friction is created to ensure the movement of the loaded cart up and out of the excavation. (Designed by Agostino Ramelli, from* The Various and Ingenious Machines of Agostino Ramelli *(1588).)* ■

INTRODUCTION

In order to understand the operation of machines, a knowledge of the basic principles of mechanics is needed. Included in the basic principles of mechanics are the laws of inertia, acceleration, and action and reaction, and the physical quantities of mass, force, linear displacement (distance), velocity (speed), and acceleration (change in velocity).

CHAPTER CONTENTS

PERFORMANCE OBJECTIVES

Once you have read and studied each section; worked through each example with pencil, paper, and calculator; worked through the section and the end-of-chapter problems; and answered the end-of-chapter questions, you should be able to

- Distinguish between force and mass.
- Apply the laws of inertia, acceleration, and action/reaction.
- Understand and use the units for physical quantities of mass, force, distance, speed, and acceleration.
- Differentiate among displacement, velocity, and acceleration.
- Understand and apply average velocity and average acceleration to the solution of problems.

2.1 MECHANICAL PRINCIPLES

Force

You no doubt have an understanding of force. If you want to slide a table across a floor, you push on it. If you want to take a child for a ride in a wagon, you pull on the handle to get the wagon to move. If you want to change the direction of a ball rolling on a floor, you must deflect it. All these life experiences are simple demonstrations of the nature of *matter* and *force*. From this discussion of force it can be stated that **force** *is a push or pull that tends to cause motion or tends to stop motion.*

Law of Inertia

All physical objects are made up of matter. Matter occupies space and has **mass**. Objects, because of their mass, resist change in motion; the property of resisting change is called **inertia**. Mass is a measure of the inertia of an object. Thus, objects with large amounts of mass have more inertia than do objects with small amounts of mass. An important law that describes force is the *law of inertia*. This law is commonly called *Newton's first law of motion*.

> ➤ **Law of Inertia** An object at rest will remain at rest and an object in motion will remain in motion at the same speed and direction unless it is acted upon by an outside force.

The law of inertia explains why your automobile slows down on a level road when you remove your foot from the accelerator. Rather than continue at a con-

FIGURE 2.1
The friction force (F_f) opposes the acceleration force (F_a), causing the moving car to slow.

stant speed, your car loses speed (decelerates). From the law of inertia we learn that the cause is due to an outside force—the force is friction. Figure 2.1 pictures the friction force acting on an automobile to oppose its forward motion.

Law of Acceleration

When a force is applied to an object and the object moves (accelerates), the movement is in the direction of the force. This concept is pictured in Figure 2.2. When an object accelerates (speeds up), then an outside force must be present. To be more specific—a *net* outside force must be present. Since there is usually more than one outside force acting on an object, the term *net force* is used to represent the total of all the forces. Since the mass of an object is an indicator of an object's opposition to a change in speed, objects with a large mass require more net force to speed them up than do objects with a small mass. In summary, when there is a change in the **velocity** (speed) of an object, then there is an **acceleration** (speeding up) or **deceleration** (slowing down) of the object, and there is a net outside force present. The *law of acceleration* describes the acceleration of an object when it is acted upon by a net outside force. This law is often called *Newton's second law of motion*.

> **Law of Acceleration** When a net outside force, F, acts on an object of mass, m, and causes it to accelerate, the acceleration may be computed by the formula $a = F/m$, and the acceleration is in the direction of the net outside force.

Mathematically, the law of acceleration is commonly stated as

$$F = ma \qquad\qquad\qquad (2.1)$$

FIGURE 2.2
The force of acceleration is applied to the ball when it is thrown, causing the ball to move in the direction of the acceleration force (F_a).

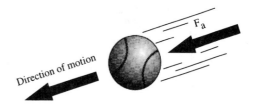

where

F = net outside force, in lb or N

m = mass of object, in slugs (slug) or kg

a = the acceleration of the object due to the force, in ft/s^2 or m/s^2

From the law of acceleration, we learn that for a constant amount of mass the acceleration is directly proportional to the net force (a ∝ F). That is, the greater the net force, the greater the acceleration of the object. In a similar fashion, when the net force is constant, then the acceleration of the object is inversely proportional to the mass of the object (a ∝ 1/m). That is, for a given net force, objects with a smaller mass will be accelerated more than objects with a larger mass. To summarize, a net force is needed to

■ Get a stationary object moving;

■ Stop a moving object;

■ Speed up or slow down (accelerate or decelerate) a moving object;

■ Change the direction of a moving object.

Example 2.1

Determine the amount of net force needed to accelerate a 1,930-lb sports car at 15.0 ft/s^2 if the mass of the car is 60.0 slugs. See Figure 2.3.

SOLUTION   Solve for force.

Given $F = ma$ (lb)

Evaluate $m = 60.0$ slug

$a = 15.0$ ft/s^2

Observation   The unit for mass in BES is the slug. The slug has no assigned unit symbol; however, the unit may be expressed as (lb-s^2)/ft. In this text, the word *slug* will be written out and will represent the unit of mass.

Substitute $F = 60.0 \times 15.0$

Solve $F = 900$ lb

The amount of force needed to accelerate the car at 15.0 ft/s^2 is 900 lb. This is a substantial amount of force compared to the weight of the car. If the 900-lb force causing the 15.0-ft/s^2 acceleration is maintained for a period of 7.0 s, then the car (starting from a stopped position) will be traveling at 105 ft/s (15.0 ft/s^2 × 7.0 s), or 72 mi/h, at the end of 7.0 s.

FIGURE 2.3
Sketch for Example 2.1.

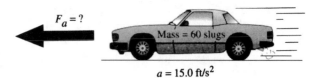

Example
2.2

Determine the acceleration of a 216 000-N (48,600-lb) aircraft that has a mass of 22 000 kg when a net force of 55 000 N is applied through the aircraft engines.

SOLUTION   Solve for acceleration. See Figure 2.4.

Given $F = ma$; solve for a:

$$a = \frac{F}{m} \quad (\text{m/s}^2)$$

Evaluate $F = 55\,\overline{0}00$ N

                $m = 22\,\overline{0}00$ kg

Substitute $a = \dfrac{55\,\overline{0}00}{22\,\overline{0}00}$

Solve $a = 2.50$ m/s^2

With a uniform acceleration of 2.50 m/s^2, the 216 000-N (48,600-lb) aircraft will travel 224 m (734 ft) during the 13.4 s it takes to reach a takeoff speed of 33.4 m/s (75 mi/h).

Example
2.3

A 250-g ball is hit with a bat, resulting in an acceleration of 22 m/s^2. Determine the net force imparted to the ball when the bat initially contacts the ball.

SOLUTION   Find the amount of net force. See Figure 2.5.

Given $F = ma$

Evaluate $m = 250$ g $\Rightarrow 0.250$ kg

                $a = 22$ m/s^2

Observation   The mass must be expressed in kilograms, since the mass in the formula $F = ma$ is defined in kilograms.

Substitute $F = 0.250 \times 22$

Solve $F = 5.5$ N

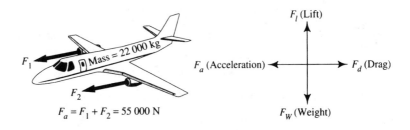

FIGURE 2.4
Sketch for Example 2.2 with the *forces of flight* indicated. Notice the pairs of opposing forces.

FIGURE 2.5
Sketch for Example 2.3.

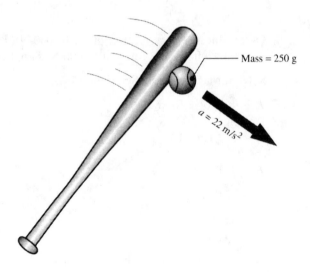

Mass = 250 g

$a = 22$ m/s²

Weight

The mass of an object is acted upon by an attractive force called gravity, which gives the object *weight*. Suppose that you are sitting on a stool and all of a sudden it is yanked out from under you. You will fall to the ground because of the downward pull of the force of gravity. The downward pull of the earth on your mass causes you to accelerate (fall) to the floor at the rate of 32.2 ft/s for each second that you fall (32.2 ft/s²). Thus, the gravitational pull on the mass of your body gives your body weight.

When the mass of an object and the acceleration due to gravity are known, then the weight of the object can be determined from the following formula:

$$F_W = mg \qquad \qquad \textbf{(2.2)}$$

where

F_W = weight of the object, in lb or N
m = mass of the object, in slug or kg
g = acceleration of gravity (32.2 ft/s² or 9.81 m/s²)

Example 2.4

Determine
(a) The weight of a 165-lb person in newtons (N).
(b) The mass of the person in kilograms (kg).

SOLUTION (a) Find the person's weight in newtons (Figure 2.6).

Given 4.448 N = 1.000 lb (Appendix A1)

Convert $165 \text{ lb} \times \dfrac{4.448 \text{ N}}{1 \text{ lb}} = 734 \text{ N}$

FIGURE 2.6
Sketch for Example 2.4.

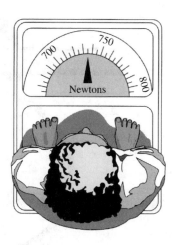

(b) Find the person's mass in kilograms.

Given $F_W = mg$; solve for m:

$$m = \frac{F_W}{g} \quad \text{(kg)}$$

Evaluate $F_W = 734$ N

$g = 9.81$ m/s^2

Substitute $m = \dfrac{734}{9.81}$

Solve $m = 74.8$ kg

   Remember that weight is a force due to the gravitational attraction of the earth. Weight is dependent upon location; that is, in space you are "weightless." On the moon you weigh about one-sixth of what you weigh on the earth at sea level. However, mass is a constant that is independent of where it is located. It is important that you use force units for force and mass units for mass. The pound and newton are force units, whereas the slug and kilogram are mass units.

Example 2.5

Determine the weight in pounds of a delivery truck that has a mass of 1250 kg.

SOLUTION Find the weight of the truck in pounds. See Figure 2.7.

Given $F_W = mg$ (lb) and 1 slug = 14.6 kg (Appendix A1)

Observation Convert the mass in kilograms to slugs.

Evaluate $m = 1250 \text{ kg} \times \dfrac{1 \text{ slug}}{14.6 \text{ kg}} = 85.6$ slug

$g = 32.2$ ft/s^2

FIGURE 2.7
Sketch for Example 2.5.

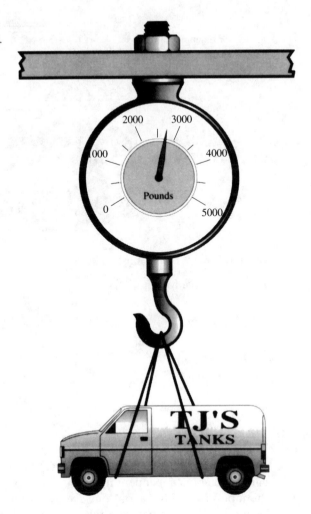

| Substitute | $F_W = 85.6 \times 32.2$ |
| Solve | $F_W = 2760 \text{ lb} \Rightarrow 2.76 \times 10^3 \text{ lb}$ |

Law of Action and Reaction

Figure 2.8 depicts a person sitting on a park bench. The person is at rest; that is, there is no motion. The weight of the person (F_W) exerts a downward force on the bench, and the bench exerts an upward force (called a reactionary force, F_R) on the person that is exactly equal to the person's weight. Since these forces are exactly equal and opposite, the *net force* is zero and the system is in equilibrium (no acceleration or change in motion). This interaction between two objects is described by the *law of action and reaction*. This law is referred to as *Newton's third law of motion*.

FIGURE 2.8
The weight of the person (F_W) exerts a downward force on the bench equal to the person's weight. The bench exerts an upward force (F_R) just equal to the weight of the person. The two forces are equal and opposite; the system is in equilibrium and the net force is zero.

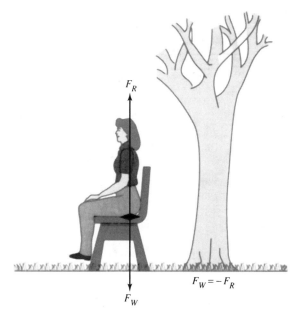

$$F_W = -F_R$$

> **Law of Action and Reaction** Forces always occur in pairs; that is, for every action there is an equal and opposite reaction with equal force but opposite direction.

When you stand on solid ground, your weight exerts a downward force on the ground and the ground, in turn, exerts an upward reactionary force on you that is exactly equal to your weight; the net force in this system of forces is zero. If this law were not so, you would sink into the ground when the force of your weight on the ground exceeded the force of the ground on your weight, or you would rise (levitate) above the ground when the force of your weight on the ground was less than the force of the ground on your weight. Remember that forces come in pairs.

Units of Force, Mass, and Acceleration

The two equations $F = ma$ (2.1) and $F_W = mg$ (2.2) relate force, mass, and acceleration to one another with relatively simple formulas. However, if the units of force, mass, and acceleration are new to you, you may have difficulty sorting them out. Table 2.1 summarizes the units of the two equations. When working with new units, try to get a conceptual understanding of the amount for each of the units. As an example:

- The weight of 1 slug of mass is approximately 32 lb when the acceleration due to gravity (BES) is 32.2 ft/s^2.
- The weight of 1 kg of mass is approximately 10 N, or 2 lb, when the acceleration due to gravity (SI) is 9.81 m/s^2.

TABLE 2.1   Force and Weight

Equation 2.1: $F = ma$			Equation 2.2: $F_W = mg$		
Force (F)	Mass (m)	Acceleration (a)	Weight (F_W)	Mass (m)	Acceleration (g)
SI					
newton (N)	kilogram (kg)	meter/second/second (m/s^2)	newton (N)	kilogram (kg)	meter/second/second (m/s^2)
BES					
pound (lb)	slug $\left(\dfrac{lb\text{-}s^2}{ft}\right)$	feet/second/second (ft/s^2)	pound (lb)	slug $\left(\dfrac{lb\text{-}s^2}{ft}\right)$	feet/second/second (ft/s^2)

- The force of 1 N is approximately equal to the force of $\frac{1}{4}$ lb.
- On a frictionless surface, the force of 1 N will give a mass of 1 kg an acceleration of 1 m/s for each second the force is applied.
- On a frictionless surface, the force of 1 lb will give a mass of 1 slug an acceleration of 1 ft/s for each second the force is applied.

When working with force or weight equations, remember to substitute the base units into the equation. That is, use newtons or pounds for force, slugs or kilograms for mass, and feet per second per second or meters per second per second for acceleration. Furthermore, remember to use units from the same system. The following examples demonstrate these concepts.

Example 2.6

Determine the force imparted to a 16-lb bowling ball if the ball has an acceleration of 45 ft/s^2.

SOLUTION   Solve for the force applied to the ball in pounds.

Given $F = ma$ (lb)

Observation   The weight of the ball (pounds) must first be changed to mass units (slugs) using $m = F_W/g$ where $F_W = 16$ lb and $g = 32.2$ ft/s^2.

Evaluate $m = \dfrac{F_W}{g} = \dfrac{16}{32.2} = .50$ slug

$a = 45$ ft/s^2

Substitute $F = .50 \times 45$

Solve $F = 23$ lb

Example 2.7	Determine the weight in newtons of a block of aluminum that has a mass of 750 g.
SOLUTION	Find the weight in newtons.
Given	$F_W = mg$ (N)
Observation	The mass must be expressed in kilograms (kg).
Evaluate	$m = 750 \text{ g} \times \dfrac{1 \text{ kg}}{1000 \text{ g}} = 0.750 \text{ kg}$
	$g = 9.81 \text{ m/s}^2$
Substitute	$F_W = 0.750 \times 9.81$
Solve	$F_W = 7.36 \text{ N}$

> ➤ **As a Rule** The base unit for mass in the SI system is the kilogram, not the gram.

Example 2.8	Determine the net force on a 15.5-Mg aircraft as it taxis down the runway at a steady acceleration of 10.0 ft/s². Express the answer in pounds.
SOLUTION	Find the force on the aircraft in pounds.
Given	$F = ma$
Observation	The mass is first converted from megagrams to slugs using the identity 1 slug = 14.6 kg. Express 15.5 Mg as 15.5×10^3 kg.
Evaluate	$m = 15.5 \times 10^3 \text{ kg} \times \dfrac{1 \text{ slug}}{14.6 \text{ kg}} = 1.06 \times 10^3 \text{ slug}$
	$a = 10.0 \text{ ft/s}^2$
Substitute	$F = (1.06 \times 10^3)(10.0)$
Solve	$F = 10{,}600 \text{ lb} \Rightarrow 1.06 \times 10^4 \text{ lb}$

The engines of the aircraft provide 10,600 lb of thrust to accelerate the aircraft down the runway. As time passes, the velocity (speed) of the aircraft increases due to the acceleration to a point where it has sufficient speed to create enough lift to enable the plane to "take off."

Exercise 2.1

Solve the following problems. Make sketches to aid in the solution of the problems and structure your work so it follows in an orderly progression and can easily be checked.

1. Determine the net force needed to accelerate a 3800-kg car at 2.5 m/s².
2. Determine the mass of an object that is accelerated to 12 ft/s² by a net force of 540 lb.
3. Determine the acceleration of a 65-kg mass when a net force of 180 N is applied to it.
4. Determine the net force needed to accelerate a 1.8-lb radio-controlled car at 10 ft/s².
5. A trash barrel weighing 250 N is slid across a level, ice-coated driveway (assume no friction) with an acceleration of 0.60 m/s². Determine the force imparted to the barrel.
6. Determine the mass of a person weighing 450 N at sea level.
7. A net force F causes a mass of 550 g to accelerate to 12 m/s². Determine the acceleration of a 2.2-kg mass that is acted upon by the same force.
8. An automobile weighing 2,600 lb picks up speed (accelerates) at 3.00 mi/h per second (3.0 mi/h/s). Determine the net force applied to the vehicle.

2.2 MOTION

Displacement

A continual change in position results when an object is in motion. Linear motion and rotary motion are the two basic types of motion. Linear motion is movement along a straight line, whereas rotary motion is rotation about a central axis. Combinations of linear and rotary motion result in complex motion.

Linear motion, the motion of an object along a straight path, produces a **displacement** of the object from one place to another. The term *displacement* is used to indicate a *change in position* of an object. Displacement is measured in feet or meters and the quantity symbol (the symbol used in formulas) is the letter *s*.

Figure 2.9 pictures an automobile moving along a highway. The car started at time zero (t_0) and finished at time 2 (t_2). During the time between t_0 and t_1, the car traveled distance 1 (d_1). During the time between t_0 and t_2, the car traveled distance 2 (d_2). The displacement (*s*) is the *change in position* of the car between d_1 and d_2.

Velocity

The *average velocity* of an object is a measure of its rate of change of position with time. Since change of position is displacement, the average velocity of travel between two places is the ratio of displacement (*s*) to the time (*t*) taken for the travel:

$$v = \frac{s}{t} \qquad\qquad \textbf{(2.3)}$$

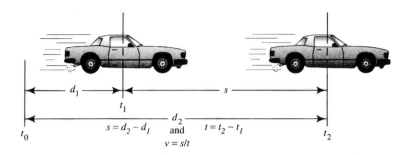

FIGURE 2.9
Displacement or distance (s) is the interval between d_2 and d_1. Time (t) is the interval between t_2 and t_1.
Time and distance are related to the average velocity by the equation $v = s/t$.

where

v = average velocity, in ft/s or m/s
s = displacement (distance), in ft or m
t = time interval in s

The average velocity is used to account for travel of fairly long duration and where the speed of travel varies from time to time. For example, you may average 15 mi/h for a 1-h trip while driving in traffic in the city. During the course of the trip, your speed may vary from 0 to 45 mi/h.

The terms *velocity* and *speed* are often used interchangeably. There is, however, a slight difference in their meaning. Speed is used to tell how fast an object is traveling and is not concerned with the direction of travel. Velocity, on the other hand, takes into consideration both how fast an object is traveling and the direction of travel. Speed and velocity are used interchangeably throughout this text, as are the terms *displacement* and *distance*.

Example 2.9	A runner is timed during the course of a 1500-m race to determine his speed. Determine the runner's average speed in **(a)** meters per second and **(b)** miles per hour if it takes him 76.0 s to complete one lap of the 500.0-m track.
SOLUTION	**(a)** Find the speed of the runner in meters per second.
Given	$v = \dfrac{s}{t}$ (m/s)
Evaluate	$s = 500.0$ m
	$t = 76.0$ s
Substitute	$v = \dfrac{500.0}{76.0}$

Solve $v = 6.58$ m/s

(b) Find the speed of the runner in miles per hour by converting the velocity in meters per second to miles per hour.

Given $v = 6.58$ m/s and 1 m/s = 2.24 mi/h (Appendix A1)

Convert $6.58 \ \cancel{\text{m/s}} \times \dfrac{2.24 \text{ mi/h}}{1 \ \cancel{\text{m/s}}} = 14.7$ mi/h

$$\therefore \quad 6.58 \text{ m/s} = 14.7 \text{ mi/h}$$

Example 2.10

The Amtrak train travels 585 mi from Los Angeles, California, to Flagstaff, Arizona, at an average speed of 50.9 mi/h. Determine its average speed for the trip in **(a)** meters per second and **(b)** feet per second. **(c)** Determine the time to make the trip.

SOLUTION   **(a)** Find the average speed in meters per second by converting the velocity from miles per hour to meters per second.

Given   50.9 mi/h and 1 m/s = 2.24 mi/h (Appendix A1)

Convert $50.9 \ \cancel{\text{mi/h}} \times \dfrac{1 \text{ m/s}}{2.24 \ \cancel{\text{mi/h}}} = 22.7$ m/s

$$\therefore \quad 50.9 \text{ mi/h} = 22.7 \text{ m/s}$$

(b) Find the average speed in feet per second by converting the velocity from miles per hour to feet per second.

Given   50.9 mi/h and 88 ft/s = 60 mi/h (exact)

Convert $50.9 \ \cancel{\text{mi/h}} \times \dfrac{88 \text{ ft/s}}{60 \ \cancel{\text{mi/h}}} = 74.7$ ft/s

$$\therefore \quad 50.9 \text{ mi/h} = 74.7 \text{ ft/s}$$

(c) Find the time needed to make the trip in hours.

Given $v = \dfrac{s}{t}$; solve for t.

$$t = \dfrac{s}{v}$$

Evaluate $s = 585$ mi

$v = 50.9$ mi/h

Substitute $t = \dfrac{585}{50.9}$

Solve $t = 11.5$ h

Acceleration

You may remember from the previous definition of velocity that velocity is the rate of change of position with time. Acceleration, like velocity, is a rate of change. *Acceleration* is the rate of change of velocity with time. Simply stated, acceleration is equal to a change in velocity divided by the time interval during which acceleration is taking place.

We experience acceleration when we drive away from the curb in our automobile or when we are in an aircraft moving down a runway to take off. In the case of an automobile, acceleration is seen in the changing velocity indication of the speedometer. Initially the reading is zero; however, as the accelerator pedal is depressed, the car begins to accelerate and the speedometer indicates 10 mi/h, then 20 mi/h, etc. The car is accelerating because the velocity, as indicated by the speedometer, is increasing (changing). Once the car is up to cruising speed, the speedometer remains steady, indicating no further increase in velocity—the car is no longer accelerating. When the brake is applied, the car slows, as indicated by a drop in the speedometer reading, and negative acceleration (deceleration) is experienced.

The average acceleration may be computed by taking the difference between the final and the initial velocity and dividing this difference by the interval of time between the two velocities. Equation 2.4 expresses average acceleration in terms of the change in velocity with time. Figure 2.10 pictures the change in velocity seen on a speedometer.

$$a = \frac{v_f - v_i}{t}$$

(2.4)

where

a = average acceleration, in ft/s^2 or m/s^2
v_f = final velocity, in ft/s or m/s
v_i = initial velocity, in ft/s or m/s
t = time interval during acceleration, in s

FIGURE 2.10
The average acceleration (a) of a vehicle may be determined by taking the change in velocity (v) over the time interval (t) during acceleration: $a = v/t$, where $v = v_f - v_i$ and $t = t_2 - t_1$.

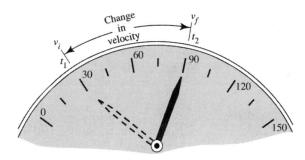

Example 2.11

Determine the average acceleration (in miles per hour per second) indicated by the race-car speedometer of Figure 2.10 when the time interval is 5.0 s.

SOLUTION

Solve for the average acceleration in miles per hour per second for an initial velocity of 30 mi/h and a final velocity of 90 mi/h.

Given

$$a = \frac{v_f - v_i}{t}$$

Evaluate

$v_f = 90$ mi/h

$v_i = 30$ mi/h

$t = 5.0$ s

Substitute

$$a = \frac{90 - 30}{5.0}$$

Solve

$a = 12$ mi/h/s

The car is gaining speed at the rate of 12 mi/h for each second of acceleration. That is, at the end of the first second of acceleration, the speed is 30 mi/h + 12 mi/h, or 42 mi/h; it is 54 mi/h at the end of the second second, 66 mi/h at the end of the third second, and 78 mi/h at the end of the fourth second. At the end of the fifth second, the speed is 90 mi/h.

Final Velocity

The *final velocity* of an object can be determined (when there is uniform acceleration and straight-line motion) if the initial velocity, average acceleration, and time interval are known. The necessary equation is stated as Equation 2.5. This is the acceleration equation (Equation 2.4), which has been solved for the final velocity:

$$v_f = v_i + at \tag{2.5}$$

where

v_f = final velocity, in ft/s or m/s
v_i = initial velocity, in ft/s or m/s
a = average acceleration, in ft/s² or m/s²
t = time interval during acceleration, in s

If the race car in the previous example has the same average acceleration of 12 mi/h/s, how fast will it be going at the end of 7.0 s when the initial velocity is 30 mi/h? Equation 2.5 can help find the answer to this question. Thus, $v_f = 30 + (12 \times 7.0) = 114$ mi/h.

Example 2.12

A freight train traveling at 30 m/s (67 mi/h) applies its air brakes and uniformly decelerates at -0.20 m/s². Determine its velocity in meters per second after 90 s.

Observation	The acceleration is given a negative sign to indicate deceleration.
SOLUTION	Find the final velocity (v_f).
Given	$v_f = v_i + at$
Evaluate	$v_i = 30$ m/s
	$a = -0.20$ m/s^2
	$t = 90$ s
Substitute	$v_f = 30 + (-0.20 \times 90)$
Solve	$v_f = 12$ m/s (27 mi/h)

Example 2.12 points out the long stopping time (2.5 min for a complete stop) and distance (7200 ft) needed by objects such as freight trains that have very large masses. The obvious implication is that, by the time the engineer of the train sees a vehicle on the track, it is too late to stop the train.

Interval of Time

When an object is accelerated, it takes an *interval of time* (the time from the start of acceleration to the end of acceleration) for the object to come up to speed. When the interval of time for acceleration is needed, then it may be computed using Equation 2.6 if it is assumed that the object is being uniformly accelerated in a straight line. Equation 2.6 may be used to find the interval of time when the initial and final velocities and the average acceleration are known.

$$t = \frac{v_f - v_i}{a} \qquad \textbf{(2.6)}$$

where

t = interval of time during acceleration, in s
v_f = final velocity, in ft/s or m/s
v_i = initial velocity, in ft/s or m/s
a = average acceleration, in ft/s^2 or m/s^2

Example *2.13*	An aircraft starts from rest at the end of a runway and uniformly accelerates at 8.80 ft/s^2 to a velocity of 132 ft/s (90.0 mi/h) at takeoff. Determine the interval of time for takeoff.
SOLUTION	Find the time interval in seconds.
Given	$t = \dfrac{v_f - v_i}{a}$
Observation	The initial velocity is 0.00 ft/s, since the aircraft starts from rest.

Evaluate	$v_f = 132$ ft/s
	$v_i = 0.00$ ft/s
	$a = 8.80$ ft/s^2
Substitute	$t = \dfrac{132 - 0.00}{8.80}$
Solve	$t = 15.0$ s

Average Velocity

The average velocity of an object moving over a course during the time interval in which average acceleration is taking place can be determined by taking the average of the initial velocity (v_i) and the final velocity (v_f).

$$v = \frac{v_i + v_f}{2} \tag{2.7}$$

where
> v = the average velocity, in ft/s or m/s
> v_i = initial velocity, in ft/s or m/s
> v_f = final velocity, in ft/s or m/s

Example **2.14**	A cyclist traveling at a constant 12 mi/h accelerates uniformly for 10.0 s to a velocity of 16 mi/h. Determine the average velocity, in feet per second, during the 10.0 s of acceleration.
SOLUTION	Find the average velocity of the cyclist in feet per second.
Given	$v = \dfrac{v_i + v_f}{2}$
Evaluate	$v_i = 12$ mi/h
	$v_f = 16$ mi/h
Substitute	$v = \dfrac{12 + 16}{2}$
Solve	$v = 14$ mi/h
Convert	$14 \text{ mi/h} \times \dfrac{3.28 \text{ ft/s}}{2.24 \text{ mi/h}} = 21$ ft/s

Distance

The last parameter to be considered is the distance traveled by an object while it is being accelerated by a net force. Assuming uniform acceleration, the distance traveled can be determined from the definition of average velocity. The average velocity was first defined as $v = s/t$ in Equation 2.3 and later as $v = (v_i + v_f)/2$ in Equation 2.7. The distance traveled during the period of acceleration can be determined by setting these two equations equal to each other and solving for the distance (s).

$$v = s/t \quad \text{but} \quad v = \frac{v_i + v_f}{2} \quad \text{so} \quad \frac{s}{t} = \frac{v_i + v_f}{2} \quad \text{and}$$

$$s = \left(\frac{v_i + v_f}{2}\right)t \qquad (2.8)$$

where

$\quad s$ = distance, in ft or m
$\quad v_i$ = initial velocity, in ft/s or m/s
$\quad v_f$ = final velocity, in ft/s or m/s
$\quad t$ = time interval during acceleration, in s

Example 2.15

A switch engine in a rail yard accelerates uniformly from 22.0 ft/s (15 mi/h) to 35.0 ft/s (24 mi/h) in 12.0 s.

(a) Determine how far the engine travels while it is accelerating.
(b) Find the average acceleration during the 12-s time period.

SOLUTION

(a) Find the distance the train travels in the 12 s.

Given

$$s = \left(\frac{v_i + v_f}{2}\right)t$$

Evaluate

$$v_i = 22.0 \text{ ft/s}$$

$$v_f = 35.0 \text{ ft/s}$$

$$t = 12.0 \text{ s}$$

Substitute

$$s = \left(\frac{22.0 + 35.0}{2}\right)12.0$$

Solve

$$s = 342 \text{ ft}$$

(b) Find the average acceleration for the 12-s period.

Given

$$a = \frac{v_f - v_i}{t} \qquad \text{(Equation 2.4)}$$

Evaluate

$$v_f = 35.0 \text{ ft/s}$$

$$v_i = 22.0 \text{ ft/s}$$

$$t = 12.0 \text{ s}$$

Substitute

$$a = \frac{35.0 - 22.0}{12.0}$$

Solve

$$a = 1.08 \text{ ft/s}^2$$

Example
   2.16   A sports car with a weight of 1,932 lb uniformly accelerates from zero to a velocity of 88.0 ft/s in 5.87 s. Determine each of the following.

   (a) Average acceleration
   (b) Average velocity during acceleration
   (c) Distance traveled during acceleration
   (d) Mass of the car
   (e) Net force needed to produce the average acceleration

SOLUTION   **(a)** Find the acceleration (ft/s^2).

Given $a = \dfrac{v_f - v_i}{t}$ (Equation 2.4)

Evaluate $v_f = 88.0$ ft/s

             $v_i = 0.00$ ft/s

             $t = 5.87$ s

Substitute $a = \dfrac{88.0 - 0.00}{5.87}$

Solve $a = 15.0$ ft/s^2

             (b) Find the average velocity during acceleration (ft/s).

Given $v = \dfrac{v_i + v_f}{2}$ (Equation 2.7)

Evaluate $v_i = 0.00$ ft/s

             $v_f = 88.0$ ft/s

Substitute $v = \dfrac{0.00 + 88.0}{2}$

Solve $v = 44.0$ ft/s

             (c) Find the distance (ft) traveled in 5.87 s.

Given $s = vt$ (Equation 2.8)

Evaluate $v = 44.0$ ft/s

             $t = 5.87$ s

Substitute $s = 44.0 \times 5.87$

Solve $s = 258$ ft

             (d) Find the mass of the car in slugs.

Given $m = \dfrac{F_W}{g}$ (Equation 2.2)

Evaluate	$F_W = 1932$ lb
	$g = 32.2$ ft/s^2
Substitute	$m = \dfrac{1932}{32.2}$
Solve	$m = 60.0$ slug

(e) Find the net force needed to accelerate the car.

Given	$F = ma$ (Equation 2.1)
Evaluate	$m = 60.0$ slug
	$a = 15.0$ ft/s^2
Substitute	$F = 60.0 \times 15.0$
Solve	$F = 900$ lb

Table 2.2 summarizes the various equations for force, weight, average acceleration, average velocity (speed), time, and distance (displacement). Included at the bottom of Table 2.2 are two equations derived from the earlier equations in Table 2.2. These equations are very useful in solving for distance and acceleration. The first provides a solution for distance when initial velocity, time, and acceleration are known. It is sometimes called the $\frac{1}{2} at^2$ *equation.*

$$s = v_i t + \tfrac{1}{2} at^2 \tag{2.9}$$

where

$\quad s$ = distance, in ft or m
$\quad v_i$ = initial velocity, in ft/s or m/s
$\quad t$ = time interval during acceleration, in s
$\quad a$ = average acceleration, in ft/s^2 or m/s^2

The second equation, commonly called the *2as equation*, is useful in solving for acceleration or distance when initial and final velocity are known along with either distance or acceleration.

$$2as = v_f^2 - v_i^2 \tag{2.10}$$

where
$\quad a$ = average acceleration, in ft/s^2 or m/s^2
$\quad s$ = distance, in ft or m
$\quad v_f$ = final velocity, in ft/s or m/s
$\quad v_i$ = initial velocity, in ft/s or m/s

Example 2.17

In a bicycle race, a cyclist overtakes her opponent by increasing her average velocity from 10.0 m/s to 12.0 m/s in a distance of 35 m.
(a) Determine the average acceleration (m/s^2).
(b) Find the time to overtake her opponent.

TABLE 2.2 Summary of Formulas Used in Chapter 2

Physical Quantity	Unit	Equation Number	Equation
Force	N or lb	2.1	$F = ma$
Weight	N or lb	2.2	$F_W = mg$
Velocity (average)	m/s or ft/s	2.3	$v = \dfrac{s}{t}$
Acceleration (average)	m/s^2 or ft/s^2	2.4	$a = \dfrac{v_f - v_i}{t}$
Final velocity	m/s or ft/s	2.5	$v_f = v_i + at$
Time	s	2.6	$t = \dfrac{v_f - v_i}{a}$
Velocity (average)	m/s or ft/s	2.7	$v = \dfrac{v_i + v_f}{2}$
Displacement (distance)	m or ft	2.8	$s = \left(\dfrac{v_i + v_f}{2}\right)t$
Displacement (distance)	m or ft	2.9	$s = v_i t + \frac{1}{2}at^2$
Acceleration or distance	m/s^2 or ft/s^2 m or ft	2.10	$2as = v_f^2 - v_i^2$

SOLUTION   **(a)** Find the acceleration using the 2*as* equation.

Given $2as = v_f^2 - v_i^2$; solve for *a*:

$$a = \frac{v_f^2 - v_i^2}{2s}$$

Evaluate $v_f = 12.0$ m/s

$v_i = 10.0$ m/s

$s = 35$ m

Substitute $a = \dfrac{12^2 - 10^2}{2 \times 35}$

Solve $a = 0.63$ m/s^2

(b) Find the time it took to overtake her opponent.

Given $t = \dfrac{v_f - v_i}{a}$ (from Equation 2.4)

Evaluate $v_f = 12.0$ m/s

$v_i = 10.0$ m/s

$a = 0.63$ m/s^2

Substitute	$t = \dfrac{12 - 10}{0.63}$
Solve	$t = 3.2$ s

Example 2.18

A carpenter working on a three-story apartment building inadvertently drops his tape measure from 35 ft above the ground. Assuming no resistance from the air, determine how long it will take the tape to hit the ground.

SOLUTION	Find the time for the tape to fall to the ground.
Given	$s = v_i t + \frac{1}{2}at^2$. However, $v_i = 0$, so $s = \frac{1}{2}at^2$. Solve for t:
	$t = \sqrt{\dfrac{2s}{a}}$
Evaluate	$s = 35$ ft
	$a = g = 32.2$ ft/s^2 (acceleration due to gravity)
Substitute	$t = \sqrt{\dfrac{2(35)}{32.2}}$
Solve	$t = 1.5$ s

Exercise 2.2

Solve the following problems. Make sketches to aid in solving the problems and structure your work so it follows in an orderly progression and can easily be checked.

1. You are at a football game with your handy stopwatch when the left end receives a pass at his 20.0-yd line; you press your stopwatch as he crosses his 30.0-yd line and runs straight down the field along the sideline. At the opponent's 20.0-yd line you stop your watch and read a time of 8.20 s. Determine the receiver's average velocity
   (a) In feet per second.
   (b) In miles per hour.
2. A high-speed interceptor aircraft is initially traveling at 1930 km/h when it is ordered to overtake an aircraft directly ahead. The interceptor uniformly increases its speed to 2420 km/h over 4.70 s. Determine
   (a) Its average acceleration in km/h/s.
   (b) Its average acceleration in m/s^2.
3. You pull your car onto the interstate and begin to accelerate uniformly up to cruising speed. As the speedometer on your vehicle passes 30.0 mi/h, you start your handy stopwatch; you continue timing for 10.0 s, at which time you note the speedometer is indicating 60.0 mi/h. Determine
   (a) The average velocity (ft/s) during acceleration.

      (b) The average acceleration (mi/h/s).

      (c) The distance traveled (ft) during acceleration.

4. The ram on a large horizontal extrusion press reduces speed uniformly from 22.0 ft/s to 6.0 ft/s in a distance of 4.0 ft. Determine

      (a) The remaining distance (ft) it will travel before coming to a complete stop, assuming the same uniform deceleration.

      (b) The total time taken for the ram to decelerate to a complete stop.

5. A 4,200-lb automobile is cruising down an interstate at 65 mi/h (95.3 ft/s) when the driver applies the brakes. If the braking force is a uniform 2,850 lb, determine

      (a) The time to bring the vehicle to a complete stop assuming uniform deceleration.

      (b) The distance traveled (ft) during the deceleration.

      (c) The velocity of the vehicle (ft/s) after 3.20 s of braking.

6. An automatic guided vehicle (AGV) in a computer-controlled warehouse is traveling at a speed of 12.0 ft/s (8.18 mi/h). When the vehicle is signaled to stop, it decelerates uniformly from the initial speed of 12.0 ft/s to a final speed of 0.00 ft/s in 18.0 ft. Determine the average deceleration.

7. The arm of an industrial robot without tooling and payload has a mass of 30.0 kg. If the arm accelerates from its initial velocity of 1.0 m/s to 1.6 m/s in a distance of 2.2 m, determine

      (a) Its average acceleration.

      (b) The force of acceleration.

      (c) The force of acceleration with 65 N of tooling and a payload of 82 N.

8. Circuit board card holders are moved along an automatic assembly line by a conveyor system, which accelerates the holders from station to station at an average acceleration of 125 cm/s^2 and decelerates the holders at an average deceleration of -125 cm/s^2. If the stations are 2.00 m apart and the holders are accelerated to a constant velocity of 0.500 m/s, determine

      (a) The time and distance for acceleration of the card holder from a stopped condition to a speed of 0.500 m/s.

      (b) The time and distance for deceleration of the card holder from a velocity of 0.500 m/s to a stopped condition.

      (c) The time and distance for movement between acceleration and deceleration at the constant velocity of 0.500 m/s.

      (d) The time it takes to move the card holder 2.00 m between workstations.

CHAPTER SUMMARY

- Force is a push or pull that tends to cause motion or tends to stop motion.
- Inertia is the tendency of an object to resist a change in position or a change in motion.
- Mass is a measure of the inertia of an object; heavy objects have more inertia than light objects.

- The law of inertia explains the fact that objects in motion stay in motion and objects at rest remain at rest.
- Force of acceleration is a net outside force applied to an object to cause it to speed up (accelerate) or slow down (decelerate).
- The law of acceleration is stated mathematically as $F = ma$.
- Weight is the result of gravity acting on the mass of an object.
- The law of action and reaction states that forces occur in pairs that are equal but opposite to one another.
- Displacement (distance) is used to indicate a change in position of an object.
- Velocity is the time rate of change of the distance (displacement) of an object.
- Acceleration is the time rate of change of speed (velocity) of an object.
- Average acceleration is the difference between the final and initial velocity of an object being accelerated divided by the interval of time during which acceleration takes place.
- Average velocity during acceleration is the average of the sum of the initial and final velocities of the object being accelerated.

SELECTED TECHNICAL TERMS

The following technical terms, abbreviations, and acronyms are defined in the glossary located after Chapter 16. You are encouraged to use the glossary to aid your understanding and to test your knowledge of these important terms.

acceleration	law of action and reaction
displacement	law of inertia
force	mass
inertia	velocity
law of acceleration	

END-OF-CHAPTER QUESTIONS

Write T if the statement is true and F if the statement is false.

1. The inertia of an object is directly related to its mass.
2. An object at rest will move only when a net outside force is applied to it.
3. An object being accelerated moves in the direction of the net outside force.
4. Pounds and kilograms are force units.
5. Forces occur in pairs.
6. A change in the position of an object is measured in units of newtons.
7. The terms *velocity* and *speed* are used interchangeably in straight-line motion.
8. When working problems with deceleration, a negative sign is placed in front of the number representing acceleration.
9. When a car speeds up from 20 mi/h to 45 mi/h, the initial velocity of the car is 35 mi/h.
10. The 2*as* equation allows for the solution of either acceleration or force.

In the following, select the word or words that make the statement true.

11. For every force there is an equal and (negative, opposite) force.

12. A change in an object's speed is due to a net outside (velocity, force, displacement).

13. The property of an object that causes it to remain at rest is due to its (velocity, acceleration, inertia).

14. Assuming a uniform acceleration force, the rate of acceleration of an object depends on its (mass, change in velocity, weight).

15. When a magician pulls a tablecloth from under the dishes resting on the cloth, the law of (acceleration, action-reaction, inertia) best explains why the dishes are not moved.

Answer each of the following questions with a short answer in the form of a complete sentence. Include a restatement of the question in your answer.

16. What forces act on a candy dish at rest on a table?

17. In terms of action-reaction, what happens between the oar face and the water when a boat is rowed?

18. A 50-lb carton of canned food is at rest on the floor of a warehouse; determine the net force on the carton when a 10-lb stray cat is asleep on top of the carton. Explain how you arrived at your answer.

19. If an object starts from rest, how far (in meters) will a heavy object fall in one second when dropped from a third story window?

20. An object is accelerated at 1.5 ft/s^2 by a net force F; determine the acceleration when the net force is doubled to $2F$. Explain how you arrived at your answer.

END-OF-CHAPTER PROBLEMS

Solve the following problems. Make sketches to aid in solving the problems and structure your work so it follows in an orderly progression and can easily be checked.

1. Determine the force of acceleration needed to move a 40.0-kg gantry robot from rest to 4.0 m/s in 12 s.

2. Determine the final velocity of a forklift that accelerates from rest for 8.0 s at 1.5 ft/s^2.

3. A 115-N automatic guided vehicle is moved with an accelerating force of 22.5 N. Determine the average acceleration.

4. Determine the time it would take to accelerate a 9400-N overhead crane from 1.5 m/s to 4.2 m/s if the average acceleration is 0.80 m/s^2.

5. Determine the distance (in feet) it would take to bring an overhead conveyor system to a stop if the conveyor was moving 5.6 ft/s at the time it was shut down. The rate of deceleration is −9.0 in/s^2.

6. A belt conveyor system is used to move product from the mine site to the processing plant. The combined weight of the belt and product is 25,000 lb. Determine
   (a) The time it takes for the belt to come up to its running speed of 15.0 ft/s from a stopped condition if it accelerates at a uniform .335 ft/s^2.
   (b) The net force on the belt during acceleration.

7. After machining, a 345-N cast iron part is removed from a holding fixture by the ram from a pneumatic cylinder located beneath the part. The completed part is pushed 10.2 cm up and out of the fixture in 820 ms, enabling a robot to grasp the part. Determine the force of the ram in newtons.

FIGURE 2.11
Dot matrix printer mechanism for
Problem 8.

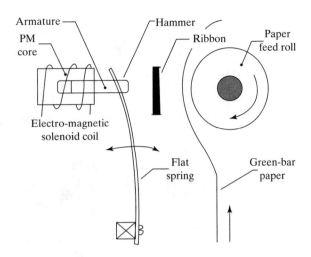

8. The armature of a hammer from a commercial dot matrix printer (pictured in Figure 2.11) is attached to a flat spring that is held under tension in the permanent magnet core of an electromagnetic solenoid. The 6.40-g hammer is released (driven forward 0.280 mm) when a pulse of electric current passes through the coil of the solenoid. The dot-shaped end of the hammer strikes the ribbon in 0.760 ms, printing a dot onto the paper. Determine
   (a) The acceleration of the hammer using Equation 2.9.
   (b) The final velocity of the hammer.
   (c) The force of the hammer on the ribbon.

Principles of Simple Machines and the Lever

Chapter **3**

About the Illustration. *This Renaissance machine utilizes a class 1 lever in conjunction with a screw (and nut) to remove large, heavy doors from their hinges. In use, the claw end of the lever is placed beneath the door. The nut (X) is then turned with a wrench (E), causing lever P to descend, thereby raising the door up and off its hinge pins. (Designed by Agostino Ramelli, from* The Various and Ingenious Machines of Agostino Ramelli *(1588).)* ■

INTRODUCTION

The law of simple machines is helpful in understanding the operation of machines. This law and the concepts of ideal mechanical advantage (IMA), actual mechanical advantage (AMA), and efficiency make up the fundamental principles of simple machines. Each of these concepts, along with force, displacement, and equilibrium, is used in the study and analysis of the first of the simple machines, the lever. The concept of a moment of force is introduced in this chapter.

CHAPTER CONTENTS

PERFORMANCE OBJECTIVES

Once you have read and studied each section; worked through each example with pencil, paper, and calculator; worked through the section and the end-of-chapter problems; and answered the end-of-chapter questions, you should be able to

- Identify the five basic machines and the two categories that divide them.
- Differentiate between a machine element and a mechanism.
- Identify the principle sources of power for machines in modern industry.
- Apply the law of simple machines.
- Use the principles of mechanical advantage, velocity ratio, and efficiency to solve for machine parameters.
- Classify levers.
- Understand how moments of force are used to bring a machine to a state of equilibrium.

3.1   INTRODUCTORY CONCEPTS

Ancient Machines

Machines have evolved through the ages from crude stone implements to the machines of today. As you look around, you may see machines that are direct descendants of people's earliest tools. Figure 3.1 pictures the crowbar (lever), block and tackle (pulley), winch (wheel and axle), ramp (wedge), and screw—all examples of modern-day simple machines adapted from ancient machines. Each machine came into being from people's need to do things that were not possible with bare hands and body strength. Machines have served people by making physical work easier.

Machine Characteristics

Machines amplify force; the claw hammer pictured in Figure 3.2 can be used to remove a nail from a board. In using the hammer for this task, you may have noticed that the effort applied to the handle is transferred across the *pivot point* to the nail and that the handle of the hammer moves much farther than the nail moves. The relationship between force and distance is one of the fundamental principles of machines.

➤ **As a Rule** Machines enable us to amplify force at the expense of distance.

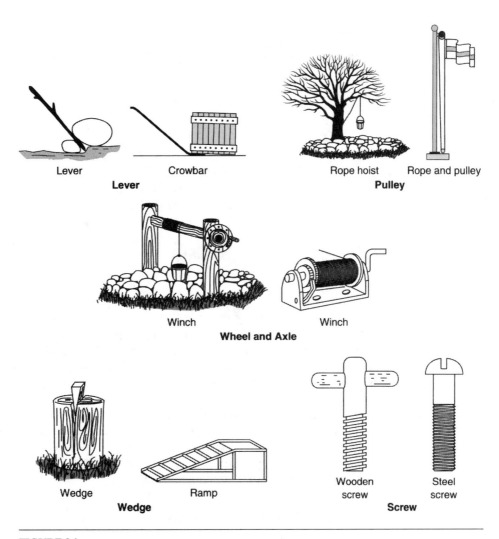

FIGURE 3.1
Ancient machines have evolved over time into today's simple machines.

Machines amplify speed; for example, a bicycle provides a means of transforming force into motion. When a bicycle is shifted into its highest gear, it takes considerable force to move the pedals. However, a gain in speed results. This concept is also a fundamental principle of machines.

➤ **As a Rule** Machines enable us to amplify speed at the expense of force.

Notice that in each of these examples speed or distance and force are traded for one another. *Machines do not allow an amplification in both speed and force or distance and force at the same time.*

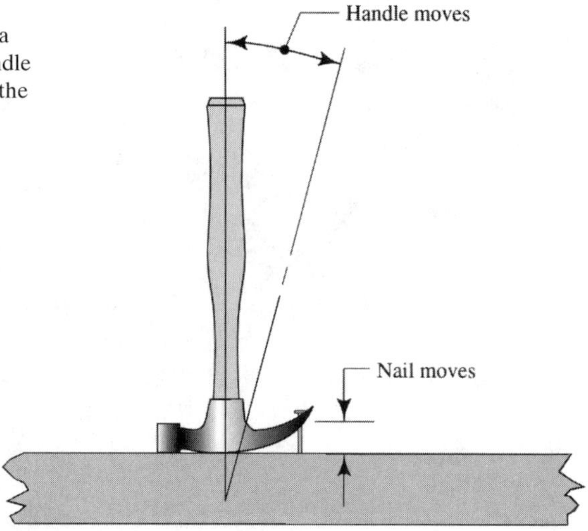

FIGURE 3.2
A claw hammer is used to remove a nail. Note that the distance the handle moves is greater than the distance the nail moves.

Machines change direction; for example, the rope and pulley on a flagpole allow us to raise a flag while standing on the ground. Without the ability of a machine to change direction, the seemingly simple task of raising a flag would require the raiser to scale the pole and attach the flag to the top of the pole—both a physically demanding and dangerous job.

The five simple machines (known to ancient civilizations) are the **lever**, the **pulley**, the **wheel and axle**, the **inclined plane**, and the **screw**. These simple machines are the basis of all compound machines. Once you have studied simple machines and become familiar with their principles of operation, then you will be able to understand the working of compound machines.

Machines and Machine Elements

A *machine* is a device that is used to do work by converting or transferring applied force from one point to another. A piano, a copier, a food processor, a lawn mower, and a can opener are only a few of the endless number of compound machines. Machines are used to change the amount of force, the direction of the applied force, or the point to which force is applied so that daily tasks are made easier.

All compound machines are made up of two or more simple machines in combination with various *machine elements*. Gears, bearings, springs, links, couplers, clutches, and escapements are examples of machine elements. The various parts of a machine, arranged to transform the input motion and force into the desired output motion and force, are collectively referred to as a *mechanism*. The internal mechanism of a piano (the machine) transforms the touch of the pianist (input energy/work) into vibrations of the strings to create the music that is heard (the output energy). Figure 3.3 pictures the mechanism used by Cristofori, the creator of the modern piano.

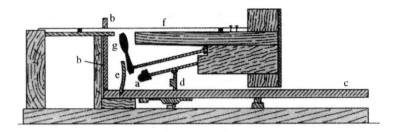

FIGURE 3.3
Early piano mechanism: (a) Under-hammer; (b) Damper; (c) Key; (d) Escapement; (e) Spring; (f) String; (g) Hammer. *(Reprinted by permission of the Oxford University Press. Musical Instruments by Karl Geiringer © 1945, New York.)*

Prime Movers

The principle sources of power for machines used in today's modern industries can be categorized into three general power groups—electrical, mechanical, and fluid. Electrical sources include the *electric motor*, which is powered by electricity in the form of either alternating current (ac) or direct current (dc). Mechanical sources include the *internal combustion engine*, powered by fossil fuel, and fluid sources include *hydraulic* and *air motors*. Each of these **prime movers** is the initial source of input power needed to drive machines.

3.2 PRINCIPLES OF SIMPLE MACHINES

Law of Simple Machines

An **ideal machine** is a machine that has no loss in energy; that is, the force applied through a distance at the input (*work in*, W_{in}) exactly equals the resulting force applied through a distance at the output (*work out*, W_{out}).

➤ **As a Rule** Work = force × distance.

Thus, for the ideal machine, there are no losses due to friction, and the work (energy) into the machine equals the work (energy) out of the machine.

➤ **Law of Simple Machines** In the ideal machine, the input work ($F_E \times s_E$) is exactly equal to the output work ($F_R \times s_R$).

Thus, $W_{in} = W_{out}$ and

$$F_E \times s_E = F_R \times s_R \qquad (3.1)$$

where

F_E = force applied at the input of the machine (effort force), in N or lb
s_E = distance over which the input force is applied (effort distance), in m or ft

F_R = force applied at the output of the machine (resistance force), in N or lb

s_R = distance over which the output force is applied (resistance distance), in m or ft

Example 3.1

Verify that the conditions of the hand-brake lever, pictured in Figure 3.4, are valid by using the law of simple machines (Equation 3.1).

SOLUTION

Show that the product of force and distance on the effort side (input) of the hand brake is equal to the product of force and distance resulting on the resistance side (output) of the ideal machine.

Given

$F_E \times s_E = F_R \times s_R$ and assume the hand brake is an ideal simple machine with no loss due to friction.

Evaluate

$F_E = 8.00$ lb

$s_E = 18.0$ in

$F_R = 72.0$ lb

$s_R = 2.00$ in

Substitute

$8.00 \times 18.0 = 72.0 \times 2.00$

Solve

$144 = 144$

Therefore, 144 in-lb of input equals 144 in-lb of output.

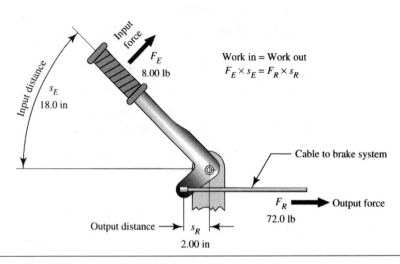

FIGURE 3.4
Hand-brake lever for Example 3.1.

Mechanical Advantage

Each machine assists the person using the machine by a specific factor. The measure of a machine's usefulness is called the machine's *mechanical advantage*. The machine's mechanical advantage is referred to as the *actual mechanical advantage* (**AMA**); it is stated as the ratio between the force of resistance (F_R) and the force of effort (F_E).

$$AMA = \frac{F_R}{F_E} \qquad (3.2)$$

where

AMA = the mechanical advantage
F_E = effort force, in N or lb
F_R = resistance force, in N or lb

Efficiency

In the ideal machine, the energy (work) into the machine cannot exceed the work (energy) out of the machine. In actual machines (less than ideal), the losses may be considerable. The major source of loss is the friction force that opposes the effort force. The losses due to friction (heat, drag, etc.) are accounted for in the **efficiency** of the machine. Efficiency, represented by the Greek letter η (eta), is the ratio of useful work out of the machine (W_{out}) to total work into the machine (W_{in}).

$$\eta = \frac{W_{out}}{W_{in}} \qquad (3.3)$$

where

η = efficiency—a unitless decimal fraction
W_{out} = work out of the machine, in J or ft-lb
W_{in} = work into the machine, in J or ft-lb

Many machines have been constructed to convert, or transform, energy from one form to another. In the process of transforming the energy, some energy is lost (usually as heat). The law of conservation of energy accounts for the energy in a system.

> ➤ **The Law of Conservation of Energy** Energy may be converted or transformed, but energy cannot be created or destroyed.

To put this mathematically, input energy = output energy + lost energy. Figure 3.5 illustrates this concept. The motor converts electrical energy (W_{in}) to mechanical energy (W_{out}) and, in the process, some energy is lost (W_{lost}) to friction and other sources in the form of heat.

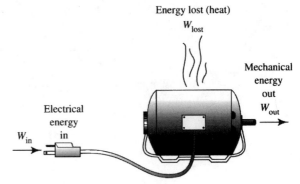

FIGURE 3.5
Energy is conserved when energy is converted from form to form: $W_{in} = W_{out} + W_{lost}$.

Example 3.2

For the motor pictured in Figure 3.5, determine
  (a) The efficiency when 150 kJ of energy is put into the motor and 95 kJ is converted to useful work at the output of the motor.
  (b) The amount of lost energy.

SOLUTION

(a) Find the efficiency of the motor.

Given

$$\eta = \frac{W_{out}}{W_{in}}$$

Evaluate

$W_{out} = 95 \text{ kJ}$

$W_{in} = 150 \text{ kJ}$

Substitute

$$\eta = \frac{95}{150}$$

Observation

Since both energies are expressed in kilojoules, then kilojoules factor out, and the pure unitless ratio, $\frac{95}{150}$ is left.

Solve

$\eta = 0.63$

(b) Find the amount of lost energy.

Given

$W_{lost} = W_{in} - W_{out}$

Evaluate

$W_{in} = 150 \text{ kJ}$

$W_{out} = 95 \text{ kJ}$

Substitute

$W_{lost} = 150 \text{ kJ} - 95 \text{ kJ}$

Solve

$W_{lost} = 55 \text{ kJ}$

Velocity Ratio

For the ideal machine with no frictional losses, the efficiency is equal to 1 ($\eta = 1$). The maximum mechanical advantage for a machine is the ratio of the amount of movement made by the effort (s_E) to the amount of movement made by the resistance (s_R). This is the *velocity ratio* and it is referred to as the *ideal mechanical advantage* (**IMA**).

$$IMA = \frac{s_E}{s_R} \qquad (3.4)$$

where

IMA = velocity ratio or ideal mechanical advantage
s_E = effort distance, in m or ft
s_R = resistance distance, in m or ft

Efficiency in Less-Than-Ideal Machines

For the ideal machines (lever, inclined plane, etc.), the law of simple machines holds true. However, for compound machines, the law does not hold true because of the frictional losses in the machines. A new equation is needed for machines where some of the force of effort is used to overcome the friction forces. If the law of simple machines is stated as a proportional equation, then the AMA is equal to the IMA (in the ideal machine). By introducing efficiency into the equation, the less-than-ideal machine can be analyzed. Thus, for the ideal machine:

$$F_E \times s_E = F_R \times s_R \qquad \text{(From Equation 3.1)}$$

Setting Equation 3.1 up as a proportional equation results in

$$\frac{s_F}{s_R} = \frac{F_R}{F_E}$$

However,

$$IMA = \frac{s_E}{s_R}$$

$$AMA = \frac{F_R}{F_E}$$

Substituting gives IMA = AMA in an ideal machine.

This expression indicates that, for ideal machines, the velocity ratio and the mechanical advantage of the machine are equal. That is, IMA = AMA when there is no loss of energy in the machine ($\eta = 1$). However, in less-than-ideal machines (the usual case), AMA is always less than IMA (AMA < IMA). So, for these machines:

$$AMA = (IMA)(\eta) \qquad \text{For less-than-ideal machines} \qquad (3.5)$$

This leads to a definition of efficiency for less-than-ideal machines in terms of their ideal and actual mechanical advantage.

$$\eta = \frac{AMA}{IMA} \qquad (3.6)$$

where

η = efficiency—a unitless decimal fraction
AMA = actual mechanical advantage
IMA = ideal mechanical advantage, or velocity ratio

Expressed as a percent, the expression of efficiency becomes

$$\eta_\% = \frac{\text{AMA}}{\text{IMA}} \times 100\% \qquad\qquad (3.7)$$

where AMA and IMA are as just defined and $\eta_\%$ is percent efficiency.

Example 3.3

The rope in Figure 3.6 is pulled a distance of 5.0 m, which causes the load to move 1.0 m. The force of resistance is 250 N and the force of effort is 85 N. Determine the efficiency of the pulley system pictured in the figure and express the answer both as a percent and as a decimal fraction.

SOLUTION   Find both the efficiency and the percent efficiency.

Given
$$\eta = \frac{\text{AMA}}{\text{IMA}}$$

Evaluate
$$\text{AMA} = \frac{F_R}{F_E}, F_R = 250 \text{ N}, F_E = 85 \text{ N}$$

$$\text{IMA} = \frac{s_E}{s_R}, s_E = 5.0 \text{ m}, s_R = 1.0 \text{ m}$$

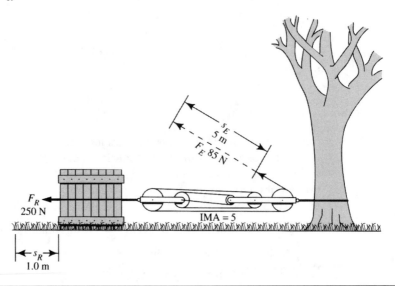

FIGURE 3.6
The pulley system for Example 3.3.

Substitute $\eta = \dfrac{250/85}{5.0/1.0}$

Solve $\eta = 0.59$

        $\eta_\% = \eta \times 100 = 0.59 \times 100 = 59\%$

Exercise 3.1

Solve the following problems. Structure your work so it follows in an orderly progression and can easily be checked.

1. Convert 235 J of energy to foot-pounds of energy. Use the identities in Table A1 of Appendix A.
2. Convert 2800 ft-lb of energy to joules of energy. Use the identities in Table A1 of Appendix A.
3. Determine the efficiency of a machine when the input energy is 1200 J and the output energy is 980 J.
4. Determine the output energy of a machine having an efficiency of 0.35 when the input energy is 2400 ft-lb.
5. Determine the amount of energy (in joules) lost in an electric motor with an efficiency of 82% when the output energy is 42 ft-lb.
6. Determine the mechanical advantage (AMA) of a machine having an efficiency of 0.72 and a velocity ratio (IMA) of 12.
7. Determine the efficiency of a machine with an AMA of 6.4 and an IMA of 9.8.
8. Determine the velocity ratio (IMA) of a machine having a percent efficiency of 35% and an AMA of 68.
9. For the conditions of the railroad switch lever shown in Figure 3.7, determine
   (a) The IMA of the lever.
   (b) The AMA of the lever.
   (c) The efficiency of the lever.
10. For the conditions of the oar shown in Figure 3.8, determine
    (a) The IMA of the oar mechanism.
    (b) The AMA of the oar mechanism.
    (c) The efficiency of the oar mechanism.

3.3 THE LEVER

Introduction

The five simple machines (lever, pulley, wheel and axle, inclined plane, and screw) may be grouped into two basic categories of simple machines—the lever and the inclined plane. All machines are composed of one or more of these types of machines. The three machines that have the lever as their basis are the lever, the pulley, and the wheel and axle.

FIGURE 3.7
The railroad switch lever used in
Problem 9 of Exercise 3.1.

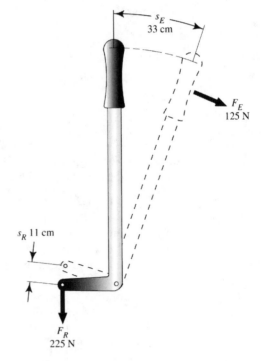

The *lever* is made up of a rigid bar and a pivot called the **fulcrum**. The force of effort (F_E) is applied to the effort arm (s_E), whereas the force of resistance (F_R) is applied to the resistance arm (s_R). This action is pictured in Figure 3.9. For the lever, the ideal mechanical advantage is equal to the ratio of the *effort arm* to the *resistance arm*.

$$\text{IMA} = \frac{\text{effort arm}}{\text{resistance arm}} = \frac{s_E}{s_R} \tag{3.8}$$

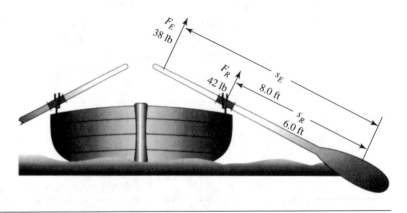

FIGURE 3.8
An oar used in Problem 10 of Exercise 3.1.

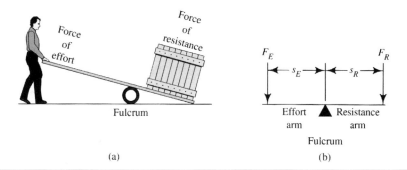

FIGURE 3.9
(a) A lever is used to raise a crate in a warehouse. (b) The free-body diagram used to represent the forces acting on the lever.

As noted in Figure 3.9(b), the effort arm is equal to the distance from the fulcrum to the point of application of the effort force. The resistance arm is equal to the distance from the fulcrum to the point of application of the resistance force.

Because the area of contact of the fulcrum with the lever is very small, the losses due to friction are also very small, and the lever is, for all practical purposes, an ideal machine. The law of simple machines can be used to solve for unknown quantities when the lever is in equilibrium (balanced).

Equilibrium

Equilibrium is achieved when the lever is *at rest*—that is, not moving when forces are applied. In Figure 3.9, the effort force tends to rotate the lever in a counterclockwise (ccw) direction, and the resistance force tends to rotate the lever in a clockwise (cw) direction. When the turning effects are equal, they balance each other, and the equilibrium is maintained. The product of the force and the corresponding distance from the fulcrum is called the *moment of force*. For the lever to be in equilibrium, the clockwise moments about the fulcrum must equal the counterclockwise moments. The law of simple machines is based on the concepts of equilibrium and moments of force. The product of F_E and s_E is the *effort moment*, whereas the product of F_R and s_R is the *resistance moment*.

Example 3.4

The simple beam balance pictured in Figure 3.10 is being designed to weigh objects with masses up to 20.0 kg. Determine the length of the balance arm if the countermass is to be 2.00 kg.

SOLUTION   Find the length of the balance arm in meters.

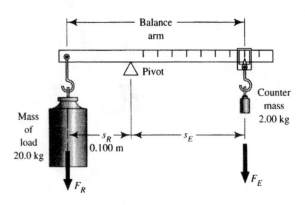

FIGURE 3.10
Beam balance for Example 3.4.

| Given | The law of simple machines, Equation 3.1. Solve for s_E: |

$$F_E \times s_E = F_R \times s_R$$

$$s_E = \frac{F_R \times s_R}{F_E}$$

Evaluate $F_R = 20.0$ kg   (mass)

$s_R = 0.100$ m

$F_E = 2.00$ kg   (mass)

Observation   Rather than convert the units from mass to force, simply substitute mass units directly into the formula. The mass units will factor, leaving only the distance in meters.

Substitute $s_E = \dfrac{20.0 \times 0.100}{2.00}$

Solve $s_E = 1.00$ m

Therefore, the balance arm is equal to $s_E + s_R$, or $1.00 + 0.10 = 1.10$ m.

Types of Levers

The three classes (types) of levers are different from each other in the placement of the fulcrum, the effort force, and the resistance force. Figure 3.11 pictures the three classes of levers.

➤ **Class 1 Lever** The fulcrum is located between the effort force (F_E) and the resistance force (F_R).

Example 3.5

Use the seesaw (class 1 lever) pictured in Figure 3.11
   (a) Find the IMA when the weight of the effort is located 2.00 m from the fulcrum and the weight of the load is located 1.60 m from the fulcrum.

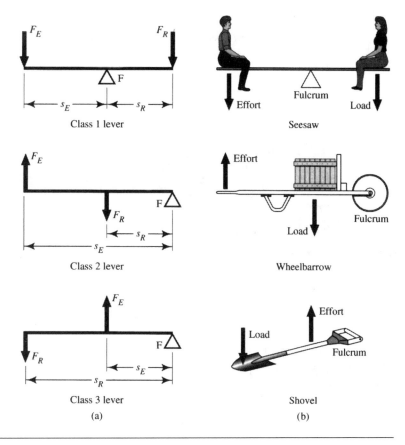

FIGURE 3.11
The three classes of levers: (a) Free-body diagrams. (b) Examples of each class of lever.

        (b) Find the effort force (weight) needed to bring the seesaw into equilibrium when the resistance force (weight) is 425 N.

        (c) Show that the effort moment is equal to the resistance moment.

SOLUTION	**(a)** Find the IMA of the class 1 lever.
Given	$\text{IMA} = \dfrac{s_E}{s_R}$
Evaluate	$s_E = 2.00 \text{ m}$
	$s_R = 1.60 \text{ m}$
Substitute	$\text{IMA} = \dfrac{2.00}{1.60}$
Solve	$\text{IMA} = 1.25$

(b) Find the weight of the effort (effort force).

Given

$F_E \times s_E = F_R \times s_R$; assume an ideal machine.

$$F_E = \frac{F_R \times s_R}{s_E}$$

Evaluate

$F_R = 425$ N

$s_R = 1.60$ m

$s_E = 2.00$ m

Substitute

$$F_E = \frac{425 \times 1.60}{2.00}$$

$F_E = 340$ N

(c) Show that effort and resistance moments are equal.

Given

$F_E \times s_E = F_R \times s_R$

Evaluate

$F_E = 340$ N

$s_E = 2.00$ m

$F_R = 425$ N

$s_R = 1.60$ m

Substitute

$340 \times 2.00 = 425 \times 1.60$

Solve

680 N·m $= 680$ N·m

➤ **Class 2 Lever** The resistance force (F_R) is located between the effort force (F_E) and the fulcrum.

Example 3.6

Use the wheelbarrow (class 2 lever) pictured in Figure 3.11.
  (a) Determine the IMA when the length of the effort arm (s_E) is 6.00 ft and the length of the resistance arm (s_R) is 1.50 ft.
  (b) Determine the effort force needed to lift the wheelbarrow when the load is 132 lb.
  (c) Show that the effort moment is equal to the resistance moment.

SOLUTION

(a) Find the IMA of the class 2 lever.

Given

$$\text{IMA} = \frac{s_E}{s_R}$$

Evaluate

$s_E = 6.00$ ft

$s_R = 1.50$ ft

Substitute $\text{IMA} = \dfrac{6.00}{1.50}$

Solve $\text{IMA} = 4.00$

(b) Find the effort force needed to lift the load.

Given $F_E = \dfrac{F_R \times s_R}{s_E}$, assume an ideal machine.

Evaluate $F_R = 132 \text{ lb}$

                $s_R = 1.50 \text{ ft}$

                $s_E = 6.00 \text{ ft}$

Substitute $F_E = \dfrac{132 \times 1.50}{6.00}$

Solve $F_E = 33.0 \text{ lb}$

(c) Show that effort and resistance moments are equal.

Given $F_E \times s_E = F_R \times s_R$

Evaluate $F_E = 33.0 \text{ lb}$

                $s_E = 6.00 \text{ ft}$

                $F_R = 132 \text{ lb}$

                $s_R = 1.50 \text{ ft}$

Substitute $33.0 \times 6.00 = 132 \times 1.50$

Solve $198 \text{ ft-lb} = 198 \text{ ft-lb}$

➤ **Class 3 Lever** The effort force (F_E) is located between the resistance force (F_R) and the fulcrum.

*Example
3.7*

Use the shovel (class 3 lever) pictured in Figure 3.11.
   (a) Find the IMA when the length of the effort arm (s_E) is 2.00 ft and the length of the resistance arm (s_R) is 4.00 ft.
   (b) Find the effort force when the load (F_R) is 10.0 lb.
   (c) Show that the effort moment is equal to the resistance moment.

SOLUTION   **(a)** Find the IMA of the class 3 lever.

Given $\text{IMA} = \dfrac{s_E}{s_R}$

Evaluate $s_E = 2.00 \text{ ft}$

$s_R = 4.00$ ft

Substitute $\text{IMA} = \dfrac{2.00}{4.00}$

Solve $\text{IMA} = 0.500$

(b) Find the effort force needed to lift the load.

Given $F_E = \dfrac{F_R \times s_R}{s_E}$; assume an ideal machine.

Evaluate $F_R = 10.0$ lb

$s_R = 4.00$ ft

$s_E = 2.00$ ft

Substitute $F_E = \dfrac{10.0 \times 4.00}{2.00}$

Solve $F_E = 20.0$ lb

(c) Show that effort and resistance moments are equal.

Given $F_E \times s_E = F_R \times s_R$

Evaluate $F_E = 20.0$ lb

$s_E = 2.00$ ft

$F_R = 10.0$ lb

$s_R = 3.00$ ft

Substitute $20.0 \times 2.00 = 10.0 \times 4.00$

Solve 40.0 ft-lb $= 40.0$ ft-lb

Figure 3.12 pictures several machines (tools) that use a double lever for their action. In addition to the tools pictured, scissors, pliers of all kinds, and tongs also utilize a double lever in their basic design. These tools (double levers) have some loss due to friction at their bearing point.

Example 3.8

Copper wire is being cut with a pair of diagonal pliers. Determine
     (a) The force exerted on the wire when 20.0 lb is exerted on the handles and the AMA is 7.80.
     (b) The IMA when the efficiency (η) is 0.975 and the AMA is 7.80.
     (c) The length of the effort arm (s_E) when the efficiency is 0.975 and the resistance arm (s_R) is .50 in.

SOLUTION   **(a)** Find the resistance force (F_R) in pounds.

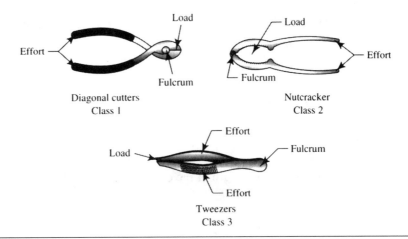

FIGURE 3.12
Examples of tools (machines) that use double levers.

Given $\text{AMA} = \dfrac{F_R}{F_E}$ (Equation 3.2) Solve for F_R:

$F_R = F_E \times \text{AMA}$

Evaluate $F_E = 20.0\ \text{lb}$

$\text{AMA} = 7.80$

Substitute $F_R = 20.0 \times 7.80$

Solve $F_R = 156\ \text{lb}$

Therefore, 156 lb of force is exerted on the wire.

(b) Find the IMA when the efficiency is 0.975.

Given $\eta = \dfrac{\text{AMA}}{\text{IMA}}$ (Equation 3.5) Solve for IMA:

$\text{IMA} = \dfrac{\text{AMA}}{\eta}$

Evaluate $\eta = 0.975$

$\text{AMA} = 7.80$

Substitute $\text{IMA} = \dfrac{7.80}{0.975}$

Solve $\text{IMA} = 8.00$

(c) Determine the length of the effort arm (s_E) when the resistance arm (s_R) is .50 in.

Given	$\text{IMA} = \dfrac{s_E}{s_R}$ (Equation 3.4) Solve for the effort arm:
	$s_E = s_R \times \text{IMA}$
Evaluate	$s_R = .50$ in
	$\text{IMA} = 8.00$
Substitute	$s_E = .50 \times 8.00$
Solve	$s_E = 4.0$ in

Therefore, the length of the effort arm is 4.0 in.

■

Exercise 3.2

Solve the following problems. Make sketches to aid in solving the problems and structure your work so it follows in an orderly progression and can easily be checked. Assume that each of the machines is an ideal machine.

1. A pair of pliers is being used to bend a steel wire. If the joint of the pliers is 2.0 in from the wire and force is applied to the handle 5.2 in from the joint, then determine the IMA of the pliers.
2. A steel bar 2.40 m long is used as a class 1 lever to lift a 430-kg boulder. If the boulder (F_R) is located 0.50 m from the fulcrum, then determine the minimum force needed at the other end of the bar (F_E) to raise the boulder.
3. Determine the class of lever pictured in each figure.
   (a) Figure 3.13   **(b)** Figure 3.14   **(c)** Figure 3.15
4. Determine the weight of the fish on the line pictured in Figure 3.13 when F_E is 3.2 lb, s_R is 8.0 ft, and s_E is 1.5 ft.

FIGURE 3.13
Fishing rod for Problem 4 of
Exercise 3.2.

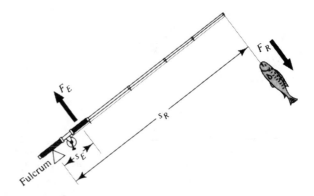

FIGURE 3.14
Micro switch for Problem 5 of Exercise 3.2.

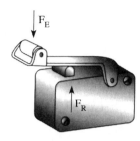

FIGURE 3.15
Tweezers for Problem 6 of Exercise 3.2.

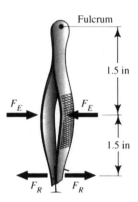

5. Determine the closure force (F_E) on the roller lever of the microswitch pictured in Figure 3.14 when F_R is 0.15 N, the distance from the pivot to the center of the switch button is 14 mm, and the distance from the pivot to the center of the roller actuator is 3.2 cm.
6. Determine the closure force (F_E) on the tweezers pictured in Figure 3.15 when the resistive force (F_R) is .15 lb.
7. Determine the class of lever pictured in each figure.
   (a) Figure 3.16   **(b)** Figure 3.17   **(c)** Figure 3.18
8. Determine the force (F_E) needed to bring the lever represented by the free-body diagram of Figure 3.16 into equilibrium.

FIGURE 3.16
Free-body diagram for Problem 8 of Exercise 3.2.

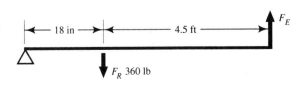

FIGURE 3.17
Free-body diagram for Problem 9 of
Exercise 3.2.

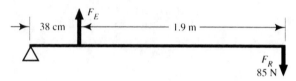

FIGURE 3.18
Clutch pedal for Problem 10 of Exercise 3.2.

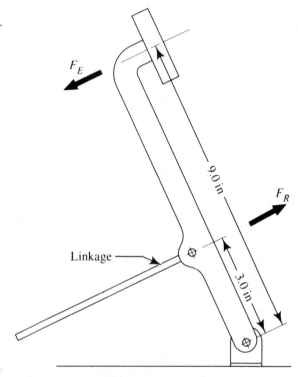

9. **(a)** Determine the force (F_E) needed to bring the lever represented by the free-body diagram of Figure 3.17 into equilibrium.
   (b) Show that the effort moment is equal to the resistance moment.
10. Determine the force (F_R) needed to disengage the clutch when the force (F_E) on the pedal of Figure 3.18 is 18 lb.

CHAPTER SUMMARY

- The lever, pulley, wheel and axle, inclined plane, and screw are the basic simple machines.
- Machines amplify force, amplify speed or distance, and change direction.
- Modern machines are made up of two or more simple machines, along with several machine elements.
- Machines get their power from prime movers.
- The law of simple machines is based on the idea that the energy (work) into an ideal machine is equal to the energy (work) out of the ideal machine.

- Mechanical advantage is referred to as the actual mechanical advantage (AMA).
- The velocity ratio (maximum mechanical advantage possible for a machine) is referred to as the ideal mechanical advantage (IMA).
- The efficiency of a machine is expressed as the ratio between the AMA and the IMA (η = AMA/IMA).
- The five simple machines may be grouped into two basic categories: the lever and the inclined plane.
- Equilibrium is achieved in a lever when the clockwise moments of force equal the counterclockwise moments of force.
- Levers are divided into three classes, each having the effort, load, and fulcrum arranged differently.

SELECTED TECHNICAL TERMS

The following technical terms, abbreviations, and acronyms are defined in the glossary located after Chapter 16. You are encouraged to use the glossary to aid your understanding and to test your knowledge of these important terms.

AMA	inclined plane
efficiency	lever
equilibrium	prime mover
fulcrum	pulley
IMA	screw
ideal machine	wheel and axle

END-OF-CHAPTER QUESTIONS

Write T if the statement is true and F if the statement is false.

1. A class 3 lever amplifies force.
2. When the AMA is equal to the IMA, the machine has frictional losses.
3. An electric motor is an example of a prime mover.
4. A single fixed pulley is used to change the direction of the effort.
5. Both bearings and springs are examples of mechanisms.
6. The velocity ratio of a machine indicates its maximum mechanical advantage.
7. The efficiency of a machine is an indicator of the losses in the machine.
8. With a lever, a moment of force results from the multiplication of a force and a corresponding distance.
9. Frictional losses are usually considered when working with ideal machines.
10. Frictional losses usually transform work to heat.

In the following, select the word that makes the statement true.

11. Machines amplify force at the expense of (acceleration, displacement, time).
12. Machine elements include (mechanisms, links, levers).
13. Sources of power for modern machines include (electrodynamic, pneumatic, hydrostatic) along with electrical and mechanical.

14. In the (best, utopian, ideal) machine, work into the machine is exactly equal to work out of the machine.

15. When lifting a 2-lb object, your arm movement at the elbow (pivot) is an example of a class $(1, 2, 3)$ lever.

Answer each of the following questions with a short answer in the form of a complete sentence. Include a restatement of the question in your answer.

16. What characteristics do machines provide to the user to make tasks easier?

17. Assuming the weight of a door is concentrated along its vertical center line, what class of lever is present when the door is opened or closed by force applied at the door knob?

18. What force is present and how does it act on the force of effort in a less-than-ideal machine?

19. What procedure would you follow to determine the IMA of a lever?

20. Mathematically, how is the actual mechanical advantage related to the ideal mechanical advantage?

END-OF-CHAPTER PROBLEMS

Solve the following problems. Make sketches to aid in solving the problems and structure your work so it follows in an orderly progression and can easily be checked. Table 3.1 summarizes the formulas used in Chapter 3.

1. Determine the efficiency of an elevator system with an actual mechanical advantage of 12.6 and an ideal mechanical advantage of 30.0. Express the answer as both efficiency (decimal fraction) and percent efficiency.

TABLE 3.1 Summary of Formulas Used in Chapter 3

Equation Number	Parameter	Equation
3.1	Law of simple machines	$F_E \times s_E = F_R \times s_R$
3.2	Actual mechanical advantage	$AMA = \dfrac{F_R}{F_E}$
3.3	Efficiency	$\eta = \dfrac{W_{out}}{W_{in}}$
3.4	Ideal mechanical advantage	$IMA = \dfrac{s_E}{s_R}$
3.5	Less-than-ideal machines	$AMA = (IMA)(\eta)$
3.6	Efficiency	$\eta = \dfrac{AMA}{IMA}$
3.7	Percent efficiency	$\eta_\% = \dfrac{AMA}{IMA} \times 100\%$
3.8	IMA of lever	$IMA = \dfrac{\text{effort arm}}{\text{resistance arm}}$

2. Determine the output energy (in joules) of a machine having a percent efficiency of 55% when the input energy is 1.74×10^3 ft-lb.

3. The simple beam balance pictured in Figure 3.10 is weighing an object (F_R) with an unknown mass. Determine the mass of the object (in kg) when the 2.00 kg counter-mass is located 72.5 cm from the pivot and the center of mass of the unknown object is 10.0 cm from the pivot.

4. Determine the amount of energy (in joules) lost in the electric motor pictured in Figure 3.5 if it has an efficiency of 0.760 and an input energy of 835 ft-lb.

5. Determine the weight (in newtons) of the fish on the line pictured in Figure 3.13 when F_E is 18.3 N, s_R is 2.6 m, and s_E is 0.46 m.

6. A 2.50 m steel bar is being used as a class 1 lever to lift one end (390 lb) of a gran-ite block on a construction site. If the granite block is located 32.0 cm from the ful-crum on one end of the bar, then determine the force in pounds needed at the other end of the bar to raise the end of the block.

7. Determine the efficiency of the electric motor pictured in Figure 3.5 when the input energy is 956 J and the output energy is 550 ft-lb.

8. For the seesaw pictured in Figure 3.11(b), determine
   (a) If the lever is in equilibrium when F_E is 800 N, s_R is 8.00 ft, s_E is 6.50 ft, and F_R is 126 lb.
   (b) The direction F_E needs to move—inward or outward—to bring the system into equilibrium.
   (c) The distance in centimeters the effort force needs to move from the current po-sition.

9. For the rowboat pictured in Figure 3.8, determine
   (a) The IMA of the oar mechanism when s_E is 2.75 m and s_R is 2.25 m.
   (b) The AMA of the oar mechanism when F_E is 116 N and F_R is 104 N.
   (c) The efficiency of the rowing system.

10. A wheelbarrow (class 2 lever) is loaded with 892 N of molding sand. The center of mass of the load is 0.250 m from the axle (fulcrum) of the wheel. If the handles are 1.62 m long from the center of the wheel axle to the point where the effort force is applied, determine
    (a) The IMA of the wheelbarrow.
    (b) The AMA when the efficiency is 0.95.
    (c) The effort force when the efficiency is 0.95.

11. A tin snip (class 1 lever) with an efficiency of 0.91 is used to cut sheet metal. If the blades contact the metal 1.0 in. from the pivot, determine the shear force (cutting force) acting on the metal when 32 lb is exerted on the handles 11.0 in. from the pivot.

12. Due to friction, the machine represented by the free-body diagram of Figure 3.16 has an efficiency of 0.82. Determine
    (a) The IMA of the machine.
    (b) The AMA of the machine.
    (c) The effort force for equilibrium.

13. If the clutch pedal mechanism of Figure 3.18 is 87% efficient and requires 44.5 N to depress the pedal, determine
    (a) The IMA of the clutch mechanism.
    (b) The AMA of the clutch mechanism.
    (c) The force transmitted through the linkage.

FIGURE 3.19
A beam suspended by a steel rod for
Problem 15 of the end-of-chapter
problems.

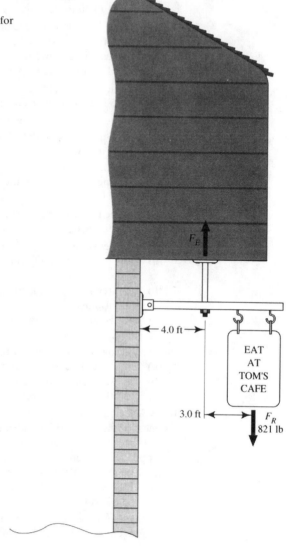

14. The closure force (F_E) on the roller lever of the microswitch pictured in Figure 3.14
is 50 mN when F_R is 0.15 N. If the distance from the pivot to the center of the switch
button is 1.4 cm and the distance from the pivot to the center of the roller actuator
is 4.5 cm, determine

(a) The IMA of the lever activator.

(b) The AMA of the lever activator.

(c) The efficiency of the lever activator.

15. A beam is suspended by a steel rod as shown in Figure 3.19. Determine
   (a) The class of the lever.
   (b) The IMA of the lever.
   (c) The force (F_E) pulling up on the beam (tension in the rod) when the percent efficiency is 100%.

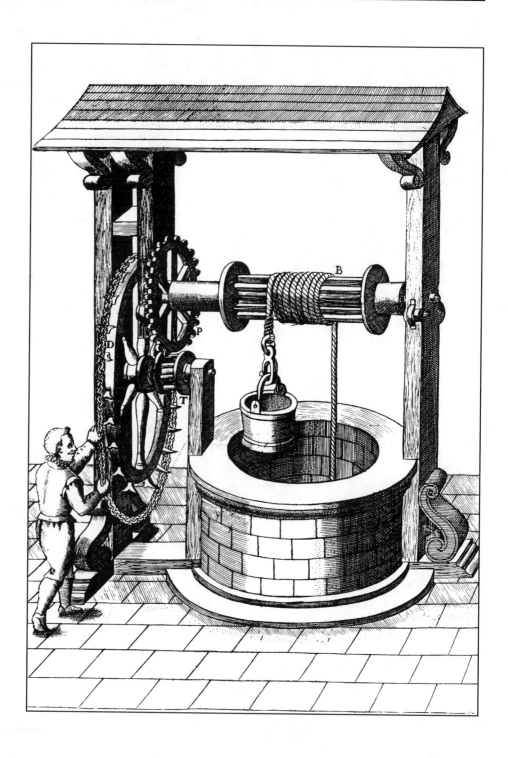

Chapter 4

The Pulley, Wheel and Axle, and Inclined Plane

■━━ ■ ■

About the Illustration. *Water is drawn from a well with this Renaissance machine that utilizes the wheel and axle. By pulling on the inertia chain drive (the chain wrapped around the sprocket wheel D), the lantern pinion (gear) T is turned. This motion is transmitted through spur gear P, which forms a wheel and axle with drum B. The shaft, with gear P and drum B, is supported by and rotates on roller shaft bearings. As illustrated, these friction-reducing bearings (consisting of two large disks whose axes are parallel to the shaft) are recessed into the supports at each end of the shaft. (Designed by Agostino Ramelli, from* The Various and Ingenious Machines of Agostino Ramelli *(1588).)* ■

INTRODUCTION

The law of simple machines and the other fundamental principles introduced in Chapter 3 are used here to study the operation of the pulley, the wheel and axle, the inclined plane, the wedge, the screw, and simple compound machines.

CHAPTER CONTENTS

4.1 Pulleys and Lifting Systems
4.2 The Wheel and Axle
4.3 Inclined Planes
4.4 Compound Machines

PERFORMANCE OBJECTIVES

Once you have read and studied each section; worked through each example with pencil, paper, and calculator; worked through the section and the end-of-chapter problems; and answered the end-of-chapter questions, you should be able to

- Identify the parts of a pulley system.
- Determine the ideal mechanical advantage of the wheel and axle.
- Understand the principle of a differential pulley.
- Calculate the IMA of the inclined plane.
- Identify the pitch of a screw thread.
- Apply the law of simple machines to the pulley, lifting system, wheel and axle, chain hoist, inclined plane, wedge, and screw.
- Calculate the IMA of a compound machine.

4.1 PULLEYS AND LIFTING SYSTEMS

Introduction

A pulley is shaped like a wheel and readily turns on the axle through its center. Its circumference is usually grooved to receive a rope or toothed to receive a chain. Pulleys of this type are often referred to as **sheaves.** When the pulley is mounted in a frame, complete with supporting rings or hooks, it forms a block. The rope that runs over the pulley is called the tackle. The complete assembly of pulley, rope, frame, and hook is called a *block and tackle.*

IMA of Simple Pulley Systems

A single fixed pulley (Figure 4.1) is a class 1 lever with equal arms and a fulcrum at its center. The effort arm is equal to the resistance arm, with the length of each equal to the radius of the pulley. Because the ideal mechanical advantage (IMA) of a single fixed pulley is equal to 1, the pulley does not amplify force. Instead, the fixed pulley changes the direction of the effort force.

When a single pulley is allowed to move (Figure 4.2), the effort arm becomes twice the length of the resistance arm (diameter versus radius), resulting in a class 2 lever with an IMA equal to 2. In this case, one end of the tackle (rope) is fixed and serves as the fulcrum; the other end is free to move.

FIGURE 4.1
(a) Pictorial of a single fixed pulley.
(b) The lever equivalent of a single fixed pulley.

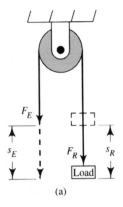
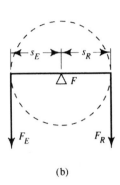

(a)   (b)

Because a single movable pulley is difficult to use by pulling upward and because pulling downward is more desirable, a system of pulleys is created to allow pulling downward. The pulley system pictured in Figure 4.3 uses the direction-changing characteristics of the fixed pulley in conjunction with the force-amplifying characteristics of the movable pulley. This system allows the user to stand on the ground and pull downward.

In the system of one movable and one fixed pulley, the distance the rope is moved is twice the distance the load is moved. For example, when the load moves 1.0 m, the rope moves 2.0 m and the IMA is equal to 2. From this example we learn that the velocity ratio (IMA) of a pulley system is equal to the ratio of the distance moved by the effort, s_E (the distance the rope moves), to the distance moved by the load, s_R.

$$\text{IMA}_{\text{pulley}} = \frac{s_E}{s_R} \tag{4.1}$$

FIGURE 4.2
(a) Pictorial of a single movable pulley. (b) The lever equivalent of a single movable pulley.

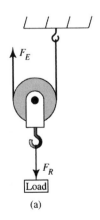
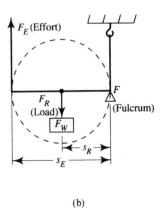

(a)   (b)

FIGURE 4.3
A system of pulleys used to change direction and amplify effort force $(s_E = 2s_R)$

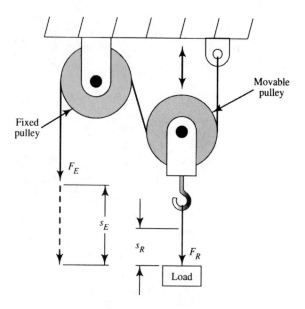

where

$\quad$ IMA = ideal mechanical advantage of the pulley system
$\quad\quad s_E$ = effort distance, the distance the rope moved in m or ft
$\quad\quad s_R$ = resistance distance, the distance the load moved in m or ft

Friction in Pulley Systems

Once a system of pulleys has more than just one movable pulley, then the frictional losses become significant and must be accounted for. Table 4.1 lists the approximate efficiencies for various systems of pulleys. The number of ropes supporting the resistance (load) is indicated by n.

TABLE 4.1   Approximate Efficiencies for Pulley Systems

n*	η
1	0.90
2	0.85
3	0.80
4	0.75
5	0.70
6	0.65

*Number of ropes pulling up on the movable pulleys.

FIGURE 4.4
Pictorial of a lifting system consisting of two blocks with one movable and two fixed pulleys; the direction of motion is indicated by arrows. The ideal mechanical advantage is 3, since three ropes are supporting the load attached to the movable pulley.

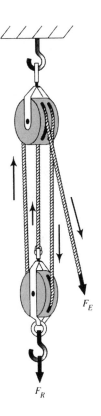

IMA of a Two-Block System

When one continuous rope is used in a lifting system made up of two blocks, as pictured in Figure 4.4, then each part of the rope supporting the load has a tension in it equal to the effort force. Because the tension is the same in each part of the rope supporting the load, the ideal mechanical advantage (IMA) can be determined by simply counting the number of ropes that support the movable pulleys.

➤ **As a Rule** Counting the number of ropes pulling up on the movable pulleys determines the IMA of the pulley system.

$$\text{IMA}_{\text{pulley}} = \text{number of ropes supporting the movable pulleys} \qquad \textbf{(4.2)}$$

Example
4.1

For the lifting system pictured in Figure 4.5, determine
    (a) The actual mechanical advantage (AMA).
    (b) The ideal mechanical advantage (IMA).
    (c) The efficiency of the system (η).

SOLUTION   **(a)** Find the AMA.

FIGURE 4.5
Lifting system for Example 4.1.

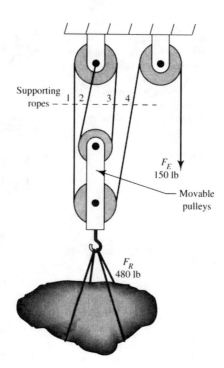

Supporting ropes

Movable pulleys

F_E
150 lb

F_R
480 lb

Given	$\text{AMA} = \dfrac{F_R}{F_E}$ and the information in Figure 4.5
Evaluate	$F_R = 480$ lb
	$F_E = 150$ lb
Substitute	$\text{AMA} = \dfrac{480}{150}$
Solve	$\text{AMA} = 3.20$

(b) Find the IMA by counting the ropes that support the movable pulleys.

Solve	$n = 4$
	$\text{IMA} = 4$

(c) Find the efficiency from the AMA and the IMA.

Given	$\eta = \dfrac{\text{AMA}}{\text{IMA}}$
Evaluate	$\text{AMA} = 3.20$
	$\text{IMA} = 4.00$

Substitute	$\eta = \dfrac{3.20}{4.00}$
Solve	$\eta = 0.800$

Example 4.2

For the lifting system pictured in Figure 4.6, determine:

(a) The IMA when the load is raised 15.0 m.

(b) The length of rope pulled through the system to raise the load 15.0 m.

(c) The effort force needed to raise a 1710-N load of building material 15.0 m when the percent efficiency is less than 100%. Select an appropriate efficiency from Table 4.1.

SOLUTION   (a) Find the number of ropes that support the movable pulleys.

Solve
$$n = 5$$
$$\text{IMA} = 5$$

(b) Find the effort distance (s_E).

Given
$$\text{IMA} = \frac{s_E}{s_R}$$

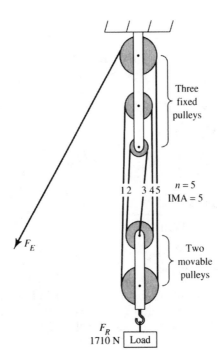

FIGURE 4.6
Lifting system for Example 4.2.

Solve for s_E:

$$s_E = \text{IMA} \times s_R$$

Evaluate $s_R = 15.0$ m

IMA $= 5$

Substitute $s_E = 5 \times 15.0$

Solve $s_E = 75.0$ m

(c) Find the effort force F_E.

Given

$$\eta = \frac{\text{AMA}}{\text{IMA}} \qquad \text{Equation 3.6}$$

$$\text{AMA} = \frac{F_R}{F_E} \qquad \text{Equation 3.2}$$

solve for AMA in Equation 3.6; then solve for F_E in Equation 3.2. Solving for AMA:

$$\text{AMA} = \eta \times \text{IMA}$$

Solve for F_E in Equation 3.2:

$$F_E = \frac{F_R}{\text{AMA}}$$

Observation   Start with the solution of AMA from AMA $= \eta \times$ IMA.

Evaluate $\eta = 0.70$ from Table 4.1 for $n = 5$

IMA $= 5$

Substitute   AMA $= 0.70 \times 5$

Solve   AMA $= 3.5$

Observation   Now find F_E from $F_E = F_R/\text{AMA}$.

Evaluate $F_R = 1710$ N

AMA $= 3.5$

Substitute $$F_E = \frac{1710}{3.5}$$

Solve $F_E = 490$ N

Due to the frictional loss in the pulley system in the previous example, the effort was greater than it would have been had the system been frictionless (ideal). The effort force was 490 N for an efficiency of 0.70 (70%), compared with 320 N for an efficiency of 1.00 (100%).

Exercise 4.1

Solve the following problems. Make sketches to aid in solving the problems and structure your work so it follows in an orderly progression and can easily be checked.

1. Draw a schematic of a lifting system with an IMA of 3 that consists of one movable and two fixed pulleys.
2. Draw a schematic of a lifting system with an IMA of 4 that consists of two fixed and two movable pulleys.
3. A single fixed pulley is used to raise a load 2.0 ft. If the pulley is .75 ft in diameter, determine
   (a) The IMA of the pulley.
   (b) The AMA of the pulley when the efficiency is 0.95.
4. A lifting system consisting of one fixed and one movable pulley is used to lift a 215-N load. Determine
   (a) The IMA of the system.
   (b) The AMA of the system when an effort force of 132 N is required to just lift the load.
   (c) The efficiency of the system.
5. A lifting system made up of one fixed and one movable pulley is used to lift a 165-lb load 7.5 ft. Determine
   (a) The effort force needed to lift the load when the system is ideal.
   (b) The length of rope that must be pulled by the person supplying the effort force.
   (c) The AMA when 92.0 lb is actually required to lift the load.
   (d) The percent efficiency of the system.
6. A lifting system consisting of three movable pulleys has a 795-lb load supported by six ropes that pull up on the movable pulleys. If 15.0 ft of rope is moved in applying the effort force, determine
   (a) The distance moved by the load.
   (b) The efficiency of the system (use Table 4.1).
   (c) The effort force needed to lift the load for the efficiency in (b).

4.2 THE WHEEL AND AXLE

Introduction

Assume that two cylindrical objects of different diameters are permanently joined to one another so they rotate together around the same axis, as pictured in Figure 4.7. The cylinder with the effort force applied is called the *wheel*, and the cylinder with the resistance force applied is called the *axle*.

FIGURE 4.7
Wheel and axle, with the radius of
each indicated.

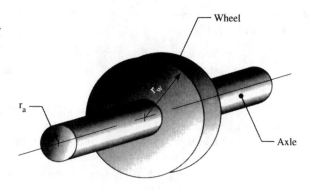

IMA of the Wheel and Axle

In Figure 4.8, a wheel and axle are shown. The effort force (F_E) is applied to the rope wrapped around the wheel, and the load (F_R) is attached to the rope wrapped around the axle. The distance moved by the rope unwinding from one turn of the wheel is equal to the circumference of the wheel, or $2\pi r_W$ ($s_E = 2\pi r_W$). Since the wheel and axle share a common axis, the rope around the axle will wind up one turn and move the load a distance equal to the circumference of the axle, or $2\pi r_a$ ($s_R = 2\pi r_a$). Thus, the velocity ratio (IMA) of the wheel and axle is equal to the ratio of the distance moved by the effort to the distance moved by the load, which may be stated as r_W/r_a, since the cylinders are fixed and rotate together through the same angle.

$$\text{IMA}_{\text{wheel\&axle}} = \frac{2\pi r_W}{2\pi r_a} = \frac{r_W}{r_a} \tag{4.3}$$

where

   IMA = ideal mechanical advantage of the wheel and axle
   r_W = radius of the wheel, which is proportional to the distance moved by the
       effort, in m or ft

FIGURE 4.8
A wheel and axle machine called a
windlass.

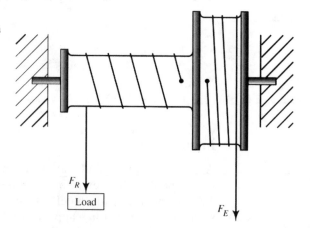

r_a = radius of the axle, which is proportional to the distance moved by the load, in m or ft

AMA and Efficiency of the Wheel and Axle

Figure 4.9 represents the **windlass** of Figure 4.8 as a lever, where s_E is the radius of the wheel ($s_E = r_W$) and s_R is the radius of the axle ($s_R = r_a$). The mechanical advantage (AMA) is found by applying the law of simple machines. For the ideal wheel and axle, the moments about the fulcrum are equal (clockwise moment equals the counterclockwise moment) and $r_W \times F_E = r_a \times F_R$. Thus, for the wheel and axle, $r_W/r_a = F_R/F_E$ and as is the case for all ideal simple machines, IMA = AMA = F_R/F_E.

As with the pulley system, the machines based on the wheel and axle have significant loss due to friction, so these machines are less than ideal; AMA = IMA $\times \eta$.

Example 4.3

The lifting system of Figure 4.10 is raising a load of 685 N. The radius of the handle is 0.60 m, and the diameter of the drum is 24 cm. Determine
  (a) The ideal mechanical advantage (IMA).
  (b) The efficiency of the machine when the effort force applied to the handle is 212 N.

SOLUTION

(a) Find the IMA.

FIGURE 4.9
Wheel and axle represented as a lever.

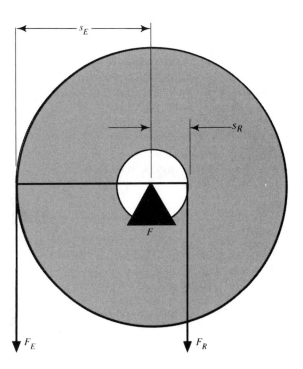

Given	$\text{IMA} = \dfrac{r_W}{r_a}$ and the information in Figure 4.10
Evaluate	$r_W = 0.60$ m
	$r_a = 0.24/2 = 0.12$ m
Observation	The radius of the drum is one-half its diameter.
Substitute	$\text{IMA} = \dfrac{0.60}{0.12}$
Solve	$\text{IMA} = 5.0$

(b) Find the efficiency.

Given	$\eta = \dfrac{\text{AMA}}{\text{IMA}}$
Evaluate	$\text{AMA} = \dfrac{F_R}{F_E} = \dfrac{685}{212} = 3.23$
	$\text{IMA} = 5.0$
Substitute	$\eta = \dfrac{3.23}{5.0}$
Solve	$\eta = 0.65$

When analyzing machines that use the wheel and axle, use the radius of both the wheel and the axle. If the diameter is given for the wheel and the radius is given for the axle, then find the radius of the wheel using diameter = 2 × radius ($d = 2r$). Even though the usual case is for the wheel to be larger than the axle, it is possible for the wheel to be smaller than the axle, in which case the IMA is less than 1.0.

FIGURE 4.10
A pictorial of the winch for Example 4.3.

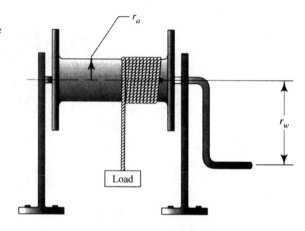

> **As a Rule** In a wheel and axle machine, the effort force is applied to the wheel and the load is attached to the axle. To identify which machine element is the wheel, simply locate where the effort force is applied.

Chain Hoist

The chain hoist, or *differential pulley*, shown in Figure 4.11, is used to lift objects up to 3 T (tons), or 27 kN. The hoist is constructed by fixing two pulleys of different diameters together. A third, movable pulley completes the assembly. An endless chain is threaded around the pulleys. The direction of movement of the several pulleys is shown in Figure 4.12. Tooth pulleys are used to accommodate the chain, which provides more strength than a rope and will not slip under very heavy loads. In using the chain hoist, the load can be left suspended at any point and it will not fall. That

FIGURE 4.11
A pictorial of a chain hoist. A chain hoist is a differential pulley.

FIGURE 4.12
A schematic of the direction of motion of the differential pulley when a load is lifted.

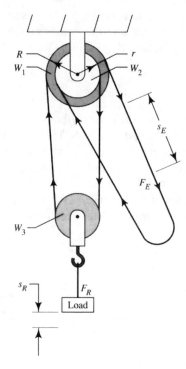

is, the chain will not run in reverse when the effort force is removed, as is sometimes the case in other pulley systems.

IMA of the Chain Hoist

The velocity ratio (IMA) of the hoist can be determined by finding the distance the effort moves (s_E) in one revolution of the fixed wheels and then comparing this distance to the distance the load moves (s_R) in one revolution. The effort moves a distance equal to the circumference of the larger of the two fixed wheels (w_1), or $2\pi R$. The load attached to the movable wheel moves a distance equal to one-half the difference between the circumferences of the two fixed wheels—the larger (w_1) and the smaller (w_2), or $(2\pi R - 2\pi r)/2$. Thus, the load moves $\pi(R - r)$.

$$IMA_{hoist} = \frac{2\pi R}{\pi(R - r)} = \frac{2R}{R - r} \tag{4.4}$$

where

$$IMA = \text{the ideal mechanical advantage of the hoist}$$
$$2\pi R = \text{distance moved by the effort, in m or ft}$$
$$\pi(R - r) = \text{distance moved by the load, in m or ft}$$
$$R = \text{radius of the larger fixed wheel } (w_1), \text{ in m or ft}$$
$$r = \text{radius of the smaller fixed wheel } (w_2), \text{ in m or ft}$$

Operation of the Chain Hoist

The operation of the chain hoist may be understood from Figure 4.12. Assume w_1 rotates through one turn; then the effort moves a distance equal to the circumference of w_1 $(2\pi R)$. The movable wheel (w_3) moves up a distance equal to one-half the circumference of w_1 (πR); remember that the IMA of a single movable pulley is 2. While w_1 is turning one turn, w_2 (the smaller of the two fixed pulleys) is also turning one turn, and it lowers the movable wheel by a distance equal to one-half the circumference of w_2 (πr). The movable pulley (w_3) does not first move up and then down; instead, it moves steadily upward a distance equal to the difference between the two distances $(\pi R - \pi r)$.

As in all machines, the mechanical advantage (AMA) is substantially less than the velocity ratio (IMA) due to the frictional losses in the machine.

Example *4.4*	A chain hoist requires an effort force of 218 lb to raise a 1.0-T load 1.0 ft when 32 ft of chain is pulled through the fixed pulleys. Determine **(a)** The ideal mechanical advantage. **(b)** The actual mechanical advantage. **(c)** The efficiency of the hoist.

SOLUTION   **(a)** Find the IMA.

Given $\text{IMA} = \dfrac{s_E}{s_R}$

Evaluate $s_E = 32$ ft

              $s_R = 1.0$ ft

Substitute $\text{IMA} = \dfrac{32}{1.0}$

Solve $\text{IMA} = 32$

              (b) Find the AMA.

Given $\text{AMA} = \dfrac{F_R}{F_E}$

Evaluate $F_R = 20\bar{0}0$ lb

              $F_E = 218$ lb

Substitute $\text{AMA} = \dfrac{20\bar{0}0}{218}$

Solve $\text{AMA} = 9.17$

(c) Find the efficiency.

Given	$\eta = \dfrac{AMA}{IMA}$
Evaluate	$AMA = 9.17$
	$IMA = 32$
Substitute	$\eta = \dfrac{9.17}{32}$
Solve	$\eta = 0.29$

Example 4.5

Determine the ideal mechanical advantage for the differential pulley pictured in Figure 4.12 when R is 8.125 in and r is 8.000 in.

SOLUTION   Find the IMA.

Given	$IMA = \dfrac{2R}{R - r}$
Evaluate	$R = 8.125$ in
	$r = 8.000$ in
Substitute	$IMA = \dfrac{2 \times 8.125}{8.125 - 8.000}$
Solve	$IMA = 130.0$

Exercise 4.2

Solve the following problems. Make sketches to aid in the solution of the problems and structure your work so it follows in an orderly progression and can easily be checked.

1. Determine the IMA of a wheel and axle when the axle diameter is 1.25 in and the wheel radius is 1.5 ft.
2. The spring balance of Figure 4.13 indicates a reading of 112 N for a load weight of 242 N. If the diameter of the flywheel is 24 cm and the radius of the axle is 5.0 cm, determine
   (a) The IMA of the wheel and axle.
   (b) The AMA of the wheel and axle.
   (c) The efficiency of the wheel and axle.
3. The windlass of Figure 4.8 raises the load of 118 N a distance of 0.85 m. If the IMA is 6.0, determine
   (a) The radius of the wheel when the diameter of the axle is 20.0 cm.
   (b) The effort force when the percent efficiency is 65%.
4. The winch of Figure 4.10 raises a load of 150 lb a distance of 2.0 ft. The radius of the handle is 2.4 ft, and the diameter of the drum is 8.0 in.

FIGURE 4.13
The flywheel used in Exercise 4.2.

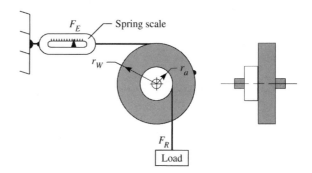

Determine
    (a) The ideal mechanical advantage (IMA).
    (b) The efficiency of the machine when the effort force applied to the handle is
        32 lb.
  5. The diameter of the wheel of a wheel and axle is 6.5 in. A weight of 82 lb is raised
    1.5 ft with an effort force of 15.5 lb. Determine
    (a) The AMA of the wheel and axle.
    (b) The diameter of the axle when the efficiency is 0.72.
    (c) The IMA of the wheel and axle.
  6. The two fixed pulleys of the chain hoist of Figure 4.11 have diameters of 16.5 in
    and 16.0 in, respectively. Determine
    (a) The IMA of the hoist.
    (b) The length of chain (in feet) pulled through the hoist to move the load 18 in.
    (c) The effort force to raise .85 T when the percent efficiency is 38%.

4.3 INCLINED PLANES

Introduction

The inclined plane is a simple machine that is used to raise and lower heavy objects.
By making the length of the inclined plane long compared with its height, the slope
is gradual, so it is possible to raise very heavy objects with the use of a relatively
small force. Figure 4.14 pictures an inclined plane with the length (l) and height (h)
noted.

IMA of the Inclined Plane

The ideal mechanical advantage is determined by dividing the distance the load is
moved along the ramp by the vertical height that the load is raised. Thus,

$$\text{IMA}_{\text{inclined plane}} = \frac{l}{h} \qquad \textbf{(4.5)}$$

FIGURE 4.14
A pictorial of an inclined plane where
IMA = l/h.

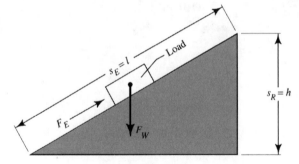

where

IMA = ideal mechanical advantage of the inclined plane
l = effort distance, which is equivalent to the distance moved by the load up the inclined plane, in m or ft
h = load distance, which is equivalent to the vertical height moved by the load, in m or ft

Effort Force

When friction is not considered, the effort force (applied parallel to the inclined plane) can be determined from the law of simple machines. For the inclined plane, this equation is expressed as:

$$(F_E \times l) = (F_W \times h)$$

Solving for F_E gives

$$F_E = \frac{F_W \times h}{l} \qquad (4.6)$$

Since l/h = IMA, then h/l = 1/IMA, and

$$F_E = \frac{F_W}{IMA} \qquad (4.7)$$

where

F_E = effort force applied to the load parallel to the inclined plane, in N or lb
F_W = weight of the load, in N or lb
h = vertical height the load is raised, in m or ft
l = distance moved by the load up the length of the inclined plane, in m or ft
IMA = the ideal mechanical advantage of the inclined plane

Example
4.6

A large crate weighing 240 lb is to be loaded into a pickup truck that has a bed 3.0 ft above the ground. Assuming no friction, determine

    (a) The ideal mechanical advantage when the loading ramp (inclined plane) is 12 ft long.

    (b) The effort force needed to move the load up the 12-ft-long inclined plane.

SOLUTION   **(a)** Find the IMA.

Given $\text{IMA} = \dfrac{l}{h}$ Equation 4.5

Evaluate $l = 12$ ft

              $h = 3.0$ ft

Substitute $\text{IMA} = \dfrac{12}{3.0}$

Solve $\text{IMA} = 4.0$

    (b) Find the force of effort.

Given $F_E = \dfrac{F_W}{\text{IMA}}$ Equation 4.7

Evaluate $F_W = 240$ lb

              $\text{IMA} = 4.0$

Substitute $F_E = \dfrac{240}{4.0}$

Solve $F_E = 60$ lb

From the previous example, you may note that the same amount of work is done (assuming no friction) whether the crate is lifted directly from the ground up 3.0 ft (vertically) to the truck bed or the crate is raised 3.0 ft by pushing it up the inclined plane. The law of simple machines holds true. Thus, $(s_E)(F_E) = (s_R)(F_R)$, $12 \times 60 = 3.0 \times 240$, and 720 ft-lb = 720 ft-lb.

Wedge

The wedge is a *double inclined plane* and is used as the basis of many cutting and splitting tools, including knives, axes, mauls, and chisels. The simple wedge is used to raise heavy loads a small distance or to split wood. Figure 4.15 pictures a splitting maul, a sledgehammer with a wedge-shaped head, which has entered a log. The ideal

FIGURE 4.15
The end of a maul is shaped like a
wedge so it can split wood. The
wedge has an IMA equal to d/s.

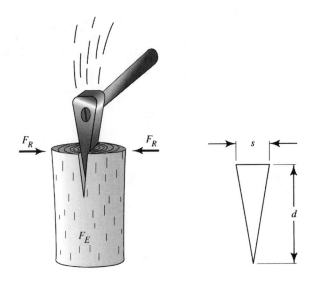

mechanical advantage is the ratio of the depth of the wedge (d) to the distance the
material is forced apart (s). From Figure 4.15, the IMA of the wedge is

$$\text{IMA}_{\text{wedge}} = \frac{d}{s} \tag{4.8}$$

Example 4.7

A 12-lb (.37 slug) splitting maul (Figure 4.15) is used to split logs into firewood. If
the maul was moving at a velocity of 28.0 ft/s when it struck the wood and the wedge
end of the maul was driven 1.5 in. (.125 ft) into the log, forcing the wood .20 in.
apart, then determine the lateral force exerted on the wood to overcome the resistive force (F_R). Assume the percent efficiency is 30%.

SOLUTION

Determine the resistive force exerted on the maul to stop its motion by applying $F = ma$. Start this solution by solving for acceleration (deceleration). If we assume that
the deceleration is uniform, since the wedge decelerates in 1.5 in, then the accelera-
tion (deceleration) can be determined from the $2as$ equation (Equation 2.10).

Given

$$2as = v_f^2 - v_i^2$$

$$a = \frac{v_f^2 - v_i^2}{2s}$$

Evaluate

$$v_f = 0$$

$$v_i = 28.0 \text{ ft/s}$$

$$s = .125 \text{ ft}$$

Substitute

$$a = \frac{-28.0^2}{2 \times .125}$$

Solve	$a = 31\overline{0}0 \text{ ft/s}^2$
Observation	The force exerted on the maul to stop its motion can be found by solving for F in $F = ma$ (Equation 2.1) when acceleration is 3100 ft/s² and mass is given as .37 slug. Note that the minus sign in –3100 ft/s² indicates deceleration.
Given	$F = ma$
Evaluate	$m = .37 \text{ slug}$
	$a = 31\overline{0}0 \text{ ft/s}^2$
Substitute	$F = .37 \times 31\overline{0}0$
Solve	$F = 1100 \text{ lb} \Rightarrow 1.1 \times 10^3 \text{ lb}$
Observation	The resistive force on the maul (F_R) of 1100 lb is equal to the effort force (law of action and reaction) applied to generate the lateral force. The lateral force exerted on the wood to split the wood apart (.20 in) is equal and opposite to the forces holding the wood together, as noted in Figure 4.15 as F_R. The effort force F_E of 1100 lb is amplified by the IMA of the wedge and reduced by the efficiency to produce the lateral force.
Given	$\text{AMA} = \text{IMA} \times \eta \quad \text{and} \quad \dfrac{F_R}{F_E} = \text{IMA} \times \eta$
	Solve for F_R:
	$F_R = \text{IMA} \times F_E \times \eta$
Evaluate	$\text{IMA} = \dfrac{d}{s} = \dfrac{1.5}{.20} = 7.5$
	$F_E = 1100 \text{ lb}$
	$\eta = 0.30$
Substitute	$F_R = 7.5 \times 1100 \times 0.30$
Solve	$F_R = 2500 \text{ lb} \Rightarrow 1.3 \text{ T}$

Even though the wedge (maul) is only 30% efficient, the lateral force splitting the wood apart is greater than 1 T (2500 lb). If the wedge were ideal, then the lateral force would be 8300 lb, or a little more than 4 T, before friction is taken into consideration. The wedge, when struck with a sledgehammer, can generate very large forces even after losses due to friction.

Screw Thread

A screw thread may be thought of as a modification of the inclined plane. By wrapping an inclined plane around a cylinder, a continuous **helix** (spiral wrapping) is formed. A screw thread is a helical groove cut into the surface of a cylindrical rod.

FIGURE 4.16
Wrapping an inclined plane around a cylinder forms a helix similar to that of a screw thread.

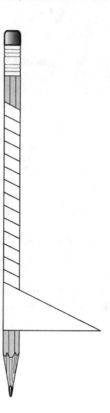

To see how a screw thread is generated, simply cut a right triangle out of paper and wrap it around a round object such as a pencil (Figure 4.16). The resulting helix represents the thread of the screw. The *pitch* of the thread is the distance between successive peaks of the thread, as indicated in Figure 4.17. The pitch of the thread may be thought of as the distance a nut would travel along a bolt when rotated one turn.

Screw threads are used to transmit motion, transmit force, and to fasten parts together (screws, bolts, and nuts). The C-clamp (Figure 4.18) is an example of a screw thread used to transmit force and hold work in place.

The *screw jack*, a machine used to lift heavy objects (pictured in Figure 4.19), uses a screw thread to generate motion and transmit force to the load. By turning the handle around one turn, the screw is advanced a distance equal to the pitch of the thread.

The effort force (F_E) is applied to the handle, which moves through a distance equal to the *circumference of the circle* ($2\pi r$) traced out by the movement of the handle. The load (F_R) moves a distance equal to the pitch of the screw thread (p), as noted in Figure 4.19. The ideal mechanical advantage of the screw jack is

$$\text{IMA}_{\text{screw jack}} = \frac{2\pi r}{p} \qquad \textbf{(4.9)}$$

FIGURE 4.17
A pictorial of a screw thread with
the pitch indicated.

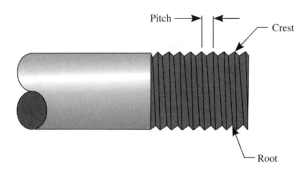

where

> IMA = ideal mechanical advantage of the screw jack
>
> $2\pi r$ = distance moved by the effort, which is equivalent to the circumference of the circle traced by the length of the handle (r), in m or ft
>
> p = distance moved by the load, which is equivalent to the pitch of the screw thread, in m or ft

The ideal mechanical advantage of the screw jack is extremely high; however, the friction between the threads of the screw and the base and the swivel cap and

FIGURE 4.18
The C-clamp uses a screw thread to
transmit force and hold work in
place.

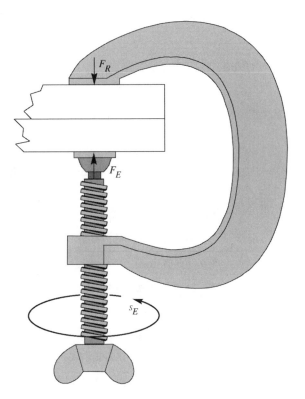

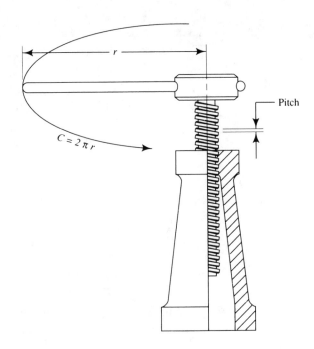

FIGURE 4.19
The parameters of the screw jack are noted. IMA = circumference/pitch, or IMA = $(2\pi r)$/pitch.

the end of the screw are substantial, resulting in a much smaller mechanical advantage (AMA). The percent efficiency for a screw jack varies from 25% to 75%.

Example 4.8

A screw jack has a screw thread with a pitch of 3.00 mm. The jack is used to lift a weight of 42.700 kN. Determine

    (a) The IMA when the radius of the jack handle is 0.625 m.
    (b) The effort force applied to the end of the 0.625-m handle when the efficiency is 0.65.

SOLUTION   **(a)** Find the IMA of the screw jack.

Given
$$\text{IMA} = \frac{2\pi r}{p}$$

Evaluate
$$r = 0.625 \text{ m}$$
$$p = 3.00 \text{ mm} = 0.00300 \text{ m}$$

Substitute
$$\text{IMA} = \frac{2\pi(0.625)}{0.00300}$$

Solve
$$\text{IMA} = 1309 \Rightarrow 1.31 \times 10^3$$

    (b) Find the force applied to the handle.

Given
The law of simple machines and the efficiency equation.

$$\text{AMA} = \text{IMA} \times \eta \quad \text{and} \quad \frac{F_R}{F_E} = \text{IMA} \times \eta$$

Solving for F_E results in

$$F_E = \frac{E_R}{\text{IMA} \times \eta}$$

Evaluate $\quad F_R = 42\ 700\ \text{N}$

$\text{IMA} = 1310$

$\eta = 0.65$

Substitute $\quad F_E = \dfrac{42\ 700}{1310 \times 0.65}$

Solve $\quad F_E = 50\ \text{N}$

An 11-lb pull on the handle (50 N) results in a lifting force of approximately 5 T (43 000 N). You can understand why the screw jack is used to raise houses off their foundations!

Exercise 4.3

Solve the following problems. Make sketches to aid in solving the problems and structure your work so it follows in an orderly progression and can easily be checked.

1. A ramp on a loading dock is 35 ft long and 4.5 ft high. A crate weighing 215 lb is pushed up the inclined plane. Determine
   (a) The IMA.
   (b) The effort force when the percent efficiency is 78%.

2. For the ramp and crate in Problem 1, determine if an effort force of 32 lb is enough to move the crate up the ramp when the efficiency is 0.82. Include your work to support your conclusion.

3. A wedge is driven 10.0 cm into a hardwood log and the log is split 4.0 cm apart. Determine
   (a) The IMA.
   (b) The resistance force when 18 kN of effort is applied to the wedge by the sledgehammer. Assume the wedge to be 25% efficient.

4. A screw jack is used to move a large 1800-lb electric motor to bring it into alignment with a pump shaft. The handle of the screw jack is 3.33 ft long from the center of the screw to the end of the handle. If the pitch of the screw is .125 in, determine
   (a) The IMA of the jack.
   (b) The effort force needed when the efficiency is 0.26.

5. The screw pitch of a screw jack is 2.0 mm and the effort is applied to the end of a handle with a radius of 30.0 cm. If the jack has a mechanical advantage (AMA) of 48 when the load is 1200 N, determine
   (a) The IMA of the jack.
   (b) The effort force applied to the handle.
   (c) The percent efficiency of the screw jack.
6. A screw jack with a 4.0-mm pitch is used to lift a 680-kg automobile. If the efficiency is 0.22 and the mechanical advantage (AMA) is 52, determine the length of the jack handle (meters).

4.4   COMPOUND MACHINES

Introduction

Compound machines are formed by combining two or more simple machines. The window latch of Figure 4.20 is an example of a compound machine. It contains two simple machines, the wheel and axle and the inclined plane. As the latch is rotated under the lock plate, the inclined plane portion of the machine amplifies the IMA of the wheel and axle and securely locks the two windows together.

IMA of Compound Machines

The ideal mechanical advantage of a compound machine is the product of the ideal mechanical advantages of the simple machines making up the compound machine:

$$\text{IMA}_{\text{total}} = \text{IMA}_1 \times \text{IMA}_2 \times \text{IMA}_3 \times \cdots \times \text{IMA}_n \qquad \textbf{(4.10)}$$

Example 4.9

Determine the total IMA of the latch of Figure 4.20 when

$$r_a = .50 \text{ in} \qquad l = .36 \text{ in}$$

$$r_W = 2.0 \text{ in} \qquad h = .18 \text{ in}$$

FIGURE 4.20
A sash lock (window lock) is a compound machine made up of two simple machines—the inclined plane and the wheel and axle.

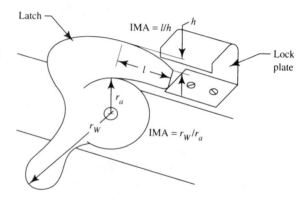

SOLUTION	Compute the IMA of each simple machine; then use Equation 4.10 to determine the total IMA of the latch.
Given	(a) $IMA_1 = \dfrac{r_w}{r_a}$ and (b) $IMA_2 = \dfrac{l}{h}$
Evaluate	Use values from the statement of the problem.
Substitute	(a) $IMA_1 = \dfrac{2.0}{.50}$ (b) $IMA_2 = \dfrac{.36}{.18}$
Solve	(a) $IMA_1 = 4.0$ (b) $IMA_2 = 2.0$
Observation	Now determine the total IMA of the compound machine.
Given	$IMA_{total} = IMA_1 \times IMA_2$
Evaluate	$IMA_1 = 4.0$
	$IMA_2 = 2.0$
Substitute	$IMA_{total} = 4.0 \times 2.0$
Solve	$IMA_{total} = 8.0$

Efficiency of Compound Machines

The efficiency of a compound machine is equal to the product of the efficiencies of the simple machines:

$$\eta_{total} = \eta_1 \times \eta_2 \times \eta_3 \times \cdots \times \eta_n \tag{4.11}$$

Example **4.10**	Determine the total efficiency of a compound machine consisting of a lever with a percent efficiency of 98%, an inclined plane with a percent efficiency of 88%, and a differential pulley with a percent efficiency of 42%.
SOLUTION	Find the total efficiency using Equation 4.11.
Given	$\eta_{total} = \eta_1 \times \eta_2 \times \eta_3 \times \cdots \times \eta_n$
Evaluate	$\eta_1 = 0.98$
	$\eta_2 = 0.88$
	$\eta_3 = 0.42$
Substitute	$\eta_{total} = 0.98 \times 0.88 \times 0.42$
Solve	$\eta_{total} = 0.36$

Cam

The **cam** is used in the design of automatic machines to produce a desired motion. The cam guides the motion of the *follower*. In Figure 4.21, the *plate cam* rotates

FIGURE 4.21
(a) A plate cam with a radial translating follower. (b) The cam is approximated by the intersection of two circles with their centers offset by a distance equal to h.

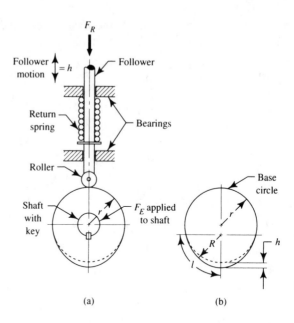

while the follower moves perpendicular to the axis of the cam, translating the rotary motion of the cam into an up-and-down motion.

The plate cam, in its simplest form, can be thought of as a compound machine consisting of a wheel and axle and a rotating wedge. The IMA and AMA may be approximated with the law of simple machines as applied to the inclined plane and the wheel and axle.

For the plate cam of Figure 4.21, the base circle is the wheel, whereas the shaft is the axle. The length of the inclined plane is approximated by the arc swept out by radius R generating distance (l). The height (h) is created by the offset between the centers of the two circles. The IMA for the inclined plane of this cam is approximated from Figure 4.21(b) as

$$\text{IMA}_{\text{cam}} = \frac{l}{h} = \frac{\pi R}{2(R - r)} \qquad \textbf{(4.12)}$$

where

$$l = \frac{C}{4} = \frac{2\pi R}{4} = \frac{\pi R}{2}$$

$\dfrac{C}{4}$ = one-fourth of the circumference (C) of the circle with radius R

$h = R - r$, the difference between the two radii

Example
4.11

Determine the IMA (from the shaft to the follower) of the cam pictured in Figure 4.21 when the shaft diameter is .500 in, $r = 1.00$ in, and $R = 1.38$ in.

SOLUTION First find the IMA of the wheel and axle; then find the IMA of the cam (inclined plane).

Given

$$\text{IMA}_{W\&a} = \frac{r_W}{r_a} \qquad \text{and} \qquad \text{IMA}_{cam} = \frac{\pi R}{2(R - r)}$$

Evaluate

$$r_a = 1.00 \text{ in} \qquad\qquad R = 1.39 \text{ in}$$

$$r_W = \frac{.500}{2} = .250 \text{ in} \qquad\qquad r = 1.00 \text{ in}$$

Substitute

$$\text{IMA}_{W\&a} = \frac{.250}{1.00} \qquad\qquad \text{IMA}_{cam} = \frac{3.14 \times 1.38}{2(1.38 - 1.00)}$$

Solve

$$\text{IMA}_{W\&a} = .250 \qquad\qquad \text{IMA}_{cam} = 5.70$$

$$\text{IMA}_{total} = \text{IMA}_{W\&a} \times \text{IMA}_{cam} = .250 \times 5.70 = 1.43$$

Exercise 4.4

Solve the following problems. Make sketches to aid in solving the problems and structure your work so it follows in an orderly progression and can easily be checked.

1. Determine the efficiency of a compound machine that is made up of levers with a combined percent efficiency of 88%, a wheel and axle with a percent efficiency of 56%, and an inclined plane with a percent efficiency of 74%.
2. The lifting system of Figure 4.22 is used to hoist ore cars up the incline so they may be dumped. The compound machine consists of three simple machines: a pulley system, an inclined plane, and a winch. If each car weighs 35 T when loaded, determine
   (a) The IMA of the pulley system.
   (b) The IMA of the inclined plane.
   (c) The IMA of the winch.
   (d) The effort force, assuming no friction.

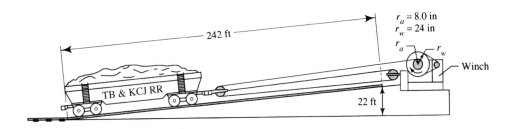

FIGURE 4.22
A compound machine used to hoist an ore car to be dumped.

FIGURE 4.23
A cam vise for Problem 5 of Exercise 4.4.

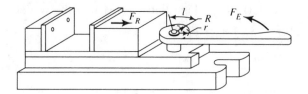

3. The lifting system of Figure 4.22 has frictional loss in each of the three simple machines. The losses, accounted for in the efficiencies of each of the simple machines, are as follows: inclined plane, 94% efficient; pulley system, 65% efficient; winch (wheel and axle), 78% efficient. Determine
   (a) The total efficiency of the compound machine.
   (b) The effort force applied to the axle of the winch to lift the loaded car with frictional forces opposing its movement.
4. The plate cam pictured in Figure 4.21 has a base circle radius of 3.0 cm and a shaft diameter of 12 mm. Determine
   (a) The IMA of the cam from the shaft to the follower when $h = 8.0$ mm.
   (b) The effort force applied to the shaft when the resistance force (including the return spring) is 8.0 N and the efficiency is 0.82.
5. The cam vise of Figure 4.23 is used to hold parts that are being machined. The effort is applied to the handle 16 in. from the center of the pivot. The radius of the base circle is 1.5 in, and the arc length is approximately equal to one-fourth of a circle with a radius equal to R. If R is 2.0 in, determine
   (a) The IMA of the wheel and axle part of the compound machine.
   (b) The IMA of the cam part of the compound machine.
   (c) The IMA of the entire compound machine.
6. Using the dimensions given in Figure 4.24, determine the following when the effort applied to the handle is 22 lb.
   (a) The IMA of the lever part of the compound machine.
   (b) The IMA of the inclined plane part of the machine.
   (c) The IMA of the entire compound machine.

FIGURE 4.24
The window latch for Problem 6 of Exercise 4.4.

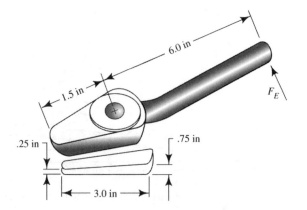

CHAPTER SUMMARY

- A block is made up of a pulley, frame, and ring or hook.
- A single fixed pulley is a class 1 lever with an IMA of 1.
- A single fixed pulley changes the direction of the effort.
- A single movable pulley is a class 2 lever with an IMA of 2.
- The frictional losses in lifting systems are sizable and cannot be ignored.
- With the wheel and axle, the wheel is not always larger than the axle. However, the effort force is applied to the wheel.
- Machines based on the wheel and axle are less than ideal.
- A chain hoist is a differential pulley, and, like all differential pulleys, it develops a very large IMA.
- An inclined plane is the basis of many machines, including the screw thread and the cam.
- A wedge is a double inclined plane, which is the basis of many cutting and splitting tools.
- The distance a bolt advances in a nut in one turn is equal to the pitch of the thread.
- A compound machine is formed when two or more simple machines are combined.
- The efficiency of a compound machine is equal to the product of the efficiency of each simple machine.

SELECTED TECHNICAL TERMS

The following technical terms are defined in the glossary located after Chapter 16. You are encouraged to use the glossary to aid your understanding and to test your knowledge of these important terms.

cam
compound machine
helix
sheaves
windlass

END-OF-CHAPTER QUESTIONS

Write T if the statement is true and F if the statement is false.

1. A single fixed pulley is a class 2 lever.
2. A single movable pulley combined with a fixed pulley has an IMA of 2.
3. A plate cam is like a rotating wedge in combination with a wheel and axle.
4. The frictional losses in a lifting system are usually very small.
5. A screw jack can have a very large AMA because of its small frictional loss.
6. The pitch of a screw is the distance between the crest of one thread and the crest of the next thread.
7. The sprocket, crank, and pedals of a bicycle are a simple wheel and axle.
8. The efficiency of a compound machine is equal to the sum of the efficiencies of each of the simple machines.
9. The rope in a pulley system is called the sheave.

10. The IMA of a pulley system can be determined by counting the number of ropes that support the fixed pulleys.

In the following, select the word that makes the statement true.

11. Compound machines are formed by combining (three, two, four) or more simple machines.
12. The combined IMA of a compound machine is equal to (the product, the sum, the quotient) of the IMA of each of the simple machines.
13. A block and tackle is made up of a sheave, a frame, a hook, and (a rope, a pulley, a wheel).
14. The IMA of a pulley system is determined by (multiplying, counting, adding) the number of ropes that support the movable pulleys.
15. In computing the IMA of machines that use the wheel and axle, use (the diameter of the wheel, the radius of both wheel and axle, the diameter of the axle and the radius of the wheel).

Answer each of the following questions with a short answer in the form of a complete sentence. Include a restatement of the question in your answer.

16. Why is the IMA of a single fixed pulley equal to 1?
17. Why is the IMA of a single movable pulley equal to 2?
18. What are some advantages of using a chain hoist over a block and tackle?
19. What are some of the characteristics of the screw jack?
20. A 4,000-lb rock is sitting flush with the ground. Using simple tools, how would you, working by yourself, initially raise the rock so a cable could be placed under it to hoist it with a crane?

END-OF-CHAPTER PROBLEMS

Solve the following problems. Make sketches to aid in solving the problems and structure your work so it follows in an orderly progression and can easily be checked. Table 4.2 summarizes the equations of IMA for the machines in this chapter.

1. Determine the IMA of a wheel and axle when the axle diameter is 4.5 cm and the wheel radius is 27 cm.
2. Determine the minimum number of movable pulleys for a lifting system with an IMA of 5. Sketch the system.
3. A compound machine is made up of two machines. One has an IMA of 8.40 and an AMA of 6.2; the other has an IMA of 12.6 and an AMA of 4.4. Determine
   (a) The efficiency of the compound machine.
   (b) The IMA of the compound machine.
   (c) The AMA of the compound machine.
4. The helix of a screw is formed by wrapping an inclined plane around a cylinder. Determine the approximate IMA of a .75-in diameter bolt with a pitch of .10 in.
5. A lifting system uses four ropes to support the movable pulleys attached to the 1180-N load. Determine
   (a) The efficiency of the system; use Table 4.1.
   (b) The IMA of the system.

TABLE 4.2 Summary of Formulas Used in Chapter 4

Equation Number	Equation
4.1	$IMA_{pulley} = \dfrac{s_E}{s_R}$
4.2	IMA_{pulley} = number of ropes supporting the movable pulleys
4.3	$IMA_{wheel\&axle} = \dfrac{r_w}{r_a}$
4.4	$IMA_{hoist} = \dfrac{2R}{R - r}$
4.5	$IMA_{inclined\ plane} = \dfrac{l}{h}$
4.6	$F_E = \dfrac{F_W \times h}{l}$
4.7	$F_E = \dfrac{F_W}{IMA}$
4.8	$IMA_{wedge} = \dfrac{d}{s}$
4.9	$IMA_{screw\ jack} = \dfrac{2\pi r}{p}$
4.10	$IMA_{total} = IMA_1 \times IMA_2 \times IMA_3 \times \cdots \times IMA_n$
4.11	$\eta_{total} = \eta_1 \times \eta_2 \times \eta_3 \times \cdots \times \eta_n$
4.12	$IMA_{cam} = \dfrac{\pi R}{2(R - r)}$

(c) The distance the effort force must move to move the load 1.0 m.

(d) The effort force needed to lift the load for the efficiency in (a).

6. The spring scale of Figure 4.13 indicates a reading of 48 lb for a load weight of 130 lb. If the radius of the flywheel is 1.0 ft and the diameter of the axle is 8.0 in, determine

(a) The IMA of the wheel and axle.

(b) The AMA of the wheel and axle.

(c) The efficiency of the wheel and axle.

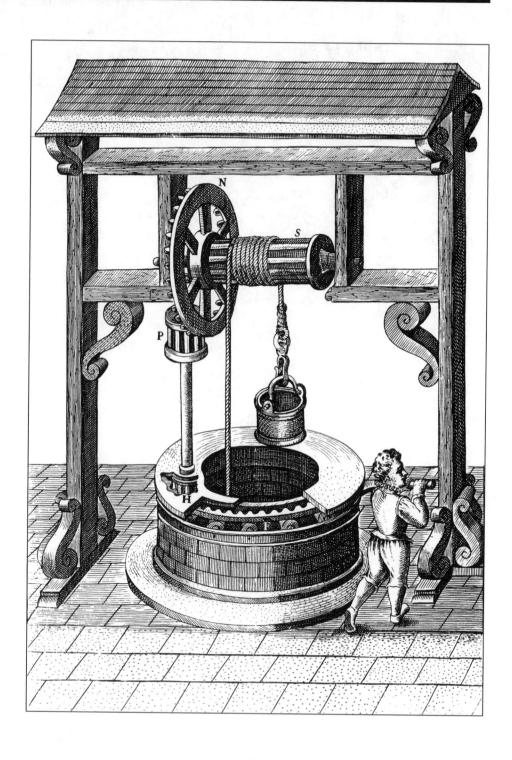

Chapter **5**

Friction, Lubrication, and Bearings

About the Illustration. *A* roller thrust bearing *supports the internal gear (F) of this Renaissance machine, which is moved by turning the external lever. The circular motion, resulting from the lever being moved, is transmitted through the lantern pinion (gear) H to pinion P, which rotates crown gear N. Gear N and drum S form a wheel and axle, which is used to wind the bucket rope up or down. The ends of the shaft (with gear N and drum S) are supported by journal bearings formed by boring holes in the support members. The bearings may be operated dry or lubricated with animal or vegetable oil. (Designed by Agostino Ramelli, from* The Various and Ingenious Machines of Agostino Ramelli *(1588).)* ■

INTRODUCTION

If machines are to operate for long periods of time without failure, they need to be engineered so that friction is at a minimum and the heat due to friction is under control. The elements of machines must operate smoothly with as little wear as possible. The key to all this is good bearing design, coupled with the correct lubricant and lubrication system. This chapter introduces friction (the force of friction) in three of its forms (static, kinetic, rolling) and studies various lubricants and bearings and how they are applied to minimize friction.

CHAPTER CONTENTS

PERFORMANCE OBJECTIVES

Once you have read and studied each section; worked through each example with pencil, paper, and calculator; worked through the end-of-chapter problems; and answered the end-of-chapter questions, then you should be able to

- Use the coefficients of friction to make calculations.
- Distinguish between plain and antifriction bearings.
- Identify the three categories of wear.
- Know the conditions under which fluid film lubrication, mixed film lubrication, and boundary lubrication take place.
- Use kinematic viscosity in selecting lubricating oil.
- Name the three classes of bearings and relate the types of plain and antifriction bearings to these classes.

5.1 FRICTION

Introduction

Friction is a good news–bad news parameter. As you know from experience, *friction forces* (friction) represent a loss of energy in the form of heat and cause inefficiency and wear in machines. That is the bad news. The good news is that friction enables V-belts to turn pulleys, brakes to stop cars, tires to grip the road, bolts and nails to hold things together, and you and me to walk without slipping. **Friction** is the resistive force that tends to keep surfaces from moving (sliding or rolling) over one another when they are in contact. An understanding of the nature of friction is necessary in order to specify lubrication and bearings.

Friction Parameters

For dry, clean surfaces in contact, the friction force depends principally on the kind of material in contact, the *profile* (roughness, waviness, lay, etc.) of the surface, and the *normal force* pressing the surfaces together. The normal force is a force that acts at right angles (perpendicular) to the surfaces of the two bodies.

FIGURE 5.1
(a) A block of wood at rest on a table top. (b) Free-body diagram with weight (F_W) and normal force (F_N) indicated; $F_W = -F_N$.

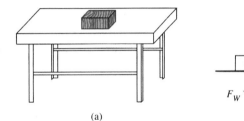

(a)   (b)

These three parameters, friction force (F_f), materials and profile (μ), and normal force (F_N), are related by an equation that approximates their interdependence:

$$F_f = \mu \times F_N \qquad (5.1)$$

where

F_f = friction force, in N or lb
μ = coefficient of friction, no units
F_N = normal force (perpendicular to the surface), in N or lb

Coefficient of Static Friction

Imagine that a wooden block is sitting on a level metal tabletop, as pictured in Figure 5.1(a). Since the block is at rest, we know that the system of forces acting on it are in equilibrium; that is, each force that could move the block has an opposite force of equal magnitude keeping the block stationary. In Figure 5.1(a), the weight of the block, F_W, acting perpendicular to the two surfaces, is pressing the surfaces together, and an opposing normal force, F_N, is pushing back. By the law of action and reaction, the two forces are equal but opposite, and $F_W = -F_N$. The free-body diagram of Figure 5.1(b) shows the normal force (F_N) equal to but opposite in direction to the weight (F_W) of the block.

Now suppose a gradually increasing effort force (F_E) is applied to the block parallel to the surface of the tabletop, as pictured in Figure 5.2(a). For small amounts of effort force, there is no movement; however, as the force is increased, there is a point at which sliding is just about to occur (called the *point of impending motion*). Once again, by the law of action and reaction, an opposing force must be present to counter the effort force and maintain equilibrium. This force is the static friction force, F_{fs}, as noted in Figure 5.2(b). Since the block is in equilibrium just be-

FIGURE 5.2
(a) An effort force (F_E) is applied to the block. (b) An opposing force, F_{fs}, keeps the system in equilibrium; $F_E = -F_{fs}$.

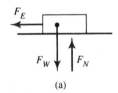

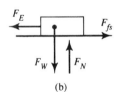

(a)   (b)

fore it slides, the effort force (F_E) must be equal but opposite to the static friction force (F_{fs}), and $F_E = -F_{fs}$.

The friction force (F_{fs}) depends on the normal force (F_N) for its magnitude; that is, as the weight of the object is made heavier, the normal force increases to counter the increase in weight, and the friction force is increased in direct proportion to the normal force. Since friction force is proportional to the normal force, a constant of proportionality may be used to relate these two forces. The constant is called the **coefficient of static friction**, denoted by the Greek letter, μ (mu). Thus

$$\mu_s = \frac{F_{fs}}{F_N} \tag{5.2}$$

where

μ_s = coefficient of static friction, no units
F_{fs} = static friction force, in N or lb
F_N = normal (perpendicular) force, in N or lb

Example 5.1

Determine the coefficient of static friction between a 4.0-lb wooden block at rest on a level metal surface when 1.2 lb of effort force is applied parallel to the metal surface to bring the block to the point of impending motion.

SOLUTION   Find the coefficient of static friction.

Given $\mu_s = \dfrac{F_{fs}}{F_N}$, $F_W = F_N$, and $F_E = F_{fs}$

Evaluate $F_{fs} = 1.2$ lb

$F_N = 4.0$ lb

Substitute $\mu_s = \dfrac{1.2}{4.0}$

Solve $\mu_s = 0.30$

Observation   For each pound of increase in weight, the friction force increases by .30 lb.

Table 5.1 lists representative values of the coefficient of static friction for several combinations of materials with clean, dry, unlubricated surfaces. The listed coefficients are dimensionless numbers and are, at best, good approximations. They should be used only as general guidelines, since the surface profile (roughness) causes the values to vary considerably.

In looking over Table 5.1, notice that the coefficient of static friction for aluminum on aluminum is twice that of aluminum on steel. This means that it would take twice the effort force to move an aluminum block, with a constant weight, across an aluminum surface than across a steel surface.

TABLE 5.1 Coefficients of Static Friction for Selected Materials (Approximate)

Materials	μ_s	Angle of Repose
Aluminum on aluminum	1.2	50°
Aluminum on steel	0.60	31°
Steel on steel	0.80	39°
Steel on brass	0.40	22°
Steel on cast iron	0.40	22°
Teflon on steel	0.040	2.3°
Nylon on nylon	0.20	11°
Wood on metal	0.20–0.60	11–31°
Clutch lining on metal	0.30–0.60	17–31°

Kinetic Friction

Once the point of impending motion is exceeded, movement occurs between the surfaces. The effort force needed to maintain movement once the friction force has been overcome is slightly less than the effort force needed to produce the point of impending motion. The friction force opposing movement is called the **kinetic force of friction** (F_{fk}). Since movement is taking place, work is being done by the effort force, which, in turn, produces heat at the surfaces and causes the surface temperatures to rise.

Once movement starts, a cause-and-effect event takes place to lessen the friction force. Under static conditions, motion is impeded by the adhesion of the materials in contact. Microscopic surface irregularities catch and form microscopic welds, which take an added force to break apart. However, once motion starts, there is less meshing of the surface irregularities and fewer adhesions (welds) to break, so the force of friction is reduced.

Kinetic friction is usually less than static friction. As an example, the coefficient of static friction for steel on bronze of 0.40 (dry) is reduced to the coefficient of kinetic friction of 0.24 (dry) once the surfaces start to slide over one another. This concept is pictured in Figure 5.3.

As the effort force (horizontal axis of Figure 5.3) is increased from 0 to 40 lb, the friction force (the reactionary force, vertical axis) also increases in step with the effort force from 0 to 40 lb. This linear relationship is reflected in the graph by the constant slope of the curve (straight line at an angle of 45° with the axis) representing the static friction. Once the effort force exceeds 40 lb, then the point of impending motion is reached, and the block begins to move. The effort force may be scaled back from 40 lb to 24 lb to meet the reactionary force resulting from the kinetic friction (the flat horizontal line). If the effort force continues to increase, then the block will be accelerated faster and faster.

Although there is wide variation in the kinetic coefficient of friction (μ_k) at high and very low surface speeds, it is safe to assume that the value of μ_k is constant for most moderate operating conditions.

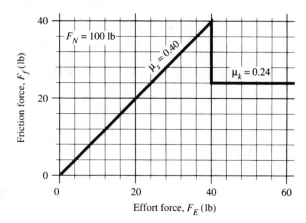

FIGURE 5.3
Effort force (F_E) versus friction force (F_f) for a constant normal force (F_N) of 100 lb and an interface of steel on bronze, where $F_E = -F_f$.

Example 5.2

Determine the force needed to produce motion by overcoming friction between two dry, cast-iron surfaces. The kinetic coefficient of friction is 0.20. One surface is the table on a milling machine and the other is the bottom of an 85-lb vise being moved along the table.

SOLUTION   Find the kinetic friction force (F_{fk}).

Given $F_{fk} = \mu_k \times F_N$

Evaluate $\mu_k = 0.20$

$F_N = F_W = 85 \text{ lb}$

Substitute $F_{fk} = 0.20 \times 85$

Solve $F_{fk} = 17 \text{ lb}$

From the preceding discussion, the following empirical observations about the nature of friction forces between dry surfaces may be made.

■ Friction forces tend to oppose the actual or impending motion of a body.
■ It takes a greater force of effort to start a body from rest than it takes to keep it in motion.
■ The force of friction (F_f) (for a moderate normal force) is directly proportional to the normal force (F_N) holding the two surfaces together and is totally independent of the area in contact.
■ The coefficients of static friction are usually greater than the coefficients of kinetic friction $(\mu_s > \mu_k)$, and for most operating conditions both are constant.

Normal Force

To understand fully the concept of a normal force (F_N), an inclined plane with a block at rest on its surface will be studied. To start with, the inclined plane will be thought of as being hinged at one end, as pictured in Figure 5.4(a), so the plane can be adjusted to various angles with the block on it.

If the plane is not adjusted at too steep an angle, then the block placed on the inclined plane will be prevented from sliding down the incline by the static friction force. Figure 5.4(b) pictures the several forces that are operating on the block as it rests on the surface of the inclined plane. The two components of the weight are the normal force (F_N) that acts perpendicular to the surfaces to press the block to the inclined plane (creating friction) and the acceleration force (F_a) that acts parallel to the surface of the plane and attempts to move the block down the inclined plane. The friction force (F_{fs}) resulting from the normal force and the coefficient of friction keep the block from sliding down the incline by opposing the acceleration force (F_a). Notice that the normal force and the weight are no longer equal. They are, however, related by the cosine function:

$$F_N = F_W \cos(\theta) \qquad (5.3)$$

where

F_N = the normal force, in N or lb
F_W = the weight of the object, in N or lb
$\cos(\theta)$ = the cosine function of the angle of inclination

As the angle of the inclined plane is slowly adjusted upward, there will be a certain angle where the block reaches the point of impending motion. This angle is called the *angle of repose*. The angle of repose is related to the coefficient of static friction by a very simple function:

$$\mu_s = \tan(\theta_r) \qquad (5.4)$$

where

μ_s = the coefficient of static friction, no units
$\tan(\theta_r)$ = the tangent function of the angle of repose
θ_r = the angle of repose, in degrees

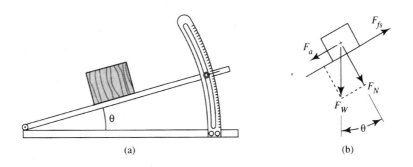

(a)   (b)

FIGURE 5.4
(a) A block at rest on an adjustable inclined plane. (b) Free-body diagram of the force components acting on the block.

Example
5.3

Determine the coefficient of static friction for a steel block on an aluminum plate if the angle of repose is determined to be 31°.

SOLUTION   Find the coefficient of static friction.

Given $\mu_s = \tan(\theta_r)$

Evaluate $\theta_r = 31°$

Substitute $\mu_s = \tan(31°)$

Solve $\mu_s = 0.60$

Industrial Friction Brake

Besides stopping cars, friction brakes are used on industrial machines to hold the machine while it is stopped. Most industrial friction brakes used to hold electric motors are of the spring-set, electromagnetic-release type. This category of brakes uses a stiff compression spring to store mechanical energy and an electromagnetic **solenoid** to control the activation of the brake shoes. Figure 5.5 pictures an electromagnetic-release, spring-set, holding brake. The braking force is changed by adjusting the spring tension on the compression spring.

When the machine is operating, the solenoid is energized, overcoming the force of the spring (compressing the spring), and the brake shoes are moved away from the brake drum, releasing the brake. Turning the motor off de-energizes the solenoid (which is wired into the motor-control circuit) and allows the spring to expand and activate the brake. Since the energy to activate the brake is stored in the compressed spring, this type of brake is automatically activated when there is a power failure. When the machine is running, the energized solenoid prevents the brake shoes from engaging the brake drum that is mounted on the shaft of the rotating motor.

5.2 LUBRICATION

Introduction

To reduce the surface wear and the heat of friction, elements of machines (bearings, gears, flexible couplings, chains, and cams) require **lubrication.** Lubrication of machine elements is facilitated by the introduction of a friction-reducing film called a lubricant that is used to support the moving load. Machine elements have machined surfaces that move relative to each other by sliding, rolling, or a combination of both motions. When loads are heavy and/or speeds of operation are high, then contact between surfaces results in high frictional forces, which produce high temperatures and rapid wear. The lubrication of the machine elements by developing and maintaining lubricating films between contact surfaces is critical for the prevention or reduction of wear between surfaces.

Without lubrication or with inadequate lubrication, modern machines would fail in a very short time due to the wearing away of the bearing surfaces. Without

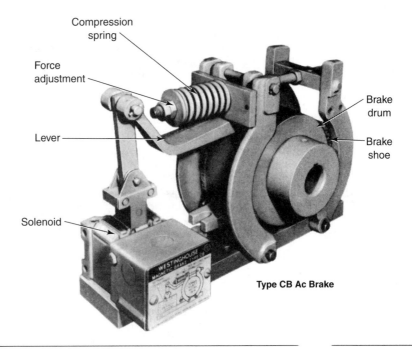

FIGURE 5.5
The friction holding brake is an electromagnetically released, spring-set type of industrial brake. (*Courtesy of the Westinghouse Electric Company*)

proper lubrication of the machine elements, machines would not run smoothly and quietly because of vibration, and they would be out of service most of the time. However, with lubricants that shear (separate) easily and with properly designed machined surfaces, friction forces that oppose motion can be reduced, thus lessening wear, reducing heat, and improving the life of the machine.

Bearing Surfaces and Wear

In general, there is no consistent relationship between the coefficient of friction and the amount of wear for well-lubricated surfaces. However, the relative hardness of the two surfaces and the relative surface profile (Figure 5.6) do directly affect the wear of the lubricated surface. Wear can be divided into three general categories: adhesive, abrasive, and pitting.

Adhesive Wear When two surfaces, as pictured in Figure 5.7, are not completely separated by a lubricant, small surface areas with peaks (known as **asperities**) weld (adhere) to each other, causing shearing and deforming of the material at these points. Material is broken away in these areas. Adhesive wear is minimized when the surface finish (roughness) is carefully controlled and the surfaces are made from hard materials or made hard by plating or case hardening. Also, adhesive

FIGURE 5.6
Surface profile produced by surface grinding.

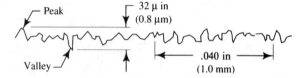

wear is lessened when unlike materials slide over each other, as when carbon steel is used against brass, bronze, or cast iron. Table 5.2 lists design requirements for surface roughness of machine element-bearing surfaces.

Abrasive Wear When a hard rough surface slides over a softer surface, material is cut away in a series of grooves. Abrasive wear may also occur when abrasive particles contaminate the area between the surfaces that slide over each other. Abrasive wear can be controlled by using very hard surfaces so the abrasive particles won't affect the surfaces. The abrasive contaminants can be filtered out when lubricant (oil) is pumped between the surfaces. Also, a hard (carbon steel) and a very soft surface (tin alloys, aluminum, or copper-lead alloys) may be used so that the abrasive particles are embedded into the softer surface and are immobilized.

Pitting When repeated heavy loads (stresses beyond the endurance limit of the material) are transmitted between two surfaces passing over one another, the surface material is *deformed* and, in time, will *fatigue* and fail, resulting in small particles falling out of the surface (*pitting*). The localized high stress that causes pitting is common during the "wear-in" period of two surfaces. Once the asperities are worn away and the surface is smoothed, then the load is distributed over a greater surface area and the process of pitting stops. However, if the pitting is due to excessive contact stress, then destructive pitting will occur, and the surface will fail due to fatigue. Destructive pitting is controlled by keeping loads below the *fatigue limit* of the materials.

Lubricating Film Formation

Lubricants may be divided into four general classes: *fluids* (commercial petroleum oils), *semisolids* (commercial greases), *solids* (molybdenum disulfide, polytetrafluoroethylene [PTFE], etc.), and *gaseous* (air). Of the four classes, only the first three will be considered. The fourth, gaseous lubrication, plays a very important role in lu-

FIGURE 5.7
Surface wear as seen through a microscope. The thin film of oil is on the order of .0003 in (0.008 mm) thick.

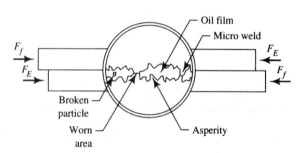

TABLE 5.2 Design Requirements for Surface Roughness for Bearing Surfaces of Machine Elements

μin	μm	Machine Element
63	1.6	Piston pin bores, brake drums, ratchet and pawl teeth, rolling surfaces
32	0.8	Bronze journal bearings, gear teeth, ways
16	0.4	Motor shafts, camshaft lobes, gear teeth
13	0.3	Cylinder bores, crankshaft bearings
8	0.2	Cam faces, hydraulic-cylinder bores
4	0.1	Ball bearing races, piston pins
1	0.03	Balls and rollers for bearings

bricating very high speed mechanisms; however, its discussion is beyond the scope of this text.

The selection of a lubricant for a given application depends upon how the lubricating film is formed between the two surfaces to be lubricated. The formation of the lubricating film depends upon several factors that affect the coefficient of friction. Included in these are the design of the bearing surface, the load per unit area, the speed, and the **viscosity** (resistance to shear or motion) of the lubricant.

Figure 5.8 pictures the three lubrication zones in which a bearing may operate. These zones result from speed, load, and viscosity, the general conditions under which the lubricating film is formed. The coefficient of friction is pictured along the vertical axis, with the maximum (μ_{max}) and minimum (μ_{min}) coefficient of friction indicated. The factors representing viscosity (Z), speed (N), and load per unit area (P) are shown along the horizontal axis.

FIGURE 5.8
The bearing operating curve divided into lubrication zones. The curve pictures the relationship between the coefficient of friction (μ) and the factors representing viscosity (Z), speed (N), and load per unit area (P).

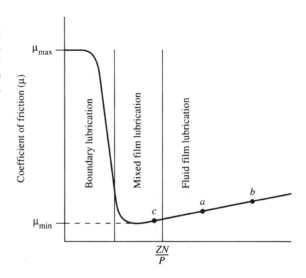

To understand the relationship between viscosity, speed, and pressure (the horizontal factors) and the coefficient of friction (the vertical factor), assume the following: First, the point of operation is point *c*; and second, the viscosity and pressure remain constant, but speed is increased. This being the case, the function *ZN/P* will be larger, causing the *point of operation* to shift from point *c* to the right toward point *a* or point *b*. Because the portion of the curve from point *c* to point *b* slopes upward, the coefficient of friction (vertical axis) will increase from the lower value at point *c* to a higher value at point *b*. The three zones of lubrication pictured from right to left in Figure 5.8 are as follows:

Fluid film lubrication, or *hydrodynamic lubrication*, produces a film that completely separates the two sliding surfaces (a very desirable condition). Because the two surfaces do not contact each other, there is no wear, and frictional losses are very low. Because the force of friction results only from the shearing of the lubrication film, the operation temperature is low. Fluid film lubrication is maintained when the surface velocity is greater than 25 ft/min (7.6 m/min), which results in a coefficient of friction on the order of 0.003. A surface speed (surface velocity) of 25 ft/min is approximately equal to the edge speed of a 1.0-in-diameter shaft rotating at 100 rev/min.

Because of the fluid film formed by the lubricant, the rules for lubricated surfaces are totally different than those for dry surfaces. With fluid film lubrication, the friction force (F_f) is virtually independent of the normal force per unit area (lb/in^2).

Mixed film lubrication, or *thin film lubrication*, as it is often called, results in a film that partially separates the sliding surfaces. Because the sliding surfaces are in contact to some degree (the fluid film carries some of the load), there is wear and the speed of operation is usually low. Mixed film lubrication, which is achieved when the surface velocity is greater than 10 ft/min (3 m/min), results in a coefficient of friction on the order of 0.05.

Boundary lubrication results when there is not enough lubricant present between the sliding surfaces, when the load is large, or when velocities are too low. Under these conditions the lubricant has little to do with supporting the two surfaces, and considerable contact is present. Because the rubbing action of the sliding surfaces generates wear and heat, the operation of machine elements is limited to intermittent, oscillatory, or very slow motion of under 10 ft/min (3 m/min). Typical coefficients of friction for boundary lubrication are on the order of 0.1.

Under fluid film conditions (point *a* of Figure 5.8), an increase in the operating speed (*N*) of the bearing would cause an increase in the coefficient of friction from point *a* up to point *b* and would require a decrease in the viscosity of the lubricant (a lighter oil) to keep the coefficient of friction in the desired operating range. On the other hand, an increase in the load (*P*) on a bearing would cause the operating point to shift from point *a* down to point *c* and would require a proportional increase in the viscosity of the lubricant (a heavier oil) to keep the coefficient of friction in the desired fluid film operating range.

If fluid film operation is desired, then a point of operation is selected to produce a film that won't rupture during a momentary increase in the load. The point of minimum coefficient of friction (μ_{min}) is not a desirable point of operation in this case, since it is in the mixed film zone and any change in operating conditions would force the operating point to the left and into the boundary zone—a very undesirable operating area when fluid film lubrication is required. Instead, a safety factor is built into the design by selecting an operating point to the right of point *c*.

5.3 LUBRICANTS

As you now know, lubricants are used to reduce friction and lessen or prevent wear. Additionally, lubricants are used to prevent surface adhesion, remove heat (cool), prevent corrosion, and help distribute the load over the bearing surface. A wide variety of materials are used as lubricants, including petroleum oil, grease, and even water and air. In addition to petroleum oil and grease, solid films are used where environmental conditions (clean room, food processing, etc.) will not permit the use of petroleum-based lubricants or where normal lubricants cannot easily be applied.

Lubricating Oils

Of the several types of lubricating oils, *mineral oils* (petroleum based) are the most versatile of all the types of oils. Their versatility comes from the inherent properties of the oil plus chemical additives that give the oil additional properties for a wide range of applications. Mineral oils are formulated with a balanced group of additives for best performance and do not require oil supplements to be added by the end user. However, if the additives are to be effective over long periods of service under high temperatures and/or high loads, then the oil must be changed at regular intervals. The following are a few of the additives used to modify mineral oil for a particular application.

Viscosity index improvers are used to improve the viscosity of an oil at elevated temperatures. Viscosity is inherently reduced as temperature increases; to counteract this, long molecules (polymers) are used to thicken the oil as temperature is increased. Oils used for internal combustion engines, transmissions, gears, and hydraulic systems all benefit from viscosity index improvers.

Detergents and dispersants cause particles in engine oil to be dispersed (stopped from clotting together), preventing harmful deposit buildup. The particles are kept in suspension and are removed when the oil is changed. Figure 5.9(a) depicts particles of grime clotted together and blocking an oil passage into a bearing. Figure 5.9(b) represents the particles in a fine suspension due to the detergent-dispersant additives in the oil.

Extreme pressure additives are added to oil such as gear oil (manual transmission) to enable gears to operate at extreme pressures. Oils with extreme-pressure additives are noted by a suffix EP added to the SAE rating of the oil, as in SAE 90 EP. Sulfur-phosphorus compounds are used to protect

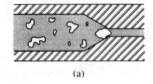 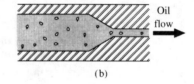

(a)   (b)

FIGURE 5.9
Oil flow in an oil channel is impeded in (a) due to the clotting of impurities in the oil; however, the oil flows unimpeded in (b) because of the detergent-dispersant additive in the oil.

against gear tooth scuffing. The EP additives react to the areas with asperities (where high temperatures are present due to the momentary welding of the sliding surfaces) to polish the surfaces, distribute the load more evenly, and form a high melting point film to lessen boundary lubrication frictional forces.

Corrosion inhibitors are used to prevent the development of corrosion on copper-lead or lead-bronze bearings due to the organic acids that develop in the oil or by contaminants carried in the oil. The corrosion inhibitor additive is either mechanically bonded onto the metal bearing surfaces or chemically deposited onto the surfaces.

Oxidation inhibitors are used to inhibit (slow down) the formation of oxide deposits on hot metal surfaces due to the reaction of the oil with air. Oxidation causes the viscosity of the oil and the formation of organic acids to increase, resulting in the formation of undesirable varnish deposits on the heated metal surfaces.

TABLE 5.3 Kinematic Viscosity System for Industrial Lubricant Oils at 40°C (Selected Listing)

Grade Identification ISO VG	Viscosity Limits (cSt [mm^2/s] at 40°C)	
	Minimum	*Maximum*
2	1.98	2.42
5	4.14	5.06
10	9.00	11.0
22	19.8	24.2
46	41.4	50.6
100	90.0	110
220	198	242
460	414	506

TABLE 5.4 SAE Kinematic Viscosity Classification of Commercial Oils (Midpoint)

SAE Grade (Viscosity)	Engine Oil (cSt at 100°C)
10W	6.0
20W	8.0
30	11.2
40	14.5
50	19.5
	Transmission (Manual) Axle (cSt at 100°C)
75	7.0
80	8.0
90	20.0
140	34.0
250	47.0

Viscosities

Mineral oils come in a wide range of *viscosities* for motor vehicle and industrial applications. Tables 5.3 and 5.4 list some of the lubricating oils available. Table 5.3 lists industrial oils, using the International Standards Organization (**ISO**) standard, by viscosity at 40°C (100°F) and grade number, and Table 5.4 lists commercial automotive oils, using the Society of Automotive Engineers (**SAE**) standard, by viscosity at 100°C (212°F) and grade number.

The viscosity of an oil is an important property in that it plays a major role in the formation of fluid films and mixed films between bearing surfaces. Viscosity is a factor in the production of heat in bearings, transmissions (gears), and internal combustion engines. It also controls the sealing effectiveness of seals in the containment of lubricating oils. Selecting oil of the correct viscosity is a very important step in getting reliable performance from any machine. The viscosity of the lubricating oil must be high enough to develop lubricating films at the designed load and speed of the machine element, but not so high that power is wasted (due to friction losses in the film) in shearing the oil film.

Viscosity Measurement

There are two systems for reporting the viscosity of lubricating oil. The first method measures the force needed to shear an oil film (overcome fluid friction in an oil film) and is called *absolute*, or *dynamic*, *viscosity*. Dynamic viscosity is given units of poise (P) or SI units of pascal-seconds (Pa·s) where 1 Pa·s = 10 P. Dynamic viscosity is

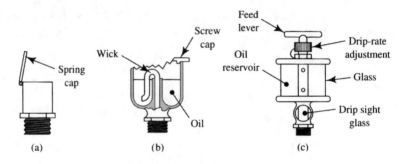

FIGURE 5.10
Gravity-feed lubricators: (a) Oil cup; (b) Wick feed; (c) Drip feed.

used in calculations relating to the design of lubricating systems for machine elements because it relates to the mass density of the lubricating oil.

The second method uses classifications from the SAE (Table 5.4), and the ISO (Table 5.3) for systems of grading lubricating oils. This method is related to the flow time of an oil through a **viscosimeter** and is called *kinematic viscosity*. Kinematic viscosity is given the units of centistokes (cSt) or SI units of square millimeters per second (mm^2/s), where 1 cSt = 1 mm^2/s. The dynamic viscosity and the kinematic viscosity are related to each other by the mass density of the oil. Specifically, dynamic viscosity is equal to the product of kinematic viscosity and fluid density.

Oiling Systems

Lubricating oil is supplied to machine bearing surfaces by three basic methods: *pressure feed*, *splash*, and *gravity feed*. Of the three, the gravity feed method is the simplest and least expensive to put into operation.

Gravity-Feed Lubricating Systems   Gravity-feed lubricating systems use a reservoir (oil cup, wick-feed oiler, or drip-feed oiler), as shown in Figure 5.10. *Hand oiling* is used to resupply oil to the reservoir of the gravity feed bearing system that has been designed to operate on boundary or mixed film lubrication. Because the oil gradually leaks away in these systems, this type of lubrication is referred as a *loss lubricant* system. Under light to moderate load conditions, wick-feed and drip-feed oilers can provide fluid film lubrication as long as the drip rate produces the required volume of oil and the reservoir is maintained with sufficient oil.

Splash-Feed Lubrication   Splash-feed lubrication is used in gearboxes and engines where the oil is contained in a leakproof gear box or engine pan. Oil is distributed by the motion of the gears or the crank of the engine. In the design of the splash system, it is important that the moving parts do not *churn* the oil, since excessive heat will develop. When refilling gear boxes and engines with oil, the recommended oil level must never be exceeded, as this will lead to excessive churning of the lubricant and elevated operating temperatures. Fluid film lubrication is maintained in splash systems as long as the machine elements are rotating. Figure 5.11

FIGURE 5.11
Transmission using splash-feed lubrication. Oil is carried from gear to gear to provide lubricant for fluid film lubrication.

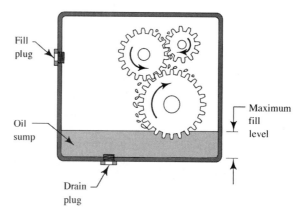

represents an enclosed gearbox with the oil being distributed by a splash feed lubrication system. Lubricating oils used in enclosed gearboxes are specified by the American Gear Manufacturers Association (**AGMA**), and they have EP ratings along with good rust and oxidation properties.

Pressure-Feed Lubrication Systems Pressure-feed lubrication systems dispense oil from a central reservoir by pumping the oil through distribution lines to all elements in the machine that require lubrication. This type of lubrication system is the most reliable method of providing lubricant to bearings. Besides providing a constant, uninterrupted supply of oil, the pressure-feed system also provides filtering and cooling of the lubricant. Because of the cooling provided, lighter grade oils can be used, which, in turn, have lower coefficients of friction and take less power from the system. Fluid film lubrication is easily achieved and maintained in pressure-feed lubrication systems.

Greases

Lubricating greases range from solids to semifluids consisting of a liquid lubricant (mineral oil) in combination with thickening agent(s) (metallic soaps) and additives. The additives commonly used in greases have the same general function as those added to lubricating oils. Included in these additives are extreme pressure agents, oxidation inhibitors, and friction-reducing agents.

Greases are classified by the thickeners used in their manufacture. Since greases are made with a variety of *metallic soaps*, many types are available. Table 5.5 lists several lubricating greases and their properties. Besides the type classification, greases are graded by consistency (the resistance to deformation by force). The National Lubricating Grease Institute (**NLGI**) consistency numbers (000 through 6) are used to grade the hardness of lubricating greases from semifluid to hard. Table 5.6 is a partial list of the grades.

TABLE 5.5 Lubricating Greases

Type	Maximum Operating Temperature (°F)	Use	Service
Barium soap	350	Multipurpose	Wide
Calcium soap	160	Chassis, cup	Limited
Lithium soap	300	Multipurpose	Wide
Sodium soap	300	Antifriction bearings	Limited

Greases are used instead of lubricating oils when relubrication would be difficult or costly, when oils can't be retained, or when contaminants need to be sealed out. Since grease can be compounded from any grade oil, greases can serve the same lubrication function as oils with a few exceptions. Because of their relatively higher coefficients of friction (typically 0.1), greases produce slightly higher heats, and because greases are not circulated and filtered, heat and contaminants are not removed.

The lubricating properties of a grease come from the release of oil from the thickener at rates (called the *bleed rate*) that are sufficient to form fluid films, mixed films, or boundary lubrication, depending on the design and application need. In some applications, where the amount of mobile lubricant required is high, the oil in the grease may be depleted to a level below 50% of the weight of the original grease. When this is the case, the grease will need renewing. In other applications, the rate of lubricant bleed is sufficiently slow that the machine element may be lubricated for life, and the grease is renewed only when the machine is dismantled for periodic maintenance or rebuilding.

Lubricating greases are applied by brush for the semifluid grades, by grease cup, by pressure gun or centralized pressure system, or by hand. Figure 5.12 pictures a grease fitting and a hand-operated grease gun used to periodically "grease" bearing surfaces.

TABLE 5.6 Grade Numbers for Lubricating Greases (Partial List)

NLGI Grade Number	Hardness/Softness
0	Very soft
1	Soft
2	Semisoft
3	Firm
4	Very firm
5	Semihard
6	Hard

FIGURE 5.12
A grease fitting (a) is attached to the machine element to be lubricated. Grease is forced through the fitting and into the bearing surface by the grease gun (b).

(a)

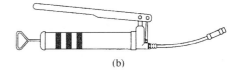

(b)

Solid-Film Lubricants

Solid-film lubricants are dry films used where greases or oils can't be used because of application difficulties, boundary lubrication, or environmental constraints. The solids, such as graphite or molybdenum disulfide, are applied in resin binders or in volatile carriers that are squeezed or evaporated out or are incorporated in plastics as the lubricating agent. In addition to powders such as graphite, plastics (polyethylene and polytetrafluoroethylene) are bonded to surfaces as solid-film lubricants for high-load, slow-speed applications. Additionally, materials such as lead and babbit are fused or thermally deposited to act as lubricants on a dry surface.

Some of the physical properties of solid lubricants include coefficients of friction of 0.020 (**PTFE**, or polytetrafluoroethylene) to 0.20 (typical), high operating temperature range (molybdenum disulfide, 750°F), and excellent boundary lubrication properties (reduced adhesive wear) due to the low shear strength of the solid materials.

Synthetic Lubricants

In industrial applications, synthetic lubricants are the lubricant of choice. Unlike traditional petroleum-based mineral oils and greases, synthetic lubricants are *man-made* by tailoring the molecular structure of the lubricant to have known and predictable properties.

Differing from mineral oils, which are composed of naturally occurring complex mixtures of hydrocarbons, synthesized lubricants start with synthetic base stocks (some made from petroleum) that are chemically combined (synthesized) to the desired viscosity and are then blended with additives to give the desired properties.

For example, a synthetic extreme pressure gear and bearing oil may combine synthesized hydrocarbons (olefin oligomers) with an organic ester as a high-viscosity, high-temperature base fluid for the lubrication of the gears and bearings in a multiple gear set speed reducer. In addition to an extreme pressure additive, other additives blended with the base fluid to form a synthetic multigrade oil (e.g., 80W – 140 SAE) would include rust inhibitors, corrosion inhibitors, and anti-foaming agents. Because synthetic oil has no aromatic (benzene) or polynuclear aromatic (naphthalene) rings, it resists oxidation that forms acids, sludge, and other deposits that degrade common mineral oils. When synthetic oils are combined with an inorganic thickener, a wide-temperature, long-life synthetic grease is formed for the lubrication of antifriction bearings used in severe duty applications. The benefits of using synthesized lubricants are noted in Table 5.7.

TABLE 5.7 Benefits of Using Synthesized Lubricants over Traditional Lubricants

Reduced energy cost	Lower operating cost
Long oil life	Reduced lubrication cost
Reduced friction	Reduced motor amperage
Reduction in wear	Reduced downtime
Lower bearing temperature	Longer equipment life
Less relubrication	Extended drain intervals
Less waste disposal	Wider operating temperature range

5.4 BEARINGS

Introduction

Bearings may be classified into three general classes: *guide bearings*, which support linear motion in machine tables and slides; *radial bearings*, which support rotational motion in shafts with radial loads; or *thrust bearings*, which support rotational motion in machine elements that have axial loads. Figure 5.13 pictures the three gen-

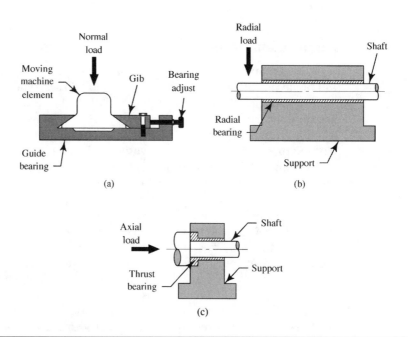

(a)

(b)

(c)

FIGURE 5.13

Classes of bearings: (a) In a guide, or flat, bearing, the load is applied normal to the linear motion of the machine element. (b) In a radial bearing, the load is applied along the radius of the rotating shaft. (c) In a thrust bearing, the load is applied along the central axis of the rotating shaft.

FIGURE 5.14
Radial bearings: (a) Plain bearing;
(b) Antifriction bearing.

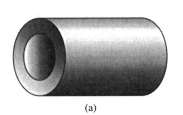

(a)   (b)

eral bearing classes with their load directions identified. Concepts related to the maintenance and lubrication of sleeve and antifriction bearings can be found in Appendix C, Section C4.

The bearing can be one of two types, a *plain bearing* (sliding contact) or an *antifriction bearing* (rolling contact), depending on the design parameters of the machine element. Figure 5.14 pictures both a plain radial bearing and an antifriction radial bearing. Each of the two types of bearings, plain and antifriction, is available for design with radial loads, axial loads, and linear motion. Also, each type of bearing has its advantages and disadvantages.

Advantages

Plain bearings: Require less space, quiet in operation, very rigid, long life.

Antifriction bearings: Less friction, no wear, sealed against contamination, large number of styles and sizes from which to choose.

Disadvantages

Plain bearings: High coefficient of friction, can be difficult to lubricate, can be damaged by reduced lubrication supply, can be damaged by contamination in lubricant.

Antifriction bearings: Life limited by the fatigue strength of the bearing material, larger size requires more space, more expensive.

Plain Bearings

Guide Bearings   Guide bearings (including *flat bearings*) provide a bearing surface on which a machine tool (lathe, mill, etc.) is moved in a linear, back-and-forth motion. Additionally, some types of cams and machines with slide valves use guide bearings in their operation. Boundary lubrication is the usual mode of operation, and the surfaces may be dry, dry film (molybdenum disulfide), oil, or grease. Figure 5.15(a) pictures a tail stock with a flat bearing on the *ways* of a lathe.

Radial Bearings   Radial bearings, called *journal bearings*, are made in various shapes and forms, ranging from a simple sleeve bearing to a complex split bearing (two parts), with oil grooves to ensure fluid film lubrication. Regardless of the variations, all radial bearings have three fundamental components in their opera-

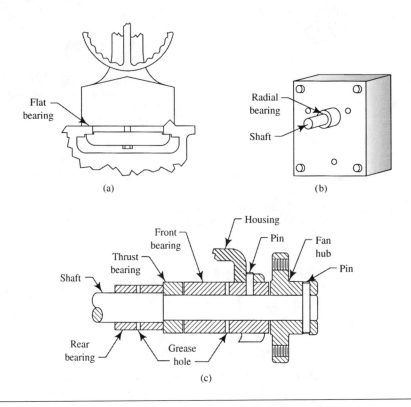

FIGURE 5.15

Plain bearings: (a) Flat bearing on the tail stock of an engine lathe. (b) Radial bearing on a speed reducer. Radial bearings are usually called journal bearings. (c) Thrust bearing on a water pump shaft. When the fan rotates, it pulls the shaft forward against the thrust bearing.

tion, namely, *shaft* (journal), *bearing* (bushing), and *lubricant*. Figure 5.15(b) shows a radial bearing being used to support the shaft of a light-duty speed reducer.

Figure 5.16(a) depicts a journal bearing with the shaft at rest. Here the shaft is resting on the bearing in the direction of the load with no lubrication under it. At this point, the center line of the shaft is slightly below the theoretical center of the bearing. Once the shaft is started, the center of the shaft rises slightly and rotates as a wedge of oil is forced under it, as pictured in Figure 5.16(b). While in operation and with sufficient speed, there are areas of high and low pressure in the oil film surrounding the circumference of the shaft, causing a slight eccentric movement in the rotation.

At start-up and at very slow speeds, the bearing operates in boundary lubrication and, as the speed picks up, the operation shifts to mixed film lubrication. At sufficient speed, fluid film is hydrodynamically developed. The hydrodynamic fluid film is formed when the motion of the bearing surfaces forms a wedge of oil that lifts the journal and separates the surfaces. As long as sufficient pressure exists in the convergence zone (high-pressure area), the surfaces will not touch or wear, and only the shearing of the oil film will produce heat. Hydrodynamic films formed in journal

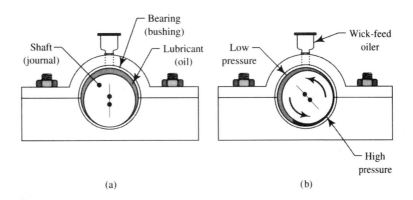

FIGURE 5.16
Journal bearing with a wick-feed oiler: (a) At rest with the journal resting on the bearing. (b) Rotating with hydrodynamic fluid film lubrication. Note the areas of low and high pressure that are developed when the shaft rotates. Oil is fed from the oiler to the low-pressure side of the bearing.

bearings and in some designs of thrust bearings are called *thick hydrodynamic films* because the film thickness is typically .001 in (25 μm).

The key to the formation of hydrodynamic films is the availability of sufficient oil to flood the gap between the bushing and the shaft. A recirculating oil system is the usual choice for supplying oil to the bearing(s). The oil wedge formed in hydrodynamic operation of the bearing depends on the speed of the surfaces (rev/min), the load per unit area (lb/in^2), and the viscosity of the oil (cP). When correctly designed, the film is thick enough to withstand variations in speed and load and still not have metal-to-metal contact. The heat due to the shearing of the fluid film is carried away in the circulating oil. To aid in heat movement and hydrodynamic operation, grooving is applied to the inside of the bearing. Grooving helps to admit the oil into the bearing and in the distribution of the oil over the load-carrying area of the bearing. Oil is introduced through a hole or holes in the low-pressure zone of the bearing. In horizontal operation, this is near the top of the bearing. Figure 5.17 pictures a bearing with grooving.

Where the loads are very heavy and/or the speed of operation is slow, hydrodynamic fluid films cannot be formed. If fluid film lubrication is required, then the fluid film is developed using hydrostatic lubrication. The bearing is specifically designed to operate with a pressurized recirculating oil feed system that provides sufficient pressure to support the bearing load and form fluid film lubrication.

Finally, both radial and axial bearings are often lubricated with grease. Depending upon design, the bearing operates as either mixed film or fluid film lubrication. Since the bearings are relubricated on a regular basis, the old grease is forced out by the new grease, so oxidation and contamination are not a factor. However, high-temperature characteristics and viscosity are important in selecting grease lubricant for plain journal and thrust bearings.

Thrust Bearings Thrust bearings are designed to resist force and movement in the direction of the axis of the shaft. Axial loads are generated by helical gears,

FIGURE 5.17
A section of a plain bearing showing an oil hole (1), feeder groove (2), and axial groove (3).

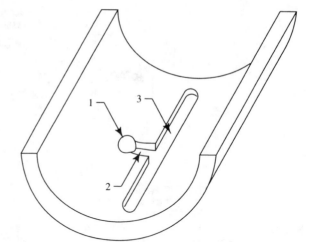

fans, blowers, worm gears, propellers, etc. The *flat plate thrust bearing* (Figure 5.15(c)) is the most frequently used thrust bearing. It is simple in design and low in cost; however, it can be used only for light or intermittent loads. The tapered land thrust bearing can handle high loads, whereas the tilting pad (Kingsbury) thrust bearing is capable of very high axial loads even when misalignment is present. The *Kingsbury pivoted-shoe thrust bearing* is designed to handle thrust from large ship propellers.

Thrust bearings, like radial bearings, operate in boundary lubrication at start-up and pass through mixed film operation as the speed picks up. Assuming sufficient oil and sufficient movement of the bearing, pressures great enough for the formation of hydrodynamic fluid films are possible. Hydrostatic lubrication is used with plain thrust bearings when hydrodynamic films are not possible. Grease lubricant is also used for mixed film and fluid film operation.

Summary of Plain Bearings

In conclusion, although plain bearings have been replaced by antifriction bearings in many applications, their use is still important due to cost and size. In addition to metal bearings, plastics and sintered metals have found an increased use in plain bearing applications. Porous metals (bronze, iron, and stainless steel) are formed from powdered metals so that oil can be forced into the small cavities (25% by volume) and held there by capillary attraction. The metals are self-lubricating and find application in light to intermediate loads, where lubrication of the bearing is difficult or where the duty of the device will not require future lubrication (lubricated for life). Boundary and mixed film lubrications are the usual modes of operation for these materials.

Of the plastics, nylon (polyamid resins), teflon (tetrafluoroethylene), and phenolic are most often used as plain bearings. Plastics can be molded into the desired shape at low cost and can be operated with no lubrication. Besides operating quietly, plastic bearings resist corrosion.

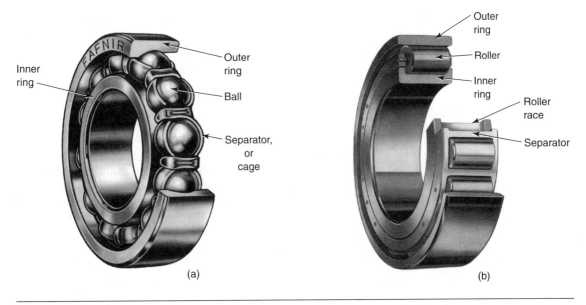

FIGURE 5.18
Parts of antifriction bearings: (a) Single-row, deep-groove ball bearing; (b) Cylindrical roller bearing. (*Courtesy of The Torrington Company*)

Antifriction Bearings

Antifriction bearings, or *roller-element bearings*, as they are often called, use a rolling element (ball or roller) between the loaded surfaces. Figure 5.18 pictures a single-row, deep-groove, ball bearing and a cylindrical roller bearing.

TABLE 5.8   Antifriction Bearing Types

	Roller Element		Load*	
Bearing Type	*Ball*	*Roller*	*Radial*	*Axial*
Guide	X		L/M	N
Single row, deep groove	X		H	L/M
Ball thrust	X		N	H
Self-aligning	X		M	L
Angular contact	X		M	H
Cylindrical		X	VH	N
Spherical thrust		X	M	VH
Spherical		X	H	H
Taper (pair)		X	VH	VH

*Note: L = light, M = moderate, H = heavy,
VH = very heavy, N = no load permitted

Antifriction bearings are divided into two categories, *ball bearings* and *roller bearings*. Ball bearings have five general types: guide, radial, thrust, self-aligning, and angular contact; roller bearings have four general types: cylindrical, thrust, spherical, and taper. Table 5.8 summarizes the types of antifriction bearings.

Roller and Ball Bearing Types

Guide Bearing   The *ball guide bearing* is used for linear motion where very low coefficients of friction (0.002 dry) and extreme smoothness in operation are desired. Ball guide bearings find application in machine tools, test instruments, business machines, and industrial processing equipment. Figure 5.19(a) pictures one style of ball guide bearing used for linear motion. In this system, the load is rolled along on ball bearings in three to six separate, continuous, oblong, recirculating circuits, as illustrated in Figure 5.19(b).

Ball guide bearings are operated on hardened and ground shafts with no lubrication when the loads are light and speeds are low, or with medium oil (ISO VG10) or soft grease (NLGI 1) when loads and/or speeds are moderate to high.

Radial Bearing   Two types of antifriction bearings are widely used as radial bearings. The first radial bearing is the *single-row, deep-groove ball bearing* (Figure

FIGURE 5.19
A standard *ball bushing*® bearing shown: (a) With the bearing installed on a hardened and ground shaft; (b) With the ball circuit sectioned for viewing. (*Courtesy of Thomson Industries, Inc.*)

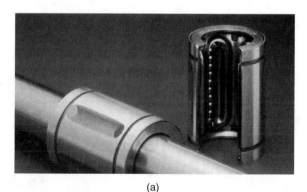

(a)

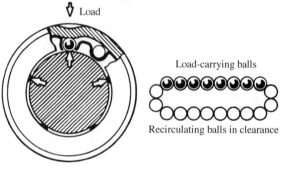

(b)

5.18(a)). This type of bearing is the most widely used antifriction bearing. When installed, it requires close alignment between the shaft and housing, since this bearing is intolerant of even small misalignment. Although its principal use is for radial loads, it can manage light to moderate axial (thrust) loads. This bearing also is available in a double-row version for greater radial loads.

The second radial bearing is the *cylindrical roller bearing* (Figure 5.18(b)). This type of bearing has a length-to-diameter ratio roughly equal to 1 and is capable of carrying larger radial loads at moderate speeds than those carried by radial ball bearings using the same size bearing. Even though it has exceptional radial load ability, no axial load is permitted with this bearing. Cylindrical roller bearings are available in a double-row version for applications requiring maximum radial rigidity.

Thrust Bearings Two types of antifriction bearings are widely used as thrust bearings (pictured in Figure 5.20). The first, the *ball thrust bearing*, is designed for axial (thrust) loads only—no radial loads. The load is transmitted parallel to the axis of the shaft, resulting in minimum axial deflection and an ability to carry heavy axial loads. The bearing is usually preloaded (has some load at all times) to ensure successful operation and is never operated at high speeds because of the excess radial forces created on the outer edge of the bearing race due to the centrifugal force.

The second, the *spherical roller thrust bearing*, is capable of very heavy axial loads as well as moderate radial loads. The single row contains barrel-shaped rollers that roll on an outer, spherical race.

Self-Aligning Bearings The two types of antifriction bearings that are self-aligning are the *spherical roller bearing* and the *self-aligning ball bearing*. Each type of bearing has a double row of rolling elements, and each is used where the shaft does not run true to the housing axis and where precise alignment cannot be maintained.

(a)   (b)

FIGURE 5.20
Antifriction thrust bearings: (a) Ball thrust bearing; (b) Spherical roller thrust bearing. (*Courtesy of The Torrington Company*)

FIGURE 5.21

The spherical roller bearing is an antifriction, self-aligning bearing, which automatically adjusts for misalignment in the shaft when the shaft does not run true to the housing axis. (*Courtesy of The Torrington Company*)

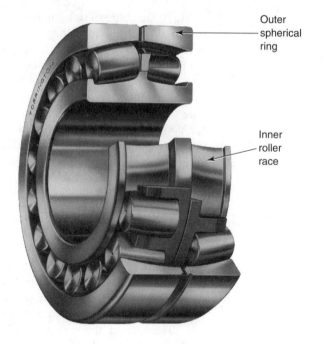

Outer spherical ring

Inner roller race

The spherical roller bearing (Figure 5.21) is free to move over the inner surface of the outer spherical ring, which has no raceway to impede its movement. Because of the *gimbling* (swiveling) action of the inner unit, the shaft is automatically adjusted even for severe misalignment. Since the barrel-shaped rollers have line contact (rather than point contact), they are able to support both heavy radial and axial loads.

The self-aligning ball bearing, like the self-aligning roller bearing, is free to move due to its two rows of balls that roll along the inner surface of the outer spherical ring, which has no grooves in the raceway. Because the balls in the self-aligning ball bearing have only point contact with the spherical surface, it is capable of supporting only moderate radial loads and light axial loads.

Angular Contact Bearings   The two types of antifriction bearings that are angular contact bearings are the angular contact ball bearing and the taper roller bearing.

The *angular contact ball bearing*, Figure 5.22(a), is characterized by high supporting shoulders. One shoulder is on the outer ring, and the other is located on the opposite side, on the inner ring. The shoulders provide for thrust (in one direction only) that is larger than the single-row, deep-radial ball bearing can handle. This bearing also can support moderate radial loads. When used in pairs with the correct orientation, the bearing can support bidirectional axial loads.

The *taper roller bearing*, Figure 5.22(b), is always used in pairs. The pairs of bearings are installed in a face-to-face or back-to-back position. A pair of taper roller bearings is capable of handling both very large axial and radial loads. Taper roller bearings are provided with a mounting system that allows for the adjustment

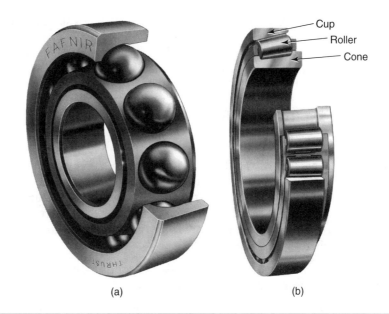

Cup
Roller
Cone

(a)   (b)

FIGURE 5.22
Antifriction angular contact bearings: (a) Angular contact ball bearing; (b) Taper roller bearing. (*Courtesy of The Torrington Company*)

of the bearings, running clearance. The bearings can be preloaded to provide for extreme rigidity in their operation.

Antifriction Bearing Life

When correctly lubricated and not subjected to extreme operation conditions, the maximum useful life of antifriction bearings (ball and roller) is dependent on the *fatigue life* of the bearing steel. When the ball or roller moves over the raceway, considerable loading per unit area is created between the rolling element and the raceway, which—besides causing *elastic deformation*—will cause *shear stresses* (force/area) in the contact surfaces. These stresses, over time, cause *microcracking* in the topmost metal (.002 in) that leads to surface *spalling* (flaking or chipping). The breaking down of the bearing surfaces signals the end of the bearing's useful life.

Antifriction Bearing Lubrication

With proper lubrication, fluid films are formed by the lubricant, which is drawn into the convergent zone just in front of the ball or roller. These areas are very small (a point for the ball and a line for the roller) and the loads per unit area are very high (>300,000 lb/in^2). It is under these conditions that an extremely thin lubricant film (1 μm, or 40 μin) is formed. Because the surface materials are undergoing elastic deformation at this point, this type of lubrication is called *elastohydrodynamic lubrication* (**EHL**). Even though this film is microscopic in thickness, it is sufficient to separate and support the two metal surfaces. EHL films are formed with oil supplied by any of the previously discussed lubricating systems as well as by the oil in grease. Today, many bearings are packed with grease for the life of the bearing and require no further lubrication.

5.5 ROLLING RESISTANCE

Introduction

When heavily loaded railcar wheels, rollers in bearings, or truck tires move over a surface, the surface is deformed around the point of contact of the rolling element (wheel, roller, etc.) and the surface. In effect, the rolling element drops down into a small depression in the surface created as the surface yields to the weight carried by the rolling element. In order for the rolling element to move forward, a force is applied to "pull it out" of the groove.

Parameters of Rolling Friction

Figure 5.23(a) pictures the mechanics of the rolling element as it passes over a surface. Here the weight of the load is seen acting normal to the surface, whereas the effort force, F_E, is acting parallel to the surface. The opposition, the rolling resistance, or rolling friction (F_{rr}), as it is often called, is shown at the point of contact of the wheel and the edge of the depression in the surface.

By taking moments about the point of contact, an expression for the force of rolling resistance (F_{rr}) may be derived for the point of impending motion of the rolling element. By applying the law of simple machines to the free-body diagram of Figure 5.23(b), the following expression results for rolling resistance:

FIGURE 5.23
The area of contact is deformed a very small amount compared to the diameter of the rolling element. For a .250-in-diameter roller, the distance of c_{rr} is .003 in.

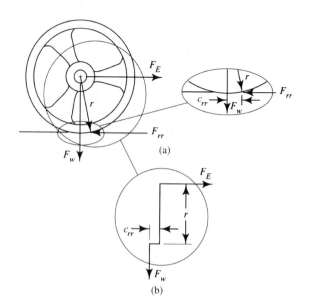

TABLE 5.9 Coefficient of Rolling Resistance (Inches)

Material	c_{rr}
Metal wheel on wood	.20
Steel wheel on mild steel	.020
Railroad wheel on rail	.0040
Roller on bearing race	.0030
Ball on bearing race	.00010
Rubber tire on asphalt	.30

$$F_E \times r = F_W \times c_{rr}$$

$F_{rr} = F_E$ when the system is in equilibrium (equal but opposite forces). Substitute $F_{rr} = F_E$ into the first equation and solve for F_{rr}:

$$F_{rr} = \frac{F_W \times c_{rr}}{r} \qquad (5.5)$$

where
F_{rr} = the rolling resistance, in lb
F_W = the weight on the rolling element, in lb
c_{rr} = the coefficient of rolling resistance, in inches
r = the radius of the rolling element, in inches

The *coefficient of rolling resistance* (also called the *coefficient of rolling friction*) has units of inches. It is unlike the coefficient of friction, a ratio of forces that has no units. Table 5.9 lists several coefficients of rolling resistance. The assumption in using these coefficients is that they are constant for a given set of materials and that the coefficients are constant for all diameters of wheels or rollers. Although these assumptions make the calculations simple, they can lead to considerable error. At best, the results are an approximation of what is actually happening.

Example 5.4

Determine the traction force (pulling force) that is needed to overcome the rolling resistance of a 72-T narrow-gauge steam locomotive (144,000 lb weight on drive wheels), which has eight 44-in diameter drive wheels.

SOLUTION

Find the rolling resistance (F_{rr}); this is equal to the traction force (F_E).

Given

$$F_{rr} = \frac{F_W \times c_{rr}}{r}$$

Evaluate

F_W = 144,000 lb

c_{rr} = .0040 in   (from Table 5.9)

$$r = \tfrac{44}{2} = 22 \text{ in} \quad (\text{diameter}/2)$$

Substitute $\quad F_{rr} = \dfrac{144{,}000 \times .0040}{22}$

Solve $\qquad F_{rr} = 26 \text{ lb}$

The traction force is 26 lb.

The kinetic (sliding) friction force for the 72-T locomotive is more than 2000 times as great as the 26-lb rolling friction force; that is, $F_k = \mu \times F_N$, or $F_k = 0.40 \times 144{,}000 = 58{,}000$ lb. Our ancestors learned very early to use rollers to move large, multiton monolithic stones. Sliding very heavy objects takes very large forces, but rolling the objects does not.

CHAPTER SUMMARY

- Static friction depends upon the kind of material, the condition of the surface, and the normal force.
- Coefficients of kinetic friction are less than coefficients of static friction.
- It takes a greater force to start a body in motion than to keep it in motion.
- The coefficient of static friction is equal to the tangent function of the angle of repose.
- Friction brakes play an important role in stopping machines.
- When machines move and friction is present, work is done to overcome friction-creating heat.
- Pitting may occur in surfaces during the wear-in period.
- Fluid film lubrication is the most desirable condition for bearing operation because the surfaces do not touch each other.
- Two kinds of viscosity are dynamic and kinematic viscosity.
- Pressure-feed lubrication systems are the most reliable means of providing oil to bearings.
- Synthetic lubricants provide superior lubrication properties.
- Lubricating greases are used to lubricate antifriction bearings and provide EHL lubricating films.
- Solid lubricants are used where conditions (environmental and load) preclude the use of petroleum lubricants.
- Plain bearings are used as flat (guide) bearings, radial (journal) bearings, and axial (thrust) bearings.
- Ball and roller bearings have application as guide, radial, axial, self-aligning, and angular contact bearings.
- Rolling resistance happens when a loaded rolling element (wheel, roller) rolls across a surface, causing the surface to deform and resulting in a miniature depression and a slight uphill condition.

SELECTED TECHNICAL TERMS

The following technical terms, abbreviations, and acronyms are defined in the glossary located after Chapter 16. You are encouraged to use the glossary to aid your understanding and to test your knowledge of these important terms.

abrasive wear
adhesive wear
AGMA
asperities
bearings
coefficient of friction
EHL
EP
friction
hydrodynamic lubrication

ISO
kinetic force of friction
lubrication
NLGI
pitting
PTFE
SAE
solenoid
viscosimeter
viscosity

END-OF-CHAPTER QUESTIONS

Write T if the statement is true and F if the statement is false.

1. Splash feed lubrication is used to lubricate gears in a transmission.
2. Axial loads are supported by journal bearings.
3. Boundary lubrication results in wear to the bearing surfaces.
4. The NLGI number system is used to grade oils.
5. Spherical roller bearings are self-aligning.
6. EHL stands for elastohydrodynamic lubrication.
7. The letters EP refer to a lubricant with extreme pressure properties.
8. The term *hydrodynamic lubrication* refers to a type of automatic transmission oil.
9. Surface roughness plays little role in adhesive wear.
10. Under certain conditions, the normal force and the weight of an object have the same magnitude of force.

In the following, select the word or phrase that makes the statement true.

11. Solid film lubricants include graphite, (lithium grease, molybdenum disulfide, sintered metal), and teflon.
12. The viscosity of lubricating oil measured in centistokes is called (kinematic viscosity, dynamic viscosity, absolute viscosity).
13. The least expensive lubricating system is the (pressure feed, splash, gravity feed).
14. The type of friction with the lowest loss of energy is (static, rolling, kinetic).
15. Surface asperities are (pits, spalling, peaks) in the bearing surface.

Answer each of the following questions with a short answer in the form of a complete sentence. Include a restatement of the question in your answer.

16. Compare the characteristics of static friction to those of kinetic friction.
17. Explain why it is possible for an object to remain at rest on an inclined plane.
18. List several advantages and several disadvantages of friction.
19. Using your knowledge of lubricants, what means could be used to detect that a bearing had too much grease applied to it when it was repacked?
20. Describe how hydrodynamic lubrication is achieved between bearing surfaces.

TABLE 5.10 Summary of Formulas
Used in Chapter 5

Equation Number	Equation
5.1	$F_f = \mu \times F_N$
5.2	$\mu_s = \dfrac{F_{fs}}{F_N}$
5.3	$F_N = F_W \cos(\theta)$
5.4	$\mu_s = \tan(\theta_r)$
5.5	$F_{rr} = \dfrac{F_W \times c_{rr}}{r}$

END-OF-CHAPTER PROBLEMS

Solve the following problems. Make sketches to aid in solving the problems and structure your work so it follows in an orderly progression and can easily be checked. Table 5.10 summarizes the equations for this chapter.

1. Using the information in Table 5.1, determine at what angle a 1.8-lb block of steel will slide down an elevated cast-iron surface.

2. A 10-N aluminum block is placed onto the middle of a 2-ft by 2-ft, $\frac{1}{2}$-in aluminum plate that has one end elevated. The elevation of the plate is gradually increased until the block just starts to slip, at which time the angle of repose is recorded as 52°. Determine the coefficient of static friction for the aluminum block on the aluminum surface.

3. A large 320-lb wooden packing crate is being pushed across the level metal floor of a moving van. Determine the effort force needed to just move the crate. *Note:* Use the average value of the coefficient of static friction for wood on metal from Table 5.1.

4. If the normal force is 56 lb, determine the maximum static friction force developed when the brake shoe of the industrial brake pictured in Figure 5.5 is pressed against the cast-iron drum by the compression spring. Assume that the brake lining has the same coefficient of friction as the clutch lining listed in Table 5.1 and use the maximum value.

5. Determine the force of deceleration of a 4200-lb automobile on a level concrete road when the driver has locked the brakes in a panic stop. The kinetic (sliding) coefficient of friction for tires on dry concrete is 1.0.

6. Determine the weight of a workbench (in newtons) being slid across a level concrete floor by a horizontal force of 204 N when the coefficient of kinetic friction is 0.34.

7. If a drill press weighs 892 N, determine the coefficient of static friction when a force of 186 N is needed to start sliding the press on a level wood floor.

8. A file box filled with books (12.7 kg) is at rest on a smooth level floor. If the coefficient of static friction between the box and floor is 0.72 and the coefficient of kinetic friction is 0.56, determine

    (a) The horizontal force (in newtons) needed to just move the box.

    (b) The horizontal force (in newtons) required to maintain the box in motion at a constant speed.

9. What force (in newtons) is needed to push a 2800-N piano across a hardwood floor? The piano is equipped with 4.0-in-diameter metal wheels.

10. It takes a combined push of 95 lb from two people to get a 2900-lb car to just begin to roll on a level parking lot. Determine the coefficient of rolling resistance (c_{rr}) of the 23-in-diameter tires on asphalt.

Chapter 6

Work, Energy, Torque, and Power

About the Illustration. *This compound machine, powered by one person aided by a flywheel, is capable of lifting heavy marble blocks used in the construction of Renaissance buildings. As illustrated, a number of simple machines are compounded to enable the power of the person turning the lever to lift the block. Included in this machine are the block (L, M, N, O) and tackle, wheel and axle (G, H), screw (worm) and wormwheel (Q), lever, gears (H, lantern gear, and S, crown gear), and wheel (sheave, G, and drums, E and F) and rope drive. Tension (friction) for the rope drive is provided by the men on either side of the machine. Notice the* lewis *(a three-piece iron tool shaped to fit into a dovetail in the marble block) used to lift and set the block. (Designed by Agostino Ramelli, from* The Various and Ingenious Machines of Agostino Ramelli *(1588).)* ∎

INTRODUCTION

When machines rotate, they do work and possess energy. In this chapter, concepts dealing with energy, work, power (the rate of doing work), angular displacement, and angular velocity (the rate of rotation) are discussed. In addition, force and torque (a twisting forcelike quantity) are studied.

PERFORMANCE OBJECTIVES

Once you have read and studied each section; worked through each example with pencil, paper, and calculator; worked through the end-of-chapter problems; and answered the end-of-chapter questions, then you should be able to

- Define work and energy and distinguish between potential and kinetic energy.
- Relate power to work and time.
- Understand the nature of torque and the units in which it is expressed.
- Distinguish between rotary motion and circular motion.
- Determine rotational work and power from torque, angular displacement, and angular velocity.

6.1 WORK AND ENERGY

Work

We usually associate work with the activities of a job. We go to work each day to do some form of physical or mental exertion, which causes us to feel exhausted by the end of the day. This type of activity, however fatiguing, does not meet the scientific definition of work.

Mechanical work is done when a *force is applied to an object and the object moves*. Movement is the key concept in the definition of work. For example, applying a force in an attempt to push a stalled car and not moving the car does not represent work, even though we may become exhausted in the attempt. In determining the amount of work done by the application of a force, two quantities must be known: first, the distance the object moves and, second, the amount of force applied in the direction of the movement. Stated as an equation,

$$W = Fs \tag{6.1}$$

where

W = work done by a force in moving a mass a distance s, in J or ft-lb
F = force applied to the object to get it to move, in N or lb
s = distance the object moved, in m or ft

TABLE 6.1 Units of Work and Energy

	Force	Distance	$W = Fs$	Unit Name
BES	lb	ft	ft-lb	foot-pound
SI	N	m	N·m	joule

Table 6.1 summarizes the units of work and energy. The derived unit for work in SI is the joule (J); that is, 1 N·m = 1 J.

The work done by a force on an object is the *product of the force in the direction of movement and the distance moved by the object.* Figure 6.1 pictures a crate being pushed across a warehouse floor by an applied force F. In this case, the motion of the crate is in the same direction as the applied force.

If the applied force is not in the same direction as the movement (force applied at an angle, as in Figure 6.2), then only the part (*component*) of the force in the direction of the movement is used in determining the work done in moving the crate. In this case, the horizontal component of force ($F \cos \theta$) is in the direction of the movement. It is this component that is used in the calculation of work, since only the horizontal component of the applied force is responsible for the movement. The equation for work (6.1) is modified to include the angle between the applied force (F) and the direction of motion, which is a more general form of the work equation.

$$W = Fs \cos(\theta) \tag{6.2}$$

where

W = work, in J or ft-lb
F = force, in N or lb
s = distance, in m or ft
θ = angle between the *line of action* of the force and the *line of movement* (direction of motion), in degrees

*Example
6.1*

The 300-lb crate of Figure 6.1 is slid 85.0 ft across the floor by pushing it with a 92.0-lb force applied parallel to the floor. Determine the work done in moving the crate.

SOLUTION   All the force is in the direction of motion, so use Equation 6.1 to find the work.

FIGURE 6.1
Work is done in moving the crate a distance s by the applied force F. The direction of motion is in the same direction as the applied force; $W = Fs$.

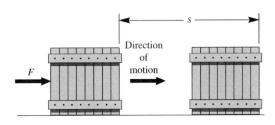

FIGURE 6.2
Only part of the force F goes into moving the crate, because the applied force is at an angle. The component of force in the direction of motion is used to compute the work in moving the crate distance s: $W = (F \cos \theta)s$.

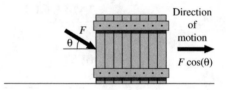

Given	$W = F \times s$
Evaluate	$F = 92.0 \text{ lb}$
	$s = 85.0 \text{ ft}$
Substitute	$W = 92.0 \times 85.0$
Solve	$W = 78\overline{0}0 \text{ ft-lb}$
Observation	The force of kinetic friction is overcome by the 92-lb force applied to the crate to get it to move. The work done in moving the crate against the frictional force is converted to heat and sound in the course of moving the wooden crate over the concrete floor.

Example 6.2

Using the 7800 ft-lb of work from Example 6.1, determine the force needed to move the 300-lb crate when the force is applied at a 30° angle, as pictured in Figure 6.2.

SOLUTION	Since the force is applied at an angle of 30° to the direction of motion, use Equation 6.2 to solve for the force.
Given	$W = Fs \cos(\theta)$; solve for F:
	$F = \dfrac{W}{s \cos(\theta)}$
Evaluate	$W = 7800 \text{ ft-lb}$
	$\theta = 30°$
	$s = 85 \text{ ft}$
Substitute	$F = \dfrac{7800}{85 \cos 30°}$
Solve	$F = 106 \Rightarrow 110 \text{ lb}$
Observation	The force needed to move the crate increased from 92 lb, when applied parallel to the direction of motion, to 106 lb, when applied at 30° to the direction of motion.

Energy

Mechanical energy has the ability to do work. Mechanical energy can be divided into two categories, *kinetic energy* and *potential energy*. Like other forms of energy, mechanical energy follows the law of conservation of energy: *Energy can neither be created nor destroyed but can only be converted from one form to another.* Both energy and work are measured in the same units, joules in SI or foot-pounds in BES.

Potential Energy

Energy possessed by an object due to its position is called **potential energy.** Objects that are elevated above a surface have stored, or potential, energy. An elevated object, such as a book resting on a table, has a potential for doing work. When the book is pushed off the table, the potential energy of the book is transformed to kinetic energy (energy of motion) in falling to the floor. Upon hitting the floor, the energy is converted into noise and heat.

The work needed to raise an object above a surface is determined by its position (height) and mass. The work done in moving an object above a surface is converted to potential energy, which may be determined by applying the following equation.

$$W = Fs = F_wh = mgh \tag{6.3}$$

where

W = work, in J or ft-lb
F = force, in N or lb
s = distance, in m or ft
F_W = weight of the object, in N or lb
m = mass of the object, in kg or slug
g = acceleration due to gravity, 9.81 m/s^2 or 32.2 ft/s^2
h = height above the reference surface, in m or ft

> **As a Rule** The potential energy of the object at rest above the surface is exactly equal to the work done in lifting the object.

Therefore, the potential energy (PE) of an object above a surface is

$$PE = mgh \tag{6.4}$$

where

PE = potential energy of the object, in J or ft-lb
m = mass, in kg or slug
g = acceleration due to gravity, 9.81 m/s^2 or 32.2 ft/s^2
h = height above the reference surface, in m or ft

Note: Remember that the potential energy is equal to the work done. That is,
PE = $W = F_wh = mgh$.

Example
6.3

Determine the potential energy in a 60.8-kg pallet of bricks that has been raised by a crane 43.0 m above the ground.

SOLUTION   Find the PE in the pallet of bricks.

Observation   Since the mass of the pallet of bricks (60.8 kg) is given, use Equation 6.4.

Given $PE = mgh$

Evaluate $m = 60.8$ kg

$g = 9.81$ m/s^2

$h = 43.0$ m

Substitute $PE = 60.8 \times 9.81 \times 43.0$

Solve $PE = 25.6$ kJ

Example
6.4

A block and tackle is used to lift a 270-lb load 3.0 ft above the ground, as shown in Figure 6.3. Determine

(a) The potential energy of the 270-lb load once it is raised 3.0 ft.
(b) The work lost to friction if the IMA is 5 and the effort force is 68 lb.
(c) The work done in moving the load 3.0 ft against gravity and friction.

SOLUTION   **(a)** Find the PE of the 270-lb load.

Observation   Since the weight of the load (270 lb) is given, use Equation 6.3. Remember that potential energy and work are equal ($PE = W$).

Given $PE = W = F_W h$

Evaluate $F_W = 270$ lb

$h = 3.0$ ft

Substitute $PE = 270 \times 3.0$

Solve $PE = 810$ ft-lb

(b) Find the frictional work loss.

Given $W_{loss} = W_{in} - W_{out}$

Evaluate $W_{in} = F_E \times s = 68 \times (5 \times 3.0) = 1020$ ft-lb

$W_{out} = F_R \times s = 270 \times 3.0 = 810$ ft-lb

Substitute $W_{loss} = 1020 - 810$

Solve $W_{loss} = 210$ ft-lb

FIGURE 6.3
A block and tackle makes it possible to raise a load with less force. However, the amount of work needed to raise the load is not reduced.

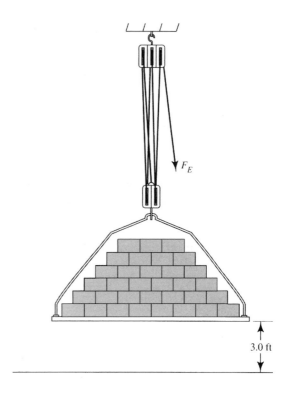

3.0 ft

(c) Find the total work done in moving the load against gravity and friction.

Observation	The total work done is equal to the sum of the potential energy and the work lost to friction.
Given	$W_T = \text{PE} + W_{\text{loss}}$
Evaluate	$\text{PE} = 810 \text{ ft-lb}$
	$W_{\text{loss}} = 210 \text{ ft-lb}$
Substitute	$W_T = 810 + 210$
Solve	$W_T = 1020 \Rightarrow 1\overline{0}00 \text{ ft-lb}$

The advantage of using a machine when doing work is not that it makes less work, but because it allows the job to be done with the force at hand. It would be virtually impossible for you to lift a 270-lb load 3.0 ft off the ground, but with the aid of a block and tackle, you can do the job with a relatively small force of 68 lb.

Kinetic Energy

Energy due to the motion of an object is called **kinetic energy.** Falling, moving, and rotating objects all have kinetic energy. This statement is true because objects that fall, move, or rotate exert force and do work on any other object that attempts to stop their motion.

When a force is applied to an object at rest, the object is accelerated, and energy is imparted to the object by the force. The amount of work done in accelerating an object and moving it a distance (s) in the direction of the force (F) is

$$W = Fs$$

From the law of acceleration, force is related to acceleration by:

$$F = ma$$

Substitute $F = ma$ into $W = Fs$ and replace the symbol for work (W) with the symbol for kinetic energy (KE):

$$W = mas \quad \text{and} \quad KE = mas$$

From the $2as$ equation (Equation 2.10), when an object is initially at rest, $v_i = 0$. Solve for as and substitute it into $KE = mas$.

$$2as = v_f^2 - 0$$

$$as = \tfrac{1}{2}v^2$$

where v is the final speed of mass m.

Substitute $\tfrac{1}{2}v^2$ into $KE = mas$ in place of as and rearrange the terms:

$$KE = \tfrac{1}{2}mv^2 \tag{6.5}$$

where

$$KE = \text{kinetic energy imparted to mass } m \text{ by the force } F, \text{ in J or ft-lb}$$
$$m = \text{mass of the object, in kg or slug}$$
$$v = \text{final speed of the object, in m/s or ft/s}$$

➤ **As a Rule** The kinetic energy of an object in motion is exactly equal to the work done in accelerating the object.

Example
6.5

A 2600-lb automobile is accelerated to a final velocity of 60 mi/h (88 ft/s). Determine the kinetic energy imparted to the car by the force of acceleration.

SOLUTION   Convert the weight to mass and then solve for the kinetic energy (KE). $m = F/a$ and $m = 2600/32.2 = 80.7$ slug, where $a = g = 32.2$ ft/s^2.

Given $KE = \tfrac{1}{2}mv^2$

Evaluate $m = 80.7$ slug

$v = 88$ ft/s

Substitute $KE = \tfrac{1}{2} \times 80.7 \times 88^2$

Solve $KE = 312{,}470 \Rightarrow 3.12 \times 10^5$ ft-lb

Exercise 6.1

Solve the following problems. Make sketches to aid in solving the problems and structure your work so it follows in an orderly progression and can easily be checked.

1. Determine the work done by an overhead differential hoist in raising a 480-lb diesel engine 2.5 ft up and out of a truck.
2. A 580-N wooden crate is pushed 12 m across a concrete warehouse floor. If the crate is moved by a 44-N force in a straight horizontal line, determine
   (a) The frictional force.
   (b) The work done in pushing the crate.
   (c) The work done to overcome the frictional force.
   (d) The work done to overcome the force of gravity.
3. Determine the force applied to the handle of a lawnmower in cutting the grass when work of 140 ft-lb is done while moving the 22-lb mower horizontally 30 ft across the lawn at a constant speed. The handle of the mower makes a 45° angle with the horizontal.
4. Determine the kinetic energy of a cyclist who weighs 710 N and is riding a 160-N bicycle at a speed of 10.0 m/s.
5. A 42-lb drum of solvent is moved 55 ft with the aid of a two-wheel hand truck. If the force in the direction of movement is 2.0 lb, determine
   (a) The work done in moving the 42-lb drum.
   (b) The coefficient of rolling friction of the 5.0-in wheels.
6. Assuming no frictional loss, determine
   (a) The work done in lifting a 240-lb electric motor vertically 9.0 in with a crow bar (class 1 lever).
   (b) The distance the effort end of the bar moves if the IMA is 3.0.
   (c) The force applied to the effort end of the bar.

6.2 POWER

Rate of Energy Conversion

The average power (*P*) delivered by doing work or expending energy is determined by computing the time rate at which work is done or energy is converted. In usage, the term *power* is sometimes used where the term *energy* is needed. To avoid confusion, remember that power is how fast energy is converted, transformed, or expended. As such, the unit of time is present along with the unit of work. Stated mathematically,

$$P = \frac{W}{t} \qquad\qquad (6.6)$$

TABLE 6.2 Units of Power

	Work/Energy	Time	P = W/t	Unit Name
BES	ft-lb	s	ft-lb/s	ft-lb/s*
SI	J	s	J/s	watt (W)

*550 ft-lb/s = 1 hp (horsepower).

where

P = average power, in W or ft-lb/s
W = energy transformed or the work done, in J or ft-lb
t = time, during which work is done or energy is transformed, in s

Table 6.2 summarizes the units of power. The derived unit for average power in SI is the watt; that is, 1 N·m/s = 1 J/s = 1 W. In BES, the horsepower (hp) is used as a unit of power: 550 ft-lb/s = 1 hp, or 33,000 ft-lb/min = 1 hp. The conversion factor between SI and BES is 746 W = 1 hp.

Example *6.6*	Determine the average power used in lifting an 8200-N load of building materials from the ground to the roof of a 110-m building under construction. If it takes 64 s to move the load, express the answer in both watts (SI) and horsepower (BES) and assume the lifting system is 42% efficient.
SOLUTION	First, find the power without losses and then compute the average power with losses in both SI and BES.
Given	$P = \dfrac{W}{t}$ and $W = Fs$; therefore, $P = Fs/t$.
Evaluate	$F = 8200$ N
	$s = 110$ m
	$t = 64$ s
Substitute	$P = \dfrac{8200 \times 110}{64}$
Solve	$P = 14\,000$ W $\Rightarrow 14$ kW
Observation	Efficiency is equal to the ratio of work out to work in ($\eta = W_{out}/W_{in}$), as stated in Chapter 3. It follows that efficiency is also equal to the ratio of power out to power in ($\eta = P_{out}/P_{in}$). The 14 000 W just determined is the power into the lifting system; the power out is determined from the efficiency equation.
Given	$P_{out} = \eta \times P_{in}$
Evaluate	$\eta = 0.42$

$P_{in} = 14\,000$ W

Substitute $P_{out} = 0.42 \times 14\,000$

Solve $P_{out} = 5900 \Rightarrow 5.9$ kW

Convert $5900 \, \cancel{W} \times \dfrac{1 \text{ hp}}{746 \, \cancel{W}} = 7.9 \text{ hp}$

The power used is 5.9 kW, or 7.9 hp.

Force Times Velocity

A useful form of the power equation expressed in terms of velocity (s/t) and force (F) is given in Equation 6.7. Look once again at the definition of power.

Power = work/time = (force × distance)/time = $(Fs)/t = F(s/t)$. Recall that distance over time (s/t) is velocity. From this, we see that power is equal to force × velocity ($P = Fv$). This assumes, of course, that the force is applied in the same direction that the object is moving. Stated mathematically,

$$P = Fv \tag{6.7}$$

where

P = power, in W or ft-lb/s
F = force, in N or lb
v = velocity, in m/s or ft/s

*Example
6.7*

Assume that it takes 85 lb to move a car on a level, paved surface. Determine
(a) The power it takes to move the car at 1.5 mi/h (2.2 ft/s), the case when the car is being pushed by hand.
(b) The power it takes to move the car at 25 mi/h (37 ft/s), the case when the car is being driven through a residential area.

SOLUTION   **(a)** Find the power needed to maintain a speed of 2.2 ft/s (1.5 mi/h) with an applied force of 85 lb.

Given $P = Fv$

Evaluate $F = 85$ lb

$v = 2.2$ ft/s

Substitute $P = 85 \times 2.2$

Solve $P = 187$ ft-lb/s

Observation   187 ft-lb/s is equivalent to .34 hp.

(b) Find the power needed to maintain a speed of 37 ft/s (25 mi/h) with an applied force of 85 lb.

Given	$P = Fv$
Evaluate	$F = 85$ lb
	$v = 37$ ft/s
Substitute	$P = 85 \times 37$
Solve	$P = 3100$ ft-lb/s
Observation	3100 ft-lb/s is equivalent to 5.7 hp.

In the previous example, notice how the power increased as the speed of the car went up. In the first case, about $\frac{1}{3}$ hp was used to move the car, whereas in the second case, about 6 hp were used to move the same car. The increase of about seventeen times in power corresponds to an increase of about seventeen times in speed. To move the car 1 mi it takes the same 450,000 ft-lb of energy (work). However, it takes about seventeen times as long to push the car 1 mi by hand at 1.5 mi/h than it does to drive the car 1 mi at 25 mi/h (40 min versus 2.4 min). Once again, the difference between work (energy) and power is noted.

Exercise 6.2

Solve the following problems. Make sketches to aid in solving the problems and structure your work so it follows in an orderly progression and can easily be checked.

1. Convert 12 hp to kilowatts.
2. Convert 4920 ft-lb/s to horsepower.
3. Assuming no losses, determine the average power expended in lifting a 420-lb electric motor 55 ft in 12 s to the top of an industrial building.
4. Assuming no losses, determine the average power delivered to a 12 500-N elevator car lifted vertically 165 m at an average velocity of 1.4 m/s.
5. Determine the efficiency of a hydraulic freight elevator that is driven by a 25-hp motor if the maximum load that can be lifted at 1.5 ft/s is 7900 lb.
6. If water weighs 8.0 lb per gallon, determine the amount of water (in gallons) that can be drawn from a 60.0-ft well each hour when the pump is driven by a 40.0-hp motor. Assume the system is 78% efficient.

6.3 TORQUE

Introduction

In machines, rotating elements are very common. Drive shafts, clutches, motor shafts, cams, gears, and sprockets all rotate. A twisting quantity called *torque* is applied to each of these elements to get them to rotate. Torque is a force applied through a distance that results in a twisting or turning effect.

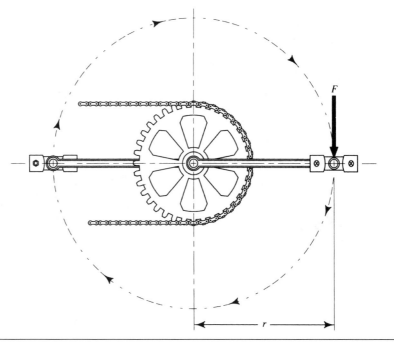

FIGURE 6.4
The force is applied tangent to the circular motion of the crank.

For example, to get the bicycle crank of Figure 6.4 to move, a force is applied to the pedals attached to the crank. The crank arm is the distance through which the force is applied. The greatest torque is achieved when the force is applied tangent to the circular motion of the crank arm. The term *applied tangent to the circular motion of the crank arm* means that the force is being applied perpendicular (at a right angle) to the crank, as pictured in Figure 6.4.

Moment Arm

The distance through which the force is applied to create a torque is referred to as the *moment arm of the force*. The length of the moment arm is determined by taking the perpendicular distance between the line of action of the force and the center of rotation (the pivot point). Figure 6.5 pictures an open-end wrench being applied to a nut to tighten it.

In Figure 6.5(a), the moment arm (r_1) is the length of the wrench. Here we can see that the force is applied tangent to the circular motion of the wrench. However, in Figure 6.5(b), the force is applied to the wrench at an angle, and the length of the moment arm is much shorter than the length of the wrench. The torque specifications for tightening steel bolts and screws are included in Appendix A, Table A4.

FIGURE 6.5
(a) Force is applied perpendicular to the wrench. The moment arm length (r_1) is equal to the length of the wrench. (b) Force is applied at an angle and the length of the moment arm, r_2, is less than r_1.

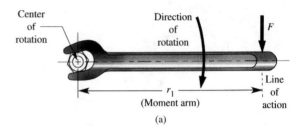

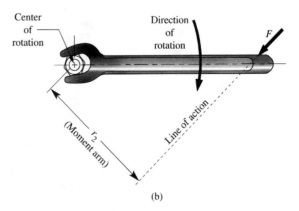

Summary

- The action of applying a force to a moment arm with a resulting rotation is called torque.
- The moment arm is the perpendicular distance (the shortest distance) from the line of action to the center of rotation.
- The units of torque, as listed in Table 6.3, are the pound-foot (BES) and the newton-meter (SI).

Notice that the units of torque are the same as the units of work and energy. However, in SI, the newton-meter (N·m) is used to emphasize the mechanical nature of torque, and the joule (J) is not used. In BES, the mechanical nature of torque is indicated by stating the force unit first and then the distance unit: pound-foot (lb-ft) rather than foot-pound (ft-lb), as with work and energy.

TABLE 6.3 Units of Torque

	Force	*Distance*	*$\tau = Fr$*	*Unit Name*
BES	lb	ft	lb-ft	pound-foot
SI	N	m	N·m	newton-meter

Torque Parameters

Stated as a formula, torque is the product of the force applied to the moment arm times the length of the moment arm.

$$\tau = Fr \tag{6.8}$$

where

τ = torque, in N·m or lb-ft
F = force applied to the moment arm, in N or lb
r = length of the moment arm, in m or ft

Example
6.8

Determine the torque (in lb-ft) applied to the nut by the wrench pictured in Figure 6.5(a) when the length of the moment arm is 9.0 in and the force is 35 lb.

SOLUTION   Convert the length of the moment arm from inches to feet and then find the torque.

Given $\tau = Fr$

Evaluate $F = 35$ lb

         $r = 9.0$ in

Convert $r = 9.0 \text{ in} \times \dfrac{1 \text{ ft}}{12 \text{ in}} = .75$ ft

Substitute $\tau = 35 \times .75$

Solve $\tau = 26$ lb-ft

Example
6.9

Determine the force that must be applied to the torque wrench of Figure 6.6 to produce a torque of 25 N·m to tighten the bolt.

SOLUTION   First express the length of the moment arm in meters and then solve Equation 6.8 for force.

Given $\tau = Fr \quad \text{and} \quad F = \dfrac{\tau}{r}$

Given $\tau = 25$ N·m

         $r = 45$ cm $= 0.45$ m

Substitute $F = \dfrac{25}{0.45}$

Solve $F = 56$ N

FIGURE 6.6
A torque wrench is used to tighten
bolts to a specified torque.

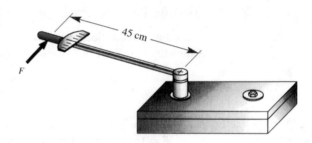

Exercise 6.3

Solve the following problems. Make sketches to aid in solving the problems and
structure your work so it follows in an orderly progression and can easily be
checked.

1. Determine the torque supplied to the pulley of Figure 6.7 in
   (a) lb-ft.   **(b)** N·m.
2. Determine the amount of torque on a spark plug when an 18-in torque wrench
   has a 25-lb force applied to the end of the wrench.
3. Determine the length of the crank, r, pictured in Figure 6.4 if the tangential force
   is 240 N and the torque is 75 N·m.
4. Determine the tangential force needed to tighten a nut to a torque of 170 N·m
   with a 54-cm wrench.

FIGURE 6.7
The net force of 95 lb carried by the
V-belt creates a torque in the shaft at-
tached to the pulley.

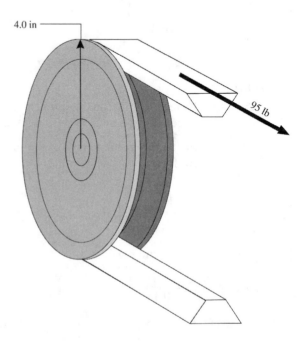

5. Determine the force on the cutting edge of a .50-in-diameter drill when the electric drill motor driving the drill produces a torque of 74 lb-ft.

6. A 1900-lb sports car is accelerated from 0.00 to 60.0 mi/h in 6.54 s. If the wheel size is 22.0 in, determine
   (a) The net force needed to accelerate the car.
   (b) The torque supplied to the drive wheels.

6.4 ROTARY MOTION

Introduction

So far the study of straight-line motion has been our principal focus. However, most prime movers (motors) rotate, and the mechanical energy developed by these machines is transmitted by rotating elements such as gears, sprockets, drive shafts, and cams. The understanding of the rotary motion of machine elements is an important consideration.

In the study of rotating objects, all the laws and principles studied in past chapters are used in understanding their rotary motion. Each of the physical quantities of straight-line motion (rectilinear motion) has a comparable physical quantity in rotational motion. Table 6.4 lists the *duals* for the various physical quantities of motion.

The term *rotary motion* implies that the motion of a turning object is around an axis, such as the wheel on the axle of an automobile. *Circular motion* is the term used to describe the motion of an object as the object travels along a circular path, such as the valve stem on a rotating tire or a race car around a circular track. Note the difference in rotary and circular motion; in rotary motion, the object rotates (*spins*) around a central axis, whereas in circular motion, the whole object travels a *circular path*. The study of rotary motion requires a system for measuring and defining *angular displacement*.

Angular Measurement

Over the ages, angles have been measured using various systems. Today three general systems are in common use:

TABLE 6.4 Physical Quantities of Motion

Translational Motion (Unit)	*Rotary Motion* (Unit)
Force (N or lb)	Torque (N·m or lb-ft)
Displacement (m or ft)	Angular displacement (rad)
Velocity (m/s or ft/s)	Angular velocity (rad/s)
Acceleration (m/s^2 or ft/s^2)	Angular acceleration (rad/s^2)
Mass, or inertia (kg or slug)	Moment of inertia (kg·m^2)

Revolutions The number of times an object revolves about an axis.

Degrees The subdivision of a circle swept out by a rotating object, where the circle is 360°.

Radians The ratio of the length of an arc of a circle to the length of the radius of the circle.

Of the three systems for measuring angular displacement (angles), radian measurement is widely used by technologists because it is compatible with all the formulas used in calculating rotational quantities. The **radian** (rad) is a derived unit in SI and is used in rotational motion calculations in BES. The simple relationship that exists among the three systems of angular measurement is:

$$1 \text{ revolution} = 360 \text{ degrees} = 2\pi \text{ radians} \tag{6.9}$$

At the onset, it is important that you understand that a radian is not a unit in the same sense as the meter or the foot. Instead, a radian has no dimension, since it is defined as a ratio of arc length to radius in a circle. Since each is measured in the same unit, the units factor out, leaving the ratio as a dimensionless quantity. When angles are substituted into the formula, the word *radian* is dropped, since the angle is dimensionless.

Figure 6.8 pictures an angle θ (theta) with its vertex at the center of a circle with radius r and an intercepted arc of length s. The angle θ *subtended* (enclosed) by the arc s is defined as the ratio of the arc length (s) to the length of the radius (r), where theta is in radians:

$$\theta = \frac{s}{r} \tag{6.10}$$

FIGURE 6.8
The angle θ subtended by the arc s is determined by the ratio of s to r. The resulting angle is stated in radians, a dimensionless quantity.

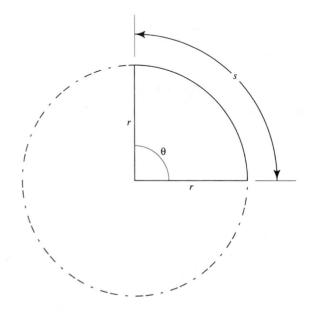

where

> θ = angular displacement, in radians (a dimensionless quantity)
> s = arc length, in any unit of linear distance (m, cm, ft, in)
> r = length of the radius, in any unit of linear distance

Circumference of a Circle

Equation 6.10 may be used to define the circumference of a circle, since one revolution of a radius generates an angle of 2π radians. Solve Equation 6.10 for s, the arc length:

$$s = r\theta \qquad\qquad (6.11)$$

where

> s = arc length, in any unit of linear distance
> r = length of the radius, in any unit of linear distance
> θ = angular displacement, in radians (a dimensionless quantity)

Now let $C = s$ and 2π radians $= \theta$ in Equation 6.11 ($s = r\theta$). Then $s = r\theta$; we substitute C for s and 2π for θ:

$$C = r2\pi$$

Rearrange the right side, and the circumference of a circle is

$$C = 2\pi r \qquad\qquad (6.12)$$

where

> C = circumference of a circle, in any unit of linear distance
> r = length of the radius of the circle, in any unit of linear distance
> 2π = angle subtended by the arc represented by the circumference, in radians (a dimensionless quantity)

Example
6.10

Using Equation 6.11 and Figure 6.8, determine the length of the arc s when
(a) $\theta = 90.0°$ and $r = 10.0$ in.
(b) $\theta = 360°$ and $r = 10.0$ in.

SOLUTION

(a) Convert $90.0°$ to radians using Equation 6.9 and then solve for s.

Convert
$$90.0° \times \frac{2\pi \text{ rad}}{360°} = \frac{\pi}{2} \text{ rad} = 1.57 \text{ rad}$$

Given
$$s = r\theta$$

Evaluate
$$r = 10.0 \text{ in}$$
$$\theta = 1.57 \text{ rad}$$

Substitute
$$s = 10.0 \times 1.57$$

Solve
$$s = 15.7 \text{ in}$$

(b) Convert 360° to radians and solve for *s*.

Convert	$360° \times \dfrac{2\pi \text{ rad}}{360°} = 2\pi \text{ rad} = 6.28 \text{ rad}$
Given	$s = r\theta$
Evaluate	$r = 10.0 \text{ in}$
	$\theta = 6.28 \text{ rad}$
Substitute	$s = 10.0 \times 6.28$
Solve	$s = 62.8 \text{ in}$

Angular Displacement

When a rigid body rotates about an axis, all the points on that body move through the same angle. The change of position of a point on the body from one place to another during rotation is called **angular displacement**; that is to say, angular displacement is the angle through which any point on the body is rotated when the body is experiencing rotary motion. Figure 6.9 pictures a laser disk rotating about a central axis. When the disk rotates one revolution (2π rad, or 360°), each of the indicated points passes through an angular displacement of 2π rad, or 360°.

Equation 6.10 may be used to determine any arc length if the angle subtended by the arc is given in radians and the radius of the arc from the axis to a point on the arc is known.

FIGURE 6.9
Each time the laser disk rotates through one revolution, each point on the disk has an angular displacement of 2π radians, or 360°.

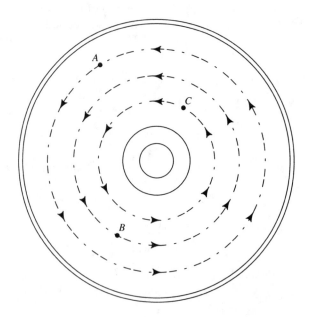

Example
6.11

The timing mark on the edge of a 2.0-m-diameter flywheel travels an arc distance of 12 cm. Determine the angular displacement

    (a) In radians.
    (b) In degrees.

SOLUTION   **(a)** Solve for θ in Equation 6.10.

Given $\theta = \dfrac{s}{r}$

Evaluate $s = 12 \text{ cm} = 0.12 \text{ m}$

    $r = \dfrac{d}{2} = \dfrac{2.0}{2} = 1.0 \text{ m}$

Substitute $\theta = \dfrac{0.12}{1.00}$

Solve $\theta = 0.12 \text{ rad}$

    (b) Convert 0.12 rad to degrees.

Convert $0.12 \text{ rad} \times \dfrac{360°}{2\pi \text{ rad}} = 6.9°$

Angular Velocity

In pure rotary motion, all parts of a rigid object rotate in circular paths around the same fixed axis of rotation. The rate of change of angular displacement with time is called the **angular velocity**; it is represented by the Greek letter ω (omega). When the rotary motion of the object is uniform, then the angular velocity is found by taking the angle through which the object turns and dividing it by the time taken to travel through the angle. Stated as an equation,

$$\omega = \frac{\theta}{t} \qquad\qquad\qquad \textbf{(6.13)}$$

where

    ω = angular velocity, in rad/s
    θ = angular displacement, in rad (a dimensionless quantity)
    t = time, in s

Angular velocity is commonly measured in revolutions per minute (rev/min, or rpm). However, the angular velocity must be expressed in radians per second (rad/s) when making calculations that use angular velocity in a formula. A simple conversion factor for revolutions per minute to radians per second is 1 rev/min = 0.1047 rad/s.

Example
6.12

Determine the angular velocity of the shaft of a motor that has had the speed reduced with the aid of a gearhead. The shaft rotates 18 times in 6.0 s. Express the answer in

  (a) Revolutions per minute.
  (b) Radians per second.

SOLUTION

(a) First find the angular velocity in revolutions per second (rev/s) and then convert to revolutions per minute.

Given

$$\omega = \frac{\theta}{t}$$

Evaluate

$\theta = 18$ revolutions

$t = 6.0$ s

Substitute

$$\omega = \frac{18}{6.0}$$

Solve

$\omega = 3.0$ rev/s

Convert

$$3.0 \, \frac{\text{rev}}{\cancel{s}} \times \frac{60 \, \cancel{s}}{1 \, \text{min}} = 180 \, \text{rev/min}$$

(b) Convert 3.0 rev/s to radians per second.

Convert

$$3.0 \, \frac{\cancel{\text{rev}}}{s} \times \frac{2\pi \, \text{rad}}{1 \, \cancel{\text{rev}}} = 6.0\pi \, \text{rad/s} = 18.9 \, \text{rad/s} \Rightarrow 19 \, \text{rad/s}$$

Linear Velocity of Rotating Objects

From time to time, we are interested in the speed of the circular motion of a point moving about a fixed axis. As you know, the angular velocity (rev/min or rad/s) is the same for all points on an object rotating around a fixed axis. However, the linear velocity (ft/s or m/s) for a point on a rotating object depends on its radial distance from the axis of rotation.

To find the linear velocity (v) of a point on an object rotating about an axis, the definition of linear velocity is combined with the definition of angular displacement, as follows:

$$v = \frac{s}{t} \quad \text{(Equation 2.3)} \quad \text{and} \quad s = r\theta \quad \text{(Equation 6.11)}$$

Substitute $r\theta$ from Equation 6.11 for s in Equation 2.3:

$$v = \frac{r\theta}{t}$$

However, θ/t is equal to ω (see Equation 6.13). Thus,

$$v = r\omega \tag{6.14}$$

where

v = linear velocity of a point on a rotating object, in m/s or ft/s

r = radial length from the center of rotation to the point on the rotating object, in m or ft

ω = the angular velocity of the rotating object, in rad/s

Example 6.13

The 24-in-diameter wheel of an automobile is rotating at an angular velocity of 565 rev/min. Determine

(a) The linear velocity of the face of the tire in feet per second.

(b) The linear velocity of the edge of the 15-in hubcap in feet per second.

SOLUTION

(a) Find v in feet per second. Start by converting 565 rev/min to radians per second by using the conversion factor 1 rev/min = 0.1047 rad/s. Then use Equation 6.14 to solve for the linear velocity (v).

Convert

$$565 \; \text{rpm} \times \frac{0.1047 \; \text{rad/s}}{1 \; \text{rpm}} = 59.2 \; \text{rad/s}$$

Given

$$v = r\omega$$

Evaluate

$$r = \frac{24 \; \text{in}}{2} = 12 \; \text{in} = 1.0 \; \text{ft}$$

$$\omega = 59.2 \; \text{rad/s}$$

Substitute

$$v = 1.0 \times 59.2$$

Solve

$$v = 59.2 \Rightarrow 59 \; \text{ft/s}$$

Observation

A linear velocity of 59 ft/s is about 40 mi/h (59/88 × 60 mi/h = 40 mi/h). As the tire rolls over the road, the car is moved forward at the same speed as the face of the tire—about 40 mi/h.

(b) Use Equation 6.14 to solve for the linear velocity of the edge of the 15-in hubcap in feet per second.

Given

$$v = r\omega$$

Evaluate

$$r = \frac{15 \; \text{in}}{2} = 7.5 \; \text{in} \times \frac{1 \; \text{ft}}{12 \; \text{in}} = .63 \; \text{ft}$$

$$\omega = 59.2 \; \text{rad/s}$$

Substitute

$$v = .63 \times 59.2$$

Solve

$$v = 37 \; \text{ft/s}$$

Observation

As expected, the speed of the edge of the hubcap is less than the speed of the edge of the tire (37 ft/s versus 59 ft/s).

FIGURE 6.10
When a stone dislodges from the tread of a tire, it leaves the tread in a straight line that is tangent to the circular path of travel.

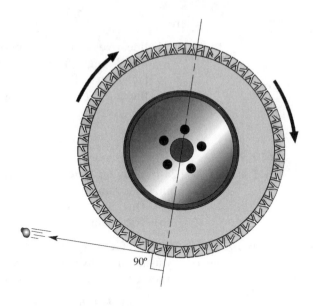

90°

Objects that are attached to bodies that are rotating, such as the valve stem on a tire or a stone lodged in the tread of a tire, are traveling in circular motion; that is, they travel along a circular path. When these objects come loose, as with a stone that dislodges itself from the tire tread, they travel in a straight line that is tangent to the circular path of travel, as shown in Figure 6.10. Their initial velocity is equal to the linear velocity ($r\omega$) of the point on the tire. Thus, a stone that leaves the tire tread of a vehicle traveling 60 mi/h is initially going at a speed of 60 mi/h (88 ft/s) relative to an observer standing still.

When operating machines, it is a very good practice to keep the guards over rotating elements in place at all times. Not only does this prevent objects from entering the machine, but it keeps objects from flying out of the machine when something comes loose.

Example
6.14

A 1-oz (.063-lb) chip of wood leaves the edge of a 10.0-in-diameter blade on a table saw and flies off, striking the machine operator's face shield. If the saw blade is rotating 5400 rev/min, determine

(a) The speed of the wood chip in feet per second as it leaves the edge of the saw blade.

(b) The work done on the face shield by the 1-oz wood chip.

SOLUTION

(a) First convert the angular velocity (ω) from revolutions per minute to radians per second and then solve for the linear velocity (v) in feet per second.

Convert

$$5400 \text{ rev/min} \times \frac{0.1047 \text{ rad/s}}{1 \text{ rev/min}} = 565 \text{ rad/s}$$

Given	$v = r\omega$
Evaluate	$r = \dfrac{10.0 \text{ in}}{2} = 5.00 \text{ in} \times \dfrac{1 \text{ ft}}{12 \text{ in}} = .417 \text{ ft}$
	$\omega = 565 \text{ rad/s}$
Substitute	$v = .417 \times 565$
Solve	$v = 236 \text{ ft/s, or } 161 \text{ mi/h}$

(b) Use Equation 6.5 to determine the kinetic energy of the wood chip.

Given	$KE = \frac{1}{2}mv^2$
Evaluate	$m = \dfrac{F}{a}$, where $F = F_W = .063 \text{ lb}$ and $a = g = 32.2 \text{ ft/s}^2$
	$m = \dfrac{.063}{32.2} = .0019 \text{ slug}$
	$v = 236 \text{ ft/s}$
Substitute	$KE = \frac{1}{2}(.0019)(236)^2$
Solve	$KE = 53 \text{ ft-lb}$
Observation	A kinetic energy of 53 ft-lb is equivalent to the energy imparted to the floor by an 18-lb object falling off a 3-ft-high table.

Angular Acceleration

Uniform angular acceleration, like linear acceleration, is the rate of change of velocity with time. Specifically, **angular acceleration** is the rate of change of angular velocity with time. Objects that rotate undergo angular acceleration when their angular velocity changes. For example, the shaft on an electric motor is accelerated as it comes up to speed. Angular acceleration is represented by the Greek letter alpha (α):

$$\alpha = \frac{\omega_f - \omega_i}{t} \tag{6.15}$$

where

α = average angular acceleration, in radians per second per second (rad/s^2)
ω_f = final angular velocity, in rad/s
ω_i = initial angular velocity, in rad/s
t = time interval during acceleration, in s

*Example
6.15*

Find the average angular acceleration (in rad/s^2) of the rotor/shaft assembly of an electric motor that is initially at rest if the final velocity of the rotor/shaft assembly is 1750 rev/min after 1.5 s.

SOLUTION   First write 1750 rev/min as an angular velocity in radians per second and then determine the average angular acceleration (α).

Convert $1750 \text{ rev/min} \times \dfrac{0.1047 \text{ rad/s}}{1 \text{ rev/min}} = 183 \text{ rad/s}$

Given $\alpha = \dfrac{\omega_f - \omega_i}{t}$

Evaluate $\omega_f = 1750 \text{ rev/min} = 183 \text{ rad/s}$

$\omega_i = 0 \text{ rev/min} = 0 \text{ rad/s}$

$t = 1.5 \text{ s}$

Substitute $\alpha = \dfrac{183 - 0}{1.5}$

Solve $\alpha = 122 \text{ rad/s}^2$

Therefore, the average angular acceleration is 120 rad/s^2.

The equations for rotational motion resulting from uniform acceleration are
of the same general form as those equations resulting from uniform linear accelera-
tion, as noted in Table 6.5. There are *duals* for distance, s and θ; velocity, v and ω; and
average acceleration, a and α.

TABLE 6.5   Equations for Uniformly Accelerated Motion

Physical Quantity	Translational Motion	Rotational Motion
Velocity	$v = \dfrac{s}{t}$	$\omega = \dfrac{\theta}{t}$
Acceleration (average)	$a = \dfrac{v_f - v_i}{t}$	$\alpha = \dfrac{\omega_f - \omega_i}{t}$
Final velocity	$v_f = v_i + at$	$\omega_f = \omega_i + \alpha t$
Time	$t = \dfrac{v_f - v_i}{a}$	$t = \dfrac{\omega_f - \omega_i}{\alpha}$
Velocity (average)	$v = \dfrac{v_i + v_f}{2}$	$\omega = \dfrac{\omega_i + \omega_f}{2}$
Displacement	$s = \left(\dfrac{v_i + v_f}{2}\right)t$	$\theta = \left(\dfrac{\omega_i + \omega_f}{2}\right)t$
Displacement	$s = v_i t + \frac{1}{2}at^2$	$\theta = \omega_i t + \frac{1}{2}\alpha t^2$
Acceleration or distance	$2as = v_f^2 - v_i^2$	$2\alpha\theta = \omega_f^2 - \omega_i^2$

Example
  6.16   A grinding wheel starts from rest and accelerates uniformly at 24.0 rad/s^2 until it reaches final velocity after 5.00 s. Determine

  (a) The angular displacement during acceleration.

  (b) The final angular velocity in radians per second and revolutions per minute.

  (c) The speed of the face of the 56-cm-diameter wheel in meters per second.

SOLUTION   **(a)** Select an equation from Table 6.5 that expresses angular displacement in terms of angular acceleration and time.

Given $\theta = \omega_i t + \frac{1}{2}\alpha t^2$

Evaluate $\omega_i = 0 \text{ rad/s}$

               $t = 5.00 \text{ s}$

               $\alpha = 24.0 \text{ rad/s}^2$

Substitute $\theta = 0(5.00) + \frac{1}{2}(24.0)(5.00)^2$

Solve $\theta = 300 \text{ rad}$

             (b) Select an equation from Table 6.5 that expresses the final angular velocity in terms of the initial velocity, average angular acceleration, and time.

Given $\omega_f = \omega_i + \alpha t$

Evaluate $\omega_i = 0 \text{ rad/s}$

               $\alpha = 24.0 \text{ rad/s}^2$

               $t = 5.00 \text{ s}$

Substitute $\omega_f = 0 + 24.0 \times 5.00$

Solve $\omega_f = 120 \text{ rad/s, or } 1150 \text{ rev/min}$

             (c) Use Equation 6.14 to determine the linear velocity (speed) of the face of the grinding wheel.

Given $v = r\omega$

Evaluate $r = \dfrac{56 \text{ cm}}{2} = 28 \text{ cm} = 0.28 \text{ m}$

               $\omega = 120 \text{ rad/s}$

Substitute $v = 0.28 \times 120$

Solve $v = 33.6 \Rightarrow 34 \text{ m/s}$

FIGURE 6.11
Information for Problem 1, Exercise 6.4.

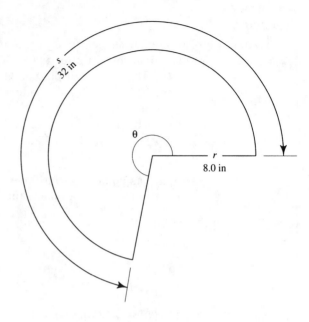

Exercise 6.4

Solve the following problems. Make sketches to aid in solving the problems and structure your work so it follows in an orderly progression and can easily be checked.

1. Given the information in Figure 6.11, determine
    (a) The angular displacement, θ, in radians.
    (b) The angular displacement, θ, in degrees.
    (c) The angular displacement, θ, in revolutions.
2. Express 360 rev/min in radians per second.
3. What is the angular velocity of a cam rotating through 14 revolutions in 1.27 min? Express the answer in
    (a) Revolutions per minute.   (b) Radians per second.
4. Determine the linear velocity of the belt pictured in Figure 6.12 when the angular velocity is 690 rev/min. Express the answer in
    (a) Feet per second.   (b) Meters per second.

FIGURE 6.12
The continuous belt of Problem 4, Exercise 6.4: (a) The belt has a 6.68-in radius; (b) The belt as it appears installed around the pulleys.

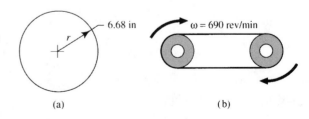

(a)   (b)

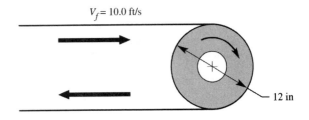

FIGURE 6.13
The conveyor of Problem 6, Exercise
6.4.

$V_f = 10.0$ ft/s

12 in

5. The drive shaft connecting an electrical motor to a magnetic clutch is rotating at
a final velocity of 1750 rev/min. Determine
  (a) The average angular acceleration during the 2.4 s start-up period. Assume
  that the shaft starts from rest.
  (b) The angular displacement (in radians) during start-up.
6. The conveyor belt of Figure 6.13 is driven by a 12-in. diameter shaft, as pictured.
If the conveyor starts from rest and accelerates uniformly to a final speed of 10.0
ft/s in an elapsed time of 18.4 s, determine
  (a) The linear acceleration of the belt in feet per second per second.
  (b) The final angular velocity of the shaft in revolutions per minute.
  (c) The angular acceleration of the shaft in radians per second per second. ∎

6.5 ROTATIONAL WORK AND POWER

Rotational Work

When a *tangential force*, F, is applied to the edge of the flywheel pictured in Figure
6.14 for a time t, the wheel rotates distance s, and it has angular displacement θ. The
work done in moving the wheel distance s is Fs ($W = Fs$). However, the torque is Fr
($\tau = Fr$); also, the distance traveled is equal to $r\theta$ ($s = r\theta$ from Equation 6.11). From
this information, the *rotational work* can be defined in terms of torque (τ) and angular displacement (θ).

$$W = Fs \quad \text{but} \quad F = \frac{\tau}{r} \quad \text{and} \quad s = r\theta$$

Substitute:

$$W = \frac{\tau}{\cancel{r}} \times \cancel{r}\theta \quad \text{Factor out } r.$$

$$W = \tau\theta \tag{6.16}$$

where

  W = work, in J or ft-lb
  τ = torque, in N·m or lb-ft
  θ = angular displacement, in rad (a dimensionless quantity)

FIGURE 6.14
The point on the rim of the flywheel moves distance s when force F is applied.

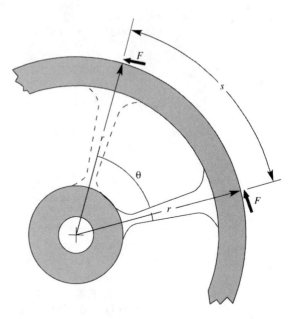

From Equation 6.16, we learn that work is done when a torque (τ) is applied to an object and the object rotates through an angle (θ). The amount of work can be determined by simply multiplying the applied torque by the resultant angular displacement.

Rotational Power

The generation of power is an important result of rotary motion from applied torque. At hydro and thermal generating sites, electrical power is generated by rotating machines (*alternators*) from rotary power taken from turbines attached to the alternators. The torque for these machines is produced by falling water, which moves the turbine in hydropower generation, and by steam, which moves the turbine blades in thermal power generation. When torque is applied uniformly to move an object, the *average power* from a rotating system is determined by dividing the rotational work by the elapsed time:

$$P = \frac{W}{t} = \frac{\tau\theta}{t}$$

However, θ/t is angular velocity (ω); therefore,

$$P = \tau\omega \tag{6.17}$$

where

P = power produced in a rotating system, in W or ft-lb/s
τ = torque, in N·m or lb-ft
ω = angular velocity, in rad/s

Example
6.17

Determine the power developed by a torque of 25 N·m causing an angular displacement of 520 rad in 2.0 s.

SOLUTION

Find the power by determining the time rate of doing work.

Given

$$P = \frac{W}{t} = \frac{\tau\theta}{t}$$

Evaluate

$\tau = 25$ N·m

$\theta = 520$ rad

$t = 2.0$ s

Substitute

$$P = \frac{25 \times 520}{2.0}$$

Solve

$P = 6.5$ kW

Observation

6.5 kW is equivalent to 8.7 hp (6500/746).

Example
6.18

Determine the power delivered by the rotating shaft of a motor when the speed of the shaft is 240 rev/min and the torque at the shaft is 5.8 lb-ft.

SOLUTION

First convert the angular velocity of 240 rev/min to radians per second and then solve for the power.

Convert

$$240 \text{ rev/min} \times \frac{0.1047 \text{ rad/s}}{1 \text{ rev/min}} = 25 \text{ rad/s}$$

Given

$P = \tau\omega$

Evaluate

$\tau = 5.8$ lb-ft

$\omega = 25$ rad/s

Substitute

$P = 5.8 \times 25$

Solve

$P = 145$ ft-lb/s $= 1\overline{5}0$ ft-lb/s

Observation

$1\overline{5}0$ ft-lb/s is .27 hp, or about $\frac{1}{4}$ hp (150/550).

Since power is usually given in horsepower, angular velocity in revolutions per minute, and torque in pound-feet in BES, the following equation simplifies the solution for power when the quantities are in these units.

$$P = \frac{\tau\omega}{5250} \qquad \textbf{(6.18)}$$

where

$$P = \text{power, in hp}$$
$$\tau = \text{torque, in lb-ft}$$
$$\omega = \text{angular velocity, in rev/min}$$

Example
6.19

Determine the torque delivered by a 10.0-hp motor having a shaft speed of 3500 rev/min.

SOLUTION   Find the torque from the rotational power equation, Equation 6.18.

Given $P = \dfrac{\tau \times \omega}{5250}$ and $\tau = \dfrac{P \times 5250}{\omega}$

Evaluate $P = 10.0 \text{ hp}$

$\omega = 3500 \text{ rev/min}$

Substitute $\tau = \dfrac{10.0 \times 5250}{3500}$

Solve $\tau = 15.0 \text{ lb-ft}$

■
_____

Exercise 6.5

Solve the following problems. Make sketches to aid in solving the problems and structure your work so it follows in an orderly progression and can easily be checked.

1. Determine the horsepower of an engine that develops 192 lb-ft of torque at 2700 rev/min.
2. Determine the power in watts of an electric motor that produces a torque of 34 N·m at an angular velocity of 120 rad/s.
3. Determine the torque developed by an electric motor that delivers .50 hp at 1750 rev/min.
4. Determine the torque transmitted by a belt drive when 35 hp is applied at 280 rev/min.
5. Determine the angular velocity in revolutions per minute of a pneumatic motor that develops 1600 W with a torque of 190 N·m.
6. The output of a transmission is rated at 440 lb-ft at 280 rev/min. Determine
    (a) The horsepower rating of the transmission at its output.
    (b) The efficiency of the transmission when the input power is 32 hp.

■

CHAPTER SUMMARY

- When machines rotate, they do work and possess energy.
- Work is done when an object moves in the direction of the applied force.
- Mechanical energy is divided into two categories, potential energy and kinetic energy.
- Power is the time rate of doing work or expending energy.
- Torque is a force applied to a moment arm that results in a twisting or turning motion.
- Rotary motion is the motion of an object spinning about an axis.
- Angular displacement results when a point on a rotating object moves from one position to another.
- The time rate of change of angular displacement is called angular velocity.
- Angular acceleration is the time rate of change of angular velocity.
- Rotational work results when a torque moves an object in the direction of the applied torque.
- Rotational power is the time rate of doing rotational work.

SELECTED TECHNICAL TERMS

The following technical terms are defined in the glossary located after Chapter 16. You are encouraged to use the glossary to aid your understanding and to test your knowledge of these important terms.

angular acceleration	potential energy
angular displacement	radian
angular velocity	torque
kinetic energy	

END-OF-CHAPTER QUESTIONS

Write T if the statement is true and F if the statement is false.

1. Work is done when an applied force results in a small movement of an object in the direction of the force.
2. The law of conservation of energy states that a small amount of energy is destroyed in the conversion process.
3. The watt is the unit of power in the SI.
4. The foot-pound is the unit of power in the BES.
5. The term *applied tangent* means that a force is directed toward the center of rotation of an object.
6. A moment arm is necessary in order for a torque to be present.
7. Angular displacement results when an object rotates.
8. The radian is a unit of angular measurement in both BES and SI.
9. Torque is one factor in rotational work.
10. The radian is a dimensionless unit.

In the following, select the word or phrase that makes the statement true.

11. The product of torque and angular displacement results in (power, work, angular velocity).

12. An object at rest on a bench top has (kinetic energy, rotational energy, potential energy).

13. The rotary motion of a machine element is due to (force, acceleration, torque).

14. The symbol that corresponds to velocity in a rotating system is (τ, ω, θ).

15. Objects that travel with a circular motion (revolve about an axis, travel in a circular path).

Answer each of the following questions with a short answer in the form of a complete sentence. Include a restatement of the question in your answer.

16. What two factors constitute work?

17. What is the difference between work and energy?

18. Why is a radian a dimensionless unit?

19. Why are torque units stated differently than work units?

20. What is the difference between rotary motion and circular motion?

TABLE 6.6 Summary of Formulas Used in Chapter 6

Physical Quantity	Unit	Equation Number	Equation
Work	J or ft-lb	6.1	$W = Fs$
Work	J or ft-lb	6.2	$W = Fs \cos(\theta)$
Work	J or ft-lb	6.3	$W = Fs = F_w h = mgh$
Potential energy	J or ft-lb	6.4	$PE = mgh$
Kinetic energy	J or ft-lb	6.5	$KE = \frac{1}{2}mv^2$
Power	W or ft-lb/s	6.6	$P = \dfrac{W}{t}$
Power	W or ft-lb/s	6.7	$P = Fv$
Torque	N·m or lb-ft	6.8	$\tau = Fr$
Angular diplacement	rad, °, rev	6.9	$1 \text{ rev} = 360° = 2\pi \text{ rad}$
Angular displacement	rad	6.10	$\theta = \dfrac{s}{r}$
Arc length	m or ft	6.11	$s = r\theta$
Circumference	m or ft	6.12	$C = 2\pi r$
Angular velocity	rad/s	6.13	$\omega = \dfrac{\theta}{t}$
Velocity	m/s or ft/s	6.14	$v = r\omega$
Angular acceleration	rad/s²	6.15	$\alpha = \dfrac{\omega_f - \omega_i}{t}$
Work (energy)	J or ft-lb	6.16	$W = \tau\theta$
Power	W or ft-lb/s	6.17	$P = \tau\omega$
Power	hp	6.18	$P = \dfrac{\tau\omega}{5250}$

END-OF-CHAPTER PROBLEMS

Solve the following problems. Make sketches to aid in the solution of the problems and structure your work so it follows in an orderly progression and can easily be checked. Table 6.6 summarizes the equations for work, torque, power, etc., in this chapter.

1. Determine the kinetic energy of a 14 200-N automobile traveling 24 m/s.

2. Determine the work done in sliding a 210-lb drill press 15 ft across a concrete floor when the coefficient of kinetic friction is 0.060.

3. Determine the power needed to lift a 485-kg load vertically through a distance of 82 m in 16 s.

4. A new 2.2-kW, 230-V, 60-Hz electric heat treating oven was purchased to replace an older gas-fired oven. If each oven has the same heating capacity, determine the heating rate of the older gas-fired oven; express the answer in Btu/h.

5. If a 4200-lb automobile is decelerated from 60.0 mi/h to a stop in 4.2 s, determine
   (a) The heat dissipated (in joules) by the braking system.
   (b) The average power dissipated (in watts) while the vehicle was being stopped.

6. The shaft of an electric motor accelerates uniformly from 330 rev/min to its final angular velocity in 4.8 s. Determine the final angular velocity if the average acceleration is 35 rad/s^2.

7. Determine the power in horsepower of an electric motor that develops a torque of 280 lb-in at 72 rev/min.

8. A speed reducer (gear box) is attached to the output shaft of an electric motor rotating at 320 rev/min. If the motor supplies .250 hp to its output shaft, determine the torque (lb-ft) produced at the output shaft of the speed reducer when the shaft's angular velocity is 6.00 rev/min. The speed reducer is 68% efficient.

9. A 460-V, three-phase, electric motor powering an elevator is rated to lift a 618-kg mass 60.0 m in 12.2 s. Determine
   (a) The horsepower rating of the three-phase motor.
   (b) The power (in watts) taken from the electric power line by the three-phase motor when it is operating at an efficiency of 0.784.

10. The 24-V direct current motor of a truck-mounted winch is operated from two series-connected 12-V industrial batteries. The 24-V dc motor takes in 20.6 amperes of current to produce 494 W of input power. Under load, the drum of the winch rotates at 12.0 rev/min. If the winch system is 34.6% efficient, determine the output torque (in lb-ft) of the drum under load.

Chapter 7

Rotational Power Transmission

About the Illustration. In this Renaissance machine, one or two people run in the vertical treadmill (A) to power the twin-drum winch (wheel and axle) attached to the block and tackle (M, N). Because of its flywheel effect *(mass at a distance from its center)*, the treadmill stores and delivers inertial energy to smooth out the operation of the machine. Torque provided by the treadmill is increased through the system of gears. The lantern pinion (B), constructed of round bars called rundles, meshes with the spur gear (C) to reduce the speed of the treadmill and increase the torque to the worm (D), which drives the winch through the wormwheel (E). (Designed by Agostino Ramelli, *from* The Various and Ingenious Machines of Agostino Ramelli *(1588).)* ■

INTRODUCTION

Mechanical rotational power is usually transmitted from the prime mover to machine mechanisms using power-transmission elements. Since the prime mover (electric motor, air motor, etc.) operates at a shaft speed greater than 1000 rev/min and because most mechanical motion needed to power machinery is well below this speed, speed reduction using *power-transmission elements* is common.

Among the power-transmission elements are couplings, clutches, gears, belts, and chains. Each of these machine elements transmits rotational power from shaft to shaft. Of the various elements used in power transmission, couplings and clutches are used to connect shafts end to end (axially), whereas gear trains, belt drives, and chain drives transmit rotary motion and torque radially from shaft to shaft. Only gearing, belt drives, and chain drives (the radial power transmitting elements) are studied in this chapter. Axial power-transmission elements are studied in Chapter 8, along with linear, reciprocating, and intermittent-motion devices.

CHAPTER CONTENTS

PERFORMANCE OBJECTIVES

Once you have read and studied each section; worked through each example with pencil, paper, and calculator; worked through the end-of-chapter problems; and answered the end-of-chapter questions, then you should be able to

- Identify various types of gears.
- Determine gear, pulley, and sprocket ratios.
- Use pitch and pitch diameter to determine the number of teeth on gears, timing belt pulleys, and sprockets.
- Differentiate among various types of belts used in belt drives.
- Specify standard roller chain size by chain number.

7.1 GEARING

Introduction

Gears provide positive transmission of rotational power (rotary motion and torque) from one shaft to another. Shafts that are parallel are connected by spur gears, helical gears, or herringbone gears (double helical gears), as pictured in Figure 7.1.

Spur gears are very popular, because they are inexpensive to manufacture and are available in a variety of sizes and materials. Because spur gears are somewhat noisy in operation, nonmetallic gears may be used where quiet operation is needed; however, nonmetallic gears are weaker than metal gears. In operation (as illustrated in Figure 7.2), only a single tooth of the spur gear carries the entire load during part of the meshing cycle, causing a very high tooth load.

Helical gears have teeth that are inclined in relation to the axis in the shape of a helix. When compared to the spur gears, helical gears are quieter (since the teeth engage gradually) and can carry a greater load (because more than one set of teeth is in contact at all times). However, the helix angle causes an **axial** load (side load), which creates a thrust on the shaft bearings. Because of this, thrust bearings are necessary with helical gears.

To avoid the thrust caused by single helical gears, double helical gears (herringbone gears) are used to neutralize the axial load and the resulting thrust. *Herringbone gears* are very smooth-running and are commonly used in heavy-duty, high-speed applications.

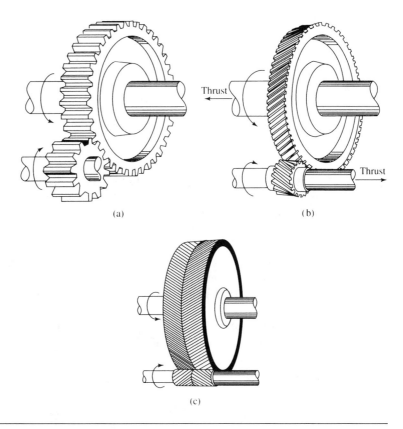

FIGURE 7.1
Parallel shafts are connected by: (a) Spur gears; (b) Helical gears; (c) Herringbone gears.

FIGURE 7.2
During part of the meshing cycle, only one tooth of the spur gear is in mesh.

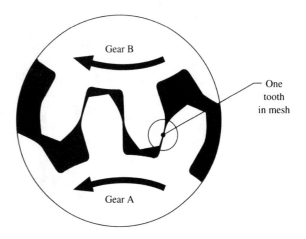

FIGURE 7.3
Types of bevel gears: (a) Straight
bevel gears; (b) Spiral bevel gears.

(a)   (b)

Shafts intersecting at right angles use *bevel gears* to transmit rotary motion and torque, as shown in Figure 7.3. Bevel gears are usually sold as matched sets, with either straight teeth or helical teeth. Bevel gears with helical teeth are called *spiral bevel gears*. Spiral bevel gears have axial thrust. *Miter gears* are bevel gears that are the same size and have the same number of teeth.

Crossed-axis shafts (shafts that do not intersect) are connected by helical gears or by *worm gears*. Crossed-axis helical gears (pictured in Figure 7.4) transmit rotary motion at right angles and, like helical gears used in connecting parallel shafts, they develop an axial thrust and require bearings that can handle both **radial** and thrust loads.

When a large torque at a greatly reduced angular velocity is needed at the driven shaft, then worm gearing is used. Worm gearing is made up of a worm and a wormwheel. The worm has a continuous helix, like the thread on a bolt, which forms one continuous tooth that engages the wormwheel. The wormwheel has concave-shaped helical teeth to ensure complete meshing with the worm. The shape of the teeth of the wormwheel allows large forces to be transmitted from the worm to the wormwheel. Unlike other gearing systems, in which the output may drive the input, the worm (the input) always drives the wormwheel (the output). The wormwheel can't drive the worm because of the large frictional forces between them. Since worm gears are unidirectional (only the worm can turn independently), they are *self-locking*.

Thus far you have learned how parallel shafts, intersecting shafts, and cross-axis shafts are interconnected by gearing. The last case is that of a pair of nonparallel shafts that do not intersect, as in the rear end of a rear-wheel-drive vehicle. Shafts of this type are connected by *hypoid gears*, which are similar in appearance to spiral bevel gears and operate somewhat like worm gears (see Figure 7.5). In operation,

FIGURE 7.4
Crossed-axis shafts
are connected by:
(a) Crossed-axis
helical gears; (b)
Worm gears.

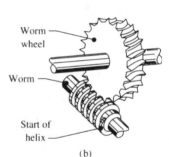

Worm
wheel

Worm

Start of
helix

(a)   (b)

FIGURE 7.5
Hypoid gears look similar to spiral bevel gears; however, their axes do not intersect.

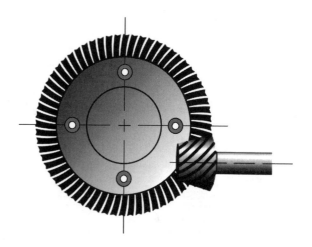

the teeth in contact do not just roll over each other; instead, the motion is a combination of sliding and rolling.

To some degree, all types of gear teeth roll and slide over one another when in operation. As you know, the force of sliding friction is much larger (about 1000 times) than rolling friction. The efficiency of a gearing system and the heat buildup in the system depend on the mix of rolling and sliding friction between the meshed teeth.

The efficiency of the parallel axis gears, pictured in Figure 7.1, can be very high—approaching 98% for gears that are precision-cut and use oil-spray lubrication. This high efficiency is the result of rolling friction with very little sliding friction. The efficiency of worm gears and hypoid gears is considerably lower than that of parallel-axis gears because of the high sliding friction. Low efficiency in worm and hypoid gears results in higher operating temperatures, which signal a need for extreme-pressure lubricants.

Terms and Definitions

The external spur gear is used in the following discussion. The terms and definitions are, in general, applicable to all other types of gears. Figure 7.6 illustrates a gear tooth, and Figure 7.7 pictures two gears in mesh. Each figure identifies some of the terms used in gear technology.

> **Pitch circle** The pitch circle is an imaginary circle that is the basis of gear design. Meshed gears meet at the point of tangency of their pitch circles.
> **Pitch diameter** Pitch diameter (PD) is the diameter of the pitch circle.
> **Pitch point** The pitch point (P) is a point on the line of centers. It is the point of tangency of the two pitch circles of gears in mesh.
> **Addendum** The addendum (the part added) is the radial distance from the pitch circle to the crest (top) of the tooth.
> **Dedendum** The dedendum (the part deducted) is the radial distance from the pitch circle to the root (bottom) of the tooth.

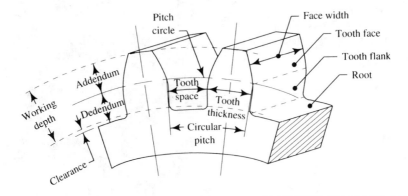

FIGURE 7.6
An involute gear tooth with several gear terms noted.

Clearance In meshed gears, clearance is the difference between the dedendum of one gear and the addendum of its mating gear. Clearance prevents the gears from binding.

Circular pitch Circular pitch (CP) is the arc distance between the centers of corresponding points of adjacent teeth, measured along the pitch circle.

Diametral pitch Diametral pitch (DP) is the ratio of the number of teeth on the gear (N) to the length of the pitch diameter (PD) (DP = N/PD). The term *diametral pitch* is so common with American standard gears that it is often shortened to pitch. Figure 7.8 relates the diametral pitch to tooth size for pressure angles of 20°.

Module The term *module* is used with metric gears. It is the ratio of the length of the pitch diameter (in millimeters) to the number of teeth on the gear (module = PD/N). Note that the value of module is the reciprocal of that of diametral pitch.

Pinion The pinion is the smaller gear in a pair of meshed gears.

Gear The gear is the larger gear in a pair of meshed gears.

FIGURE 7.7
Two spur gears meshed. The smaller of the gears is called the pinion, whereas the larger is called the gear. Externally meshed gears rotate in opposite directions to one another.

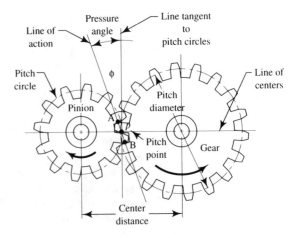

FIGURE 7.8
Relative sizes of gear teeth for selected diametral pitches with 20°
pressure angle. Diametral pitches
range as follows: (a) Coarse, $\frac{1}{2}$ to 10;
(b) Medium, 12 to 18; (c) Fine, 20 to
148; (d) Ultra-fine, 150 to 200 (not
pictured).

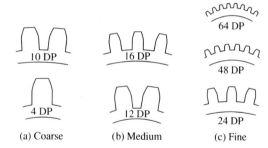

(a) Coarse   (b) Medium   (c) Fine

Line of centers The line of centers passes through the centers of two meshed
gears as well as the pitch point.

Center distance The center distance (CD) is the distance measured along the
line of centers between the centers of the pitch circles. The center distance
is equal to the sum of the pitch diameters of the pinion (PD$_p$) and gear
(PD$_g$) divided by 2 (CD = (PD$_p$ + PD$_g$)/2).

Involute The shape of modern gear teeth. The involute curve is generated by a
point on a straight line as it rolls onto a cylinder. This concept is shown in
Figure 7.9.

Base circle The base circle is the circle used to generate an involute tooth
curve. The diameter of the base circle (BD) is equal to the pitch circle diameter (PD) times the cosine of the pressure angle (ϕ) (BD = PD cos ϕ).

Pressure angle The angle between the line of action and the line tangent to the
pitch circles of meshed gears is the pressure angle ϕ (pictured in Figure 7.7).
The pressure angles are

14.5°   used for replacement only; not recommended in new design.

20°   established standard for new design; smooth and quiet-running with
good tooth strength.

25°   used when greater strength and surface durability are needed.

FIGURE 7.9
An involute curve is generated when
a taut string attached to a cylinder is
wound onto the cylinder.

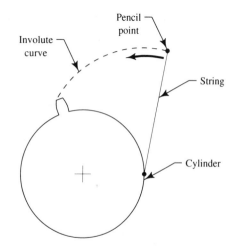

Line of Action The line of action is the contact path of the involute gear teeth in a meshed set of gears. It passes through the pitch point (point P in Figure 7.7) and it is tangent to the base circles of the meshed gears (points A and B in Figure 7.7). *Note:* The base circles are not shown in Figure 7.7.

Gear Ratio (IMA)

Suppose that two cylinders representing the pitch circles of two gears in contact without their teeth are held together with some force, as pictured in Figure 7.10. When the first cylinder, called the driver (D) because it is attached to a prime mover, rotates in a clockwise (cw) direction, then the second cylinder, called the driven (d), rotates in a counterclockwise (ccw) direction. The two cylinders in contact with one another make up a friction drive, since power is moved from the driver cylinder to the driven cylinder by the frictional forces between them.

If r_D is the radius of the driver and ω_D is its angular velocity, then the linear velocity (v) at any point on the edge of the cylinder is $v_D = r_D\omega_D$ (Equation 6.14). Assuming no slippage between the cylinders, then the linear velocity at any point on the edge of the driven cylinder must be $v_d = r_d\omega_d$.

Because the two cylinders are in contact without slipping and they share a common point of tangency, the linear velocity (v) at each edge must be equal:

$$v_D = v_d$$

Thus,

$$r_D\omega_D = r_d\omega_d$$

Dividing both sides of the equation by $r_D\omega_d$ results in

$$\frac{\omega_D}{\omega_d} = \frac{r_d}{r_D} \tag{7.1}$$

However, the velocity ratio of ω_D/ω_d in Equation 7.1 is the ideal mechanical advantage (IMA). *Note:* The velocity ratio of a set of meshed gears is referred to as the *gear ratio*.

$$\mathrm{IMA} = \frac{\omega_D}{\omega_d} \tag{7.2}$$

FIGURE 7.10
Two cylinders pressed together have equal linear velocities (v) at their edges.

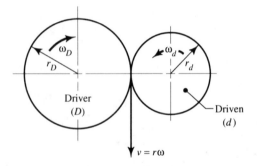

where

> IMA = ideal mechanical advantage
> ω_D = angular velocity, in rev/min or rad/s, of the driver, which supplies the effort
> ω_d = angular velocity, in rev/min or rad/s, of the driven, which moves the load

Because the number of teeth on a gear is related to the diameter of the pitch circle, the IMA (gear ratio) can be computed from the number of teeth in each meshed gear or from the pitch diameters of the gears. Thus,

$$\text{IMA} = \frac{N_d}{N_D} = \frac{\text{PD}_d}{\text{PD}_D} \tag{7.3}$$

where

> IMA = ideal mechanical advantage (gear ratio)
> N_d = number of teeth on the driven gear
> N_D = number of teeth on the driver gear
> PD_d = pitch diameter of the driven gear, in inches
> PD_D = pitch diameter of the driver gear, in inches

Torque Ratio (AMA)

When looking at two meshed gears, it can be seen that the turning moment (torque) is the result of force being applied over a lever having its pivot at the center of the gear and its length equal to the radius of the pitch circle, as shown in Figure 7.11. Since the gears are in mesh (no slippage), and assuming no frictional losses, the force acting on the teeth of the driver (F_D) must be equal to the force acting on the teeth of the driven (F_d). Since the pitch diameters of the two gears are not equal, the radii are not equal. The resulting torques ($F \times r$) are also not equal. Solving for the torque at the driver gear gives

$$\tau_D = F_D \times r_D$$

and, for the torque at the driven gear,

$$\tau_d = F_d \times r_d$$

FIGURE 7.11
The turning moment (torque) is the result of force applied over a lever with its pivot at the center of the gear and its length equal to the radius of the pitch circle.

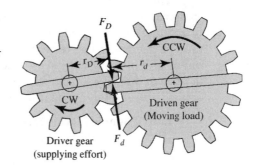

Since the force between the two gears is equal, then each equation may be solved for force and equated to each other:

$$F_D = \frac{\tau_D}{r_D}, \quad F_d = \frac{\tau_d}{r_d}, \quad \text{and} \quad F_D = F_d$$

Then

$$\frac{\tau_D}{r_D} = \frac{\tau_d}{r_d}$$

Multiplying both sides by r_d and dividing both sides by τ_D results in

$$\frac{r_d}{r_D} = \frac{\tau_d}{\tau_D} \tag{7.4}$$

However, the *torque ratio* of τ_d/τ_D is the mechanical advantage (AMA) of the meshed set of gears. Thus,

$$\text{AMA} = \frac{\tau_d}{\tau_D} \tag{7.5}$$

where

$$\begin{aligned} \text{AMA} &= \text{actual mechanical advantage} \\ \tau_d &= \text{torque, in lb-ft or N·m, at the driven gear} \\ \tau_D &= \text{torque, in lb-ft or N·m, at the driver gear} \end{aligned}$$

Law of Simple Machines

From the discussion so far for ideal meshed gears, you may have noticed that the velocity ratio is equal to the ratio of the radii of the pitch circles (Equation 7.1) and that the torque ratio is also equal to the ratio of the radii of the pitch circles (Equation 7.4). Thus,

$$\frac{\omega_D}{\omega_d} = \frac{r_d}{r_D} = \frac{\tau_d}{\tau_D}$$

or, simply,

$$\frac{\omega_D}{\omega_d} = \frac{\tau_d}{\tau_D} \tag{7.6}$$

This proportional equation indicates that the velocity ratio is inversely related to the torque ratio. That is, when the angular velocity (rev/min) at the output of a set of rotating objects (gears, pulleys, sprockets) is increased, then the torque (lb-ft) must be decreased, and conversely. What this means, simply, is that rotational power out of a set of gears may have an increase in torque with an accompanying decrease in angular velocity or an increase in angular velocity with an accompanying decrease in torque, but not an increase in both.

> ➤ **Law of Simple Machines for Rotational Motion** As in other ideal machines (frictionless), the rotational power into the machine is equal to the rotational power out of the machine.

Using Equation 7.6, this law may be stated mathematically as

$$\tau_D \omega_D = \tau_d \omega_d$$

Since power is equal to the product of torque and angular velocity,

$$P_D = P_d$$

That is, power into the driver gear (P_D) equals power out of the driven gear (P_d).

Efficiency in Radial Power Drives

When the frictional losses must be considered in gearing, belt drives, and chain drives, then—as in any other machine—the efficiency (η) is equal to the actual mechanical advantage (AMA) divided by the velocity ratio (IMA), or

$$\eta = \frac{\text{AMA}}{\text{IMA}} \quad \text{and} \tag{7.7}$$

$$\text{AMA} = \text{IMA} \times \eta \tag{7.8}$$

Example 7.1

The angular velocity of each gear shaft in a pair of meshed helical gears, as represented in Figure 7.12, is measured with a *mechanical tachometer*. The driver, which has 18 teeth, is rotating at 1800 rev/min, and the driven gear, which has 72 teeth, is rotating at 450 rev/min. Determine

    (a) The gear ratio (IMA) using the angular velocity.
    (b) The gear ratio (IMA) using the number of teeth.

SOLUTION   **(a)** Find the IMA using the angular velocity.

Given $\text{IMA} = \dfrac{\omega_D}{\omega_d}$

Evaluate $\omega_D = 1800$ rev/min

            $\omega_d = 450$ rev/min

FIGURE 7.12
Schematic of driver and driven gears. To simplify the representation of gears, only the pitch diameters are shown, with much of the detail omitted.

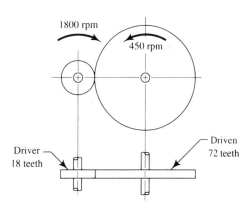

Substitute	$\text{IMA} = \dfrac{1800}{450}$
Solve	$\text{IMA} = 4.0$
Observation	The gear ratio is written as 4:1, a ratio.

(b) Find the IMA using the number of teeth.

Given	$\text{IMA} = \dfrac{N_d}{N_D}$
Evaluate	$N_d = 72$
	$N_D = 18$
Substitute	$\text{IMA} = \dfrac{72}{18}$
Solve	$\text{IMA} = 4.0$ and the gear ratio is 4:1

Example 7.2

Two spur gears are meshed. The driver gear, the larger of the two, has a pitch diameter of 3.500 in, whereas the driven gear has a pitch diameter of .875 in. If the gears are 92% efficient and the power at the driven gear (P_{out}) is 183 ft-lb/s, determine

(a) The gear ratio (IMA).
(b) The mechanical advantage (AMA).
(c) The power at the driver gear (P_{in}) in ft-lb/s.
(d) The torque at the driver gear ($\tau_D = \tau_{\text{in}}$) in lb-ft when the driver gear is rotating 180 rev/min (18.8 rad/s).

Observation	Make a sketch and label it with the given information, as in Figure 7.13.
SOLUTION	**(a)** Find the IMA using Equation 7.3 and express the answer as a gear ratio.
Given	$\text{IMA} = \dfrac{\text{PD}_d}{\text{PD}_D}$

FIGURE 7.13
Sketch for Example 7.2.

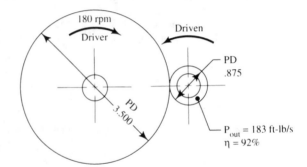

Evaluate	$PD_d = .875$ in
	$PD_D = 3.500$ in
Substitute	$IMA = \dfrac{.875}{3.500}$
Solve	$IMA = 0.250$

The gear ratio is 0.250:1, or 1:4

Observation	Since the driver gear is larger than the driven gear, the angular velocity of the driven gear will be increased by a factor of 4. The torque at the driven gear will be diminished by a factor greater than 4 once the frictional forces are considered.

(b) Find the AMA using Equation 7.8.

Given	$AMA = IMA \times \eta$
Evaluate	$IMA = 0.250$
	$\eta = 0.92$
Substitute	$AMA = 0.250 \times 0.92$
Solve	$AMA = 0.230 \Rightarrow 0.23$

The mechanical advantage (AMA) is 0.23.

(c) Find the power at the driver gear (P_{in}) using the efficiency equation (3.3) with power instead of work. Solve for the input power in foot-pounds per second.

Given	$P_{in} = \dfrac{P_{out}}{\eta}$
Evaluate	$P_{out} = 183$ ft-lb/s
	$\eta = 0.92$
Substitute	$P_{in} = \dfrac{183}{0.92}$
Solve	$P_{in} = 199 \Rightarrow \overline{200}$ ft-lb/s

The power into the driver gear is $\overline{2}00$ ft-lb/s.

Observation	The power into the driver gear is $\overline{2}00$ ft-lb/s, the power out of the driven gear is 183 ft-lb/s, and the difference of 17 ft-lb/s is lost to heat caused by the friction.

(d) Find the torque at the driver gear (τ_D-τ_{in}) when the driver gear is rotating 180 rev/min. Remember to express the angular velocity in radians per second when working with the formula. Using Equation 6.17 ($P = \tau\omega$), solve for torque.

Given	$\tau_{in} = \dfrac{P_{in}}{\omega}$

Evaluate	$P_{in} = 2\bar{0}0$ ft-lb/s
	$\omega = 180 \; \cancel{\text{rev/min}} \times \dfrac{0.1047 \quad \text{rad/s}}{1 \quad \cancel{\text{rev/min}}} = 18.8$ rad/s
Substitute	$\tau_{in} = \dfrac{2\bar{0}0}{18.8}$
Solve	$\tau_{in} = 10.6 \Rightarrow 11$ lb-ft

Therefore, the torque at the driver gear is 11 lb-ft.

Observation	The torque at the driven gear (τ_d) is found using Equation 7.5 (AMA $= \tau_d/\tau_D$) and solving for τ_d. In this example, the torque at the driven gear, which includes the frictional losses, is 2.4 lb-ft.

Gear Trains

When two external gears are in mesh they rotate in opposite directions. However, when a third gear is introduced between the two gears, then the first and last gears rotate in the same direction. The intermediate gear between the driver gear and the driven gear is called an **idler gear.** Figure 7.14 pictures a simple **gear train** using an idler gear. The idler gear does not change the gear ratio between the driver and driven gears; instead, it serves to maintain the same direction of rotation from the driver gear to the driven gear.

> ➤ **As a Rule** When the number of shafts in a gear train is an even number, then the first and last gears rotate in opposite directions. When the number of shafts in a gear train is an odd number, then the first and last gears rotate in the same direction.

Figure 7.15 pictures a compound gear train in which two sets of gears are in mesh. The two intermediate gears attached (fixed) to the center shaft are not idler gears. Gear B is driven by gear A, and gear C drives gear D. Unlike an idler gear that simply reverses rotational direction without changing the gear ratio, these intermediate gears affect the gear ratio and must be included in its calculation.

> ➤ **As a Rule** When two or more gears are attached to the same shaft, all gears are rotating in the same direction at the same angular velocity (revolutions per minute).

In compound gear trains, the angular velocity of the first and last gear is needed in order to determine the IMA (gear ratio) and the AMA (torque ratio) of

FIGURE 7.14
The idler gear is the intermediate gear between the driver gear and the driven gear. It does not affect the gear ratio, since it acts as both a driven gear and a driver gear.

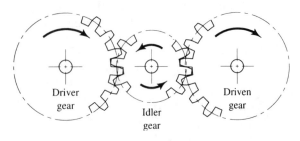

Driver gear

Idler gear

Driven gear

FIGURE 7.15
In a compound gear train, (1) gears on the same shaft rotate in the same direction at the same speed, and (2) gears in mesh must be from the same pitch series; that is, they must have the same diametral pitch.

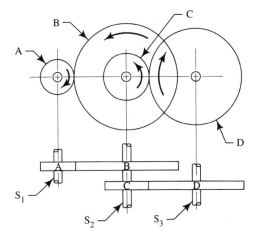

Gear	PD	N	DP	rpm	Center Distance
A	1.500	18	12	200	3.000
B	4.500	54	12	66.7	
C	2.000	36	18	66.7	3.000
D	4.000	72	18	33.3	

the gear train. The following equation, based on the relationship of the velocity ratio and the number of teeth on a gear, is used to determine the angular velocity of the last gear in a gear train:

$$\omega_d = \frac{\omega_D N_1 N_2 \cdot \cdot \cdot N_n}{n_1 n_2 \cdot \cdot \cdot n_n} \tag{7.9}$$

where

ω_d = angular velocity of last driven gear, in rev/min or rad/s
ω_D = angular velocity of first driver gear in rev/min or rad/s
N_1 = number of teeth on the first driver gear
N_2 = number of teeth on the second driver gear
N_n = number of teeth on the Nth driver gear
n_1 = number of teeth on the first driven gear
n_2 = number of teeth on the second driven gear
n_n = number of teeth on the nth driven gear

Example 7.3

In Figure 7.15, if gear A rotates at an angular velocity of 200 rev/min in a clockwise (cw) direction, determine

(a) The direction of rotation of gear D.
(b) The angular velocity of gear D.
(c) The gear ratio of the gear train.

SOLUTION

(a) Since there are three shafts, an odd number, then the first and last gears rotate in the same direction.

Therefore, gear D is rotating in a cw direction.

(b) Using Equation 7.9, find the angular velocity of gear D.

Given

$$\omega_d = \frac{\omega_D(N_1 N_2)}{n_1 n_2}$$

Evaluate

$\omega_D = 200$ rev/min

$N_1 = 18$ teeth

$N_2 = 36$ teeth

$n_1 = 54$ teeth

$n_2 = 72$ teeth

Substitute

$$\omega_d = \frac{200 \times 18 \times 36}{54 \times 72}$$

Solve

$\omega_d = 33.3$ rev/min

(c) Find the gear train's gear ratio (IMA) using the ratio of the angular velocity of gear A to the angular velocity of gear D and then express the solution as the gear ratio.

Given

$$\text{IMA} = \frac{\omega_A}{\omega_D}$$

Evaluate

$\omega_A = 200$ rpm

$\omega_D = 33$ rpm

Substitute

$$\text{IMA} = \frac{200}{33.3}$$

Solve

$\text{IMA} = 6$

Therefore, the gear ratio is 6:1.

> ➤ **As a Rule** An idler gear acts as both a driver gear and a driven gear. As such, it does not affect the gear ratio and may be omitted from the gear ratio calculation.

Example 7.4

Determine the angular velocity (revolutions per minute) of gear D in Figure 7.16 when gear A is rotating at 1750 rev/min.

Observation

Since gear B is both a driver and a driven gear (an idler), it may be omitted from the calculation.

FIGURE 7.16
Gear train for Example 7.4.

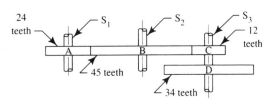

SOLUTION

Find the angular velocity of gear D. Since gears C and D are on the same shaft (S_3), then both gears are rotating at the same angular velocity. Thus, by determining the angular velocity of gear C, the angular velocity of gear D is also determined.

Given

$$\omega_d = \frac{\omega_D N_1}{n_1}$$

Evaluate

$$\omega_D = \omega_A = 1750 \text{ rev/min}$$

$$N_1 = 24 \text{ teeth}$$

$$n_1 = 12 \text{ teeth}$$

Substitute

$$\omega_d = \frac{1750 \times 24}{12}$$

Solve

$$\omega_d = 3500 \text{ rev/min}$$

Therefore, the angular velocity of gear *D* equals 3500 rev/min.

Center Distance

From time to time, a technician may need to construct a simple gear train for an instrument or may need to specify a replacement gear in the course of maintaining an electromechanical system. One of the key points of information in constructing a gear train is the center distance—that is, the center-to-center distance between the shafts used to support meshed gears.

To find the center distance between parallel shafts, add the pitch diameters of the meshed gears and divide by two.

$$CD = \frac{PD_1 + PD_2}{2} \qquad (7.10)$$

where

CD = center distance between two meshed gears, in inches
PD_1 = pitch diameter of the first gear, in inches
PD_2 = pitch diameter of the second gear, in inches

Gear Parameters

When replacing a gear, the new gear must be of the same material and type (spur, helical, etc.) as the old gear and have the same number of teeth, pitch diameter, and diametral pitch. In addition, the hub diameter must be compatible to the shaft that supports the gear. A general equation for gearing that relates the number of teeth (N) to the pitch diameter (PD) and the diametral pitch (DP) is

$$N = PD \times DP \qquad\qquad \textbf{(7.11)}$$

where

N = the number of teeth on the gear
PD = the pitch diameter, in inches
DP = the diametral pitch, in teeth/in (see Figure 7.8)

Example 7.5

Gear A in the compound gear train pictured in Figure 7.17 has a diametral pitch (DP) of 32 and a pitch diameter (PD) of .7500 in. Determine
    (a) The number of teeth on gear A.
    (b) The pitch diameter of gear B if it has 68 teeth.
    (c) The center distance between S_1 and S_2 (CD_{1-2}).

SOLUTION   **(a)** Using Equation 7.11, find the number of teeth on gear A.

Given $N = PD \times DP$

Evaluate   PD = .7500 in

             DP = 32

Substitute $N = .7500 \times 32$

Solve $N = 24$ teeth

Therefore, the number of teeth on gear A is 24 teeth.

(b) Find the pitch diameter (PD) of gear B using Equation 7.11.

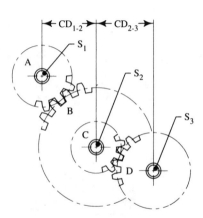

FIGURE 7.17
Compound gear train with the center distance indicated (Example 7.5).

Given	$N = PD \times DP$ and $PD = \dfrac{N}{DP}$
Observation	Two gears in mesh must have the same diametral pitch. Since gear A is 32 pitch, then gear B must also be 32 pitch.
Evaluate	$N = 68$ teeth
	$DP = 32$
Substitute	$PD = \dfrac{68}{32}$
Solve	$PD = 2.125$ in

Therefore, the pitch diameter of gear B is 2.125 in.

(c) Using Equation 7.10, find the center distance between shaft 1 (S_1) and shaft 2 (S_2).

Given	$CD_{1-2} = \dfrac{PD_1 + PD_2}{2}$
Evaluate	$PD_1 = .7500$ in
	$PD_2 = 2.125$ in
Substitute	$CD_{1-2} = \dfrac{.7500 + 2.125}{2}$
Solve	$CD_{1-2} = 1.438$ in

Therefore, the center distance between S_1 and S_2 is 1.438 in.

IMA of the Worm Gear

Virtually all the concepts presented thus far are applicable to all types of gears. The one exception is the worm gear. Because the worm gear has one continuous tooth, which looks like the thread on a bolt, the IMA (gear ratio) for the worm gear based on the teeth ratio is slightly different than that for other types of gearing.

$$IMA_{wg} = \frac{N_{wheel}}{N_{worm}} \tag{7.12}$$

where

IMA_{wg} = ideal mechanical advantage of the worm gear
N_{wheel} = number of teeth on the worm wheel (driven gear)
N_{worm} = number of leads (starts) of the helix, as seen from the end of the worm as pictured in Figure 7.4(b).

Note: Most worms are single-lead; however, double and triple leads are available.

Example
    7.6

A single-lead worm is used with an 80-tooth worm wheel in a gear reducer (a speed reducer). The worm gearing is 35% efficient. If the torque at the worm (input) is .625 lb-ft, determine

  (a) The IMA and gear ratio.
  (b) The AMA.
  (c) The torque at the shaft of the worm wheel (output torque).

SOLUTION   **(a)** Find the IMA using Equation 7.12.

Given $$\text{IMA}_{wg} = \frac{N_{wheel}}{N_{worm}}$$

Evaluate $N_{wheel} = 80$

             $N_{worm} = 1$

Substitute $$\text{IMA}_{wg} = \frac{80}{1}$$

Solve $\text{IMA}_{wg} = 80$

Therefore, the gear ratio is 80:1.

(b) Find the AMA using Equation 7.8.

Given $\text{AMA} = \text{IMA} \times \eta$

Evaluate $\text{IMA} = 80$

             $\eta = 0.35$

Substitute $\text{AMA} = 80 \times 0.35$

Solve $\text{AMA} = 28$

(c) Find the torque at the shaft of the worm wheel (driven). Use Equation 7.5.

Given $\text{AMA} = \dfrac{\tau_d}{\tau_D}$ and $\tau_d = \text{AMA} \times \tau_D$

Evaluate $\tau_D = .625$ lb-ft

             $\text{AMA} = 28$

Substitute $\tau_d = 28 \times .625$

Solve $\tau_d = 17.5 \Rightarrow 18$ lb-ft

Therefore, the torque at the shaft of the worm wheel is 18 lb-ft.

Observation The speed is reduced by a factor of 80 and the torque is increased by a factor of 28 when the percent efficiency of the gear reducer is 35%.

Summary of Rules for Gear Trains

- When the number of shafts in a gear train is odd, then the first and last gears are rotating in the same direction.
- When the number of shafts in a gear train is even, then the first and last gears are rotating in opposite directions.
- Gears attached to the same shaft rotate in the same direction at the same angular velocity.
- An idler gear may be omitted from the gear ratio calculation, since it is both a driver and a driven gear.
- Two gears in mesh must have the same diametral pitch.

Exercise 7.1

Solve the following problems. Make sketches to aid in solving the problems and structure your work so it follows in an orderly progression and can easily be checked.

1. Determine the gear ratio for a pair of meshed gears when the driver rotates at 270 rev/min and the driven rotates at 1350 rev/min.
2. For the gears in Problem 1, determine the mechanical advantage (AMA) if the gears are 87% efficient.
3. A pair of spur gears is 92% efficient. The driver gear has 48 teeth, and the driven gear has 16 teeth. Determine the torque at the driven gear when the torque at the driver gear is 8.00 lb-ft.
4. For the gears in Problem 3, determine the power into the driver gear when the shaft attached to the driven gear is rotating at 328 rev/min.
5. Determine the revolutions per minute of gear E in the gear train of Figure 7.18 when gear A is rotating at 560.0 rev/min.
6. A double lead worm is used with a 64-tooth worm wheel in a worm gear torque converter. If the efficiency of the system is 0.44, determine the torque at the worm wheel when the worm is rotating 360 rev/min with a torque of .250 lb-ft.
7. Complete the table in Figure 7.19.
8. Complete the table in Figure 7.20.

FIGURE 7.18

The gear train for Problem 5 in Exercise 7.1.

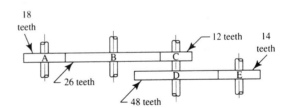

FIGURE 7.19

Gear train and table for Problem 7 of Exercise 7.1.

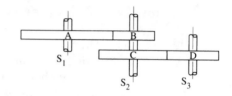

Gear	PD	N	DP	Direction	rpm	CD
A	.8125		32	S_1 CW	82.0	
B		18		S_2		
C	.6042			S_2		
D		18	48	S_3		

7.2 BELT DRIVES

Introduction

Belt drives transmit power between parallel shafts through the use of belts and **pulleys** (also called sheaves). Like gears, pulleys have a pitch diameter (PD) that is used in calculating the pulley ratio (IMA) of the system. The IMA equations for gears may be used with belt drives. The pitch diameter is also used, along with the center distance (CD), to determine the belt length.

Figure 7.21 pictures a pulley-and-belt system. Figure 7.21(a) shows that the driver pulley, the driven pulley, and the belt are all rotating in the same direction; this is an *open-belt* system. However, in Figure 7.21(b) the pulleys rotate in opposite directions; this is a *crossed-belt* system. Compared to gears, belt drives require no lubrication, are quiet, can operate with very long center distances, and can tolerate some misalignment.

Most belt drives rely on friction between the belt and sheaves to transmit power from the driver pulley to the driven pulley. The factors that permit power transmission in belt drives are these:

FIGURE 7.20

Gear train and table for Problem 8 of Exercise 7.1.

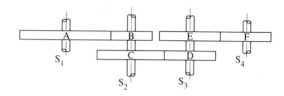

Gear	PD	N	DP	Direction	rpm	CD
A	.7500		48	S_1 CCW	180.0	
B		18		S_2		
C	1.0000			S_2		
D	.3200	16		S_3		
E	.8750		32	S_3		
F		40		S_4		

FIGURE 7.21
Pictorial and schematic representations of belt drive rotation direction for: (a) An open-belt drive; (b) A crossed-belt drive.

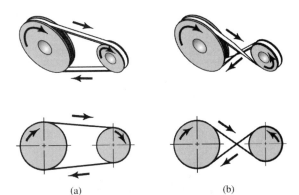

(a)   (b)

- The coefficient of friction between the belt and pulley
- The frictional force developed by tensioning the belt with some initial tension (force)
- The size of the smaller pulley
- The pulley ratio, which affects the arc of contact
- The speed of the belt

Belt Types

Belts may be divided into four general categories: round, flat, synchronous, and V-belts. Of the four categories, only the synchronous belt drive system is not a *friction drive* system; it is classed as a *positive drive* system. Each of these belt types, along with its pulley, is pictured in Figure 7.22.

Round Belts   Round belts made from neoprene and polyurethane are used with light loads in applications requiring severe bends (>90°) and twists. The belts are self-tensioning. Applications using round belts include sewing machines, video-cassette recorders, and motion picture projectors.

Flat Belts   Small, flat belts are applied in low-power, high-speed commercial equipment requiring small pulley diameters with very low inertia and noise levels. The woven nylon or dacron rubber-impregnated fabric belts require tensioning with *idlers* or spring devices to maintain sufficient tension to create the frictional forces that keep the belt from slipping. It should be understood that all belts (except toothed, synchronous belts) creep due to tension and flexure stresses. In high-power applications, large, flat belts are used as conveyor belts to move products from place to place.

FIGURE 7.22
Four types of belts and pulleys: (a) Round belt; (b) Flat belt; (c) Timing belt; (d) V-belt.

(a)   (b)   (c)   (d)

Synchronous Belts Synchronous belts, commonly called *timing belts*, are flat belts with evenly spaced, trapezoidal teeth that are used to transmit power and maintain position. Because power is transmitted by the teeth and not by friction, there is no slippage or creep, and the belt requires only a small tensioning force, resulting in low bearing loads. The information for installing and tensioning timing belts may be found in Appendix C, Section C1.

The load-carrying elements, embedded in the belt, are steel cables for low-speed–high-torque applications and polyester for high-speed–low-torque stepper motor applications requiring smooth running over small pulleys. Neoprene teeth and backing are bonded to the load-carrying elements, and a nylon facing is placed over the teeth to improve wear and lower friction. Figure 7.23(a) pictures a standard **RMA** (Rubber Manufacturers' Association) timing belt with trapezoid-shaped teeth, Figure 7.23(b) pictures a metric timing belt with stronger curvilinear-shaped teeth, and Figure 7.23(c) pictures a metric timing belt with trapezoid-shaped teeth used in Europe and found on equipment imported from Europe and Asia. In each of the three timing belts, the circular pitch (CP) has been indicated.

In most designs, a minimum of six teeth are engaged on the smaller pulley at all times. This ensures that the teeth on the belt will not shear off under load. Because of their light weight, some timing belts can operate at very high linear speeds (up to 260 ft/s).

Like gears, the grooved pulleys designed for use with timing belts are designed using an equation that relates the teeth (grooves), pitch diameter, and pitch. With timing belt pulleys, the pitch is circular pitch, not diametral pitch as with gears. The

FIGURE 7.23
Timing belt types with circular pitch (CP) noted: (a) Standard RMA; (b) Metric ISO (international); (c) Metric (European).

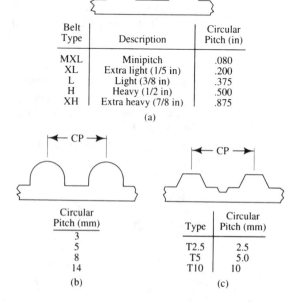

Belt Type	Description	Circular Pitch (in)
MXL	Minipitch	.080
XL	Extra light (1/5 in)	.200
L	Light (3/8 in)	.375
H	Heavy (1/2 in)	.500
XH	Extra heavy (7/8 in)	.875

(a)

Circular Pitch (mm)
3
5
8
14

(b)

Type	Circular Pitch (mm)
T2.5	2.5
T5	5.0
T10	10

(c)

following formula relates the circular pitch (CP) and the pitch diameter (PD) to the number of teeth (N) on the timing pulley:

$$N = \frac{\pi(\text{PD})}{\text{CP}}$$ (7.13)

where

N = number of teeth on the timing pulley
PD = pitch diameter, in inches
CP = circular pitch, in inches

V-Belts Of all the drive systems, the V-belt drive is by far the most common nonsynchronous method of driving loads at moderate rotational speeds (1000 to 6000 rev/min). The most significant feature of the V-belt drive is the *wedging action* of the belt against the sides of the groove in the pulley, resulting in an increase in friction force as the tension in the belt increases. Thus, a greater load simply causes a greater frictional force, resulting in a tighter belt grip. Because the V-belt drive system relies on friction to transmit power, there is a small slippage or creep between the belt and the sheaves, resulting in wear on both the belt and the sheaves.

An important consideration when working with V-belt drives is that the belt should never touch the bottom of the groove in the pulley, as excess slippage and heating will result. There must always be a clearance between the bottom of the belt and the bottom of the groove. Appendix C, Section C1, has information on the installation, maintenance, and tensioning of V-belts.

Besides supplying a wedging action, V-belts are quiet in operation; protect bearings by absorbing start and stop shock; require little maintenance; and are available in single belts, multiple matched-belt sets, or connected (banded) multiple belts.

To ensure uniformity and replaceability, V-belts, like most belts, are manufactured to RMA standards. V-belts are classified by group as industrial, automotive, or agricultural. Each group has its own set of standards. The following are the three categories of V-belts in the industrial group.

- The *classical series* of V-belts, for use with heavy-duty multiple belt drives transmitting high horsepower (up to 300 hp), is the oldest series. It has five belt sizes.
- The *narrow groove series* of V-belts, like the classical series, is designed for use in heavy-duty multiple belt drive systems transmitting high horsepower. Compared to the classical series, the modern narrow groove series of V-belts is more compact and lighter in weight for comparable horsepower capacity. It has only three belt sizes (versus five) to cover the power range.
- The *light duty or fractional horsepower (FHP) series* of V-belts is designed for single belt drive systems transmitting less than three horsepower. Designed for intermittent use, the series consists of four belt sizes.

Figure 7.24 pictures the three types of V-belts in the industrial group along with their identifiers. Additional information on the size range and codes for V-belts can be found in Appendix C, Section C1.

FIGURE 7.24
Industrial V-belts are available in three types of "sections" (cross-sections) as: (a) Classical multiple belts; (b) Narrow groove belts; (c) Light duty, or fractional horsepower (FHP), belts.

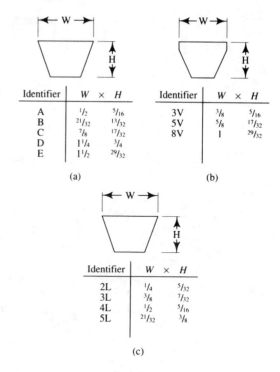

Identifier	W	×	H
A	$1/2$		$5/16$
B	$21/32$		$13/32$
C	$7/8$		$17/32$
D	$1 1/4$		$3/4$
E	$1 1/2$		$29/32$

(a)

Identifier	W	×	H
3V	$3/8$		$5/16$
5V	$5/8$		$17/32$
8V	1		$29/32$

(b)

Identifier	W	×	H
2L	$1/4$		$5/32$
3L	$3/8$		$7/32$
4L	$1/2$		$5/16$
5L	$21/32$		$3/8$

(c)

Belt Drive Parameters

Unless you are responsible for the design of drive systems, you will never be required to specify a belt initially. However, as a technician, you will, from time to time, order replacement belts. It is very important that you select the proper belt. The following discussion covers some of the factors that go into the design of a belt drive. With these concepts, you will have a better understanding of why drive components (including belts) have the characteristics that they have and why they must be replaced with compatible components when they wear out.

➤ **As a Rule** In any machine, the motor has greater power capacity than the load to be driven. The drive system must, as a matter of course, have a greater power capacity than the motor.

To use the preceding rule, several factors must be applied in sizing the motor, belt, and pulleys. Included in these are the center distance and belt length, the arc-of-contact correction factor, the pulley ratio (IMA), and the drive system's service overload factor, commonly called the *service factor* (SF). Each of these factors will be discussed.

Center Distance and Belt Length

The belt length needed to connect two pulleys in a belt drive may be approximated by the following formula:

$$l = 2\text{CD} + \frac{\pi}{2}(D + d) \qquad \qquad \textbf{(7.14)}$$

where

l = pitch length (pitch circumference) of the belt, in inches
CD = center distance between the shafts supporting the driver and driven pulleys, in inches
D = pitch diameter of the larger pulley, in inches
d = pitch diameter of the smaller pulley, in inches

This formula results in the pitch length of the belt—that is, the length of the belt along the pitch line, as pictured in Figure 7.25. Notice in the flat belt and timing belt drives (Figure 7.25) that the pitch circle of the pulley lies outside the body of the pulley, as does the pitch line of the belt. However, this is not the case with the V-belt drive, since the belt is wedged into the groove of the pulley. In this case, the pitch circle of the pulley and the pitch line of the belt lie within the body of the pulley. In V-belt drives, the pitch diameter (PD) of the pulley is always slightly smaller than the outside diameter (OD) of the pulley. Equation 7.14 may be used to calculate the length of a V-belt along the outside of the belt (outside length) by using the outside diameters (OD) of the pulleys instead of the pitch diameters (PD) of the pulleys.

Example 7.7

A .10-hp stepper motor operating at 240 rev/min is connected to the shaft of a high efficiency, zero backlash, ball bearing screw on an X-Y table by a type XL timing belt. The ball bearing screw, pictured in Figure 7.26, is operated at 96.0 rev/min for best performance. If the timing pulley attached to the table (driven pulley) has a pitch diameter of 1.593 in and the center distance is approximately 5.00 in, determine

FIGURE 7.25
Flat timing belt pulley and belt geometry.

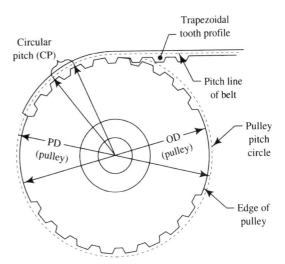

FIGURE 7.26
The precise positioning of the X-Y table is ensured by the use of a synchronous belt drive and a zero backlash ball bearing screw.

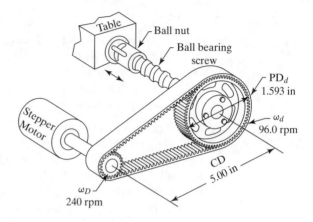

(a) The pulley ratio of the system.
(b) The pitch diameter of the timing pulley attached to the motor shaft (driver pulley).
(c) The number of teeth on the smaller of the two grooved pulleys.
(d) The approximate pitch length of the timing belt.

SOLUTION
(a) Find the pulley ratio using the revolutions per minute of the motor shaft and the table shaft. Substitute these values into Equation 7.2 and solve for the IMA. Express the result as the pulley ratio.

Given

$$\text{IMA} = \frac{\omega_D}{\omega_d}$$

Evaluate

$$\omega_D = 240 \text{ rev/min}$$

$$\omega_d = 96.0 \text{ rev/min}$$

Substitute

$$\text{IMA} = \frac{240}{96.0}$$

Solve

$$\text{IMA} = 2.50$$

Therefore, the pulley ratio is 2.50:1.

(b) Find the pitch diameter of the motor pulley, which is the driver pulley, using Equation 7.3 and solving for PD_D.

Given

$$\text{IMA} = \frac{\text{PD}_d}{\text{PD}_D} \quad \text{and} \quad \text{PD}_D = \frac{\text{PD}_d}{\text{IMA}}$$

Evaluate

$$\text{PD}_d = 1.593 \text{ in}$$

$$\text{IMA} = 2.50$$

Substitute

$$\text{PD}_D = \frac{1.593}{2.50}$$

Solve	$PD_D = .6372$ in

Therefore, the motor pulley has a pitch diameter of .6372 in.

(c) Find the number of teeth on the smaller of the two pulleys, which, in this case, is the drive pulley. Once this is known, determine if enough belt teeth will be engaged during operation to carry the load.

Given	$N = \dfrac{\pi(PD)}{CP}$
Evaluate	$PD = .637$ in
	$CP = .200$ from Figure 7.23(a)
Substitute	$N = \dfrac{\pi \times .637}{.200}$
Solve	$N = 10$ teeth
Observation	In use, the belt will wrap about 180° around the pulley and engage 5 teeth. If the load is light to moderate, then 5 teeth will be adequate.

(d) Approximate the pitch length of the type XL timing belt using Equation 7.14.

Given	$l = 2CD + \dfrac{\pi}{2}(D + d)$
Evaluate	$CD = 5.00$ in
	$D = 1.593$ in
	$d = .637$ in
Substitute	$l = 2 \times 5.00 + \dfrac{\pi}{2}(1.593 + .637)$
Solve	$l = 13.5$ in

Therefore, the belt length is selected as 13.6 in, the nearest standard length.

Arc of Contact

The arc of contact varies from an ideal 180° when the pulley ratio is 1:1, as shown in Figure 7.27(a), to a poor contact arc of less than 150° when the pulley ratio is greater than 4:1. Higher pulley ratios with medium to heavy loads cause slippage, which results in overheating and a decrease in efficiency. Figure 7.27(b) shows that the arc of contact is critical on the smaller pulley. Most single-step drives are designed with pulley ratios less than or equal to 3:1 to ensure good drive characteristics. When higher pulley ratios are needed, then a two-step drive using a countershaft is used.

Once the pitch diameters of the pulleys are known and the center distance is determined, then the arc of contact can be calculated and the horsepower correction factor can be determined. Each belt has a horsepower rating at a given number of

FIGURE 7.27
Pulley ratio versus contact arc. (a)
Ideal pulley ratio of 1:1 with 180°
contact arc. (b) Poor pulley ratio of
greater than 4:1 with less than 150°
contact arc. (c) Satisfactory pulley
ratio of less than or equal to 3:1
with equal to or greater than 150°
contact arc.

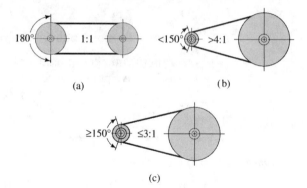

revolutions per minute for a minimum-size pulley. For example, a 4L V-belt operating at 1750 rev/min can transmit .12 hp when the pulley ratio is 1:1 and the pulley diameters are 2.00 in. However, the same belt operating with a 1:1 pulley ratio at the same 1750 rev/min can transmit 1.3 hp when the pulley diameters are 4.00 in. When the pulley ratio changes, then the power capacity of the belt is modified because the arc of contact on the smaller pulley is less than 180°. The following formula and the values in Table 7.1 are used to determine the horsepower correction factor:

$$\text{AC} + 180° - \left(\frac{D - d}{\text{CD}}\right) \times 57.3° \qquad (7.15)$$

where

AC = arc of contact, in degrees
D = the pitch diameter of the larger pulley, in inches
d = the pitch diameter of the smaller pulley, in inches
CD = the center distance, in inches

Note: The factor 57.3° converts the diameter–to–center distance ratio from radians to degrees. The ratio results from 360°/(2π) = 57.3°.

TABLE 7.1 Horsepower Correction Factor Versus Arc of Contact

Arc of Contact (Degrees)	Correction Factor	Arc of Contact (Degrees)	Correction Factor
180	1.00	130	0.86
175	0.99	125	0.84
170	0.97	120	0.83
165	0.96	115	0.81
160	0.95	110	0.79
155	0.93	105	0.76
150	0.92	100	0.74
145	0.91	95	0.72
140	0.89	90	0.70
135	0.87		

Example
7.8

A belt drive uses a driver pulley with a 3.00-in pitch diameter and a driven pulley with a 12.0-in pitch diameter. The pulleys have a 12.0-in center distance and a 4:1 pulley ratio. The belt selected for the system is rated at 1.82 hp when operated over the 3.00-in-diameter driver pulley. Determine the horsepower rating of the belt after it has been corrected for the arc of contact.

SOLUTION

Step 1: Find the arc of contact using Equation 7.15 and then select an appropriate correction factor from Table 7.1.

Given

$$AC = 180° - \left(\frac{D - d}{CD}\right) \times 57.3°$$

Evaluate

$$D = 12.0 \text{ in}$$

$$d = 3.00 \text{ in}$$

$$CD = 12.0 \text{ in}$$

Substitute

$$AC = 180° - \left(\frac{12.0 - 3.00}{12.0}\right) \times 57.3°$$

Solve

$$AC = 137°$$

Observation

Select 0.88 from Table 7.1. This is halfway between the values for 135° and 140°.

Step 2: Corrected power = rated power × correction factor

Evaluate

Rated power = 1.82 hp

Correction factor = 0.88

Substitute

Corrected power = 1.82 × 0.88

Solve

Corrected power = 1.6 hp

Observation

Belt drives work best when the pulley ratio is smaller than 3:1 and the angular velocity of the fastest pulley is greater than 1000 rev/min. When heavy V-belts turn faster than 4500 rev/min, centrifugal force tends to expand the belt and lessen the friction force, causing the belts to lose traction.

Drive System Service Factor

To ensure that the drive system has sufficient power capacity, the drive system's design power rating is increased based on the nature of the load. For example, when the load is a washer, a reciprocating pump, a food grinder, or a mixer, then the drive system design power rating is increased to 140% by multiplying by a service factor (SF) of 1.4. The drive system design power rating is increased an additional 40% (0.4) when a high-torque motor is used and/or the load is frequently started and stopped. Thus, a food grinder driven by a high-torque motor would have a total service factor of 1.8 (1.4 + 0.4). The following example uses these ideas to develop a drive system's design power capacity.

Example
7.9

A .50-hp high-torque (capacitor start) 3450-rev/min motor is used to power a recip-rocating pump operating at 980 rev/min with a 16-in center distance belt drive. The pump pulley has a pitch diameter of 10.6 in, and the motor pulley has a pitch diam-eter of 3.0 in; the 37-in belt is rated at 1.1 hp. Determine

    (a) The service factor (SF) for the system.
    (b) The drive system's design power rating.
    (c) The drive-belt loss to the arc of contact.

SOLUTION

(a) Find the total service factor (SF) of the drive system by combining the service factor of the reciprocating pump with the high-torque service factor of the mo-tor.

Given

The service factor of the reciprocating pump equals 1.4. The service factor of the high-torque motor equals 0.4.

Solve

SF = 1.4 + 0.4 = 1.8, or 180%

(b) Find the drive system's design power rating by multiplying the motor's power rating (MPR) by the service factor (SF).

Given

Drive-system power rating = MPR × SF

Evaluate

MPR = .50 hp

    SF = 1.8

Substitute

Drive-system power rating = .50 × 1.8

Solve

Drive-system power rating = .90 hp

Therefore, the motor is rated at .50 hp and the drive system is rated at .90 hp.

Observation

Due to the high starting torque of the motor and the high inertia load of the recip-rocating pump, the drive system is designed to have a greater power capacity than the motor.

(c) Step 1: Find the drive-belt loss due to the arc of contact and evaluate whether the belt rating of 1.1 hp will be sufficient for this application.

Given

$$AC = 180° - \frac{D - d}{CD} \times 57.3°$$

Evaluate

$D = 10.6$ in

$d = 3.0$ in

$CD = 16$ in

Substitute

$$AC = 180° - \left(\frac{10.6 - 3.0}{16}\right) \times 57.3°$$

Solve   AC = 153°

The correction factor from Table 7.1 for 153° is selected as 0.93.

Step 2: Corrected power = rated power × correction factor

Evaluate   Rated power = 1.1 hp

Correction factor = 0.93

Substitute   Corrected power = 1.1 × 0.93

Solve   Corrected power = 1.0 hp

Therefore, the belt loss is .1 hp (1.1 hp − 1.0 hp). Since the corrected power rating of the belt is 1.0 hp, there is still sufficient horsepower to carry the designed drive system load of .90 hp.

Two-Step Belt Drive

When higher pulley ratios (greater than 5:1) coupled with large torques are needed, then a two-step drive using countershafts, as pictured in Figure 7.28, is used. The following equation may be used to determine the angular velocity of driver or driven shafts, the diameter of the pulleys in two-step drives, or the IMA (pulley ratio) of the drive system.

$$\frac{\omega_D}{\omega_d} = \frac{d_1 d_2}{D_1 D_2} = \text{IMA} \tag{7.16}$$

where

ω_d = angular velocity of the last driven pulley, in rad/s
ω_D = angular velocity of the first driver pulley, in rad/s
d_1 = diameter of the first driven pulley, in inches
d_2 = the diameter of the second driven pulley, in inches
D_1 = the diameter of the first driver pulley, in inches
D_2 = the diameter of the second driver pulley, in inches

FIGURE 7.28
A two-step belt drive with the drive, counter, and driven shafts noted.

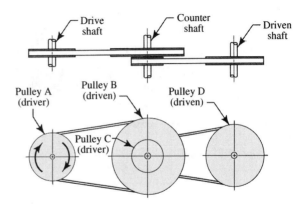

Example
7.10

The angular velocity of drive pulley A in Figure 7.28 is 1175 rev/min, and its diameter is 3.25 in, whereas the diameter of the driven pulley D is 9.5 in. If the diameter of pulley B is 10.0 in and pulley C is 2.5 in, determine:

(a) The angular velocity of the driven shaft.
(b) The pulley ratio (IMA) of the drive system.
(c) The angular velocity of the countershaft.
(d) The direction of rotation of the driven shaft.

SOLUTION

(a) Find the angular velocity of the driven shaft (ω_d) using Equation 7.16.

Given

$$\omega_d = \frac{\omega_D(D_1 D_2)}{d_1 d_2}$$

Evaluate

$\omega_D = 1175$ rev/min

$D_1 = 3.25$ in

$D_2 = 2.5$ in

$d_1 = 10.0$ in

$d_2 = 9.5$ in

Substitute

$$\omega_d = \frac{1175(3.25 \times 2.5)}{10.0 \times 9.5}$$

Solve

$\omega_d = 1\overline{0}0$ rev/min

(b) Find the pulley ratio of the drive system using Equation 7.2.

Given

$$\text{IMA} = \frac{\omega_D}{\omega_d}$$

Evaluate

$\omega_D = 1175$ rev/min

$\omega_d = 1\overline{0}0$ rev/min

Substitute

$$\text{IMA} = \frac{1175}{1\overline{0}0}$$

Solve

$\text{IMA} = 11.7 \Rightarrow 12$

Therefore, the pulley ratio is 12:1.

(c) Find the angular velocity of the countershaft by modifying Equation 7.16 and solving for the angular velocity of pulley B, a driven pulley.

Given

$$\omega_d = \frac{\omega_D D_1}{d_1}$$

Evaluate

$\omega_D = 1175$ rev/min

$D_1 = 3.25$ in

$d_1 = 10.0$ in

Substitute	$\omega_d = \dfrac{1175(3.25)}{10.0}$
Solve	$\omega_d = 382$ rev/min

Therefore, the countershaft, which is attached to pulley B, is rotating 382 rev/min.

(d) Find the direction of rotation of the driven shaft.

Observation Because the belts in the two-step drive system are open, all pulleys in the system are rotating in the same direction. Since the drive pulley is rotating clockwise, the other pulleys must also rotate clockwise.

Therefore, the driven shaft is rotating in a clockwise direction.

Exercise 7.2

Solve the following problems. Make sketches to aid in solving the problems and structure your work so it follows in an orderly progression and can easily be checked.

1. Determine the pitch diameter (PD) of a 14-tooth timing pulley designed for use with an L-type belt.
2. Determine the pulley ratio of a timing belt system when the driver pulley has 14 teeth and the driven pulley has 38 teeth.
3. Determine the approximate pitch length, in millimeters, of a flat belt used on pulleys with pitch diameters of 44.0 mm and 13.0 mm and a center distance of 105 mm.
4. Determine the power rating of a V-belt used to power a circular saw after the belt has been corrected for the arc of contact. The center distance is 30.0 in and the pulleys have pitch diameters of 3.125 in and 6.50 in. The belt is rated at 1.72 hp for an arc of contact of 180°.
5. A $\frac{1}{3}$-hp motor is used to power a warehouse package-conveyor system that is started and stopped many times an hour. Assuming an overload factor of 120% for the conveyor system, specify
   (a) The service factor (SF) for the drive system.
   (b) The drive system's design power rating.
6. Pulley A in Figure 7.28 is rotating at 3450 rev/min in a cw direction. The pitch diameters of the V-belt pulleys are pulley A, 3.60 in; pulley B, 12.0 in; pulley C, 3.00 in; and pulley D, 9.00 in. Determine
   (a) The angular velocity of pulley D.
   (b) The angular velocity of pulley C.
   (c) The pulley ratio of the two-step drive system.
   (d) The direction of rotation of the driven shaft when the belt between pulleys C and D is crossed.

7.3   CHAIN DRIVES

Introduction

Chain drives, like gearing and belt drives, transmit power from the driver element to the driven element. In chain drives, the driving sprocket is connected by way of an endless chain to the driven sprocket. Chain drives are used in three principal applications: the transmission of power, the conveyance of materials, and the synchronizing of movement. Unlike belt drives, which rely on friction, chain drives require little or no pretensioning, are more compact in size for the same design power rating, and don't creep or slip. There are several types of chain drives; however, the standard roller and the inverted-tooth (silent) types are most often used in power transmission in industrial, commercial, and agricultural applications. Information about the installation, inspection, and adjustment of roller chains may be found in Appendix C, Section C2.

Standard Roller Chain

The standard roller chain, Figure 7.29, is available in single-strand or multiple-strand format. In the multiple-strand case, two or more chains are joined by common pins to ensure alignment during operation. As noted in Figure 7.29, the stan-

FIGURE 7.29
Standard roller chain assembly. (a) Roller link with roller, bushing, and link plates noted. (b) Pin link with pin and link plates noted. (c) Assembled links forming the roller chain.

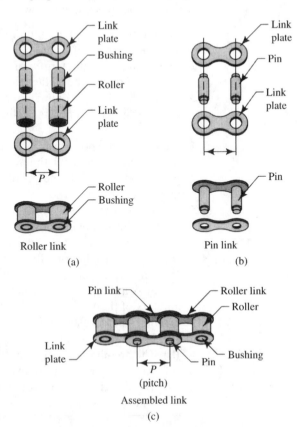

TABLE 7.2 Standard Roller Chain Sizes

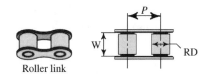

Roller link

Standard Chain Number	Pitch (P) (in)	Width (W) (in)	Roller Diameter (RD) (in)
25	.250	.125	.130
35	.375	.188	.200
40	.500	.312	.312
41	.500	.250	.306
50	.625	.375	.400
60	.750	.500	.469
80	1.000	.625	.625
100	1.250	.750	.750
120	1.500	1.000	.875

dard roller chain is made up of alternating roller links connected by pin links. The rollers, which are evenly spaced, rotate as they pass over the teeth in the sprocket, providing for low-friction, high-efficiency (98%) operation. Roller chains usually have a removable link called the *connecting link* to enable the installation and removal of the chain from the sprocket. Table 7.2 is a partial list of standard roller chain sizes from $\frac{1}{4}$-in pitch to $1\frac{1}{2}$-in pitch. The numbering system is coded as follows:

Right-hand single digit The right-hand single digit is used as follows: 0 for roller general duty; 1 for lightweight chain; 5 for chain without rollers.

Left-hand digits The left-hand digits represent the pitch in $\frac{1}{8}$-in increments measured in a straight line from the center of one roller to the center of the next roller.

Suffix letter H The suffix letter H indicates heavy series of chain.

Suffix dash number The suffix dash number is −2 for double strand, −3 for triple strand, etc.

For example, No. 25 is read as single-strand, rollerless (the right-hand 5), $\frac{1}{4}$- (.250-) pitch (the left-hand 2 represents $\frac{2}{8}$-, or $\frac{1}{4}$-in) roller chain.

Chain Lubrication

Lubrication must be provided to the chain to ensure low friction and reliable operation of the large number of precision journal bearings formed when a roller is placed over a bushing in the assembly of the roller link. The type of lubrication system used depends on the speed and torque transmitted to the load. For light loads or slow speeds, periodic manual oiling, producing boundary lubrication, is used. With moderate loads or speeds, drip lubrication from a drip lubricator, producing mixed film lubrication, is satisfactory. As speeds or loads increase, oil bath or oil slinger lubrication, producing fluid film lubrication, is applied. Oil stream lubrication, which produces fluid film lubrication, is utilized when the chain is operated at high speed or under heavy load.

Chordal Effect

Roller chain drives have one inherent disadvantage that gearing or belt drives do not have. That disadvantage is the "ripple" present in the speed of the chain due to the **chordal effect.** This effect results from the pitch radius of the chain changing size as each link engages a tooth on the sprocket and is advanced. In Figure 7.30 the nature of this phenomenon is seen in the pitch radii r_1 and r_2. When the chain is in the position shown in the figure, the pitch radius is equal to r_1. However, when the roller advances half the distance to the next groove, the pitch radius is equal to r_2, which is slightly smaller than r_1. The result of this variation in the pitch radius is an acceleration that produces a variation (ripple) in the speed of the chain. A serious consequence of the chordal effect is vibration in the chain at the point where the roller meets the sprocket. This effect is very noticeable in systems where the smallest sprocket has a low tooth number. To minimize the unevenness in the speed of the chain, the smallest sprocket in a system is designed with a minimum of 12 teeth for low speeds and 21 teeth for high speeds. Chain drives using sprockets with high tooth count have virtually no chordal effect and are smooth-running.

Sprockets

Sprockets are traditionally made from steel. However, filled nylon sprockets are finding some use today because of their long-wear characteristics and quiet operation. By using steel in combination with nylon, the life for both sprocket and chain is increased. This is due to the nylon's cushioning property, which lessens both the impact when the roller engages the sprocket tooth and the vibration due to the chordal effect.

Because of the gearlike nature of the sprocket, the sprocket ratio (IMA) of a chain drive is calculated with the same equations that are used with gearing.

$$\text{IMA} = \frac{\omega_D}{\omega_d} \qquad \text{(repeated)} \qquad \textbf{(7.2)}$$

FIGURE 7.30
The chordal effect results in a slight ripple in the output sprocket in the roller chain drive.

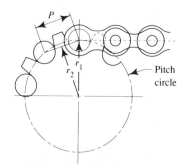

$$\text{IMA} = \frac{N_d}{N_D} \quad \text{(repeated)} \tag{7.3}$$

Also, the number of teeth (N) on the sprocket is approximated by using the same equation used with the timing belt, except that the circular pitch (CP) is replaced with the pitch of the chain:

$$N = \frac{\pi(\text{PD})}{P} \tag{7.17}$$

where

N = number of teeth on the sprocket, expressed as an integer number
PD = pitch diameter of the sprocket, in inches
P = pitch of the chain, in inches

Epilogue

Because drive systems are formed by combining two or more machine elements, they are treated like compound machines (Section 4.4 in Chapter 4). Gear trains, multiple-step belt drives, and multiple-step chain drives are just compound machines. This means that the gear, pulley, or sprocket ratio of an entire drive system is simply the product of the individual gear, pulley, or sprocket ratios.

$$\text{IMA}_{\text{total}} = \text{IMA}_1 \times \text{IMA}_2 \times \cdots \times \text{IMA}_n \tag{7.18}$$

where

$\text{IMA}_{\text{total}}$ = overall gear, pulley, or sprocket ratio
IMA_1 = gear, pulley, or sprocket ratio of the first set
IMA_2 = gear, pulley, or sprocket ratio of the second set
IMA_n = gear, pulley, or sprocket ratio of the nth set

Exercise 7.3

Solve the following problems. Make sketches to aid in solving the problems and structure your work so it follows in an orderly progression and can easily be checked.

1. By reading the code, state the characteristics of a roller chain with a 140H-2 chain number.
2. Number 25 single-strand roller chain is used a great deal in applications where the electric motor is fractional horsepower. What single characteristic sets this chain apart from most other roller chains?
3. Determine the sprocket ratio for a roller chain drive if the driver sprocket has 52 teeth and the driven sprocket has 13 teeth.
4. Using the parameters of Problem 3, determine the angular velocity in revolutions per minute of the driven sprocket if the motor shaft connected to the driver sprocket is rotating at 775 rev/min.
5. Determine the pitch diameter (PD) of a sprocket used with a number 25 roller chain if the sprocket has 27 teeth.

6. Determine the sprocket ratio for a number 41 roller chain drive if the driven sprocket has 32 teeth and the pitch diameter of the driver sprocket is 1.91 in.

CHAPTER SUMMARY

- Parallel shafts are connected by spur, helical, or herringbone gears.
- Worm gears are used for speed reduction and they are self-locking.
- The IMA of meshed gears may be determined from the velocity ratio, the tooth ratio, or the ratio of the pitch diameters.
- The torque ratio is used to determine the AMA.
- An idler gear is used in between two gears to ensure that the first and last gear rotate in the same direction.
- In an open-belt system, the pulleys and belt rotate in the same direction.
- Belts are tensioned to develop a frictional force in order to transmit rotary power from the drive to the driven pulley.
- Timing belts are used to maintain position while transmitting rotary power.
- The drive system must have a greater power capacity than the motor, and the motor must have a greater power capacity than the load.
- An ideal arc of contact of 180° results when the pulley ratio is 1:1.
- Chain drives are smaller in size than belt drives for a given design power rating.
- Chain drives do not have a uniform speed ratio between drive sprocket and driven sprocket due to the chordal effect.

SELECTED TECHNICAL TERMS

The following technical terms, abbreviations, and acronyms are defined in the glossary located after Chapter 16. You are encouraged to use the glossary to aid your understanding and to test your knowledge of these important terms.

axial	pulley
chordal effect	radial
gear	RMA
gear train	SF
idler gear	sprocket

END-OF-CHAPTER QUESTIONS

Write T if the statement is true and F if the statement is false.

1. Chain drives are used to synchronize movement.
2. Only gears with the same diametral pitch can be meshed.
3. The shape of the teeth on a gear is an involute curve.
4. Because of the chordal effect, the speed of a roller chain is very uneven when sprockets with high numbers of teeth are used.
5. Timing belts require considerable tensioning in order to prevent slippage.

6. A 5 as the right-hand digit in a roller chain size number indicates that the pitch is $\frac{1}{2}$ in.

7. Circular pitch is the same parameter as diametral pitch when making calculations with belt-drive systems.

8. Two gears on the same shaft rotate in opposite directions at the same angular velocity.

9. Occasional lubrication of a roller chain is satisfactory when the load is light and the chain speed is fast.

10. A gear train increases either the output torque or the output speed but not both at the same time.

In the following, select the word or phrase that makes the statement true.

11. The center distance between gear shafts is determined by adding the (pitch diameter, circular pitch, diametral pitch) and dividing by 2.

12. Besides ensuring positive drive with no slip, gears also change both direction and (power, angular velocity, sprocket ratio) of the output.

13. In a (closed-belt drive, open-belt drive, crossed-belt drive), the driven and the driver pulley are rotating in opposite directions.

14. The IMA of a worm gear is determined by dividing the (teeth on the worm by the leads on the wheel, pitch diameter of the worm by the pitch diameter of the wheel, teeth on the wheel by the starts in the worm).

15. To compensate for the chordal effect at moderate speeds, sprockets are used with (as few teeth as possible, a minimum of 10 teeth, 16 or more teeth).

Answer each of the following questions with a short answer in the form of a complete sentence. Include a restatement of the question in your answer.

16. What three parameters may be changed between the input and the output in a gear train?

17. What is the purpose of an idler gear?

18. What features in V-belt drives make them the most common type of belt-drive system?

19. What load and speed conditions would require fluid film lubrication of a roller chain drive?

20. Using Table 7.2, what is the difference between a number 40 chain and a number 41 chain?

END-OF-CHAPTER PROBLEMS

Solve the following problems. Make sketches to aid in solving the problems and structure your work so it follows in an orderly progression and can easily be checked. Table 7.3 summarizes the formulas used in Chapter 7.

1. Determine the AMA of a worm gear with a 64-tooth wheel and a two-lead worm when the percent efficiency is 44%.

2. Determine the pitch diameter of a sprocket with 35 teeth used with a number 41 roller chain.

3. Complete the table in Figure 7.20 when the diametral pitch (DP) of gear A is changed from 48 to 16.

4. Determine the overall gear ratio of the gear train shown in Figure 7.18.

TABLE 7.3 Summary of Formulas Used in Chapter 7

Equation Number	Equation
7.1	$\dfrac{\omega_D}{\omega_d} = \dfrac{r_d}{r_D}$
7.2	$\text{IMA} = \dfrac{\omega_D}{\omega_d}$
7.3	$\text{IMA} = \dfrac{N_d}{N_D} = \dfrac{\text{PD}_d}{\text{PD}_D}$
7.4	$\dfrac{r_d}{r_D} = \dfrac{\tau_d}{\tau_D}$
7.5	$\text{AMA} = \dfrac{\tau_d}{\tau_D}$
7.6	$\dfrac{\omega_D}{\omega_d} = \dfrac{\tau_d}{\tau_D}$
7.7	$\eta = \dfrac{\text{AMA}}{\text{IMA}}$
7.8	$\text{AMA} = \text{IMA} \times \eta$
7.9	$\omega_d = \dfrac{\omega_D N_1 N_2 \cdots N_n}{n_1 n_2 \cdots n_n}$
7.10	$\text{CD} = \dfrac{\text{PD}_1 + \text{PD}_2}{2}$
7.11	$N = \text{PD} \times \text{DP}$
7.12	$\text{IMA}_{\text{wg}} = \dfrac{N_{\text{wheel}}}{N_{\text{worm}}}$
7.13	$N = \dfrac{\pi(\text{PD})}{\text{CP}}$
7.14	$l = 2\text{CD} + \dfrac{\pi}{2}(D + d)$
7.15	$\text{AC} = 180° - \left(\dfrac{D - d}{\text{CD}}\right) \times 57.3°$
7.16	$\dfrac{\omega_D}{\omega_d} = \dfrac{d_1 d_2}{D_1 D_2} = \text{IMA}$
7.17	$N = \dfrac{\pi(\text{PD})}{P}$
7.18	$\text{IMA}_{\text{total}} = \text{IMA}_1 \times \text{IMA}_2 \\ \times \cdots \times \text{IMA}_n$

5. A 5.0-hp three-phase high-torque induction motor is used to power a conveyor on an engine transfer line that is subject to frequent starts and stops. If the drive system design power rating is increased 130% for the conveyor system, determine
   (a) The service factor for the entire system.
   (b) The drive system's design power rating.

6. Determine the pitch length of a belt needed to connect two pulleys with a center distance of 15.0 in. The driven pulley has a pitch diameter of 3.25 in, whereas the driver pulley has a pitch diameter of 9.75 in.

7. Determine the angular velocity (revolutions per minute) of the driven gear of two meshed spur gears if the shaft attached to the 18-tooth driver gear is rotating 875 rev/min. The driven gear has a diametral pitch of 6 and a pitch diameter of 10.0 in.

8. Determine the corrected power rating of a drive belt due to the arc of contact for a belt rated at 1.6 hp. The system's drive pulley is rotating at 1250 rev/min and the driven pulley is rotating at 250 rev/min. The pulleys have a center distance of 16.0 in and the drive pulley has a pitch diameter of 2.50 in.

9. Two helical gears are meshed. The driver gear, the smaller of the two, has a pitch diameter of 1.125 in and the driven gear has a pitch diameter of 2.700 in. If the gears are 92% efficient and the power at the driver gear is 37.0 ft-lb/s, determine the torque at the driven gear in pound-feet when the driver gear is rotating 438 rev/min.

10. A floor-mounted roller conveyor system, driven by a continuous chain, is used on the outfeed of a cardboard box fabricating machine to transport stacks of completed boxes to the banding station. At any given time, ten rollers are beneath a stack of boxes. If each of the ten rollers turns 96.0 rev/min and each requires 15.0 lb-in of torque when moving a stack of boxes, determine

    (a) The power (hp) out of the shaft of the speed reducer attached to the drive sprocket when the efficiency of the roller conveyor system is 0.48.

    (b) The linear velocity (ft/s) of a stack of boxes as it moves over the 4.50-in diameter rollers.

Chapter 8

Rotation, Linear, and Intermittent-Motion Devices

INTRODUCTION

Rotational, linear, and intermittent motion and its associated power is transmitted from the prime mover to machine mechanisms using machine elements that include couplings, clutches, drive systems (gears, belts, and chains), cams, linkages, ratchets, block and screw, rack and pinion, etc. Each of these elements receives rotational power, changes the torque-speed characteristics of the power, and produces the desired motion (rotational, linear, reciprocating, or intermittent), enabling the machine to perform its designed task.

CHAPTER CONTENTS

PERFORMANCE OBJECTIVES

Once you have read and studied each section; worked through each example with pencil, paper, and calculator; worked through the end-of-chapter problems; and answered the end-of-chapter questions, then you should be able to

- Identify axial power-transmission elements.
- Differentiate between a positive clutch and a friction clutch.
- Understand the operation of a rack and pinion, block and screw, cam and follower, and a Scotch yoke.
- Describe the operation of the two classes of four-bar mechanisms.
- List the characteristics of the ratchet mechanism.
- Explain the terms *constant acceleration* and *constant velocity* as they relate to the motion of mechanisms.

8.1 AXIAL POWER-TRANSMISSION ELEMENTS

Introduction

The elements that provide power transmission from one shaft to another connected end to end (connection of *axial shafts*) include couplings (both rigid and flexible), universal joints, and clutches. The choice of coupling depends upon the relative alignment of the two shafts to be connected. The need for *axial movement* (end play), *torsional flexibility* (twist) in the transmission of torque, and/or the need for disconnecting and reconnecting the shafts without removing or reinstalling any component is also an important consideration when selecting the power-transmission elements.

When two shafts are being considered for axial connection to one another, several alignment conditions must be considered. These conditions, pictured in Figure 8.1, include *lateral misalignment* (center line offset), *angular misalignment*, an allowance for *axial movement*, and an allowance for *torsional movement*.

Couplings

Couplings are generally divided into two groups, rigid and flexible. *Rigid couplings* can be used only to connect aligned shafts (shafts that are in line) as any lateral

FIGURE 8.1

Shaft-alignment conditions: (a) Lateral misalignment; (b) Angular misalignment; (c) Allowance for axial movement; (d) Allowance for torsional movement.

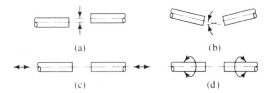

(a) (b)

(c) (d)

(right-angle) force or axial force due to misalignment will lead to failure of the bearings or bending of the shafts. Lateral forces create a bending moment on the drive shaft, causing a radial load to be produced on the support bearings. Axial forces, due to the lack of provision for end play, cause thrust loading on the support bearings.

The application of rigid couplings (pictured in Figure 8.2) is usually limited to connecting solid-line shafts together, where accurate linear alignment is guaranteed or where allowance for separating two shafts is needed. Rigid couplings are seldom used to connect prime movers to their loads, since they do not have an ability to compensate for shaft misalignment.

Since it is virtually impossible to maintain perfect alignment between the end of a motor shaft and the end of a speed reducer shaft, a pump shaft, etc., *flexible couplings* are used. The principal task of flexible couplings is to control the forces acting on the shafts and bearings because of misalignment. The use of flexible couplings allows for the following:

■ Small, unavoidable misalignments, both angular and lateral
■ Torsional movement (twist) and the absorption of vibration in the transmission of torque
■ Axial movement in the shafts from *thermal growth*
■ Electrical isolation between the motor and mechanical load when nonmetallic materials are used for the intermediate connection elements

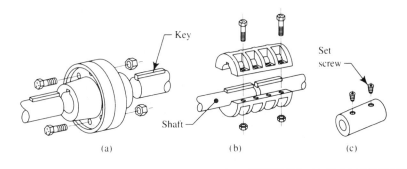

(a) (b) (c)

FIGURE 8.2

Rigid couplings: (a) Safety flange couplings, which transmit up to 150 hp, are designed so the bolt heads and nuts used for fastening the flanges together are covered by the undercut in the flange. (b) Split, or muff, coupling is used when ease of separating the two shafts is needed. The coupling is disassembled by simply removing the connecting bolts. (c) Small one-piece sleeve couplings are used to transmit torque to light loads. Setscrews are used for connecting shafts up to $\frac{1}{2}$ in. in diameter.

A very large number of flexible couplings are available to accommodate various types of load conditions and shaft misalignment. A few of these couplings are presented here to give you an understanding of the application of flexible couplings. At the onset, you should understand that flexible couplings do not give complete freedom of movement in all axes of motion; instead, a rather limited freedom of movement is provided, depending upon the parameter or parameters that must be satisfied to ensure minimal increase in bearing loading when machines are coupled together. The alignment, inspection, and maintenance of flexible couplings is included in Section C3 of Appendix C.

Of the couplings used with moderate to heavy amounts of torque, the *Oldham*, or *floating center, coupling* of Figure 8.3 is representative of the design of this type of flexible coupling. Couplings of this type have three members: two outer-flange members with protrusions that engage and trap the third member between them. The Oldham coupling (named for its inventor) is used to accommodate

- Axial movement due to temperature variation;
- Lateral misalignment (up to 10% of the shaft diameter);
- Small, angular misalignment (up to 3°) between the shafts of medium to large motors (3 to 150 hp) and their loads.

Because of the rigid nature of the center element in the Oldham coupling, there is little allowance for torsional movement, which results in high-torque capacity, zero **backlash** (side play), and no variation in angular velocity. When the center disk is made of metal, the sliding surfaces between the two flanged halves and the lugs must be lubricated.

Several types of light-duty flexible couplings are available for use with automated manufacturing and processing equipment to interconnect the shafts of stepper motors, ball screws, robotic drives, servo systems, encoders, etc. Two of these flexible couplers, the *helical coupling* and the *bellows coupling* (Figure 8.4), are representative of this family of couplings, which have been designed to handle small amounts of torque (up to 300 oz-in) and transmit small amounts of power at constant angular velocities with zero backlash. These two representative couplings have the following general characteristics:

- Are compliant enough to manage lateral shaft misalignment of 3 to 30 mils maximum (.003 to .030 in)

FIGURE 8.3
(a) The Oldham flexible coupling is used for moderate- to high-torque applications where the two shafts are laterally misaligned. (b) The center disk may be made of steel, brass, nylon, polyurethane, or rubber, and it can be spring loaded to dampen axial movement.

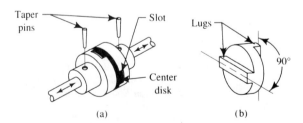

(a)   (b)

FIGURE 8.4
Light-duty flexible couplings. (a)
The helical coupling is made from
one piece of metal by cutting a
helical groove around its outside
diameter and through its central
part, thus creating a compliant and
flexible coupling. (b) The all-metal
bellows coupling in this picture is
constructed by attaching two clamp-
style hubs to the central flexible
bellows unit.

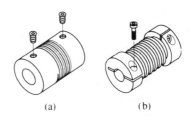

(a)   (b)

- Are compliant enough to manage angular shaft misalignment of 1.5° to 7° maximum
- Allow *axial deformation* (end float) from nil to 60 mils maximum for drive-shaft movement due to thermal growth
- Allow torsional movement (twist) from nil to 3° for cushioning shock in the drive shaft and bearings
- Have no need for lubrication
- Are small, lightweight, low inertia, high speed

The last types of flexible couplings to be considered are highly flexible and yield when pulsations of power are present in the drive shaft. These couplings are made in three parts, two end fittings and an intermediate element between the two flanges. The intermediate element is made from a flexible material such as rubber, neoprene, polyurethane, or steel springs. Figure 8.5 pictures several flexible couplings (gear grip, "k" type, and disk type) that yield to variations in torque and speed.

FIGURE 8.5
Flexible couplings designed to yield
to changes in torque and speed
include the gear grip, "k" type, and
disk type. (a) Gear grip flexible
coupling uses a setscrew key system
for attaching the hubs to the shafts and
a neoprene intermediate element for
absorbing shock. (b) The "k"-type
flexible coupling uses an elastomeric
(polyurethane) intermediate element.
(c) Disk-type flexible coupling uses a
rubber composite intermediate
element.

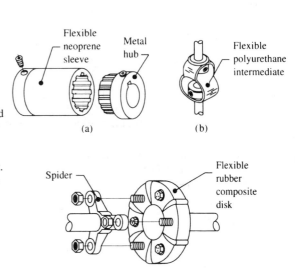

Because these couplings are made with materials that have considerable torsional resilience (shock resistance), shafts that are seriously out of alignment or operated under heavy pulsating and shock loads can be accommodated. By using couplings of this type, even drive shafts subject to severe conditions can be operated
with minimum vibration, bearing loads, and wear on the bearings and couplings.

The coupling, because of the property of its materials and the design, absorbs
impact energy caused by the irregular loading (sudden change of torque or speed)
of the drive shaft. The intermediate element momentarily stores the energy in its
material for return at a later time and/or converts the energy to heat by the internal
friction in the material. This feature holds for each of the flexible couplings illustrated in Figure 8.5, which both store and dissipate energy when they yield to variations in power in the drive shaft. The representative couplings of Figure 8.5 have the
following general characteristics:

- Can manage lateral shaft misalignment up to 190 mils ($\frac{3}{16}$ in)
- Can manage angular shaft misalignment up to 15°
- Allow shaft movement (end float) up to 125 mils ($\frac{1}{8}$ in)
- Allow torsional movement (twist) greater than 10° for cushioning shock in
  the drive shaft and bearings
- Operate at temperatures up to 140°F
- Have no need for lubrication

*Example
8.1*

A gear-grip flexible coupling is rated at a maximum torque of 108 lb-in. Determine
if the coupling can be used to link a 2.0-hp electric motor operating at 1750 rev/min.

Observation

Equation 6.18 is repeated here to assist in the solution of this example.

$$P = \frac{\tau \omega}{5250} \qquad \text{(repeated)} \qquad \textbf{(6.18)}$$

where

P = power, in hp
τ = torque, in lb-ft
ω = angular velocity, in rev/min

SOLUTION

Find the torque of the 2.0-hp motor at an angular velocity of 1750 rev/min and compare it with the 108 lb-in torque of the coupling. Use Equation 6.18 to determine the
torque.

Given

$$P = \frac{\tau \omega}{5250} \quad \text{and} \quad \tau = \frac{5250P}{\omega}$$

Evaluate

$P = 2.0$ hp

$\omega = 1750$ rev/min

Substitute	$\tau = \dfrac{5250 \times 2.0}{1750}$
Solve	$\tau = 6.0$ lb-ft
Convert	$6.0 \text{ lb-ft} \times \dfrac{12 \text{ in}}{1 \text{ ft}} = 72 \text{ lb-in}$

Therefore, the 72 lb-in of the 2.0-hp motor can be accommodated by the 108 lb-in rating of the gear-grip flexible coupling.

Universal Joints

Universal joints are used to connect two shafts that intersect at an angle to each other and/or whose operating angle can vary while the shafts are rotating. Figure 8.6 pictures a *Hooke's universal joint*, the simplest and most common type of universal joint.

As illustrated, the universal joint consists of two yokes, one attached to the input shaft and the other to the output shaft. The forks (openings) in the two yokes are joined by the spider, the cross-shaped central member. The universal joint, which is found in all types of equipment, including aircraft, medical, and industrial, is limited to a maximum operating angle (β) of 20° (Figure 8.6) at a typical operating speed of less than 650 rev/min. In general, the speed of rotation may increase as the operating angle becomes closer to zero. At operating angles greater than 30°, the speeds are very slow, requiring manual operation.

When two shafts are connected by a single universal joint at an operating angle (β), there is a cyclic fluctuation in the angular velocity of the output (driven) shaft, even though the input (driving) shaft is rotating at a constant angular velocity. The output shaft may appear to *lope* due to the irregularity in the speed (revolutions per minute) of the output shaft.

Table 8.1 lists several values of operating angles and the resulting variation in revolutions per minute of the output shaft for a constant input angular velocity of 1000 rev/min. As noted in the table, the variations in speed increase with larger operating angles. Figure 8.7 pictures the *cyclic fluctuations* in the output angular velocity for several values of operating angles. The figure shows that the output angular velocity is never constant but varies in a **sinusoidal** pattern, alternately accelerating

FIGURE 8.6
(a) Hooke's type of universal joint with the yokes shown attached to the central spider. (b) A representation of the spider from the center of the universal joint.

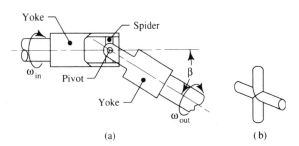

(a)

(b)

TABLE 8.1 Variation in Output Angular Velocity of a Single Universal Joint as a Function of Operating Angle (β) for a Constant Input Angular Velocity of 1000 Revolutions per Minute

Operating Angle ($\beta°$)	Minimum Revolutions/Minute	Maximum Revolutions/Minute	Variation (Rev/Min)
0	1000	1000	0
5	1004	996	8
10	1015	985	30
15	1035	966	69
20	1064	940	124
25	1103	906	197
30	1155	866	289

and then decelerating between the maximum and minimum values of the output angular velocity twice during each 360° input cycle. If uniformity of motion is essential in the output shaft, then connecting two shafts with a single universal joint should be avoided.

Since cyclic fluctuations in the velocity of the drive shaft may be unsuitable for the operation of some mechanisms, two universal joints and an intermediate shaft, as shown in Figure 8.8, may be used to bring the output shaft into phase with the input shaft and maintain a constant output angular velocity. In order to achieve this state of operation, two conditions must be met.

1. The operating angles, α and β, must make the same angle with the intermediate shaft, as illustrated in Figure 8.8.
2. The orientation of the two yokes attached to the intermediate shaft must be in the same direction. The universal joints pictured in Figure 8.9(a) are correctly installed on the intermediate shaft, but the universal joints shown in Figure 8.9(b) are incorrectly installed on the intermediate shaft.

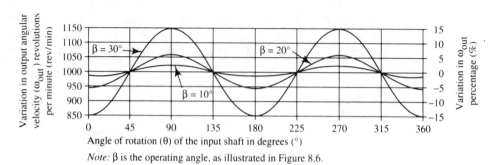

Note: β is the operating angle, as illustrated in Figure 8.6.

FIGURE 8.7

The speed (revolutions per minute) of the output shaft is continually changing in a periodic manner. The input shaft is rotating at a constant (uniform) speed of 1000 rev/min, whereas the output shaft fluctuates (accelerating and decelerating), as noted in the sinusoidal pattern in the graph.

FIGURE 8.8
Fluctuations in the output angular velocity (ω_{out}) are canceled out when the two universal couplings are correctly mounted on the intermediate shaft with the yokes oriented in the same direction and the operating angles are equal ($\angle\alpha = \angle\beta$).

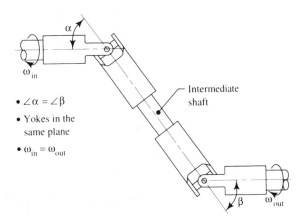

- $\angle\alpha = \angle\beta$
- Yokes in the same plane
- $\omega_{in} = \omega_{out}$

The basic design of the Hooke's universal joint was modified into the *ring-and-trunnion universal joint* for use in high-torque automotive drive trains for rear-wheel drive vehicles. Like the Hooke's universal joint, the ring-and-trunnion universal joint is used in pairs along with an intermediate shaft, called a propeller shaft, to cancel out variations in the output angular velocity.

When space won't permit the use of two universal joints and an intermediate shaft, then a single **constant-velocity joint** is used. The constant-velocity joint permits the smooth transmission of torque through large angles that are constantly changing, as in the drive train of a front-wheel-drive vehicle. The joint, which uses a ball-and-socket design, is driven by several ball bearings that circulate in specially engineered channels within the joint between the ball and socket. The constant-velocity joint operates smoothly, thus permitting the driven shaft (output) to rotate at a constant angular velocity.

Clutches

A **clutch** is used in the drive train to start and stop a machine by connecting and disconnecting the drive shaft (rotating shaft) from the driven shaft (either stationary or rotating). The two principle types of clutches are the positive clutch and the friction clutch.

FIGURE 8.9
(a) Universal joints correctly installed with the yokes on the intermediate shaft oriented in the same direction; $\angle\alpha = \angle\beta$. (b) Universal joints incorrectly installed as the yokes on the intermediate shaft are oriented 90° out of phase; $\angle\alpha \neq \angle\beta$.

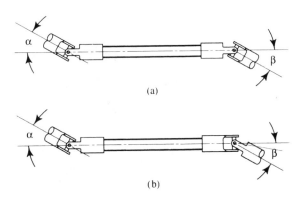

(a)

(b)

Positive Clutch The positive clutch connects the driving and driven shafts by interlocking the teeth on the *clutch plates* attached to the drive and driven shafts. These teeth are pressed together with force applied along the axis of the connecting shafts by a lever-engagement mechanism or a spring mechanism.

Figure 8.10(a) pictures a spiral-jaw positive clutch that has been engaged, through the aid of an operating lever, to provide drive to the output shaft. Because of the design, the spiral-jaw clutch gives positive drive in only one direction. Like the rear wheel on a bicycle equipped with a coaster brake, the crank drives the rear wheel in only one direction and sets the brake in the other. Positive clutches are classified by the form (shape) of the teeth, as illustrated in Figure 8.10(b).

Because a positive clutch may cause impact loading if engaged while the drive shaft is rotating, it is used only when sudden starting is tolerable, when the inertia of the driven load is small, or if the drive shaft can be brought to a stop before engaging the clutch.

Friction Clutch The friction clutch is designed to couple the driven shaft (load) to the rotating drive shaft smoothly with little or no shock, thereby overcoming the shortcoming of the positive clutch. The friction clutch is used in applications where the prime mover (internal combustion engine) must continue to rotate when the load is connected or disconnected. These desired functions are performed through various clutch systems, each of which uses the friction between engaged surfaces in their design.

Magnetic Clutch The magnetic clutch is a friction-disc clutch commonly used in industrial applications in which the engagement of the friction surfaces is controlled by energizing a magnetic coil. When the magnetic clutch is *shaft-mounted* to the motor, the drive-shaft torque may be transmitted either to an axially driven shaft (a split-shaft application), as illustrated in Figure 8.11, or to a parallel-driven shaft through a belt drive, chain drive, or gearing, as illustrated in Figure 8.12. When used to drive a parallel shaft, the clutch module may be either a *single-ended unit*, as pictured in Figure 8.12(a), or a *double-ended unit*, as pictured in Figure 8.12(b).

Figure 8.13 pictures a cutaway of a *shaft-mounted magnetic clutch*. The motor shaft enters the double-ended clutch unit on the right and is attached to the clutch rotor by setscrews and a square key. A sprocket or pulley is attached to the *hub* with bolts that thread into the pretapped holes. When in operation, the armature is en-

FIGURE 8.10

(a) A spiral-jaw positive clutch. (b) Several styles (types) of clutch teeth: (1) Straight-tooth; (2) Modified straight-tooth; (3) Sawtooth.

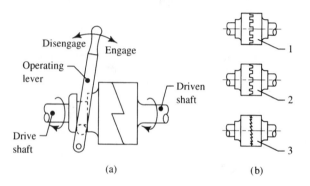

FIGURE 8.11

A shaft-mounted magnetic clutch used to couple the drive shaft from the motor to an axially driven shaft connected to the load in an in-line split-shaft application.

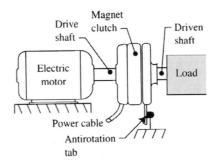

gaged (pulled against the rotor) by the force of magnetic attraction created when the dc electric current passes through the magnetic coil located next to the rotor. The air gap (.002 to .015 in) between the armature and the rotor serves as mechanical clearance when the clutch is disengaged.

Moment of Inertia

The torque that a machine element must handle (torque rating) may be approximated using the following equation:

$$\tau = \frac{5250P \times \text{SF}}{\omega} \tag{8.1}$$

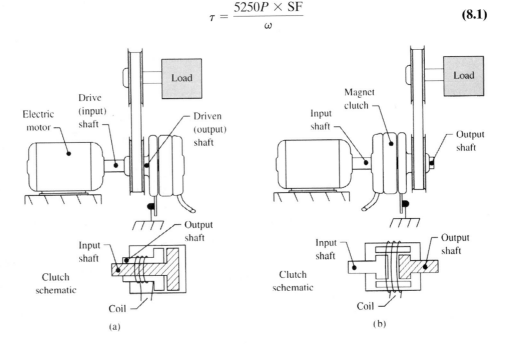

FIGURE 8.12

A shaft-mounted magnetic clutch used to connect the drive shaft of the motor to the pulley driving the parallel shaft of the load. (a) Single-ended clutch type, where the input and output shafts are on the same side of the clutch unit. (b) Double-ended clutch type, where the input and output shafts are on opposite sides of the clutch unit.

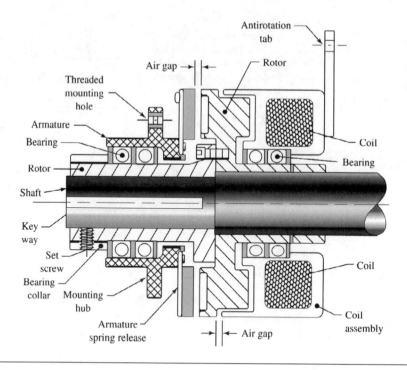

FIGURE 8.13
Sectional view of a shaft-mounted magnetic clutch with the armature, rotor, and coil assemblies shown.

where

τ = the transmitted torque, in lb-ft
P = the power into the clutch, in hp
SF = the service factor, which accounts for the moment of inertia of the load:
1.5 for low-inertia loads
3.0 for high-inertia loads
ω = the angular velocity of the load, in rev/min

Before using Equation 8.1 in an example, we will explore the concept of the moment of inertia. The **moment of inertia** (I) has the same relationship to angular acceleration in rotational motion as mass has to linear acceleration in linear motion. That is, $F = ma$ in a linear-motion system and $\tau = I\alpha$ in a rotating system.

In order to determine how fast a rotating load will be brought up to speed or accelerated (α) by an applied torque (τ), the moment of inertia (I) of each rotating part of the system must be known. Standard formulas for each element are developed using the concept that the moment of inertia is found by taking the product of mass and the square of the distance from the axis of rotation:

$$I = mr^2 \tag{8.2}$$

where

I = moment of inertia of mass, in slug-ft^2 (also written as ft-lb-s^2) or kg·m^2
m = mass, in slug or kg
r = radius from the axis of rotation, in ft or m

Since each formula is dependent on the shape of the object and the location of the reference axis, the moments of inertia are calculated separately and then combined. The formulas for moments of inertia of most common geometric shapes are listed in engineering handbooks; Figure 8.14 pictures several of these.

In summary, the moment of inertia appears in formulas that involve rotational motion when torque and/or acceleration are parameters. For rotary motion, the moment of inertia is proportional to the mass of the object being rotated times the square of its distance from the axis of rotation.

*Example
8.2*

Determine the torque rating for a single-ended magnetic clutch that is used to couple a $\frac{3}{8}$-in-diameter drive shaft supplying .375 hp at 1750 rev/min to a pulley in a belt drive that powers a low-inertia material-handling system.

SOLUTION   Find the torque using Equation 8.1.

Given
$$\tau = \frac{5250P \times \text{SF}}{\omega}$$

Evaluate
$P = .375$ hp

SF = 1.5 (low-inertia)

$\omega = 1750$ rev/min

Substitute
$$\tau = \frac{5250 \times .375 \times 1.5}{1750}$$

Solve
$\tau = 1.69$ lb-ft

Observation   Small amounts of torque are often expressed in pound-inches rather than pound-feet. In this example, 1.69 lb-ft is approximately 20 lb-in.

FIGURE 8.14
Moments of inertia (I) for selected geometric shapes: (a) A disk or cylinder; (b) A hollow cylinder; (c) A thin ring.

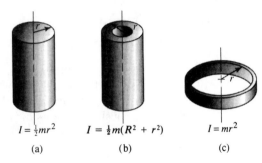

$I = \frac{1}{2}mr^2$ $I = \frac{1}{2}m(R^2 + r^2)$ $I = mr^2$

(a)   (b)   (c)

Magnetic Clutch Response Time

Magnetic clutches are applied in automated assembly lines where mechanisms are turned on and off by switching the magnetic clutch on and off while the electric motor continues to run at a constant angular velocity.

When fast response time for on-off operation of the magnetic clutch is needed because of the cycle time of the product being produced on the assembly line, several parameters must be considered to minimize the delay time. The following are included in these:

- The number of cycles on or off per minute
- The heat dissipation and the permitted temperature rise of the clutch at the desired cycle rate
- The *weight moment of inertia*, the amount of weight in the load, and the distance the center of weight is located from the axis of rotation, expressed in units of lb-ft^2
- The amount of torque needed to move the load
- The type of arc suppression used in conjunction with the electric coil of the clutch
- That the clutch electromagnet be switched from the dc side of the power supply
- That the armature plate and rotor surface of a new clutch be properly burnished (worn in) before being put into service

In general, fast cycle rates result in greater wear and reduced operating life of the magnetic clutch. This reduction in operating life is a direct result of the increase in temperature due to the heat produced by the *engagement slippage* between the armature plate and the rotor surface each time the unit is cycled on and off.

Increased temperature causes the resistance of the wire in the coil of the electromagnet to increase, which lessens the magnetic attraction force and increases the engagement slippage. Elevated temperature also causes bearings to release oil at a greater rate due to the reduced viscosity of oil (grease) at elevated temperature, which, in turn, may lead to early bearing failure. Section C6 of Appendix C has information for the general maintenance of magnetic clutches.

8.2 LINEAR, RECIPROCATING, AND OSCILLATORY MOTION MECHANISMS

Rack and Pinion

Rack-and-pinion gearing consists of a pinion gear that operates in a straight-toothed bar called the *rack*, as shown in Figure 8.15. The rack, which is essentially a straight gear with an infinite pitch diameter, is made in both spur and helical types and will only mesh with gears having the same pressure angle and diametral pitch. As noted in Figure 8.16, the circular pitch of the pinion is equal to the linear pitch of the rack.

FIGURE 8.15
A rack and pinion converts rotational motion to linear motion.

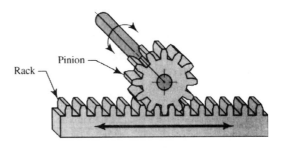

FIGURE 8.16
Rack-and-pinion parameters: (a) The linear pitch of the rack is equal to the circular pitch of the pinion. Circular pitch (CP) is equal to π divided by the diametral pitch (DP): $CP = \pi/DP$. (b) The pinion with the circular pitch (CP) and pitch diameter (PD) noted.

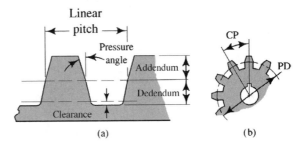

(a)   (b)

The rack and pinion converts rotary motion to linear motion and linear motion to rotary motion in the following ways:

- With the pinion rotating about a fixed center, the rack may be moved back and forth in a straight line, as shown in Figure 8.17.
- Moving the rack back and forth in a straight line causes the pinion to rotate about a fixed center in a clockwise and counterclockwise direction.
- With the rack fixed and the pinion rotating, the pinion and its shaft will move along the rack, with the center of the shaft moving in a straight line.

The distance moved by the rack or pinion is determined by the circular pitch of the pinion teeth (the linear pitch of the rack equals the circular pitch of the pin-

FIGURE 8.17
A rack and pinion is used with machine tools to move the tooling into the work, as illustrated here with the drill press.

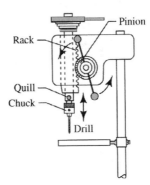

TABLE 8.2 Diametral Pitch of the Pinion Versus Circular Pitch of the Rack for Selected Values

Diametral Pitch (DP) (teeth/inch)	Circular Pitch (CP) (inches)
8	.3927
12	.2618
18	.1745
24	.1309
32	.0982
48	.0654

ion), the number of teeth in the pinion, and the angle of rotation of the pinion. These parameters are related by the following formula.

$$s = CP \times N \times \theta \qquad (8.3)$$

where

s = distance traveled by the rack or pinion, in inches
CP = circular pitch of the rack, from Table 8.2
N = number of teeth on the pinion
θ = the angular displacement of the pinion, in revolutions

Example 8.3

A 32-tooth, 18-diametral-pitch (DP) pinion is rotated about a fixed axis, causing the rack and the table attached to the rack to move in a straight line. Determine the distance the table travels when the pinion rotates 2.260 rev.

SOLUTION Find the distance using Equation 8.3 and the information in Table 8.2.

Given $s = CP \times N \times \theta$

Evaluate $CP = .1745$ in

$N = 32$ teeth

$\theta = 2.260$ rev

Substitute $s = .1745 \times 32 \times 2.260$

Solve $s = 12.62$ in

Observation If the circular pitch is not available in Table 8.2, then it may be determined from the pinion's diametral pitch with the following formula.

$$CP = \frac{\pi}{DP} \qquad (8.4)$$

where

CP = circular pitch of the rack, in inches
DP = diametral pitch of the pinion

Example 8.4

A 96-tooth, 64-diametral-pitch (DP) pinion attached to the shaft of an optical encoder is rotated about a fixed axis by a rack attached to the moving table of a machine tool. Determine the angle of rotation of the pinion when the table travels 8.375 in. Express the angle of rotation both in revolutions and in degrees.

SOLUTION

Find the rotational distance traveled by the pinion using Equation 8.3.

Given

$$s = CP \times N \times \theta, \quad \text{and} \quad \theta = \frac{s}{CP \times N}$$

Evaluate

$$s = 8.375 \text{ in}$$

$$CP = \frac{\pi}{DP} = \frac{\pi}{64} = .0491 \text{ in}$$

$$N = 96 \text{ teeth}$$

Substitute

$$\theta = \frac{8.375}{.0491 \times 96}$$

Solve

$$\theta = 1.777 \text{ revolutions}$$

Convert

$$1.777 \text{ rev} \times \frac{360°}{1 \text{ rev}} = 639.7°$$

Block and Screw

The block-and-screw mechanism (Figure 8.18) is used extensively in machine tools to move tables precisely past the cutters and move tooling past a workpiece. It is also applied in equipment used in automated assembly lines for converting rotary motion into linear motion.

FIGURE 8.18
The mechanism used to move a machine work table is made up of a block (nut) and screw. When the screw shaft is rotated, the table attached to the block will move in a longitudinal direction.

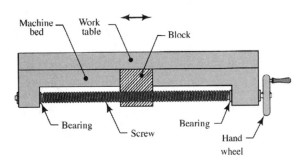

For single-lead threads, the block will move a distance equal to the pitch of the thread of the screw for each complete rotation of the screw shaft. This concept is noted in the following equation:

$$s = p\theta \tag{8.5}$$

where

s = distance the block travels, in inches
θ = angular displacement of the screw shaft, in rev
p = pitch of the thread of the screw, in inches; equal to the reciprocal of the threads per inch listed in Table 8.3 and in Table A5 of Appendix A.

Note: For double-lead threads, the block will move a distance of twice the pitch for each complete rotation of the screw shaft.

Example 8.5

The work table of a milling machine is moved into the cutter by the power-feed mechanism of Figure 8.19. The block-and-screw mechanism is engaged by rotating a handle to allow the split nut (block) to be connected to the lead screw. Determine

(a) The distance in inches that the table advances each time the 1.00-in-diameter, 12-thread/in lead screw makes one complete revolution.

(b) The speed of the table in inches per minute (in/min) when the 1.00-in-diameter, 12-thread/in (1.00-12) lead screw is rotating at an angular velocity of 6.00 rev/min.

SOLUTION

(a) Using Equation 8.5, find the linear distance traveled in inches for each revolution of the lead screw.

Given $s = p\theta$ and   Table 8.3

TABLE 8.3   Selected Thread Specifications*

| Size | American Standard Unified Screw Thread | | Constant Pitch Screw Threads | | | |
| | UNC | | 8-Thread Series | | 12-Thread Series | |
	Threads/Inch	Pitch	Threads/Inch	Pitch	Threads/Inch	Pitch
$\frac{1}{4}$	20	.050	—	—	—	—
$\frac{1}{2}$	13	.077	—	—	12	.083
$\frac{3}{4}$	10	.100	—	—	12	.083
1	8	.125	—	—	12	.083
$1\frac{1}{2}$	6	.167	8	.125	12	.083
2	$4\frac{1}{2}$	.222	8	.125	12	.083
$2\frac{1}{2}$	4	.250	8	.125	12	.083
3	4	.250	8	.125	12	.083

*Pitch = 1/(threads/inch).

FIGURE 8.19
Linear motion results when the split nut is engaged with the lead screw in this block-and-screw mechanism.

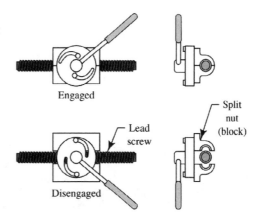

Engaged

Disengaged

Split nut (block)

Lead screw

Evaluate $p = \dfrac{1}{12 \text{ threads/in}} = .083 \text{ in}$

$\theta = 1 \text{ rev}$

Substitute $s = .083 \times 1$

Solve $s = .083 \text{ in}$

Therefore, the table advances .083 in for each revolution of the lead screw.

(b) Find the speed of the table in inches per minute when the 12-thread lead screw is rotating at 6.00 rev/min.

Given $v = \dfrac{s}{t}$ and $s = p\theta$; substituting and solving gives

$v = \dfrac{p \times \theta}{t}$

Evaluate $p = \dfrac{1}{12 \text{ threads/in}} = .083 \text{ in}$

$\theta = 6.00 \text{ revolutions}$

$t = 1.00 \text{ min}$

Substitute $v = \dfrac{.083 \times 6.00}{1.00}$

Solve $v = .50 \text{ in/min}$

Therefore, the lead screw moves the block with the table attached $\frac{1}{2}$ in (.500 in) each minute when the lead screw is rotating at 6 rev/min.

The *nut* (block) can drive the screw. When the nut is held in the frame of the mechanism and rotated, the screw shaft will move through the nut in a longitudinal direction. The *linear actuator*, an electromechanical device, uses a motor to drive a rotating nut to move a screw back and forth to open and close valves, etc.

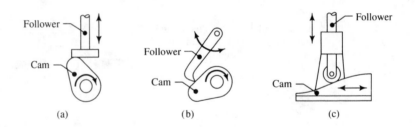

(a)   (b)   (c)

FIGURE 8.20

Cam mechanisms: (a) Plate cam with a flat-face follower. The rotary motion of the cam results in a reciprocating motion of the follower. (b) Plate cam with an edge follower. The rotary motion of the cam results in an oscillating motion of the follower. (c) Linear cam with a roller follower. The reciprocating motion of the cam results in a reciprocating motion of the follower.

Cam and Follower

A *cam* is a rotating or reciprocating machine element that has a shape designed to impart a desired motion to the follower, the component that rides against the curved surfaces of the cam and is guided by the motion of the cam. Figure 8.20 pictures several types of cams. Cams are widely applied in both manual and automatic machines used in the manufacture, assembly, and processing of all types of goods and products. Cams may be made to impart virtually any type or pattern of motion, including simple oscillatory motion and linear motion as well as precisely controlled movements that result in very complex shapes.

    The *profile*, or shape, of the cam's surface or groove determines the motion of the follower. Cams in general use may be divided into two categories, *radial cams* or *cylindrical cams*. The follower of the radial, or plate, cam moves in a direction perpendicular (90°) to the axis of rotation of the camshaft, whereas the follower of the cylindrical, or drum, cam moves in a direction parallel to the axis of rotation of the cam. These concepts are illustrated in Figure 8.21(a), where an *eccentric plate cam* is shown with the follower moving perpendicular to the axis of the cam's rotation, and

FIGURE 8.21

Two types of cam mechanisms: (a) An eccentric plate cam (radial cam) with the motion of the follower perpendicular to the axis of rotation. (b) A drum cam (cylindrical cam) with the motion of the follower parallel to the axis of rotation.

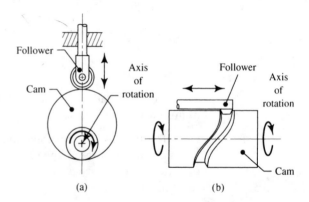

(a)   (b)

FIGURE 8.22
A displacement diagram of the plate cam with the displacement (s) of the follower related to the angular displacement (θ) of the cam.

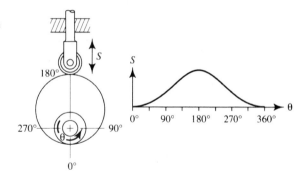

in Figure 8.21(b), where a drum cam is pictured with the follower moving parallel to the axis of the cam's rotation.

Cams of all types are characterized by their ability to have the follower remain stationary (motionless) for periods of time. The period of time that the follower is stationary is called **dwell.** Dwell occurs in a radial cam when the shape is an arc with a constant radius. Figure 8.22 represents the displacement (motion) of the follower of the plate cam of Figure 8.21(a) as it relates to the position in its cycle. In the graph, called a *displacement diagram*, the movement (displacement) of the follower is shown on the vertical axis, and the position of the cam in its cycle is shown on the horizontal axis. These types of diagrams are used in the design of the cam.

Virtually all cam mechanisms are made up of three parts, as noted in Figure 8.23:

- The frame, which supports the cam
- The cam, with its unique shape
- The follower, whose motion depends on the shape of the cam

In summary, the purpose of all cam mechanisms, regardless of their shape, is to provide motion and force to other mechanisms so that the machine with which they are associated can do its designed task or function.

Springs

Although not a mechanism, springs play an important role in the operation of many mechanisms by storing potential energy in their coils. Springs are used to control

FIGURE 8.23
A cam mechanism with the roller follower, cam, and frame noted.

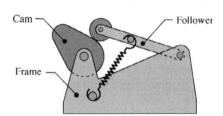

FIGURE 8.24
The force from the spring keeps the valve closed when the cam is in its lowest position.

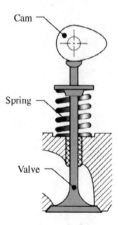

shock in load applications, reduce vibration, and return an element to its original position after displacement, as with the return spring in the cam-operated valve system of Figure 8.24.

Of the many types of springs, the three shown in Figure 8.25 have many applications in mechanisms. The compression and extension helical coil springs, made from round wire, are compared (specified) based on their *spring constant* (k), which relates how the spring reacts to the application of a load. Specifically, the spring constant is equal to the applied force divided by the resulting deflection (extension or compression) δ (delta), as indicated in the following formula:

$$k = \frac{F}{\delta} \tag{8.6}$$

where

k = spring constant, in lb/in or N/m
F = applied load (force), in lb or N
δ = deflection (extension or compression) of the helical spring, in m or in

The force a helical spring represents in a mechanism can be approximated by Equation 8.6 when the spring is in equilibrium and the spring constant and amount of deflection are known.

Example 8.6

The compression spring in the valve system of Figure 8.24 is deflected a distance of .310 in during operation of the opening of the valve. Determine the force of closure

FIGURE 8.25
Three standard spring types: (a) The extension spring; (b) The compression spring; (c) The torsion spring.

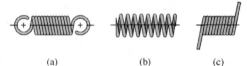

(a)   (b)   (c)

acting on the valve due to the spring once the cam reaches its lowest point and is no longer restraining the compressed spring. The spring constant is specified as 48.5 lb/in.

SOLUTION   Find the force using Equation 8.6.

Observation   When the spring is in equilibrium, the force compressing the spring is equal to the reactionary force (force keeping the spring from moving any further). For every action there is an equal and opposite reaction.

Given $k = \dfrac{F}{\delta}$; solving for F,

$$F = k\delta$$

Evaluate $k = 48.5$ lb/in

$\delta = .310$ in

Substitute $F = 48.5 \times .310$

Solve $F = 15.0$ lb

Scotch Yoke

The *Scotch yoke*, pictured in Figure 8.26, uses a yoke as a follower. A crank with a roller is rotated, imparting motion to the yoke by moving within the groove. The resulting motion is dependent on the shape of the groove in the yoke. Because the roller is the driver and the yoke with the groove is driven, the Scotch yoke is called an *inverse cam*. The positive motion of the inverse cam is used with applications re-

FIGURE 8.26
The Scotch yoke (inverse cam) is used to convert rotary motion to reciprocating motion.

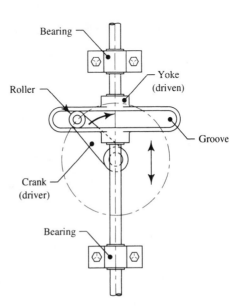

FIGURE 8.27
Four-bar mechanism with the links
and joints identified.

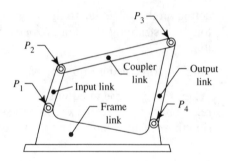

quiring reciprocating motion, as in the drive head of a sewing machine or in a simple pump.

Four-Bar

A *four-bar mechanism* is used in machines to transmit motion and torque at a greater velocity or at a greater mechanical advantage. The mechanism is made up of four rigid links joined together with movable joints to provide relative motion between the links when one of the links is fixed (held stationary). Pictured in Figure 8.27, the four rigid links (input, coupler, output, and frame) are connected by pivots (P_1 through P_4) made up of bearings and pins.

Not all combinations of four rigid links can be joined together to form the desired class of four-bar mechanism. Certain combinations of lengths are too short and will not allow the joints of the links to be joined.

> **As a Rule** Four rigid links can be assembled into a four-bar mechanism only when the longest link is shorter than the sum of the lengths of the other three links.

The two classes (kinds) of four-bar mechanisms are

- The crank-lever mechanism, pictured in Figure 8.28(a)
- The double-lever mechanism, pictured in Figure 8.28(b)

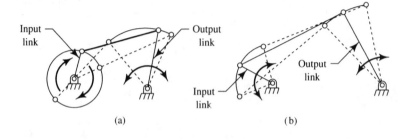

FIGURE 8.28
Two classes of four-bar mechanisms represented by kinematic symbols: (a) Crank-lever mechanism; (b) Double-lever mechanism.

FIGURE 8.29
A crank-lever mechanism with the common names of the four links indicated.

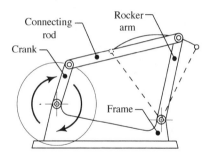

Crank-Lever Mechanism The crank-lever mechanism of Figure 8.28(a) is characterized by the 360° travel of the input link (crank) and the limited oscillatory motion of the output link (rocker arm).

When the *crank* of the crank-lever mechanism of Figure 8.29 is rotated through a complete revolution, the *rocker arm*, which is connected to the crank by the *connecting rod*, oscillates (rocks) back and forth through an arc.

The crank-lever mechanism is also operated with an input oscillatory motion at the rocker arm, which is connected to the crank by the connecting rod, producing a rotary motion at the crank. This configuration of the crank-lever mechanism was widely applied in foot-powered machines of the nineteenth and early twentieth century prior to the electric motor. Lathes, grinders, sewing machines, etc., all used foot power. Figure 8.30 pictures the crank-lever mechanism used to power a sewing machine.

Double-Crank Mechanism The double-crank, or drag-link, mechanism of Figure 8.31 is an inversion of the crank-lever mechanism. The inversion takes place when the shortest link, the crank, is fixed (held stationary) and the frame link is al-

FIGURE 8.30
A treadle drive used on a foot-powered sewing machine. A person's foot provides the necessary force and oscillatory motion. The motion is transformed into rotary motion and torque to power the sewing machine.

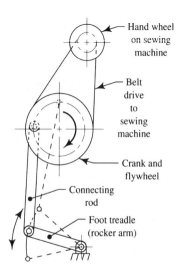

FIGURE 8.31

The double-crank mechanism is an inversion of the crank-lever mechanism.

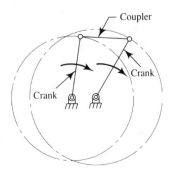

lowed to become movable as one of the cranks. In the double-crank mechanism, both cranks (input and output) make complete revolutions.

Slider-Crank Mechanism   The slider-crank mechanism, pictured in Figure 8.32, is a crank-lever mechanism in which the rocker arm (output link) is a point on the slider element that has its reciprocating motion constrained by the bearing in the frame of the mechanism. The slider-crank mechanism is used to convert rotary motion into reciprocating motion, as in an air compressor, or it may be used to convert reciprocating motion into rotary motion, as in the internal combustion engine.

When used in an air compressor to convert rotary motion into reciprocating motion, the slider element is the piston and an electric motor is connected to the crank through a belt-drive system. In operation, the rotary motion of the electric motor is converted by the slider-crank mechanism into the reciprocating motion of the piston, resulting in compression of the air inside the cylinder.

In the most common slider-crank mechanism, reciprocating motion is converted into rotary motion by the piston (slider) in the block (frame) of an internal combustion engine (Figure 8.33). The reciprocating motion of the piston is connected to the engine's crank shaft through the connecting rod to produce rotary motion to power the vehicle.

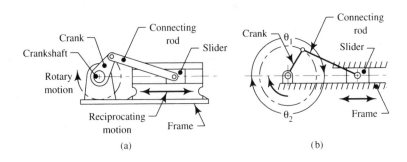

FIGURE 8.32

(a) A slider-crank mechanism with the pin joint of the slider located on the center line with the center of the crankshaft. (b) Graphic representation of the slider-crank mechanism illustrating that the forward displacement of the slider (θ_1) is equal to the reverse displacement (θ_2). $\theta_1 = \theta_2$ when the slider is on the same center line with the crank.

FIGURE 8.33
In the reciprocating motion of the slider-crank mechanism, the slider (piston in an engine) has two positions where its velocity is zero: top dead center (TDC) and bottom dead center (BDC).

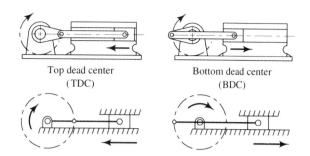

Top dead center
(TDC)

Bottom dead center
(BDC)

Double-Lever Mechanism The double-lever mechanism of Figure 8.34 is characterized by the fact that neither lever (input or output) makes a complete revolution. The double-lever mechanism is used when oscillatory motion is to be transmitted from one mechanism to another and the direction of the output lever follows the direction of the input lever.

8.3 INTERMITTENT-MOTION MECHANISMS

Ratchet Mechanism

The *ratchet mechanism* is the least complex of the devices used to produce an *intermittent output motion* (motion with dwell) from an oscillating or reciprocating input motion for use in *indexing*, or feed, devices. The ratchet mechanism consists of a driving **pawl** that is attached to the oscillating *input link* (rocker arm) and a *ratchet wheel* that is attached to the *output shaft*, which provides intermittent motion to the load. The design of the ratchet mechanism permits

■ The development of an intermittent motion from a rotary motion
■ The transmission of motion in only one direction

The ratchet may be locked (arrested) in place by an additional pawl and spring to prevent the wheel from moving backward. Figure 8.35(a) pictures an external ratchet mechanism with the drive pawl and the spring-loaded arresting pawl indicated.

When used for feeding material into a machine, a rachet mechanism is coupled with a crank-lever mechanism to form a mechanism with an adjustable stroke length. Figure 8.36 illustrates such a mechanism. By varying the crank length, the os-

FIGURE 8.34
Double-lever mechanism.

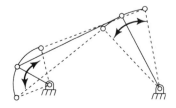

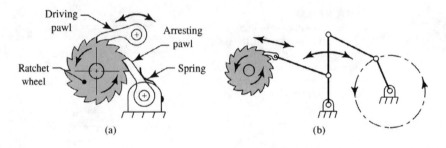

FIGURE 8.35

Ratchet mechanism: (a) An external ratchet mechanism with the drive pawl and spring-loaded arresting pawl shown; (b) Graphic representation of the ratchet mechanism combined with a four-bar driving mechanism.

cillatory motion of the rocker arm can be changed from near zero to maximum, which, in turn, permits the angular displacement of the ratchet wheel to be variable.

The *silent ratchet,* named for its lack of noise while operating, uses friction for its operation. Figure 8.37 shows a silent ratchet being used as an *overrun clutch,* which permits rotation in only one direction and locks when stopped to prevent reverse motion. One safety application of the overrun clutch is to prevent backward motion on an inclined conveyor when it is stopped.

Geneva Drive Mechanism

The *Geneva mechanism,* like the ratchet mechanism, provides intermittent motion at regular intervals for driving feeder and indexing mechanisms used in automated assembly lines. Unlike the ratchet mechanism, which can produce intermittent motion with very small angular displacements, the Geneva drive is used to produce motion with large angular displacement (up to 120°). The Geneva mechanism is classi-

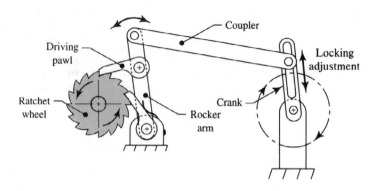

FIGURE 8.36

Ratchet-feed mechanism formed by combining a four-bar crank-lever mechanism with an external ratchet mechanism.

FIGURE 8.37
The silent ratchet uses frictional forces to couple the driver (internal member) to the driven (external member) with the wedging action of the balls. With the driver turning clockwise, the balls are moved toward the narrower section.

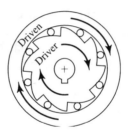

fied by the position of the driver in relation to the Geneva wheel. External, internal, and spherical are the three types of Geneva drive mechanisms.

The name of the mechanism comes from Geneva, Switzerland, where it was incorporated into watch mechanisms to prevent overwinding of the drive spring. The externally driven Geneva mechanism, pictured in Figure 8.38, is made up of a four-point drive *Geneva wheel* and a *driver* (crank), with a ball bearing roller that engages the wheel and drives it 90° for each revolution of the driver. Geneva drive mechanisms are commonly available in 4, 5, 6 and 8 *points of drive*, with one or two drive rollers on the driver. The displacement angle (θ) of the Geneva wheel is equal to 360° divided by the number of points of drive on the wheel. That is, a 5-point drive (5 slots) Geneva wheel will produce 72° of angular displacement for each 360° cycle of the driver when one drive roller is attached.

8.4 MOTION CHARACTERISTICS

Introduction

This section will help clarify and classify the movements that are observed when a machine is in operation. It will also give a qualitative understanding of **kinematic** aspects as related to select machine elements.

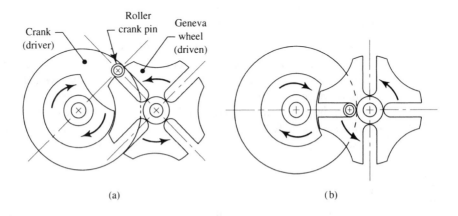

(a)   (b)

FIGURE 8.38
An externally driven Geneva mechanism with four points of drive on the Geneva wheel. (a) Roller crank pin on the crank engages the Geneva wheel to start the cycle. (b) Roller crank pin has moved the Geneva wheel 45° (halfway) through its cycle.

One of the principal concerns in the design of a mechanism is that it operates with no sudden or abrupt change in movement. For this criterion to be met, the change in acceleration (rate of change in the speed of the motion) must be within a measurable limit. It cannot be infinite; that is, the acceleration cannot be allowed to change instantaneously with time, because this creates sudden or abrupt spikes in the motion and causes dramatic increases in loading ($F = ma$) on the machine element, thus producing noise, vibration, wear, and excessive stress.

Jerk

The average time rate of change of acceleration ($\Delta a/\Delta t$), called *jerk*, is a key parameter in evaluating the performance of a mechanism. Jerk can be thought of in human terms as a shove, as opposed to a hit, that thrusts you forward.

When designing a machine, the designer attempts to keep the jerk under control as a finite rate of change of acceleration. A finite change in acceleration is a desirable condition, since it keeps force and stress low and controllable. However, when the change in acceleration is instantaneous (an undesirable condition), then the jerk is infinite, as is the instantaneous force on the machine elements.

Characteristic Curves

Motion can be classified from the *characteristic curves* of a machine element as having zero acceleration (no jerk), constant acceleration (no jerk), finite change in acceleration (finite jerk), or infinite change in acceleration (infinite jerk). Of these four conditions, only infinite change (very sudden change) in acceleration must be avoided.

The three characteristic curves used to describe motion, pictured in Figure 8.39, are

- The *displacement diagram* (*st* curve)
- The *velocity diagram* (*vt* curve)
- The *acceleration diagram* (*at* curve)

These three curves are used to interpret motion and are treated as a set with a common time base. The velocity-time curve (*vt* curve) is developed from the displacement-time curve (*st* curve) and the acceleration-time curve (*at* curve) is developed from the velocity-time curve.

To interpret these curves, first determine the *slope* of the characteristic curves and then calculate the representative parameter. The following example explores these concepts.

Example 8.7

Given the fragment of the displacement curve at the top of Figure 8.40, determine
 (a) The slope of the curve in the vicinity of each labeled point.
 (b) The velocity diagram.

Observation

The slope of a curve is approximated by first constructing a tangent line to a point in the curve and then forming a right triangle using the tangent line as the hypotenuse.

FIGURE 8.39
Motion is described by using three characteristics: displacement (s), velocity (v), acceleration (a).

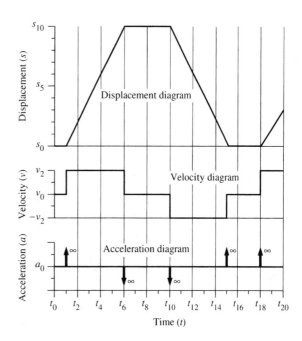

The slope is determined by dividing the altitude of the triangle by the base of the triangle (rise over run). The slope of the st curve is the average velocity at that point, and it is equal to

$$v = \frac{\Delta s}{\Delta t} \qquad\qquad (8.7)$$

where

v = average velocity at the tangent point, in ft/s or m/s
Δs = change of displacement, in ft or m
Δt = change of time, in s

SOLUTION   **(a)** Find the slopes at points $A, B, C,$ and D.

Given $v_A = \dfrac{\Delta s}{\Delta t}$ $v_B = \dfrac{\Delta s}{\Delta t}$

Evaluate $\Delta s = 1.6$ ft $\Delta s = 4.0$ ft

$\Delta t = 0.35$ s $\Delta t = 0.30$ s

Substitute $v_A = \dfrac{1.6}{0.35}$ $v_B = \dfrac{4.0}{0.30}$

$v_A = 4.6$ ft/s $v_B = 13$ ft/s

Solve $v_A = 4.6$ ft/s $v_B = 13$ ft/s

Observation   Since the portions of the curve around points A and D of Figure 8.40 (top) have approximately the same slope, they also have the same velocities. This statement is also true for the portions of the curve around points B and C and their velocities.

FIGURE 8.40
Characteristic curves for Examples
8.7 and 8.8.

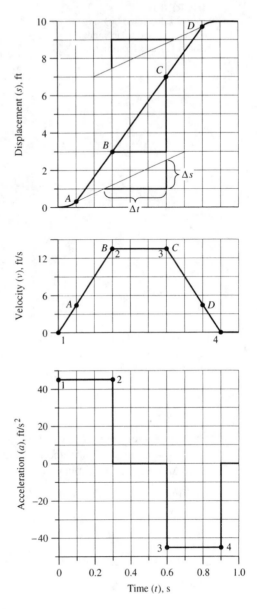

(b) The computed values of velocity are placed on the velocity diagram (middle
curve of Figure 8.40) at points that coincide with time in the displacement diagram. The vertical placement of each point corresponds to the calculated value
of velocity found in part **(a).**

➤ **As a Rule** The value of the average slope at a point in time on one of the motion diagrams is equal to the vertical height at that same point on the next lower diagram.

Example 8.8

Verify the average acceleration between the numbered points in the acceleration diagram of Figure 8.40 using the three line segments of the velocity diagram (1,2; 2,3; 3,4).

SOLUTION

Find the slope of the *vt* curve for the three line segments making up the curve.

Given

$$a = \frac{\Delta v}{\Delta t} \tag{8.8}$$

where

a = average acceleration, in f/s^2 or m/s^2
Δv = change of velocity, in ft/s or m/s
Δt = change of time, in s

Evaluate

$v_{1,2} = 13.3 - 0$ ft/s $\quad v_{3,4} = 0 - 13.3$ ft/s

$\Delta v = 13.3$ ft/s $\qquad \Delta v = -13.3$ ft/s

$t_{1,2} = 0.3 - 0$ s $\qquad t_{3,4} = 0.9 - 0.6$ s

$\Delta t = 0.3$ s $\qquad\quad \Delta t = 0.3$ s

Substitute

$a_{1,2} = \dfrac{13.3}{0.3} \qquad a_{3,4} = \dfrac{-13.3}{0.3}$

Solve

$a_{1,2} = 44$ ft/s$^2 \quad a_{3,4} = -44$ ft/s^2

Observation

The slope of line segment 2,3 in the acceleration diagram is zero because the velocity did not change with time during this period.

➤ **As a Rule** The slope of a straight line is

Positive if a point on the line rises as the point moves from left to right along the line, as pictured in Figure 8.40 for the *vt* curve from point 1 to point 2;
Negative if a point on the line falls as the point moves from left to right along the line, as pictured in Figure 8.40 for the *vt* curve, from point 3 to point 4.
Zero if a point on the line stays horizontal from left to right along the line, as pictured in Figure 8.40 for the *vt* curve from point 2 to point 3.

Motion

Uniform Motion The motion of a worktable (milling machine, X-Y table, etc.) driven by a rack-and-pinion gear is an example of uniform motion resulting from constant velocity. The pinion, which may be part of a gear train or attached to the shaft of a gearhead or a stepping motor, is rotating at a constant angular velocity, moving the table at a constant speed.

The displacement diagram of Figure 8.41 represents the motion of the work-table into the work zone as a straight line with a constant positive slope starting at t_1 and stopping at t_6. During the 5 units of time from t_1 to t_6, the table moves from position s_0 to s_{10}, covering a distance of 10 units with a smooth uniform motion (except for the start and stop) resulting from a constant velocity of two units (indicated in the velocity diagram between t_1 and t_6).

The remainder of the displacement diagram pictures the worktable stopping (dwelling) in the work zone (t_6 through t_{10}) where the workpiece is worked on by the machine, the table returning from the work zone during the time t_{10} through t_{15}, and the table dwelling for unload/load from time t_{15} through t_{18}. Recall that *dwell* is the period of time that the motion of a machine element (cam, table, link, etc.) is momentarily stopped (zero velocity) to allow time for the completion of a function, such as the automatic loading or unloading of a workpiece.

Because the movement is carried out at a constant velocity (no increase or decrease in speed), the acceleration is zero. This is confirmed by the acceleration diagram, where, except for the points in time where the motion starts or stops, the acceleration is pictured as zero. From a mathematical point of view, since there is no change in velocity over time, there is no acceleration.

In theory, the table drive is instantaneously started at the beginning of the travel period and is instantaneously stopped at the end of the travel period. In both cases, the acceleration-deceleration would be infinite (as would be the force). This state of acceleration is noted by the arrows and the infinity symbol (∞) on the acceleration-time diagram of Figure 8.41. The condition of instantaneous acceleration is comparable to being in a very low-mass, high-performance car and revving the 1500-

FIGURE 8.41

Constant-velocity motion (uniform motion) is characterized by a displacement curve made up of straight lines and an acceleration curve with zero amplitude except for starting and stopping points.

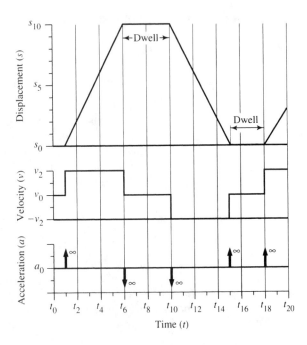

hp engine to full revolutions per minute and then popping the clutch. This scenerio should be avoided unless, of course, the machine is a drag racer.

Because constant-velocity motion has high starting and stopping forces associated with it when coming out of periods of dwell, uniform motion (constant velocity) is limited to very low operating speeds, where the machine elements cannot be accelerated instantaneously due to several factors, including the moment of inertia of the drive train, the moment of inertia of the rotating load elements, and the elastic deformation of the surfaces in contact (gear teeth, cam and follower, etc.). Even at moderate speeds, a potentially destructive jerk may be present each time the mechanism is started or stopped. However, in low-speed operation, the operating speed (ω) limits the power $(P = \omega\tau)$ in the system to a safe amount, thereby ensuring finite angular acceleration at start and stop during periods of dwell.

As a final thought, it is poor reasoning to assume that, because a machine is working satisfactorily at a slow or moderate speed, it will also work satisfactorily at a high speed. Unless the machine has been specifically designed for high-speed operation, speeding up a machine will result only in excessive wear and vibration, resulting in shorter machine life. In most applications where equipment operating speed is increased without careful evaluation, the increase in the cost of operating the machine is not paid for by the increase in productivity.

Modified Uniform Motion   To alleviate the jerk associated with uniform motion when periods of dwell are introduced into the work cycle, the displacement characteristic of uniform motion (constant velocity) is modified by gradually increasing the speed during the starting periods of time and gradually reducing the speed during the stopping periods of time, thereby producing a motion profile that is a combination of *constant acceleration* (for smooth start/stop) and *constant velocity* (for smooth travel).

The modification to the motion is done in the motor-control software for mechanisms under computer control or in the design of the hardware when dedicated hard automation is used. Regardless of how or where the modification is done, the process utilizes certain harmonic or polynomial mathematical functions (sine wave or parabolic curve) to blend the constant acceleration needed for smooth starting and stopping into the linear (straight-line) function of constant velocity for smooth running. This modification can be seen in the top displacement diagram of Figure 8.42, where part of a parabolic curve (resulting from the expression $s = \frac{1}{2}at^2$) is added to the curve (straight line) of Figure 8.41 to produce the characteristic S-curve of modified uniform motion.

The curves of Figure 8.42 picture the worktable of Figure 8.41 being moved over the same pathway. However, in this set of motion curves, the control has been modified to produce constant acceleration-deceleration for start or stop at times $t_{0.5}$–t_2 (start), t_4–t_7 (stop), t_{11}–$t_{12.5}$ (start), and $t_{14.5}$–$t_{17.5}$ (stop). Constant velocity has been retained for movement at times t_2–t_4 (movement into the work zone), and $t_{12.5}$–$t_{14.5}$ (movement out of the work zone).

With the motion curve modified, all the accelerations are finite, and the mechanism operates smoothly without jerk, even when starting or stopping after dwell.

FIGURE 8.42
Constant-acceleration/constant-velocity motion (modified uniform motion) is characterized by periods of either constant acceleration or constant velocity while the machine element is moving.

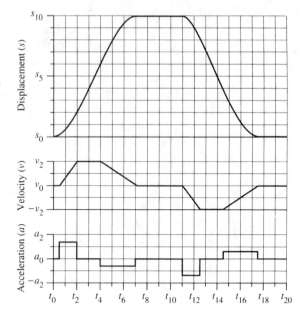

The one apparent disavantage to the modified curve (Figure 8.42) is the increase in cycle time. In the uniform velocity curve of Figure 8.41, the cycle time is 17 time units $(t_{18} - t_1)$, compared with 20 time units for the modified motion curve. However, the 18% increase in cycle time is more than offset by the increase in operating speed made possible by the smooth operating characteristics of the modified uniform motion.

Harmonic Motion   Of the various motions having a continually varying acceleration, *simple harmonic motion* (**SHM**) is the most common. The reciprocating motion characteristic of SHM is used with the Scotch yoke (Figure 8.26) to power the needle in sewing machines and to turn the crank shaft of slider-crank mechanisms in reciprocating engines (internal combustion engines), pumps, and compressors (Figure 8.33).

Harmonic motion is depicted in Figure 8.43 with cosine waveforms for the displacement and acceleration diagrams, whereas the velocity diagram is shown as a sine waveform. The operation of the Scotch yoke mechanism and the slider-crank mechanism is characterized by smoothness in the acceleration and velocity during the displacement (stroke) of the slider or yoke. However, the acceleration is maximum at the beginning (t_0) and end (t_{12}) of the stroke, which may cause vibration due to the instantaneous change in acceleration. A *flywheel* (an **inertial load**) is attached to the crankshaft of the slider-crank mechanism to increase the moment of inertia and lessen the instantaneous loads, thereby smoothing out the flow of power by storing kinetic energy in the rotating wheel for use at the top and bottom dead center of the stroke.

FIGURE 8.43
Simple harmonic motion (SHM).

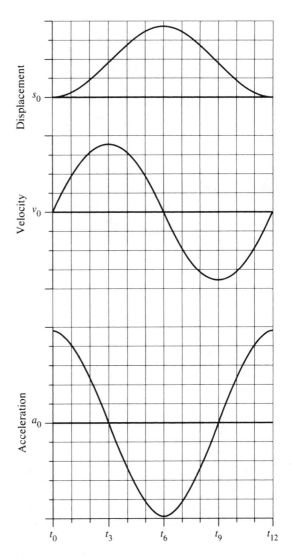

The motion of the characteristic curves of these mechanisms is periodic in that it is repeated over and over. The period (T) of the motion is the time (in seconds) taken to complete one cycle; the frequency of operation (f) is the reciprocal of the period ($f = 1/T$), or

$$f = \frac{\omega}{2\pi}$$ (8.9)

where

f = frequency, or number of cycles per second, in hertz (Hz)
ω = the angular velocity of the crank, in rad/s
2π = 2π rad

Example
8.9

The crank in the head of a sewing machine rotates 300 times a minute (300 rev/min). Determine the number of times the needle moves up and down in a second (the frequency of the needle). Express the answer in units of hertz.

SOLUTION

Find the frequency of the needle using Equation 8.9 by converting the angular velocity (rev/min) of the crank to radians per second.

Observation

Each time the crank completes one cycle, the Scotch yoke mechanism with the needle attached to one end moves down and up once.

Given

$$f = \frac{\omega}{2\pi}$$

Evaluate

$$\omega = 300 \ \text{rpm} \times \frac{0.1047 \ \text{rad/s}}{1 \ \text{rpm}} = 31.4 \ \text{rad/s}$$

Substitute

$$f = \frac{31.4}{2\pi}$$

Solve

$$f = 5.0 \ \text{Hz}$$

Therefore, the needle moves up and down 5 times a second.

CHAPTER SUMMARY

- Couplings are divided into two groups, rigid and flexible.
- Flexible couplings control the forces acting on bearings and shafts.
- Coupling are designed to accommodate misalignment and sudden changes in torque and angular velocity.
- When used alone (not in pairs), the universal joint produces a fluctuation in angular velocity in the output.
- The constant-velocity joint is a specially designed universal joint that provides a constant output angular velocity at all angles.
- A magnetic clutch is a friction-disc clutch commonly used in industrial applications.
- The moment of inertia found in rotary motion is the dual of mass in linear motion.
- A block-and-screw mechanism may be used for very precise movement.
- A cam and follower may be designed to impart virtually any type of motion.
- Springs store energy in their coils, thereby controlling shock.
- A slider-crank mechanism may be used to change reciprocating motion to rotary motion.
- A Geneva drive mechanism is used to change rotary motion to intermittent motion in indexing and feeding mechanisms.
- Uniform motion is characterized by constant velocity.
- Modified uniform motion uses constant acceleration for starting and stopping.

SELECTED TECHNICAL TERMS

The following technical terms, abbreviations, and acronyms are defined in the glossary located after Chapter 16. You are encouraged to use the glossary to aid your understanding and to test your knowledge of these important terms.

backlash
clutch
constant-velocity joint
coupling
dwell
inertial load
kinematic

lope
moment of inertia
pawl
SHM
sinusoidal
universal joint

END-OF-CHAPTER QUESTIONS

Write T if the statement is true and F if the statement is false.

1. The helical coupling is highly compliant and allows for shaft misalignment up to 12°.
2. A moment of inertia (I) appears in formulas for rotary motion.
3. Clutch armature plate-and-rotor surfaces must be properly burnished before the clutch is put into service.
4. Jerk, the average time rate of acceleration, is not a concern in the motion of a mechanism.
5. The drag-link mechanism is an inversion of the double-lever mechanism.
6. The Oldham coupling allows a fairly large torsional movement in its operation.
7. The rack and pinion must have the same diametral pitch and pressure angle if they are to mesh properly.
8. A slider-crank mechanism may be used in an air compressor to compress air.
9. A ratchet mechanism produces an intermittent motion between periods of dwell.
10. Because uniform motion has a high starting force, it is limited to very slow operating speeds when used with mechanisms.

In the following, select the word that makes the statement true.

11. Mechanisms with recriprocating motion include the (block and screw, ratchet mechanism, slider-crank).
12. The period of time that the follower of a cam remains stationary is called (pause, dwell, cycle).
13. The three characteristic curves used to describe motion share a common time base and include velocity and (displacement and inertia, jerk and acceleration, acceleration and displacement).
14. Machine elements used in the axial transmission of power include (gears, clutches, ratchets).
15. Rotary motion is changed to linear motion with the (rack and pinion, four-bar mechanism, inverse cam).

Answer each of the following questions with a short answer in the form of a complete sentence. Include a restatement of the question in your answer.

16. List three mechanisms that change rotary motion into linear motion.

17. Explain how the displacement angle of the Geneva wheel is determined and then determine the displacement angle of a Geneva wheel that has eight points of drive on the wheel.

18. Explain how jerk is lessened in uniform motion by using modified uniform motion.

19. Describe the desirable properties of motion as they relate to the movement of mechanisms in general.

20. A conveyor is used to move small work pallets along an automated assembly line. The motion of the conveyor is such that the pallets move a short distance, stop for a period of time while work is done, and then advance once again. This motion is repeated over and over, with equal time and equal displacement.

(a) List two mechanisms that could be used to produce this motion.

(b) Describe why you would select one over the other for this application.

END-OF-CHAPTER PROBLEMS

Solve the following problems. Make sketches to aid in solving the problems and structure your work so it follows in an orderly progression and can easily be checked. Table 8.4 summarizes the equations for torque, power, frequency, etc., used in this chapter.

TABLE 8.4   Summary of Formulas Used in Chapter 8

Equation Number	Equation
8.1	$\tau = \dfrac{5250P \times SF}{\omega}$
8.2	$I = mr^2$
8.3	$s = CP \times N \times \theta$
8.4	$CP = \dfrac{\pi}{DP}$
8.5	$s = p \times \theta$
8.6	$k = \dfrac{F}{\delta}$
8.7	$v = \dfrac{\Delta s}{\Delta t}$
8.8	$a = \dfrac{\Delta v}{\Delta t}$
8.9	$f = \dfrac{\omega}{2\pi}$
6.18 (repeated)	$P = \dfrac{\tau \omega}{5250}$

FIGURE 8.44
Drive system for end-of-chapter
problem 6.

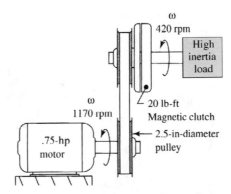

1. A light-duty helical coupling rated at 2.5 lb-in is used to couple a stepper motor to a load. The load requires .065 hp at a torque of 1.2 lb-in. Determine
   (a) The minimum rotational speed (revolutions per minute) that the load can turn and not exceed the coupling torque rating.
   (b) The speed of rotation needed to develop .065 hp at a torque of 1.2 lb-in.

2. A rack-and-pinion system is used to provide power from a small gearhead motor to a positioning table. Determine
   (a) The angular displacement of the 36-tooth, 48-DP pinion in degrees when the rack attached to the table travels 5.310 in.
   (b) The time taken (in seconds) for the table to travel 5.310 in when the angular velocity of the pinion is 42 rev/min.
   (c) The pitch diameter (PD) of the pinion.

3. The .138-in-diameter screw of a linear actuator has 32 threads per inch. Determine the distance traveled by the screw in 3.00 s when the nut is rotating 185 rev/min.

4. A helical return spring is compressed 2.80 mm while in operation. Determine the applied force (in newtons) when the spring constant is 3717 N/m.

5. The follower of a cam has a constant reciprocating motion of 125 cycles/min (rev/min) that is simple harmonic motion. Determine
   (a) The angular velocity of the follower in radians per second.
   (b) The frequency of the follower in hertz.
   (c) The period of the cycle in seconds.

6. A magnetic clutch, rated at 20 lb-ft, is used with a belt-drive system to couple a high-inertia load to the output shaft, as pictured in Figure 8.44. The load rotates at 420 rev/min and requires .50 hp for its operation. The output shaft of the motor, which rotates at 1170 rev/min, has a 2.5-in-diameter pulley attached to it. Determine
   (a) The torque out of the clutch with provision for the high-inertia load.
   (b) The diameter of the driven pulley, assuming no loss.
   (c) The power (hp) at the shaft of the motor when the drive system is 87% efficient.

Dam

Reservoir

A

Generator

Penstock

Spillway

C

Dam

Generator

B

River

C

Shaft

Draft tube

Turbine

Chapter **9**

The Electric Circuit

About the Illustration. *One source of electrical energy is from the generators within the powerhouses of dams across (energy-storage dams, A) and along rivers (run-of-river dams, B) throughout the nation. In the energy-storage dam (A) (such as Hoover Dam on the Colorado River, pictured here), the river is blocked and a lake of water containing a huge reserve of potential energy is formed behind the dam. The seventeen generating units in the powerhouses (C) of Hoover Dam have a combined nameplate capacity of 2 000 000 kW (2 GW). (Courtesy of Bureau of Reclamation.)* ■

INTRODUCTION

Of all the sources of electrical energy, 60-cycle alternating current is responsible for virtually all of the nation's energy needs. The alternating current (ac) available from hydroelectric and thermoelectric generation sites is available in two general formats:

- Single-phase, for households, to power lights, heating devices, and fractional-horsepower motors
- Three-phase, for commercial and industrial plants, to power large, multihorse-power motors; heavy-duty heaters and dryers; and large lighting systems

When direct current (dc) is needed in industrial applications or elsewhere, it is obtained simply by rectifying alternating current. Unlike direct current, alternating current, with its simple harmonic motion (sine wave), has a continually changing magnitude, which allows it to be transformed.

The nature of alternating current from the source to the load is studied in this chapter, as are conductor size and control devices. Additionally, load impedance, resistance, inductance, and capacitance are introduced, along with power factor and circuit phase angle. Electric power and efficiency are presented.

CHAPTER CONTENTS

PERFORMANCE OBJECTIVES

Once you have read and studied each section; worked through each example with pencil, paper, and calculator; worked through the end-of-chapter problems; and answered the end-of-chapter questions, then you should be able to

- Understand the nature of the electric circuit and how alternating current is energized and conducted through the load.
- Use the AWG to assist in determining the voltage drop across the circuit conductors.
- Explain the operation of manual, automatic, and electromagnetic switches.
- Solve for the inductive and capacitive reactance and the impedance in the circuit load.
- Compute the power in the electric circuit using the source voltage, circuit current, and the load power factor.
- Determine efficiency in an electric circuit.

9.1 THE ELECTRIC CIRCUIT

The electric circuit, in its simplest form, is a complete pathway through which electric charge carriers (electricity) travel, carrying energy from the source to the load. The four basic components of the electric circuit pictured in Figure 9.1 include

- The source of electric energy, where work is done and energy is put onto electric charge carriers;
- The conductive path (wiring) for the movement of the electric charge (electricity) between the source and load;
- The control device to control the flow of electricity in the conductive path;
- The load, where energy is converted to useful work.

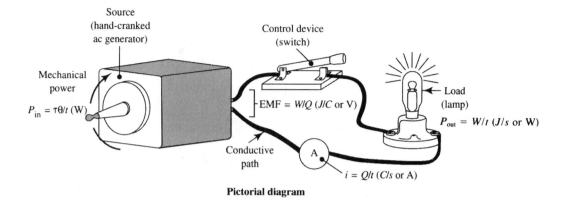

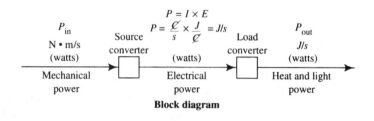

Block diagram

FIGURE 9.1

The mechanical work of turning the crank is converted to electrical power by the source conversion device (ac generator). The electrical energy is carried by the flow of electric charge (electricity) moving in the conductive path (wires) to the load (lamp). The moving energized electric charge performs work in the material of the lamp to convert electrical energy to heat and light energy.

9.2 THE SOURCE

Alternating Voltage

The hand-cranked alternating current (ac) generator of Figure 9.1 provides mechanical work through the application of a torque over an angular distance ($W = \tau\theta$). When the handle is rotated, a force is applied through a radial distance to create a torque that moves a coil of wire along a circular path through a magnetic field, as illustrated in Figure 9.2.

The electric charge carriers (electrons) present in the copper wire of the coil and the conductors have energy added to them from the mechanical work being done at the generator. The amount of energy added to each unit of electric charge (**coulomb**) is called the electromotive force (**EMF**), measured in volts (V). Each volt of electromotive force is equivalent to the energy of 1 J (joule) for each coulomb of electric charge (J/C):

$$\text{EMF} = \frac{W}{Q} \tag{9.1}$$

FIGURE 9.2

As the coil of wire rotates through the magnetic field, work is performed and energy is imparted to the electric charge carriers in the wire of the coil, inducing a voltage at the ends of the coil that is called an electromotive force (EMF).

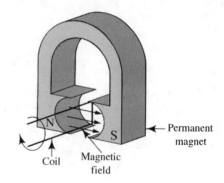

Permanent magnet

Coil   Magnetic field

where

EMF = electromotive force, in V
W = work done on the charge carriers, in J
Q = unit quantity of electric charge, in C

An electromotive force (voltage) is set up between the ends of the generator coil when two conditions are met. First, there must be relative motion between the magnetic field and the metallic coil of wire, and second, the conductors in the coil must cut through the magnetic field. The magnitude of the voltage from the generator depends on

■ The relative speed of motion between the coil conductors and the magnetic field;
■ The number of conductors in the coil that pass through the magnetic field;
■ The strength of the magnetic field.

Alternating Current

When an external circuit is connected to the ends of the coil, as pictured in Figure 9.3(a), an alternating current flows through the coil, the conductors, and the load. The current, which moves back and forth (alternates) and varies with time through the load, is measured in amperes (A). The current at any time (t) is the rate of movement of the charge carriers:

$$I = \frac{Q}{t}$$

(9.2)

where

I = current, in A
Q = unit quantity of electric charge, in C
t = time, in s

If the switch in the circuit of Figure 9.3 is opened, the flow of current stops; however, an electromotive force is still present at the ends of the coil. The electromotive force created across the moving coil, called an *induced voltage*, is the

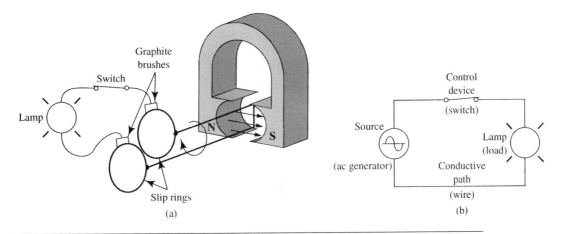

FIGURE 9.3
(a) A representation of an alternating-current (ac) generator with the rotating coil of wire attached to slip rings. The external circuit of conductors, switch, and lamp is connected to the generator through graphite brushes that ride on the generator slip rings. (b) An electrical schematic of the ac generator with an external circuit connected to it.

source voltage, and the induced current that flows out of the coil of the generator and through the external circuit is the *circuit current*.

The movement of the coil through the magnetic field, converting mechanical work to electrical energy, is opposed by a reactionary force caused by the induced electric current that circulates through the coil and the external circuit. The induced current in the coil sets up a magnetic field with the same polarity as the polarity of the magnetic field from the permanent magnet. Since like magnetic poles oppose each other, a reactionary force is set up that opposes the motion of the moving coil.

The reactionary force is predictable from the law of action and reaction. If there were no reactionary force, then the current moving through the external circuit would bring energy to the load to produce heat and light with no work being done—a clear violation of the law of conservation of energy. The presence of a reactionary force is summarized in a statement called Lenz's law.

> ➤ **Lenz's Law** The induced current resulting from the induced voltage caused by the relative motion between a magnetic field and an electric conductor will always flow in a direction so as to set up a magnetic field to oppose the motion.

Frequency of Alternating Current

The rotational motion of the ac generator is *simple harmonic motion*, and the current and voltage resulting from this motion is *sinusoidal alternating current and voltage*. As pictured in Figure 9.4, the waveforms representing the current and voltage have amplitudes that change with time. The resulting *sinusoidal*

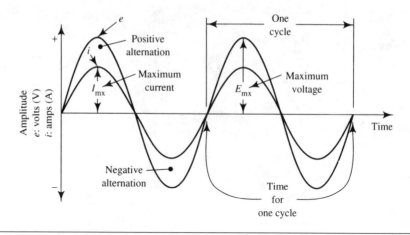

FIGURE 9.4

A sine wave of ac voltage and current is supplied from an ac generator when a complete external resistive circuit is attached to the generator terminals.

current and voltage waveforms are *periodic* (repetitive) and have a constant *frequency*.

Each revolution of the coil in the generator produces one cycle of alternating current and voltage. Each cycle consists of two alternations—one positive and one negative. The number of cycles produced in 1 s is the *frequency* of the alternating current and voltage waveforms:

$$f = \frac{cyc}{t} \tag{9.3}$$

where

f = frequency, in Hz
cyc = number of cycles of the waveform
t = time to complete the cycles, in s

The *period* of the waveforms is the time taken to complete one cycle:

$$T = \frac{1}{f} \tag{9.4}$$

where

T = period of the waveform, in s
f = frequency of the waveform, in Hz

Example 9.1

Determine

(a) The frequency of an alternating current that completes 300.0 cycles in 5.00 s.

(b) The amount of time to complete one cycle.

SOLUTION

(a) Find the frequency. The answer is expressed in hertz, the standard unit of frequency.

Given

$$f = \frac{\text{cyc}}{t}$$

Evaluate

$$\text{cyc} = 300.0 \text{ cycles}$$

$$t = 5.00 \text{ s}$$

Substitute

$$f = \frac{300.0}{5.00}$$

Solve

$$f = 60.0 \text{ Hz}$$

(b) Find the period of the 60.0-Hz sine wave. The period is the amount of time needed to complete one cycle of the wave.

Given

$$T = \frac{1}{f}$$

Evaluate

$$f = 60.0 \text{ Hz}$$

Substitute

$$T = \frac{1}{60.0}$$

Solve

$$T = 0.0167 \text{ s} \Rightarrow 16.7 \text{ ms}$$

Alternating current for electric power transmission has been standardized at 60 Hz in the United States. Power is supplied to

- Residences at 120 V and 240 V single-phase.
- Commercial and light industries at 120 V single-phase and 208 V three-phase.
- Industries at 115–120 V single-phase, 240 V three-phase, and 480 V three-phase.

Figure 9.5(a) represents a coil rotating in a two-pole (one north and one south), one-pair generator. Figure 9.5(b) pictures the same coil rotating in a four-pole machine. When one side of the coil in Figure 9.5(a) makes one complete revolution ($\theta = 360°$), then one sine wave of EMF is generated. However, when the coil in Figure 9.5(b) makes one complete revolution ($\theta = 360°$), then two sine waves of EMF are generated, one cycle for each 180° movement of the coil.

FIGURE 9.5
The frequency of ac current and voltage depends on the number of poles in the generator when the speed of rotation (rev/min) is constant.

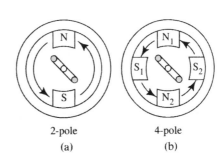

2-pole

(a)

4-pole

(b)

In a generator, each time the coil is moved through a pair of poles (one north and one south), one cycle of electromotive force is developed. The greater the number of pairs of poles, the greater is the frequency for a given angular velocity of rotation. That is, the frequency of the alternating current and voltage from a generator depends on the speed of rotation of the coil (revolutions per minute) and the number of pairs of poles:

$$f = \frac{n_c p}{60} \qquad\qquad (9.5)$$

where

f = frequency, in Hz, of the ac current and voltage
n_c = speed of rotation of the coil, in rev/min
p = number of pairs of magnetic poles
60 = conversion factor to change minutes to seconds

Example 9.2

Determine the frequency of the generators pictured in Figure 9.5 when the coil in each is rotated at 1800 rev/min.

SOLUTION   Find the frequency of each using Equation 9.5.

Given $f = \dfrac{n_c p}{60}$

Evaluate   2-pole generator: 4-pole generator:

$n_c = 1800$ rev/min $n_c = 1800$ rev/min

$p = 1$ pair of poles $p = 2$ pairs of poles

Substitute $f = \dfrac{1800 \times 1}{60}$ $f = \dfrac{1800 \times 2}{60}$

Solve $f = 30$ Hz $f = 60$ Hz

Observation   To generate the same frequency, a two-pole machine must rotate at twice the speed of a four-pole machine.

Exercise 9.1

Solve the following problems. Make sketches to aid in solving the problems and structure your work so it follows in an orderly progression and can easily be checked.

1. Determine the source voltage (induced voltage) at the terminals of an ideal ac generator when a torque of 12 N·m moves the coil through 25 rev. Assume an electric charge of 15.7 C.

2. Determine the quantity of electric charge moved in an electric circuit in 1.00 h when the current is measured at 6.25 A.

3. Determine the frequency and the period of an ac-voltage sine wave that completes 1440 alternations in 12.0 s.

4. A 4-pole ac generator is driven from a 5.0-hp gasoline engine using a V-belt drive. The engine has a governor that may be adjusted to set the revolutions per minute of the engine. Determine the speed of the engine for 60.0-Hz operation of the generator if a 3.00-in-diameter pulley on the engine drives a 6.00-in-diameter pulley on the generator.

5. An ideal electrical ac generator is rotated for 5.00 min. Determine the work done in converting the mechanical motion to electrical energy if the source voltage (voltage at the generator) is 120.0 V and the circuit current is 3.00 A.

9.3 ELECTRICAL CONDUCTORS

Introduction

Electrical conductors (wires) are made from copper and aluminum; both are extremely good conductors of electricity. Metals, with copper being the principal conductor in homes and industries, have an abundance of *free electrons*, which can support a substantial electric current when an EMF is applied.

Copper, because of its mechanical properties (strength, **ductility**, etc.), physical properties (electrical conductivity, solderability, etc.), and cost ($\frac{1}{100}$ that of silver), is the conductor (wire) of choice for all electromagnetic devices (motors, relays, etc.) and most wiring. Copper is 95% as conductive as the best conductor, silver.

Copper and aluminum, like all metals at room temperature, are not perfect conductors but have some resistance. The amount of resistance—and, consequently, the power loss due to the resistance—depends upon several factors, including the length of the conductor, the diameter of the conductor, the conductor material, and the temperature.

Ampacity

The *National Electric Code*® (NEC®)* is the principal source of information for selecting and installing electrical conductors. The tables of Section 310 of the NEC are used to specify conductors for general wiring. An example of an *ampacity table* is found in Table A7 of Appendix A. Table A7 is a reproduction of NEC Table B-310-1 listing the **ampacity** of up to three insulated conductors within an overall covering (multiconductor cable) in a raceway in free air at an ambient temperature (surrounding air temperature) of up to 30°C (86°F). This table, like the others in the series of tables in section 310 of the NEC, is used to select a conductor gauge number (AWG wire size) based on the parameters (current, temperature, etc.) of the circuit.

National Electrical Code® and NEC® are registered trademarks of the National Fire Protection Association, Inc., Quincy, MA 02269.

Table A7 of Appendix A contains information that relates conductor temperatures (60°C, 75°C, 90°C) to insulation types, thereby specifying the amount of current a copper or aluminum conductor is able to carry. Depending on heat tolerance, different insulation types (TW, RH, THHN, etc.) permit different amounts of current to be carried in the conductor.

In addition to heat tolerance, the insulation material (nonconductive covering around the wire) and its thickness also control the voltage rating of the conductor. When the temperature of the air surrounding the conductor is greater than 30°C (86°F), then the specified ampacity must be reduced by a *deration factor* specified in the lower portion of Table A7.

Example 9.3

Using Table A7 of Appendix A, specify the maximum ampacity of a number 6 AWG for both copper and aluminum conductors operating at an ambient temperature of 115°F (46°C). The insulation on the copper wire is type TW and that on the aluminum wire is type RH.

SOLUTION | For copper | For aluminum

Given | 6 AWG, 115°F, and TW | 6 AWG, 115°F, and RH

Observation | Locate copper with type TW insulation in the first column and aluminum with type RH insulation in the fifth column of Table A7. Locate the temperature deration factor for copper in the first column and aluminum in the fifth column of Table A7 (bottom).

Given | From Table A7 48 A derate by .58 | From Table A7 45 A derate by .75

Substitute | Ampacity = 48 × .58 | Ampacity = 45 × .75

Solve | Ampacity = 28 A | Ampacity = 34 A

Observation | Ampacity depends on conductor material, insulation type, conductor size, and ambient temperature.

The American Wire Gauge

The resistance per thousand feet (Ω/1000 ft), the gauge number, and the diameter in mils (.001 in) of copper and aluminum electrical conductors are specified by the American Wire Gauge (**AWG**). In this system, as noted in Table 9.1, the gauge numbers range from the largest diameter 0000 (also noted as 4/0 and pronounced "four-aught"), to 40, the smallest diameter.

Although the ampacity tables from section 310 of the NEC are usually consulted first to determine wire size, occasionally a wire run is long, resulting in a significant increase in conductor resistance that is not accounted for by the ampacity tables. When this is the case, the wire size (gauge) selected from the NEC ampacity table needs to be confirmed by computation of the voltage drop made in conjunction with the resistance taken from the AWG.

TABLE 9.1 American Wire Gauge (20°C/68°F)

Gauge No.	Diam. (mil)*	Ω/1000 ft		Gauge No.	Diam. (mil)	Ω/1000 ft	
		Copper	Aluminum			Copper	Aluminum
4/0	460	0.0490	0.0804	19	36	8.05	13.2
3/0	410	0.0618	0.101	20	32	10.2	16.7
2/0	365	0.0779	0.128	21	28.5	12.8	21.0
1/0	325	0.0983	0.161	22	25.3	16.1	26.5
1	289	0.124	0.203	23	22.6	20.4	33.4
2	258	0.156	0.256	24	20.1	25.7	42.1
3	229	0.197	0.323	25	17.9	32.4	53.1
4	204	0.249	0.408	26	15.9	40.8	67.0
5	182	0.313	0.514	27	14.2	51.5	84.4
6	162	0.395	0.648	28	12.6	64.9	106
7	144	0.498	0.817	29	11.3	81.8	134
8	128	0.628	1.03	30	10.0	103	169
9	114	0.792	1.30	31	8.9	130	213
10	102	0.999	1.64	32	8.0	164	269
11	91	1.26	2.07	33	7.1	207	339
12	81	1.59	2.61	34	6.3	261	428
13	72	2.00	3.29	35	5.6	329	540
14	64	2.53	4.14	36	5.0	415	681
15	57	3.18	5.22	37	4.5	523	858
16	51	4.02	6.59	38	4.0	660	1080
17	45	5.06	8.31	39	3.5	832	1360
18	40	6.39	10.5	40	3.1	1049	1720

*1 mil = .001 in.

In an electrical circuit, it is customary to allow up to 4% of the source voltage as a voltage drop across the length of the conductors (wiring). This loss of voltage is due to the resistance in the conductors and depends on the operating temperature (ambient), the material (copper or aluminum), the conductor size (diameter), and the length of the conductor.

When computing the resistance of the conductors (wiring), the conductor length is determined by measuring the distance from the source to the load and then applying one of the following rules.

> ➤ **As a Rule** In a *single-phase circuit*, each conductor carries the same amount of current from the source to the load and back. The length of the conductor for resistance calculation is twice the distance from the source to the load.

> ➤ **As a Rule** In a *three-phase circuit*, each conductor carries an unequal part of the total current from source to load and back. This results from the currents being out-of-phase with one another. The length of the conductor for resistance calculation is the square root of three ($\sqrt{3}$) times the distance from the source to the load.

Example 9.4 will use these rules along with the specifications of Table 9.1, the ampacity Table A7 of Appendix A, and Ohm's Law to determine the ampacity and an appropriate conductor size (AWG gauge number) and to insure that the voltage drop along the length of the wiring is within the allowed 4% drop. A complete set of Ohm's Law formulas for dc circuits is given in Table A10 of Appendix A.

$$E_{con} = IR_{con} \tag{9.6}$$

where

E_{con} = voltage drop across the conductors, in V
I = current in the load, in A
R_{con} = resistance of the conductors, in Ω (ohms)

Example 9.4

A 230/460-V three-phase motor is to be installed 310 ft from the power panel, as illustrated in Figure 9.6. The motor current is 28.0 A when the motor is operating from a 240-V source and 14.0 A when it is operating from a 480-V source. If the ambient temperature is 38°C and copper-clad aluminum with THW insulation is used, determine

(a) The gauge of the conductors for each case.
(b) The voltage drop (E_{con}) across the conductors.
(c) That E_{con} is within the allowed 4% voltage drop.

SOLUTION

(a) Using Table A7 of Appendix A, determine the ampacity.

Given

38°C, THW, 240 V/28.0 A   38°C, THW, 480 V/14.0 A

From Table A7, #8,   From Table A7, #12,
33 A derate by .88   18 A derate by .88

Substitute

Ampacity = 33 × .88   Ampacity = 18 × .88

Solve

Ampacity = 29 A   Ampacity = 16 A

Observation

For 240 V operation, an 8 gauge, 1.03 Ω/1000 ft copper-clad aluminum wire with an ampacity of 29 A at 38°C is specified. For 480 V operation, a 12 gauge, 2.61 Ω/1000 ft copper-clad aluminum wire with an ampacity of 16 A at 38°C is specified. Each wire meets the amperage requirements of the example.

FIGURE 9.6
Three-phase 3-wire system used to supply 480/240 V to a 230/460-V motor.

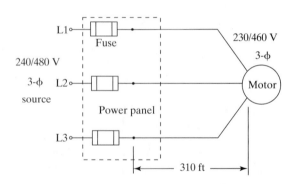

SOLUTION

(b) Using the AWG (Table 9.1), determine the voltage drop (E_{con}) across each of the aluminum conductors for a conductor length of 310 ft times the square root of three ($\sqrt{3}$).

Given

Length = $\sqrt{3} \times 310$ ft = 537 ft

$E_{con} = IR_{con}$ (for No. 8) $E_{con} = IR_{con}$ (for No. 12)

Evaluate

$I = 28.0$ A $I = 14.0$ A

$R_{con} = \dfrac{1.03\ \Omega}{1000\ ft} \times 537\ ft$ $R_{con} = \dfrac{2.61\ \Omega}{1000\ ft} \times 537\ ft$

$R_{con} = 0.553\ \Omega$ at 20°C $R_{con} = 1.40\ \Omega$ at 20°C

Calculate the resistance at 38°C   Calculate the resistance at 38°C

$R_{con} = 0.553(38.0 + 236)/256$ $R_{con} = 1.40(38.0 + 236)/256$

$R_{con} = 0.592\ \Omega$ at 38°C $R_{con} = 1.50\ \Omega$ at 38°C

Substitute

$E_{con} = 28.0 \times 0.592$ $E_{con} = 14.0 \times 1.50$

Solve

$E_{con} = 16.6$ V $E_{con} = 21.0$ V

SOLUTION

(c) The voltage at the terminals of the motor is:

240–16.6 = 223 V, a voltage drop of 7%
480–21.0 = 459 V, a voltage drop of 4%

Observation

For 240-V operation, the ampacity rating of the 8 gauge conductor is adequate, but the conductor has too much resistance, causing the voltage drop across the conductors to be too high. A 6 gauge conductor is needed to lower the resistance to 0.372 Ω at 38°C and the E_{con} to 10.4 V, resulting in a voltage drop of 4%.

When given a choice, it is always best to operate electrical loads at the highest recommended voltage because the current in the conductors will be less, as will the voltage drop across the conductors. Also, the voltage across the terminals of the load (the *load voltage*) will not vary as much from the specified source voltage. Finally, the smaller current (due to higher voltage) results in a considerable cost savings by permitting the use of smaller, less costly conductors, fuses, conduit, and disconnects.

Temperature Effect

Both copper and aluminum conductors respond to an increase in temperature by increasing their resistance. That is to say, copper and aluminum wire have a *positive temperature coefficient*, which means that the resistance of the wire varies directly with temperature: Temperature increases, resistance increases; temperature decreases, resistance decreases.

Current flow in a conductor has a power loss that may be computed by the I^2R form of the power equation:

$$P = I^2R \qquad (9.7)$$

where

P = average power dissipated by the resistance, in W
I = current in the conductor, in A
R = resistance of the conductor, in Ω

The complete set of power law equations for dc circuits is located in Table A10 of Appendix A.

The resistance of a copper (Cu) conductor due to the heating of the conductor caused by the power loss in the conductor can be determined with the following equation:

$$R = \frac{R_{\text{AWG}}(T + 234.5)}{254.5} \qquad (9.8)$$

where

R = resistance at temperature T, in Ω
R_{AWG} = resistance of the Cu conductor in the AWG at 20°C, in Ω
T = temperature of the operating conductor (conductor with current flowing through it), in °C

Example 9.5

Determine the following for a copper conductor operating at 60.0°C:

(a) The resistance of 250 ft of number 8 (AWG) wire.
(b) The power loss in the conductor when carrying 26.0 A of current while installed in a $\frac{3}{4}$-in-diameter conduit.

SOLUTION

(a) First find the resistance of the 250 ft of conductor at 20°C by using the resistance per 1000 ft from the AWG, and then determine the resistance at 60.0°C using Equation 9.8.

Given

AWG resistance at 20°C for number 8 wire is $(0.628\ \Omega)/(1000\ \text{ft})$.

$$\frac{0.628\ \Omega}{1000\ \text{ft}} \times 250\ \text{ft} = 0.157\ \Omega$$

$$R = \frac{R_{\text{AWG}}(T + 234.5)}{254.5}$$

Evaluate

$R_{\text{AWG}} = 0.157\ \Omega$

$T = 60.0°C$

Substitute

$$R = \frac{0.157(60.0 + 234.5)}{254.5}$$

Solve

$R = 0.182\ \Omega$

Observation

The information found in Table A2 in Appendix A may be used to convert from degrees Fahrenheit (°F) to degrees Celsius (°C).

(b) Find the I^2R power loss using Equation 9.7.

Given $P = I^2R$

Evaluate $I = 26.0$ A

                $R = 0.182$ Ω

Substitute $P = 26.0^2 \times 0.182$

Solve $P = 123$ W

Exercise 9.2

Solve the following problems. Make sketches to aid in solving the problems and structure your work so it follows in an orderly progression and can easily be checked.

1. The coil of a relay is specified to operate with 0.200 A of current. Determine the power loss in the coil when the operating resistance is 120 Ω.
2. Determine the resistance of 250 ft of number 14 (AWG) copper wire operating at 55°C.
3. The secondary of a transformer is wound from 200.0 ft of number 28 copper wire. Determine
   (a) The resistance of the wire at 20°C.
   (b) Its operating temperature of 74°C.
4. Determine the power loss in 640 ft (total length) of number 4 (AWG) copper conductor when carrying 48 A of current in a single-phase circuit at an operating temperature of 45°C.
5. The resistance of the windings (coil) of an armature in an electric motor was measured before operation and found to be 1.52 Ω at 20°C. Directly after the motor had been in operation for 20 min, the armature coil was once again measured and found to be 1.88 Ω. Determine the operating temperature of the windings.
6. Using Table A7 in Appendix A, determine the maximum ampacity of a number 10 AWG copper wire operating at an ambient temperature of 98.0°F when the insulation is type THHN. Also specify the maximum overcurrent protection for the conductor.

9.4 CONTROL DEVICES

Categories of Control Devices

Control devices play an important role in electric circuits by governing the application of power to the load. The control device is often referred to as a *pilot device* because it directs the electric current

- From the source to the load by enabling the completion of the electric circuit.

■ To stop flowing by opening the electric circuit.
■ To a different load by changing connections in the electric circuit.

Included in the control devices are manually operated switches, mechanically operated switches, and electromagnetically operated switches, which are covered in this section. In addition to these devices, a wide variety of solid-state switches, including the solid-state relay (Chapter 13), are in everyday use. These switches are based on semiconductors such as thyristors (SCR, triac, etc.) and metal-oxide-semiconductor field-effect transistors (MOSFETs). Unlike manually, mechanically, and electromagnetically operated switches, which have moving parts, the solid-state switch has no moving components in its structure.

Manually Operated Switches   Manually operated switches are used to make (close) and break (open) an electric circuit. That is, they are used to turn current *on* and *off* when someone pushes a button, throws a lever, or steps on a lever attached to the switch. Figure 9.7 pictures several types of manually operated switches used in controlling electric circuits along with their electrical symbols and designation letter(s).

Mechanically Operated Switches   Mechanically operated switches are used to make and break an electric circuit when the switch mechanism is actuated by non-human means. For example, a limit switch is activated by a stop attached to a moving machine table or a temperature switch is activated by a change in temperature.

Mechanically operated switches are "automatic" in that they change states when conditions around them change. Figure 9.8 illustrates some of the *automatic switches* (mechanically operated switches), along with their electrical symbols and designation letters. Included in this category of switch are the following:

■ The *limit switch* has a wedge symbol added to the contact arm symbol to represent the source of force used to activate the switch. The force is from the movement of the wedge-shaped stop that acts on the switch lever.
■ The *temperature switch* has a heating element symbol added to the contact arm symbol to represent the source of force used to activate the switch. The force is from the motion created by the change in the vapor pressure of the liquid in the immersion bulb. When the bulb temperature changes, the resulting change in vapor pressure of the liquid is transmitted to the bellows, causing the switch to operate.
■ The *pressure switch* has a diaphragm symbol added to the contact arm symbol to represent the source of force used to activate the switch. The force is from the change in pneumatic (air) or hydraulic (water, oil, etc.) pressure on the diaphragm (bellows) in the switch.
■ The *float switch* has a ball float symbol added to the contact arm symbol to represent the source of force used to activate the switch. The force is from the motion of the float in response to the change in the liquid level.

Electromagnetically Operated Switches   Electromagnetically operated switches are used to make and break an electric circuit with a mechanism called a relay. The *contact set* (switching unit), as pictured in Figure 9.9, is attached to the steel arma-

FIGURE 9.7
Manually operated switches: (a) Toggle switch (TGS); (b) Two-position selector switch (SS); (c) Push-button switch (PBS) *(Courtesy of the Allen Bradley Company, Inc.)*; (d) Foot switch (FTS). *(Courtesy of the Square D Company)* Note: NO is normally open; NC is normally closed.

ture of the relay, which is moved by the force of attraction between it and the electromagnet formed by the relay coil. The magnetic field and the resulting attractive force are created when electricity passes through the relay coil.

In summary, control devices (pilot devices) may be operated either as *momentary-contact devices* or as *maintained-contact devices*.

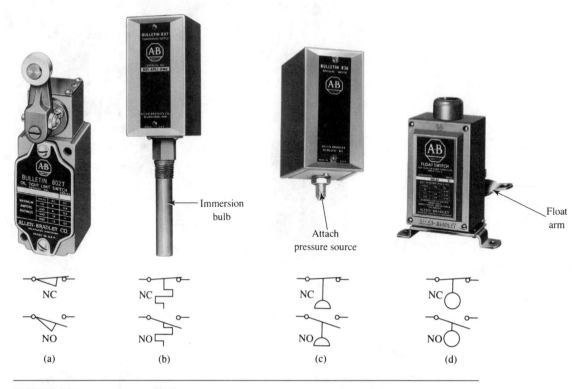

FIGURE 9.8
Mechanically operated switches (automatic switches): (a) Limit switch (LS); (b) Temperature switch
(TAS); (c) Pressure switch (PS); (d) Float switch (FS). *(Courtesy of the Allen Bradley Company, Inc.)*

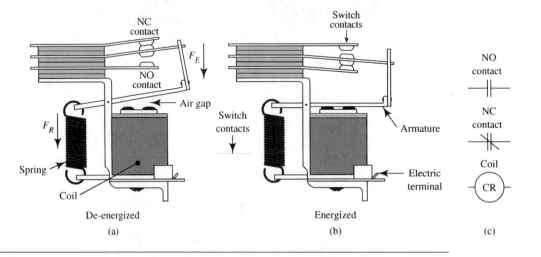

FIGURE 9.9
Electromagnetically operated switch such as a control relay (CR) shown: (a) De-energized with switch
contacts shown in their normal state; (b) Energized with switch contacts shown in their alternate state;
(c) As electrical symbols for coil and contacts.

Momentary-contact switches are characterized by a spring-loaded mechanism that returns the switch to its nonactivated position once the force of activation is removed. Manually activated push-button switches, used for START/STOP circuits, are good examples of momentary-contact devices. The normally open pushbutton switch, when pushed, will close. However, when the force on the pushbutton is released, the switch reverts to its normally open condition.

Maintained-contact switches are characterized by the fact that the switch mechanism changes state each time it is activated. The switch does not revert back to a previous state after activation. A manually activated toggle switch used to turn lights on and off is an example of a maintained-contact switch.

Contact Arrangements

Switches are defined in terms of *poles* and *throws*. The pole in a switch is the movable contact arm used to make or break the circuit; the throw is the number of positions the contact arm may move to complete a circuit connection. Figure 9.10 pictures the common contact arrangements used with manual, automatic, and electromagnetic (relay) switches.

The single-pole, single-throw (SPST) contact arrangement has one movable pole and one position to which the pole can be moved to complete the circuit connection. The SPST contact arrangement is used in simple on-off applications.

Additional contact arrangements include the single-pole, double-throw (SPDT), the double-pole, double-throw (DPDT), and the three-pole, double-throw (3PDT). The SPDT contact arrangement is designed to direct the circuit current in one of two possible directions by closing one of two circuits. The DPDT contact arrangement can connect two circuits in one of two directions, because the two poles move simultaneously from one position to the other. This action occurs because the poles are mechanically linked together, as indicated by the dashed lines in Figure 9.10.

FIGURE 9.10	Symbol	Operation	Abbreviation
Toggle switch electrical symbols with contact arrangements noted.		Single-pole, single-throw (shown closed)	SPST
		Single-pole, single-throw (shown open)	SPST
		Single-pole, double-throw	SPDT
		Double-pole, double-throw	DPDT
		Three-pole, double-throw	3PDT

9.5 CIRCUIT LOAD

Makeup of the Load

The load in an ac electric circuit is virtually any device that requires electric energy for its operation. The load, as shown in Figure 9.11, can be the induction motor of a machine, a dc power supply of a computer, or an electric toaster. Each of these loads uses electric energy to perform the task it has been designed to do.

The induction motor uses alternating current and voltage to create magnetic flux that causes the rotor and shaft to rotate and produce torque to drive a machine. The dc power supply changes the alternating current and voltage into direct current and voltage (dc) to power the electronic circuits of the computer. The electric toaster uses electric power to heat the heating elements so they produce radiant energy to toast bread.

The design of each of these load devices serves intended purposes that are very different from each other. However, when these devices are looked at from an electrical point of view (Figure 9.11), they all share something in common. Each of these loads may be represented as a *resistance* (R), an *inductance* (L), a *capacitance* (C), or any combination of these electrical elements. The load may also be represented in its simplest form as an impedance (Z).

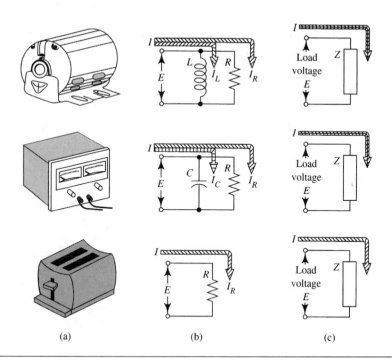

(a) (b) (c)

FIGURE 9.11
Typical electric loads: (a) Physical devices—induction motor, dc power supply, electric toaster; (b) Electrical equivalent load—resistance (R) and inductance (L), resistance (R) and capacitance (C), resistance (R); (c) Load represented as an impedance (Z).

Modeling the Electrical Load

In planning the conductor sizes for the load, the amount of current required for operation of the load is one of the first parameters that is needed. Once known, the current is then used to specify the conductor size by consulting the *National Electric Code* (NEC). The line drop is determined by using the current required by the load and the information in the American Wire Gauge (AWG).

To aid in the design of the conductive path (wiring), the load is looked at with various electrical models. These include a *current model* (Figure 9.11(b)) and an *impedance model* (Figure 9.11(c)). The current model, in which the load is divided up into resistive and reactive elements, is used to analyze the makeup of the current in the load. The impedance model is used to calculate the current in the load, which is the circuit current.

Electrical Elements in the Current Model

Three electrical elements are used to define the *parallel equivalent circuit* of the load in the current model. These elements are resistance (R), inductance (L), and capacitance (C). Each of these three quantities will be defined and their relation to alternating current will be studied.

Resistance **(R)**   The resistive element of the model of the load is that part of the load that *opposes the movement of current* and *dissipates power*. That is, the rate of the transformation of electrical energy (power) into other forms of work-producing power may be determined from the amount of **resistance** in the model of the load. It should be noted that resistance is the only part of the parallel equivalent circuit that dissipates power. Unlike resistance, inductance and capacitance do not dissipate power; instead, they momentarily store and then release energy.

The resistance in the load may be computed once the resistive component (I_R) of the circuit current is known. This current (I_R), along with the voltage (E) across the load (the load voltage), is used to compute the resistance in the load:

$$R = \frac{E}{I_R} \tag{9.9}$$

where

    R = resistance in the load, in Ω
    E = load voltage, in V
    I_R = resistive component of circuit current, in A

Inductance **(L)**   The inductive element of the model results from alternating current passing through the windings of wire in the coils of electric motors, solenoids, or relays. The resulting varying magnetic field (from the alternating current) sets up an induced voltage in the coil that *opposes any change* in the alternating current. The generation of an induced voltage in the coil by the changing current is called *self-induction*, and the induced voltage (which opposes any change in the current) is called the *counterelectromotive force* (**CEMF**).

The **inductance** of the coil is a measure of the induced voltage (CEMF) produced by the varying magnetic field of the alternating current as the current passes through the coil. The inductance is measured in units of henrys (H), the quantity of which depends on the number of turns in the coil and several physical parameters of the magnetic core, including the length of the magnetic path and the size of the cross-sectional area of the core. These parameters are related in the following equation.

$$L = \frac{\mu N^2 A}{l} \tag{9.10}$$

where

    L = inductance of the coil in the load, in H
    μ = permeability of the core, in H/m (Wb/A·m)
    N = number of turns in the coil, a unitless quantity
    A = cross-sectional area of the core, in m^2
    l = length of the magnetic pathway of the core, in m

Capacitance **(C)** The capacitance element of the model of the circuit load results from the use of a capacitor (condenser), an electrical device, or the capacitive properties resulting from the use of a synchronous motor. A *capacitor* is constructed of two metal, parallel plates separated by a dielectric, an insulating material. The **capacitance** of a capacitor is the measure of the potential difference (voltage) produced by the movement of electric charge (electric current) from one plate to the other. The capacitance is measured in units of farads (F), the quantity of which depends on several physical parameters of the metal, parallel plates and the kind of insulating material (dielectric) between the plates, as related in the following equation:

$$C = \frac{8.85 \times 10^{-12} \, kA}{d} \tag{9.11}$$

where

    C = capacitance in the load, in F
    k = dielectric constant (relative permittivity (ε_r)) of the insulating material (unitless)
    A = interface plate area, in m^2
    d = plate spacing, in m

Resistance and Reactance

Each of the three electrical elements contributes to the makeup of the model of the load of the electrical circuit, and each element responds to the movement of alternating current through it by causing work to be done.

    In the case of inductance and capacitance, the response to alternating current is called **reactance.** Reactance is the result of the inductor or capacitor storing electrical energy during part of the ac cycle and returning energy during another part of the same cycle. Since this action requires work to be done by the moving electric current, the presence of inductance or capacitance is seen by the alternating current as an opposition to its flow. This opposition, measured in ohms, is determined by the following equations.

Inductive Reactance

$$X_L = 2\pi f L \quad \text{or} \quad X_L = \omega L \tag{9.12}$$

where

X_L = inductive reactance of the inductance in the load due to the flow of alternating current, in Ω

f = frequency of the ac current, in Hz

L = inductance in the load, in H

ω = angular velocity of the ac current, measured in rad/s, equal to $2\pi f$.
Note: $\omega = 377$ rad/s when $f = 60$ Hz.

Capacitive Reactance

$$X_C = \frac{1}{(2\pi f C)} \quad \text{or} \quad X_C = \frac{1}{\omega C} \tag{9.13}$$

where

X_C = capacitive reactance of the capacitance in the load due to the flow of alternating current, in Ω

f = frequency of the ac current, in Hz

C = capacitance in the load, in F

ω = angular velocity of the ac current, measured in rad/s, equal to $2\pi f$.
Note: $\omega = 377$ rad/s when $f = 60$ Hz.

The *resistance* of the load has work done on it by the movement of alternating current, and this work is converted to other forms of energy. Since work is done on the resistance by the alternating current, the resistance is seen as an opposition to the current. This opposition is measured in units of ohms and is determined by the following equation:

$$R = \frac{E}{I_R} \quad \text{(repeated)} \tag{9.9}$$

where

R = resistance in the load, in Ω

E = load voltage, in V

I_R = resistive component of circuit current, in A

Example 9.6

Determine the resistance, inductive reactance, and the capacitive reactance in the current model of an electrical load when the following ac circuit parameters are present:

$E = 230$ V	$I_R = 8.3$ A
$L = 0.11$ H	$C = 0.20\ \mu$F
$f = 60$ Hz	$\omega = 377$ rad/s

SOLUTION   Find R, X_L, and X_C with Equations 9.9, 9.12, and 9.13.

Given	$R = \dfrac{E}{I_R}$	$X_L = 2\pi f L$	$X_C = \dfrac{1}{2\pi f C}$
Evaluate	$E = 230$ V	$f = 60$ Hz	$f = 60$ Hz
	$I_R = 8.3$ A	$L = 0.11$ H	$C = 0.20\ \mu\text{F}$
Substitute	$R = \dfrac{230}{8.3}$	$X_L = 2\pi(60)(0.11)$	$X_C = \dfrac{1}{377 \times 0.20 \times 10^{-6}}$
Solve	$R = 28\ \Omega$	$X_L = 41\ \Omega$	$X_C = 13\ \text{k}\Omega$

Reactive Components of Circuit Current

Even though the inductive and capacitive reactance of the load do not dissipate power, they do represent part of the load, and they do require current flow in the load for their operation, as was noted in Figure 9.11. Current in the inductive or capacitive element of the load may be determined with the following equation.

$$I_L = \frac{E}{X_L} \quad \text{and} \quad I_C = \frac{E}{X_C} \tag{9.14}$$

where

I_L = inductive component of circuit current, in A
I_C = capacitive component of circuit current, in A
E = load voltage, in V
X_L = inductive reactance of the inductive element of the load, in Ω
X_C = capacitive reactance of the capacitive element of the load, in Ω

*Example
9.7*

Determine the resistive and inductive components of the circuit current in the equivalent circuit of Figure 9.12 when the 60-Hz load voltage is 230 V. The resistive element of the circuit load model is 26.4 Ω and the inductive element of the load model is 0.467 H.

SOLUTION   Determine the inductive reactance (X_L) and then solve for I_L and I_R using Equations 9.14 and 9.9.

FIGURE 9.12
Circuit for Examples 9.7 and 9.8.

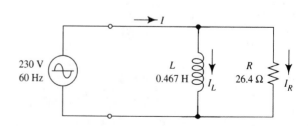

Given	$I_L = \dfrac{E}{X_L}$	$I_R = \dfrac{E}{R}$
Evaluate	$E = 230 \text{ V}$	$E = 230 \text{ V}$
	$X_L = 377 \times 0.467 = 176 \ \Omega$	$R = 26.4 \ \Omega$
Substitute	$I_L = \dfrac{230}{176}$	$I_R = \dfrac{230}{26.4}$
Solve	$I_L = 1.3 \text{ A}$	$I_R = 8.7 \text{ A}$

Vector Addition of Current

When the various components of the current model (I_R, I_L, I_C), are known, the currents may be added together to determine the total circuit current (I). The method for adding the currents requires the use of *vector addition* (rather than scaler addition) through the application of a form of the *Pythagorean theorem* to the current triangle (a right triangle), as noted in Figure 9.13. The information in Figure 9.13, along with the following equation, is used to compute the magnitude (amount) of circuit current (I):

$$I = \sqrt{I_R^2 + (I_C - I_L)^2} \tag{9.15}$$

where

I = circuit current, in A
I_R = resistive component of circuit current, in A
I_C = capacitive component of circuit current, in A
I_L = inductive component of circuit current, in A

This equation will be used in the following example to determine the circuit current (I) when several components of current are present in the load.

Example 9.8

Using the currents ($I_R = 8.7 \text{ A}, I_L = 1.3 \text{ A}$) determined in Example 9.7 and pictured in Figure 9.12, compute the circuit current (I) using vector addition.

FIGURE 9.13
Currents in the load elements of the current model are added using vector addition. (a) Vector diagram of the components of the circuit current (I). (b) The current triangle with the resistive (I_R) and reactive component (I_X) of the circuit current (I) noted.

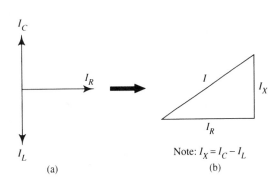

SOLUTION   Find the sum of I_R and I_L by applying Equation 9.15.

Given $I = \sqrt{I_R^2 + (I_C - I_L)^2}$

Evaluate $I_R = 8.7$ A

             $I_L = 1.3$ A

             $I_C = 0$

Substitute $I = \sqrt{8.7^2 + (-1.3)^2}$

Solve $I = \sqrt{77.38} = 8.8$ A

Impedance

The **impedance** of the circuit represents the total opposition to the flow of alternating current in the electrical circuit. The load impedance, expressed in ohms, can be determined by simply dividing the circuit current (I) into the load voltage, E, as noted in the first equation of the following set of equations.

$$Z = \frac{E}{I} \text{ and } I = \frac{E}{Z} \text{ and } E = IZ \qquad (9.16)$$

where

> Z = impedance of the load, in Ω
> E = load voltage, in V
> I = circuit current, in A

Example
9.9

Using Equation 9.16, determine the impedance of a circuit load given the following information: $E = 460$ V, $I = 23.0$ A.

SOLUTION   Find Z with voltage and current known.

Given $Z = \dfrac{E}{I}$

Evaluate $E = 460$ V

             $I = 23.0$ A

Substitute $Z = \dfrac{460}{23.0}$

Solve $Z = 20.0 \ \Omega$

The load impedance is an indicator of the magnitude (size) of the load. The circuit current, the current flowing from the source through the conductors into the load, can be quickly determined once the impedance is known. Since the circuit current is inversely related to the impedance of the load, the circuit current will be large when the impedance is low and small when the impedance is high.

Example
9.10

Determine the current in the following loads connected to a 208-V source when

(a) The impedance of fluorescent lights in an office is 16.6 Ω and the line drop is 2.0% of the source voltage.

(b) The impedance of a 3-hp motor is 13.8 Ω with a line drop of 3.0 V.

SOLUTION

(a) Find the current in the fluorescent lights by first determining the load voltage and then solving for the current using Equation 9.16.

Given

The source voltage of 208 V and a line drop of 2%,

$$I = \frac{E}{Z}$$

Evaluate

$E = 208 - (0.020 \times 208) = 204 \text{ V}$

$Z = 16.6 \text{ Ω}$

Substitute

$$I = \frac{204}{16.6}$$

Solve

$I = 12.3 \text{ A}$

(b) Find the current in the motor by first determining the load voltage and then solving for the current using Equation 9.16.

Given

The source voltage of 208 V and a line drop of 3 V,

$$I = \frac{E}{Z}$$

Evaluate

$E = 208 - 3 = 205 \text{ V}$

$Z = 13.8 \text{ Ω}$

Substitute

$$I = \frac{205}{13.8}$$

Solve

$I = 14.9 \text{ A}$

Exercise 9.3

Solve the following problems. Make sketches to aid in solving the problems and structure your work so it follows in an orderly progression and can easily be checked.

1. Determine the inductive reactance of the inductance of a motor of 0.12 H when the motor is operating from a 120-V, 60-Hz source.

2. Determine the capacitive reactance of 0.68 μF of capacitance resulting from a power supply operating from a 240-V, 60-Hz source.

3. An electric circuit is operating at 60 Hz with a line drop of 8.3 V when the circuit current is 7.3 A. The load impedance is 36.8 Ω. Determine

FIGURE 9.14
The equivalent circuit of the load of
Problem 5 in Exercise 9.3.

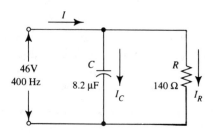

    (a) The load voltage.
    (b) The source voltage.
  4. The currents in an electric motor (the load) are 12.6 A resistive and 2.80 A inductive. Determine the impedance of the load when the source voltage is 240.0 V and the line drop is 5 V.
  5. The load voltage of an airborne electric load is 46 V, 400.0 Hz. Determine the circuit current (I) in the parallel equivalent circuit pictured in Figure 9.14 when the resistive element is 140 Ω and the capacitive element of the current model is 8.2 μF.

9.6 POWER IN THE LOAD

When determining the electrical power into the load of an electrical circuit, several parameters must be known. Included in these parameters are the source voltage (E), the circuit current (I), and the circuit phase angle used to determine the load power factor ($\cos \theta$).

Effective Value of Sinusoidal Current and Voltage

It is the effective current and voltage (indicated with capital letters I and E) that are used in computing the power in the load of an ac circuit. The effective current or voltage may be determined by measurement with an ac meter. The term *effective* results from the "effectiveness" of the current and voltage in producing power in the load. For a sinusoidal waveform (current or voltage), the effective value is $1/\sqrt{2}$ (or 0.707) times the maximum amplitude of the waveform (I_{max} or E_{max}).

Circuit Phase Angle

When a *resistive load* (R) is placed across the terminals of an alternating source, an alternating current will pass through the resistance, causing an ac voltage drop across the resistance, as illustrated in Figure 9.15(a). If the two waveforms, one representing the *load current* (the current in the load) and the other representing the *load voltage* (the voltage across the load), are placed over each other on a graph of amplitude (vertical axis) versus angular displacement (horizontal axis), the two waves are seen to start and finish together. Also, both waves reach their maximum positive amplitude as well as their maximum negative amplitude together. When

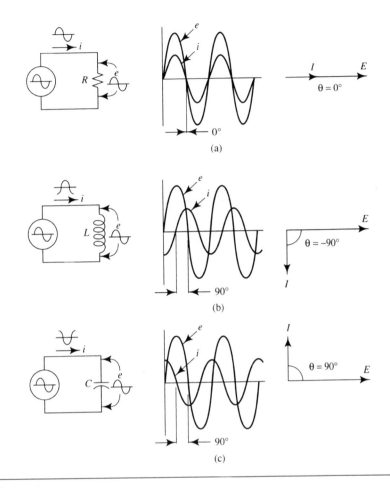

FIGURE 9.15
Circuit phase angle (θ). (a) The current is in phase ($\theta = 0°$) with the voltage when the load is resistive. (b) The current lags the voltage by 90° when the load is inductive ($\theta = -90°$). (c) The current leads the voltage by 90° when the load is capacitive ($\theta = 90°$).

this is the case (each wave passing through corresponding points in their waves at the same time), the two waves are *in phase* with each other.

The angular displacement between the voltage and current waves is the *circuit phase angle* (θ), and for a resistive load the phase angle is zero ($\theta = 0°$). The power factor for the resistive load is equal to 1 because cos 0° = 1.00.

> ➤ **As a Rule** Resistive loads have unity power factors (cos θ = 1) because the voltage and current waves are in phase.

When an *inductive load* (L) is placed across the terminals of an alternating source voltage, an alternating current passes through the inductance, creating an in-duced voltage (CEMF) that opposes the change in current. The opposition to the change in current in an inductor is the inductive reactance of the inductor. Besides opposing the change in current, the inductive reactance also causes the load current to *lag* behind the load voltage (a shift in phase), as illustrated in Figure 9.15(b). Thus, when the curves representing the load current and voltage are placed on the same axis, they are seen *out of phase* with each other. The load current lags the load voltage by 90° when the load is inductive.

Since the phase angle for an inductive load is −90° ($\theta = -90°$), then the power factor for the inductive load is zero because $\cos(-90°) = 0$.

> ➤ **As a Rule** Inductive loads have a zero power factor ($\cos \theta = 0$) because the load current lags the load voltage by 90°. The two waves are 90° out of phase with each other.

When a *capacitive load* (C) is placed across the terminals of an alternating source voltage, an alternating current passes through the capacitance, creating a capacitive reactance that opposes the change in the load voltage. Besides opposing the change in voltage, the capacitive reactance also causes the load current to *lead* the load voltage (a shift in phase), as illustrated in Figure 9.15(c). Thus, when the curves representing the load current and voltage are placed on the same axis, they are seen *out of phase* with each other. The load current leads the load voltage by 90° when the load is capacitive.

Since the angular displacement between the voltage and current waves is 90° ($\theta = 90°$), the power factor for the capacitive load is zero ($\cos 90° = 0$).

> ➤ **As a Rule** Capacitive loads have a zero power factor ($\cos \theta = 0$) because the load current leads (has a more positive amplitude at the origin than the voltage wave) the load voltage by 90°.

Power

The phase relationship between the source voltage and the circuit current plays an important role in the amount of average power (watts) delivered from the ac source to the load. This is noted in the formula for average power, which contains the power factor ($\cos \theta$) as part of the equation.

$$P = EI \cos \theta \qquad\qquad (9.17)$$

where

$$P = \text{average power, in W}$$
$$E = \text{effective voltage, in V}$$
$$I = \text{effective current, in A}$$
$$\cos \theta = \text{power factor, where } \theta \text{ is the circuit phase angle, the phase relationship}$$
between the source voltage and the circuit current, in degrees

Example
9.11

Using Equation 9.17, determine the average power (watts) in each of the three loads pictured in Figure 9.15 when the 60-Hz ac voltage across each of the loads is 480 V and

(a) Resistance (R) is 22.0 Ω.
(b) Inductance (L) is 0.22 H.
(c) Capacitance (C) is 22 μF.

SOLUTION (a) Find the current in the resistive load and then use Equation 9.17 to determine the average power.

Given
$$I = \frac{E}{R} = \frac{480}{22.0} = 21.8 \text{ A}$$

$$P = EI \cos (\theta)$$

Evaluate
$$E = 480 \text{ V}$$
$$I = 21.8 \text{ A}$$
$$\theta = 0°$$

Substitute $\quad P = 480 \times 21.8 \cos 0°$

Solve $\quad P = 10\ 464 = 10.5 \text{ kW}$

Observation The current of 21.8 A carries energy from the source to the load, where it is used to do work and produce average power (watts).

(b) First determine the inductive reactance and the current in the inductive load; finally, solve for the average power (watts).

Given $\quad X_L = \omega L = 377(0.22) = 82.9 \text{ Ω}$

$$I = \frac{E}{X_L} = \frac{480}{82.9} = 5.8 \text{ A}$$

$$P = EI \cos \theta$$

Evaluate
$$E = 480 \text{ V}$$
$$I = 5.8 \text{ A}$$
$$\theta = -90° \qquad \text{(inductive load)}$$

Substitute $\quad P = 480 \times 5.8 \cos(-90°)$

Solve $\quad P = 0 \text{ W}$

Observation For each alternation of the ac voltage, the inductor stores energy in its magnetic field and then releases the energy back to the circuit. Because no work is done over the period of the cycle, there is no average power (watts) developed in the inductive load. The inductive current of 5.8 A carries energy back and forth in the conductors between the source and the load, doing no work. However, the inductive current takes up current-carrying capacity in the conductors (copper wire).

(c) First determine the capacitive reactance and the current in the capacitive load; then solve for the average power (watts).

Given
$$X_C = \frac{1}{\omega C} = \frac{1}{377 \times 22 \times 10^{-6}} = 120.6 \text{ Ω}$$

$$I = \frac{E}{X_C} = \frac{480}{120.6} = 4.0 \text{ A}$$

$$P = EI \cos \theta$$

Evaluate $E = 480 \text{ V}$

$I = 4.0 \text{ A}$

$\theta = 90°$ (capacitive load)

Substitute $P = 480 \times 4.0 \cos (90°)$

Solve $P = 0 \text{ W}$

Observation   For each alternation of the ac voltage, the capacitor stores energy in its electric field and then releases the energy back to the circuit. Because no work is done over the period of the cycle, there is no average power (watts) developed in the capacitive load. The capacitive current of 4.0 A carries energy back and forth in the conductors between the source and the load, doing no work. However, the capacitive current takes up current-carrying capacity in the conductors.

The previous example showed that a purely resistive load dissipates (uses) power, does useful work, and has a unity power factor. Also, it showed that a purely inductive load and a purely capacitive load dissipates no power, does no work, and has a zero power factor. Although there is no power dissipated in reactive loads (inductors/capacitors), there is power moving back and forth from the source to the load and from the load to the source.

The power that does no work but whose current uses the *ampacity* (current-carrying capacity) of the conductor is termed *wattless power*. Wattless power is called *reactive power*, and it is symbolized in formulas by P_q, where the subscript q stands for quadrature, or 90°. This means that the reactive power is at quadrature (90°) with the work-producing average power. This concept is illustrated in Figure 9.16(a), where average power (P) is expressed in units of watts (W) and the reactive power (P_q) is expressed in units of *volt-amperes-reactive* (**VAR**).

The vector addition of average power and reactive power is shown in Figure 9.16(b) as the *apparent power*, the hypotenuse of the *power triangle*. The unit of

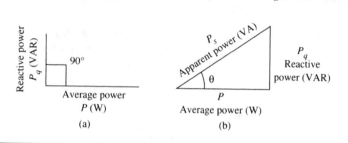

(a)   (b)

FIGURE 9.16
(a) The reactive power is at quadrature (90°) with the average power. (b) The power triangle shows the results of the vector addition of the average power (P) and the reactive power (P_q). The result is the apparent power (P_s). Theta (θ) is the circuit phase angle.

apparent power is the volt-ampere (VA), and its formula symbol is P_s. Both work-producing power and wattless power go out from the source, and the vector addition of these two powers is called the *apparent power* (P_s). Apparent power is used to specify electrical equipment, including transformers. The apparent power can be determined by multiplying the source voltage (E) and the circuit current (I), as noted in the following equation:

$$P_s = EI \qquad\qquad (9.18)$$

where

P_s = apparent power, in VA
E = effective source voltage, in V
I = effective circuit current, in A

Average Power

The average power measured in units of watts is often referred to as true power, real power, active power, or simply, power. The term *average power* comes from the process of averaging the power curve to obtain the power. The word *power* will be used to indicate average power (watts) in this text.

Figure 9.17(a) pictures the *instantaneous power curve* of a resistive load (voltage and current in phase). The curve is formed by multiplying the amplitudes of the

FIGURE 9.17
The average power (watts) developed in the load of the electric circuit may be determined from the instantaneous power curve by averaging the height of the curve. The average is indicated by the dashed line in each of the curves. (a) The average of the resistive power curve is at its midpoint, which is one-half of its maximum amplitude (height). (b) The average of the inductive power curve is zero, since the midpoint of the curve is the zero reference. (c) The average of the capacitive power curve is zero, since the midpoint of the curve is at the zero reference.

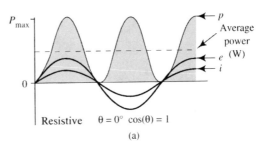

(a)

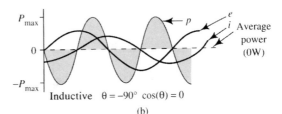

(b)

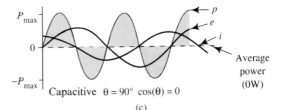

(c)

current and voltage waves at various points and plotting the results. The instantaneous power curve, as this curve is called, is the result of this process. The average of the power curve for the resistive load is one-half the maximum height (maximum power) of the curve ($P = \frac{1}{2}P_{max}$).

Example 9.11 showed that the power of inductive or capacitive loads is zero. This is verified by the instantaneous power curve of the inductor pictured in Figure 9.17(b) and the capacitor pictured in Figure 9.17(c). The average of each of these curves is zero since an equal amount of the curve is above and below the reference axis.

Loads made up of resistance and inductance or resistance and capacitance are quite common, and each has a power factor between 0 and 1. The power curves of these loads do not average to zero. The amount of power depends on the circuit phase angle, the angle between the voltage and current, and the resulting power factor ($\cos \theta$).

Example 9.12	Determine the load power in the circuit pictured in Figure 9.18 when the circuit phase angle is $-30.0°$, resulting in a lagging power factor of 0.866.
Observation	An inductive load has a *lagging power factor* because the load current lags the load voltage. A capacitive load has a *leading power factor* because the load current leads the load voltage.
SOLUTION	Find the power by applying Equation 9.17.
Given	$P = EI \cos \theta$
Evaluate	$E = 115 \text{ V}$
	$I = 11.6 \text{ A}$
	$\cos \theta = \cos(-30°) = 0.866$
Substitute	$P = 115 \times 11.6 \times 0.866$

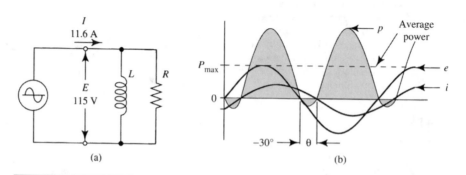

(a)   (b)

FIGURE 9.18
Circuit of Example 9.12. (a) Parallel equivalent circuit of the load. (b) Instantaneous power in the load. The average power above the axis is taken from the load, whereas the average power below the axis is returned to the load. The lagging power factor is equal to $\cos(30°) = 0.866$.

Solve	$P = 1155 = 1.16 \text{ kW}$
Observation	The power factor ($\cos \theta$) may be used as an *indicator* to understand the makeup of the load. As the power factor approaches unity, the load becomes more resistive, and as the power factor approaches zero, the load becomes more reactive.

Example 9.13

Determine the apparent power in the load of the circuit pictured in Figure 9.18.

SOLUTION	Find the apparent power by applying Equation 9.18.
Given	$P_s = EI$
Evaluate	$E = 115 \text{ V}$
	$I = 11.6 \text{ A}$
Substitute	$P_s = 115 \times 11.6$
Solve	$P_s = 1330 = 1.33 \text{ kVA}$
Observation	The average power and the reactive power may be determined by applying the trigonometric functions to the power triangle of Figure 9.16(b). The following equations result from the application of the trigonometric functions to the power triangle:

$$P_s = EI$$

$$P = P_s \cos \theta = EI \cos \theta$$

$$P_q = P_s \sin \theta = EI \sin \theta$$

$$\cos \theta = \frac{P}{P_s}$$

Efficiency

When a machine is powered from an electric motor, the power into the motor (P_{in}) is equal to the product of the circuit current, source voltage, and the power factor:

$$P_{in} = EI \cos \theta \qquad \text{(repeated)} \qquad \textbf{(9.17)}$$

where

P_{in} = average power, in W
E = effective voltage, in V
I = effective current, in A
$\cos \theta$ = power factor with phase angle in degrees (°)

The power out of the motor that is used to power the machine is equal to the horsepower (or watts) developed by the motor in converting the electric power into mechanical power. The maximum power from an electric motor may be determined by reading the power specification from the *motor nameplate*. When the shaft speed and torque are specified, the horsepower can be computed using the following equation.

$$P_{out} = \frac{\tau\omega}{5250} \quad \text{(equation 6.18)} \tag{9.19}$$

where

P_{out} = mechanical power out of the motor, in hp
τ = torque, in lb-ft
ω = angular velocity of the shaft of the motor, in rev/min

Note: 550 ft-lb/s = 1 hp and 746 W = 1 hp.

When power is converted from electrical to mechanical, some power is lost to the process and usually goes off as heat. This results in the input power being greater than the output power by the amount of lost power. The efficiency (η) of the conversion process is a mathematical ratio that indicates how well the motor performs in converting the input power to output power.

The motor nameplate, pictured in Figure 9.19, contains information needed to wire the motor. It also contains important information on the limits, operating range, and characteristics for operation of the motor. The nameplate power rating (horsepower, watts) is the maximum power that may be delivered to the shaft by the motor. In operation, the actual power delivered to the shaft by a motor depends upon the load placed on the shaft.

Figure 9.20 illustrates the process of converting power from current and voltage at the input to the motor into torque and angular velocity at the output of the machine, with losses accounted for by the efficiency:

$$\eta = \frac{P_{out}}{P_{in}} \tag{9.20}$$

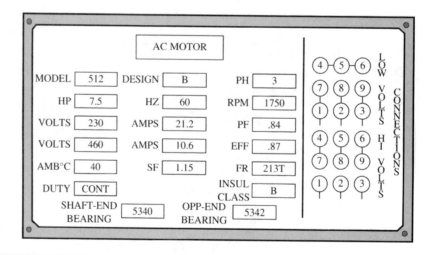

FIGURE 9.19

Motor nameplate data include horsepower (HP), angular velocity (RPM), frequency (HZ), current (AMPS), voltage (VOLTS), phase (PH), power factor (PF), efficiency (EFF), service factor (SF), ambient temperature (AMB°C), duty rating (DUTY), frame (FR), and design (DESIGN).

FIGURE 9.20
In a machine, power is converted from power into the motor of the machine to mechanical motion of the machine's mechanisms. Power is lost in the conversion process due to electrical resistance and mechanical friction.

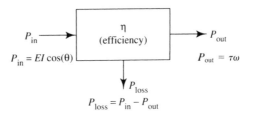

$P_{in} \longrightarrow$ η (efficiency) $\longrightarrow P_{out}$

$P_{in} = EI \cos(\theta)$ $\qquad P_{out} = \tau\omega$

P_{loss}

$P_{loss} = P_{in} - P_{out}$

where

η = efficiency, a ratio expressed as a decimal fraction with no units

P_{out} = power out of the motor or machine expressed in the same units (W, hp, ft-lb/s) as the power into the motor or machine

P_{in} = power into the motor or machine expressed in the same units (W, hp, ft-lb/s) as the power out of the motor or machine

Example 9.14

A 72-rev/min synchronous stepping motor is rated at 6.2 lb-ft when powered from a 120-V, 60-Hz source. Determine the efficiency of the motor when the current in the motor is 0.740 A and the power factor is 0.88.

SOLUTION   First determine the *power out* with Equation 9.19 and *power in* using Equation 9.17; then compute the efficiency using Equation 9.20.

Given $\eta = \dfrac{P_{out}}{P_{in}}$

Evaluate $P_{out} = \dfrac{\tau\omega}{5250} = \dfrac{6.2 \times 72}{5250} = 0.085 \text{ hp} \times 746 \dfrac{W}{hp} = 63 \text{ W}$

$P_{in} = EI \cos\theta = 120 \times 0.740 \times 0.88 = 78 \text{ W}$

Substitute $\eta = \dfrac{63}{78}$

Solve $\eta = 0.81$

Observation   One watt into the motor produces 0.81 W out of the motor to do useful work over time; 0.19 W is lost to heat resulting from mechanical friction and electrical resistance.

■

Exercise 9.4

Solve the following problems. Make sketches to aid in solving the problems and structure your work so it follows in an orderly progression and can easily be checked.

1. An inductive load has a circuit phase angle of $-18°$. Determine
   (a) The power factor.
   (b) Whether the power factor is leading or lagging.

2. Determine the power in a lighting circuit operating at 277 V, 60 Hz when the load current is 12.4 A and the power factor is 0.92.
3. Determine the power in a 230-V, 60-Hz motor circuit when the inductive component of load current is 3.6 A and the resistive component of load current is 8.7 A.
4. Determine the power lost to the conversion process when a 48-V, 400-Hz alternator provides 66 A while operating at 78% efficiency. The circuit power factor is 1.0.
5. The 480-V, 15-hp drive motor with a power factor of 0.90 is used to power a machining center. The motor has an efficiency of 0.85. Determine
   (a) The power out of the motor in watts when delivering 15 hp.
   (b) The electrical power into the motor.
   (c) The electric current in the motor.

CHAPTER SUMMARY

- The electric circuit in its simplest form is made up of four basic components: source, conductors, control device, and load.
- An alternating current in a conductor will set up a magnetic field around the conductor.
- Alternating current is measured in amperes.
- The relative motion between the conductor and a magnetic field induces an electromotive force across the ends of the conductor.
- Electromotive force (EMF) is measured in units of volts.
- The current passing through the conductors of an external circuit is directly proportional to the voltage across the circuit and inversely proportional to the impedance in the circuit.
- The total opposition to the flow of alternating current in the electrical circuit is called impedance.
- When a conductor carrying current is moved into a magnetic field, a reactionary force is developed to oppose the conductor's movement.
- The frequency of alternating voltage and current is measured in hertz (Hz).
- Electrical conductors are made from copper (Cu) and aluminum (Al), and each has a positive temperature coefficient.
- The power loss to the conductor depends on the length, diameter, material, and temperature of the conductor.
- The American Wire Gauge (AWG) is used to size conductors (wire).
- Manual, mechanical (automatic), and electromagnetic are all categories of switches.
- The electric circuit may be modeled as a current model or an impedance model, both of which are useful in the design of the circuit.
- The generation of an induced voltage in the coil by the changing current is called self-induction.
- The circuit phase angle, the angle between the voltage and current, is greater than zero when the current is out of phase with the voltage.
- The power factor is the cosine of the circuit phase angle—i.e., $\cos \theta$.
- Power in the load of an electrical circuit expressed in watts may be determined when the current, voltage, and power factor are known.
- Apparent power, measured in volt-amperes, is the vector summation of average power and reactive power.

● Efficiency is an indicator of the performance of the conversion from one form of power to another form; it is the ratio of power out to power in.

SELECTED TECHNICAL TERMS

The following technical terms, abbreviations, and acronyms are defined in the glossary located after Chapter 16. You are encouraged to use the glossary to aid your understanding and to test your knowledge of these important terms.

AWG	impedance
capacitance	inductance
CEMF	NEC
coulomb	reactance
ductility	resistance
EMF	VAR

END-OF-CHAPTER QUESTIONS

Write T if the statement is true and F if the statement is false.

1. The electric current carries energy from the source of EMF to the load.
2. The electric current in an electric circuit comes from the free electrons in the copper wire of the conductors.
3. Lenz's law explains how energy is moved through the electric circuit without work being done.
4. Copper is the best conductor of electricity.
5. Metal conductors (copper, aluminum, etc.) have resistance that increases with temperature due to the negative temperature coefficient of the metals.
6. The I^2R loss of energy in the conductor causes the resistance of the conductor to decrease in ohmic value.
7. The limit switch is a manually operated switch.
8. A dashed line in a schematic symbol indicates a mechanical link.
9. Vector addition uses a form of Lenz's law to combine the components of the circuit current.
10. The circuit phase is cos θ.

In the following, select the word that makes the statement true.

11. A six-pole generator has (two, three, four) pairs of poles.
12. Power is dissipated by (inductance, capacitance, resistance) in the electric load.
13. The angular velocity of a 60-Hz alternating current is (127, 187, 377) rad/s.
14. The toggle switch is a(n) (electromagnetically, automatically, manually) operated switch.
15. Impedance, the total opposition to alternating current, is measured in units of (henrys, ohms, farads).

TABLE 9.2 Summary of Formulas
Used in Chapter 9

Equation Number	Equation
9.1	$EMF = \dfrac{W}{Q}$
9.2	$I = \dfrac{Q}{t}$
9.3	$f = \dfrac{cyc}{t}$
9.4	$T = \dfrac{1}{f}$
9.5	$f = \dfrac{n_c p}{60}$
9.6	$E_{con} = IR_{con}$
9.7	$P = I^2 R$
9.8	$R = \dfrac{R_{AWG}(T + 234.5)}{254.5}$
9.9	$R = \dfrac{E}{I_R}$
9.10	$L = \dfrac{\mu N^2 A}{l}$
9.11	$C = \dfrac{8.85 \times 10^{-12} kA}{d}$
9.12	$X_L = 2\pi fL = \omega L$
9.13	$X_C = \dfrac{1}{2\pi fC} = \dfrac{1}{\omega C}$
9.14	$I_L = \dfrac{E}{X_L}$ and $I_C = \dfrac{E}{X_C}$
9.15	$I = \sqrt{I_R^2 + (I_C - I_L)^2}$
9.16	$Z = \dfrac{E}{I}, \quad I = \dfrac{E}{Z}, \quad E = IZ$
9.17	$P = EI \cos \theta$
9.18	$P_s = EI$
9.19	$P_{out} = \dfrac{\tau \omega}{5250}$
9.20	$\eta = \dfrac{P_{out}}{P_{in}}$

Answer each of the following questions with a short answer in the form of a complete sentence. Include a restatement of the question in your answer.

16. What is the definition of the circuit phase angle?
17. When given the choice, why is it best to operate loads at the higher of two recommended voltages?
18. What do the terms *leading* and *lagging power factor* indicate?
19. What effect does an increase in temperature have on the resistance of electrical conductors?
20. How is the term *in phase* related to current and voltage in an electric circuit?

END-OF-CHAPTER PROBLEMS

Solve the following problems. Make sketches to aid in solving the problems and structure your work so it follows in an orderly progression and can easily be checked. Table 9.2 summarizes the formulas used in Chapter 9.

1. Determine the current (in amperes) in an electric circuit when an electric motor is operated from a 230-V, 60-Hz source for 6.00 min. During the 6.00 min of operation, the motor takes in 720 kJ of energy.
2. Determine the minimum size AWG copper wire that may be used in a 120-V single-phase circuit when the distance from source to load is 100 ft and the load current is 12.3 A. The permitted line drop is 4.0 V (3.3%) at 20°C.
3. A 400-Hz, 8-pole generator is connected through a V-belt drive to an electric drive motor that is rotating at 1750 rev/min. Determine
    (a) The speed (revolutions per minute) of the generator shaft to produce 400.0 Hz alternating current.
    (b) The diameter of the pulley on the generator shaft when the diameter of the pulley on the motor shaft is 12.0 in.
4. Using the equivalent circuit of Figure 9.12 with the source voltage of 230-V, 60-Hz, determine
    (a) The current in the resistive element (I_R) when the resistance is 48.0 Ω.
    (b) The current in the inductive element (I_L) when the inductance is 0.250 H.
    (c) Combined load current (I) using vector addition.
5. An electric motor is operated from a 120-V, 60-Hz source for 18.0 min. During that time the motor takes in 370 kJ of electric energy to produce 212,000 ft-lb of work. Determine
    (a) The efficiency of the motor.
    (b) The horsepower out of the motor.
6. Determine the horsepower at the shaft of an electrical motor that requires 17.2 A at 240 V for its operation. The motor is 0.85 efficient, and the electric circuit power factor is 0.80.
7. A 460-V, three-phase motor is located 146.3 m from the power source. If the motor current is 32.0 A at an ambient temperature of 36°C and the motor is operating from a 3-ϕ, 480-V source, determine
    (a) The resistance of the copper conductors at 36°C.
    (b) The Ω/1000 ft of the copper conductor at 20°C.
    (c) The minimum gauge to insure no more than a 20-V (4%) drop at 36°C.
    (d) The ampacity from Table A7 of Appendix A for the minimum gauge at 36°C for type THW-2 insulation.

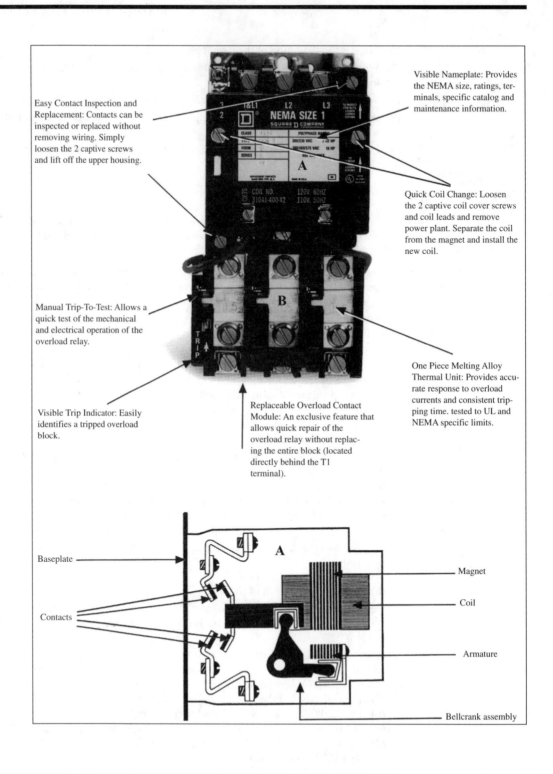

Easy Contact Inspection and Replacement: Contacts can be inspected or replaced without removing wiring. Simply loosen the 2 captive screws and lift off the upper housing.

Visible Nameplate: Provides the NEMA size, ratings, terminals, specific catalog and maintenance information.

Quick Coil Change: Loosen the 2 captive coil cover screws and coil leads and remove power plant. Separate the coil from the magnet and install the new coil.

Manual Trip-To-Test: Allows a quick test of the mechanical and electrical operation of the overload relay.

Visible Trip Indicator: Easily identifies a tripped overload block.

Replaceable Overload Contact Module: An exclusive feature that allows quick repair of the overload relay without replacing the entire block (located directly behind the T1 terminal).

One Piece Melting Alloy Thermal Unit: Provides accurate response to overload currents and consistent tripping time. tested to UL and NEMA specific limits.

Baseplate

Contacts

Magnet

Coil

Armature

Bellcrank assembly

Chapter 10   *Electromagnetic Circuits and Devices*

About the Illustration. *The motor starter is an electromagnetic device made up of two separate assemblies: the* contactor *module (A) and the* thermal overload *module (B).* (Courtesy of the Square D Company.) ■

INTRODUCTION

The electromagnetic circuit is analogous to the electric circuit in that each has a pathway through which an energy carrier circulates and each is energized by a source. The electromagnetic circuit uses iron and other ferromagnetic materials as a pathway for magnetic flux (the energy carrier) to pass. The electromagnetic circuit is energized by passing electric current through turns of wire wrapped around the closed metal pathway.

The nature of the electromagnetic series circuit is studied in this chapter, as are the magnetic parameters of flux, flux density, magnetomotive force, magnetic field strength, and permeability. Once electromagnetism is developed, the concept of the air gap in the magnetic pathway is introduced; this concept leads to an understanding of the solenoid principle (force results when an air gap is present in a magnetic circuit). Three sections dealing with the operation and application of devices based on the solenoid principle follow. The solenoid, contactor, and the control relay are introduced in this chapter.

CHAPTER CONTENTS

PERFORMANCE OBJECTIVES

Once you have read and studied each section; worked through each example with pencil, paper, and calculator; worked through the end-of-chapter problems; and answered the end-of-chapter questions, then you should be able to

- Apply the appropriate SI unit to magnetic quantities.
- Calculate with magnetic quantities.
- Understand the role that electric current plays in electromagnetism.
- Identify various kinds of solenoids.
- Describe the purpose of a shading coil.
- Understand the operation and use of a magnetic contactor.
- Differentiate between the load circuit and the coil circuit in a control relay.

10.1 ELECTROMAGNETIC CIRCUIT

Introduction

A series electromagnetic circuit, as illustrated in Figure 10.1, is formed when a closed ferromagnetic pathway, called the *core*, is wrapped with a coil of insulated wire, through which electric current is passed. The magnetic field resulting from the movement of electric current in the coil of wire is conducted through the core in the form of *magnetic flux*.

Magnetic Flux, ϕ

Magnetic flux, symbolized by the Greek letter ϕ (phi), circulates in the core of the magnetic circuit when electric current passes through the coil. The **flux** (ϕ) alternates back and forth in the core when alternating current is passed through the coil. However, when direct current is passed through the coil, the flux moves in only one direction in the core. Magnetic flux is measured in units of webers (Wb) in SI. Our discussion of electromagnetics will use only SI units.

Flux Density, B

For a given **ferromagnetic** material (iron, cobalt, nickel, and their alloys), the ability of the magnetic core to conduct flux is limited by the size of the cross section of the

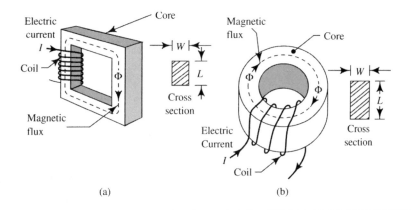

FIGURE 10.1

Examples of a series magnetic circuit with the core, coil, flux, and cross section indicated. (a) Laminated core made up of thin, insulated steel pieces used to lessen energy loss in the core. (b) Ring core, called a *toroid*.

pathway. Like electrical conductors, the ferromagnetic material of the core that conducts magnetic flux is limited in its capacity to carry flux. The term *saturated* is used to indicate that the core is operating at its maximum capacity and no more flux can be carried in the core.

Figure 10.2 pictures the magnetic characteristic curve of a simple alloy of iron and carbon called mild steel. Notice how the material's **flux density** (B) increases in a linear manner from point A to point B. Between points B and C, the flux density does not increase much, indicating that the material is becoming saturated. Above point C, the curve has become horizontal, indicating that the material is saturated.

The magnetic flux density, the amount of flux per unit cross-sectional area, is used as an indicator of the force of the magnetic flux. The force of the flux is directly proportional to the concentration (density) of the flux in the core.

The flux density in the core of a magnetic circuit is measured in teslas (T). Stated mathematically,

$$B = \frac{\phi}{A} \tag{10.1}$$

where

B = flux density, in Wb/m^2 or T
ϕ = flux in the core, in Wb
A = area, in m^2

Magnetomotive Force, mmf

In order for magnetic flux to be present in the core, a **magnetomotive force** (mmf) must be applied to the magnetic circuit. Magnetomotive force is the result of electric current circulating in the coil wrapped around the core. Unlike electromotive force

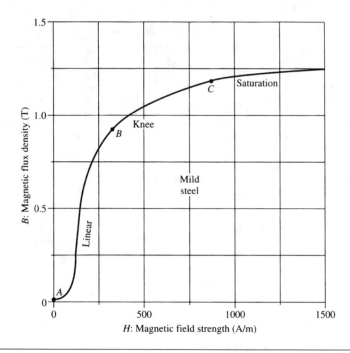

FIGURE 10.2

For a given cross-sectional area, ferromagnetic materials have a limited capacity for carrying magnetic flux. The mild steel core material easily carries flux between 0 and 0.9 T (points A to B); however, the capacity of the material begins to fill up in the knee region (points B to C), and the material is saturated (full) above point C.

in the electric circuit, which has a fixed value, the magnetomotive force in the magnetic circuit is multivalued, because its magnitude (amount) increases with each turn of wire added to the coil.

Thus, a one-turn coil with 2 A of electric current circulating in it has a magnetomotive force of "2 amp-turns" (2 A), whereas a ten-turn coil with the same 2 A of current has a magnetomotive force of "20 amp-turns" (20 A). Thus,

$$\text{mmf} = NI \qquad\qquad (10.2)$$

where

mmf = magnetomotive force, in A (commonly read "amp-turns")
N = number of turns traveled by the electric current in the coil, a unitless quantity
I = electric current in the coil, in A

| *Example* |
| *10.1* |

The magnetic circuit of Figure 10.1(a) has a core cross-sectional area of 24.0 cm² ($A = L \times W = 4.00$ cm $\times$ 6.00 cm) and a flux of 1.80 mWb circulating in it. The flux

results from a 60-Hz alternating current of 0.750 A passing through a 200-turn coil wrapped around the core of the magnetic circuit. Determine
   (a) The flux density in the core.
   (b) The magnetomotive force used to set up the flux in the core.

SOLUTION   (a) Find the flux density using Equation 10.1.

Given $B = \dfrac{\phi}{A}$

Evaluate $\phi = 1.80 \, \text{mWb}$

                $A = 24.0 \, \text{cm}^2 = 4.00 \times 10^{-2} \times 6.00 \times 10^{-2} = 24.0 \times 10^{-4} \, \text{m}^2$

Substitute $B = \dfrac{1.80 \times 10^{-3}}{24.0 \times 10^{-4}}$

Solve $B = 0.75 \, \text{T}$

                (b) Find the magnetomotive force using Equation 10.2.

Given $\text{mmf} = NI$

Evaluate $N = 200$

                $I = 0.750 \, \text{A}$

Substitute $\text{mmf} = 200 \times 0.750$

Solve $\text{mmf} = 150 \, \text{A (amp-turns)}$

In the magnetic circuit of the previous example, a flux of 1.8 mWb circulates back and forth in the core, changing directions 60 times a second. This action is caused by the magnetomotive force of 150 A, which results from a 60-Hz alternating current passing through the 200-turn coil of the magnetic circuit.

Magnetic Field Strength (H)

In addition to the flux/cross-sectional area (related by the flux density) and the turns of wire/electric current (related by the magnetomotive force), the electromagnetic circuit has length over which the flux must travel.

The magnetic field strength takes into consideration the length of the path taken by the magnetic flux as it circulates in the core. In Figure 10.3, there are two magnetic circuits pictured. Each circuit is made from the same ferromagnetic material, each has the same cross-sectional area, and each is initially magnetized with the same magnetomotive force. The only physical difference between them is the average length of the magnetic path, as illustrated in the figure. Because the flux in the smaller core travels a shorter distance when compared with the distance traveled in the larger core, the magnetic field strength (H) is greater in the shorter path. This concept is verified by the mathematical statement for magnetic field strength.

FIGURE 10.3
Two toroid cores with equal cross-sectional area are magnetized with equal magnetomotive force. The magnetic field strengths (H) are different because the path lengths of the circuit are unequal. The flux densities (B) are also unequal, since $B \propto H$.

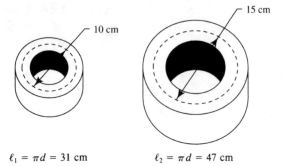

$\ell_1 = \pi d = 31$ cm $\ell_2 = \pi d = 47$ cm

$$H = \frac{NI}{\ell} = \frac{\text{mmf}}{\ell} \qquad\qquad (10.3)$$

where

H = magnetic field strength of the magnetic path, in A/m (amp-turns per meter of path length)

mmf = magnetomotive force, in A (amp-turns)

ℓ = average length of the magnetic path, in m

I = electric current, in A

N = number of turns, unitless quantity

Example 10.2

The two cast-iron magnetic circuits of Figure 10.3 are each magnetized by winding 50 turns of wire around the core and passing 4.0 A of electric current through the turns. Determine the magnetic field strength of each electromagnetic circuit.

SOLUTION   Find the magnetic field strength with Equation 10.3.

Given

$H_1 = \dfrac{NI}{\ell_1}$ $\qquad\qquad$ $H_2 = \dfrac{NI}{\ell_2}$

Evaluate

$N = 50$ $\qquad\qquad\qquad$ $N = 50$

$I = 4.0$ A $\qquad\qquad\quad$ $I = 4.0$ A

$\ell_1 = 31$ cm $= 0.31$ m $\qquad$ $\ell_2 = 47$ cm $= 0.47$ m

Substitute

$H_1 = \dfrac{50 \times 4.0}{0.31}$ $\qquad$ $H_2 = \dfrac{50 \times 4.0}{0.47}$

Solve

$H_1 = 645 \Rightarrow 6\overline{5}0$ A/m $\qquad$ $H_2 = 426 \Rightarrow 4\overline{3}0$ A/m

Observation

All physical parameters being equal in the magnetic circuit except length, the shorter magnetic path will have the greater magnetic field strength.

Permeability of the Core Material

You now know that the magnetic field strength (H) varies inversely with the length of the magnetic circuit when a constant magnetomotive force is applied to the magnetic circuit. However, when the field strength varies, the flux density (B) also varies, because the flux density (B) is related to the magnetic field strength (H) through the permeability of the core material (μ).

$$B = \mu H \qquad\qquad (10.4)$$

where

B = magnetic flux density, in T
μ = permeability of the material in the magnetic core, in Wb/A·m
H = magnetic field strength, in A/m

The **permeability** of a substance is an indication of its ability to carry magnetic flux when acted on by a magnetomotive force. Because the permeability of the ferromagnetic core material is not linear, a magnetization curve that relates flux density (B) to the magnetic field strength (H) must be used when computing circuit parameters of the series magnetic circuit. The following examples demonstrate the use of the *B-H magnetization curve*.

Example 10.3

Determine the permeability of the mild steel core of the series magnetic circuit pictured in Figure 10.1(a) when the magnetic field strength of the circuit is 250 A/m.

SOLUTION
Find the permeability using Equation 10.4 and the *B-H* curve for mild steel (Figure 10.2).

Given
$B = \mu H$; solve for μ.

$$\mu = \frac{B}{H}$$

Evaluate
$H = 250$ A/m

$B = 0.8$ T   (read from Figure 10.2)

Observation
Find B from the vertical axis of the mild steel *B-H* curve (Figure 10.2). Enter the curve at $H = 250$ A/m and more vertically up until the magnetization curve is intersected. Then move horizontally to the left and read the value of flux density ($B = 0.8$ T) from the vertical axis of the curve.

Substitute
$$\mu = \frac{0.8}{250}$$

Solve
$\mu = 3.2 \times 10^{-3} \Rightarrow 3$ mWb/A·m

FIGURE 10.4

B-H magnetization curve for cast iron, a ferromagnetic material. Note the nonlinear magnetization characteristic of cast iron.

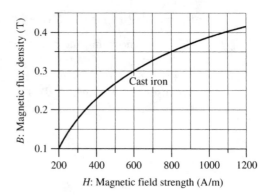

Example 10.4

Using the magnetization curve for cast iron pictured in Figure 10.4 and the computed values of magnetomotive force of Example 10.2, determine

   (a) The flux density in each of the cast-iron cores of Figure 10.3.

   (b) The flux in each of the cores when the cross-sectional area is 20.0 cm².

SOLUTION **(a)** Find the flux density using the magnetomotive force determined in Example 10.2 and the magnetization curve in Figure 10.4.

Given $H_1 = 6\overline{5}0$ A/m $H_2 = 4\overline{3}0$ A/m

Solve   Enter the curve at the value of $H_1 = 6\overline{5}0$ A/m. Move vertically up to intersect the magnetization curve, then move horizontally to the left. Read the value of flux density ($B = 0.31$ T) from the vertical axis of the curve. Repeat the process for H_2: $B_1 = 0.31$ T and $B_2 = 0.25$ T.

   (b) Find the flux using Equation 10.1 ($B = \phi/A$); solve for ϕ.

Given $\phi_1 = B_1 \times A$ $\phi_2 = B_2 \times A$

Evaluate $B_1 = 0.31$ T $B_2 = 0.25$ T

   $A = 20.0$ cm² $= 20.0 \times 10^{-4}$ m²

Substitute $\phi_1 = 0.31 \times 20.0 \times 10^{-4}$ $\phi_2 = 0.25 \times 20.0 \times 10^{-4}$

Solve $\phi_1 = 0.62$ mWb $\phi_2 = 0.50$ mWb

Exercise 10.1

Solve the following problems. Make sketches to aid in solving the problems and structure your work so it follows in an orderly progression and can easily be checked.

   1. Determine the magnetic flux (ϕ) in a ferromagnetic circuit with a cross-sectional area of 0.45 mm² when the flux density (B) is 280 mT.

2. Determine the flux density (*B*) in a mild-steel torroid core when the flux is 170 μWb and the cross section is round with a diameter of 1.8 cm.
3. Determine the magnetomotive force (mmf) of a 620-turn coil wound on a cast-iron core when the electric current in the coil is 120 mA.
4. Determine the number of turns of wire needed to produce a magnetic field strength of 250 A/m in a 33-cm long, closed, mild-steel magnetic circuit. The current in the coil around the core is 330 mA.
5. Determine the permeability of a cast-iron core made from the material in Figure 10.4 when the magnetic field strength (*H*) is 600 A/m.

10.2 SOLENOID

Air Gap in the Magnetic Circuit

When an air gap is introduced into the magnetic circuit, the magnetomotive force must be greatly increased to maintain a given flux density. Thus, more turns, an increased current, or both are needed in the coil to keep flux in the core at a constant level.

The effect of even a very small air gap (.004 in, 0.102 mm) is to alter the shape of the *B-H* magnetization curve for the ferromagnetic core material. In effect, the permeability of the material is lessened, which prevents the core from being saturated by excessively large currents in the coil.

Besides air gaps purposely designed into a magnetic circuit, an air gap is needed for mechanical clearance between moving parts in electric motors, analog meter movements, and loudspeakers and for the operation of solenoids and relays. Thus, out of necessity, most magnetic circuits include an air gap.

Force Resulting from an Air Gap

One of the properties of magnetic lines of force is that they form paths that are as short as possible. With the addition of an air gap in a magnetic circuit, the magnetic lines of force exert a tractive force on the magnetic surfaces (poles) on either side of the air gap in an effort to shorten the path by closing the gap. Assuming that the flux in the magnetic circuit is constant, then this force may be determined by the following formula, which relates the flux density (*B*), the cross-sectional area (*A*) of the air gap, and the permeability of the air gap (μ_0):

$$F = \frac{B^2 A}{2\mu_0} \tag{10.5}$$

where

F = force, in N
B = flux density, in T
A = cross-sectional area of the air gap, in m^2
μ_0 = permeability of free space (air); a constant equal to $4\pi \times 10^{-7}$ Wb/A·m

Solenoid

The tractive force resulting from an air gap in a magnetic circuit is the basis of operation of the **solenoid** and is commonly referred to as the *solenoid principle*. The solenoid principle is used in the design and operation of electromagnetic switching devices, including the contactor, the motor starter, and the control relay. The solenoid is a basic electromagnetic device constructed of three main parts, as illustrated in Figure 10.5:

- The coil, which carries current and energizes the magnetic circuit with magnetomotive force (amp-turns)
- The iron frame, or case, which provides the ferromagnetic pathway of the magnetic circuit for the movement of flux
- The plunger, the movable element that converts magnetic attraction through the closing of the air gap into rectilinear or rotary motion

Example 10.5

Determine the force (N) between the faces of the poles (plunger and stationary pole) of the C-frame solenoid pictured in Figure 10.5 when it is energized. When energized, the face of the plunger is drawn into the coil and rests on the stationary pole. The flux density in the mild-steel magnetic pathway is 1.1 T, and the diameter of the mild-steel plunger is 8.50 mm.

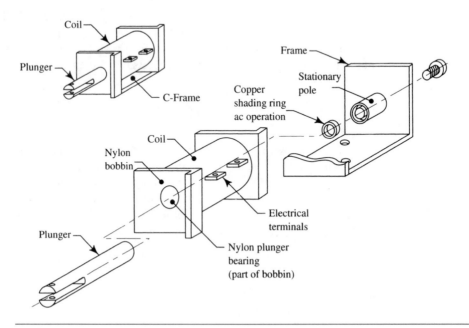

FIGURE 10.5
Essential parts of the open frame solenoid, C-frame type.

| SOLUTION | Find the force in newtons using Equation 10.5. |

Given

$$F = \frac{B^2 A}{2\mu_0}$$

Evaluate

$$B = 1.1 \text{ T}$$

$$A = \frac{\pi d^2}{4} = \frac{\pi (8.50 \times 10^{-3})^2}{4} = 56.7 \times 10^{-6} \text{ m}^2$$

$$\mu_0 = 4\pi \times 10^{-7} \text{ Wb/A·m}$$

Substitute

$$F = \frac{1.1^2 \times 56.7 \times 10^{-6}}{2 \times 4\pi \times 10^{-7}}$$

Solve

$$F = 27 \text{ N} \quad (6 \text{ lb})$$

The principle of operation of the solenoid calls for the transformation of electrical power resulting from an electric current and voltage into mechanical power in the form of motion (velocity) and force or motion (angular velocity) and torque. The motion is **rectilinear motion** in push-pull types of *linear solenoids* and rotational in *rotary solenoids*. Each type of motion serves to actuate a mechanism.

Linear Solenoid

Figure 10.6 pictures three types of linear solenoids (open frame, tubular, and low profile), along with a rotary solenoid.

- *Open-frame* types are less powerful and are the least expensive solenoids. The D-frame type is more rugged than the less costly C-frame type. The D-frame type also produces more force than the C-frame type for a given size. The open-frame solenoid is available for ac or dc operation with a wide choice of operating voltages.
- *Tubular* types are available for dc operation only in either a push or pull format. The tubular solenoid is characterized by quiet operation, long stroke length, high force per volume, high resistance to shock and vibration, and very long life.
- *Low-profile* types are faster with greater force than any other type when a short stroke is needed. They are also very rugged and dependable. Low-profile solenoids are available for dc operation only.

The ac operation of the solenoid and other ac-powered electromagnetically actuated devices requires the use of a *shading ring* (shading coil) to prevent chatter. Chatter occurs when the magnetic flux drops to zero as the alternating current in the coil passes through zero, allowing the plunger in the solenoid to drop out. The shading ring or coil (see Figure 10.5) is a one-turn copper coil that has current induced in it by magnetic induction. The resulting current in the shading coil produces a magnetic flux in the core that is out of phase with the principal magnetic flux produced by the multiturn coil. The shading coil, because of its phase difference,

Open-frame solenoid
(D-frame type)

Pull

Push
Tubular solenoids

Low-profile solenoids
(push/pull)

(a)

Rotary solenoid

(b)

(c)

FIGURE 10.6
Types of solenoids and the electrical symbol for a solenoid: (a) Linear type (open frame, tubular, and low profile); (b) Rotary type; (c) Electrical symbol. *(Courtesy of Lucas Automation and Control Engineering, Inc.)*

prevents the flux in the magnetic pathway from ever reaching zero, thereby keeping the plunger firmly seated and preventing chatter between the plunger and the stationary pole.

Electromagnetic devices operated from an ac source will inherently have hum. Some devices come with a mechanical adjustment to minimize the "buzz" associated with ac operation. When quiet operation is needed, then dc-operated electromagnetic devices must be used. General solenoid maintenance is included in Appendix C, Section C5.

Force and Stroke

The operation of the solenoid for a given stroke length requires that the initial flux density in the magnetic circuit be sufficient to provide enough closing force to accelerate (move) the mass of the plunger against the reactionary forces (load, return

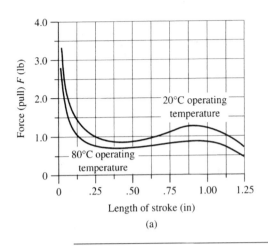 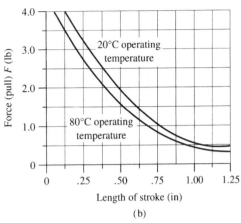

FIGURE 10.7

Graph of the closing force in pounds versus the stroke length of the plunger in inches for a D-frame solenoid operating at 20°C (cold) and at 80°C (hot) for: (a) A continuous-duty, alternating-current solenoid; (b) A continuous-duty, direct-current solenoid.

spring, gravity, and friction). Figure 10.7 pictures the operating curve of force versus stroke length, and Table 10.1 lists typical values of force versus stroke length for a family of pull-type, 120-V, 60-Hz, continuous-duty linear solenoids.

This information shows that the force of the solenoid is inversely proportional to the stroke length. That is, the longer the stroke, the smaller the closure force. This, of course, is due to the increase in the *reluctance* (the opposition to the flow of magnetic flux) resulting from an increase in the length of the air gap between the poles.

Current and Stroke

An *ac solenoid* with the plunger fully extended (maximum stroke) will have a coil current (inrush current) that is *five to ten times* as great as its corresponding seated current (often called the sealed current), as noted in Figure 10.8(a). The large inrush current for maximum stroke length is due to the lack of self-induction in the

TABLE 10.1   Specifications for a 120-V, 60-Hz Family of Continuous-Duty AC Solenoids

Coil Resistance (Ω)	Maximum Current (A)	Seated Current (A)	Force (lb) Versus Stroke Length				
			.25 in	.50 in	.75 in	1.0 in	1.25 in
21	4.9	0.4	6.4	6.4	6.0	4.0	—
15	6.5	0.5	9.0	9.1	8.4	5.8	—
6.5	12	0.7	12	12	15	16	12

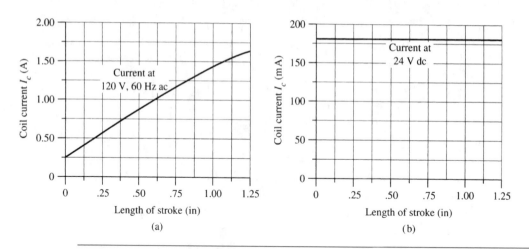

FIGURE 10.8

Graph of the core current versus the plunger stroke length for comparable open, D-frame, pull-type solenoids operated from: (a) A 120-V, 60-Hz alternating-current (ac) source with a current of 1.6 A at a maximum stroke length and 0.25 A with the plunger seated (the *sealed current*); (b) A 24-V direct-current (dc) source with a maximum stroke length current of 180 mA and a sealed current of 180 mA (no change).

coil, resulting in little to no inductive reactance to oppose the current in the coil of the magnetic circuit. Thus, the initial current is large in ac solenoids with the plunger extended. The current in the coil is small when the plunger is seated due to the full effect of the inductive reactance (opposition) to the flow of current.

The variation in the current in an ac solenoid works to its advantage, since the mmf is increased with an increase in current in the coil. The increase in mmf serves to overcome the increase in reluctance in the magnetic circuit when the plunger is extended, and the mmf tends to maintain a reasonable flux density as the plunger moves from its open to its closed position. This, however, is not the case with *dc solenoids*, since the current is limited by the resistance of the copper wire in the coil and not by the impedance (reactance and resistance) of the coil, as with ac operation. The current is constant in dc solenoids, as noted in Figure 10.8(b).

Temperature Effects

The copper coil of an electromagnetic device increases in resistance with an increase in temperature. The increase in resistance reduces the current in the coil of the solenoid, which, in turn, causes the amp-turns (mmf) and the output force to decrease. The 80°C curve of Figure 10.7 pictures the lessening of force due to an increase in the operating temperature.

As noted in Figure 10.7, the temperature increase is a direct result of the operation of the device. The wire of the coil has resistance that dissipates power (I^2R) and heats the coil. Since the coil is wound around the magnetic core, it is also heated by the heat generated in the core by the changing magnetic flux in ac operative devices.

The core losses in the magnetic circuit are due to hysteresis and eddy currents. **Eddy currents** are electrical currents induced in the core that create magnetic flux opposing the flux in the core. In ac operative devices, the core is laminated (made up of thin sheets of metal) and insulated to minimize eddy current losses.

Hysteresis is caused by the change in the direction of the alternating current and the inability of the magnetic flux in the core to change direction in step with the current. The flux in the core cannot change instantaneously, so when the electric current and its magnetic field change direction, the flux in the core is still moving in the previous direction. The two fields are out of step with each other, and they interfere with one another. Hysteresis is the name given to the lag between the coil magnetic field and the flux in the core. The "magnetic friction" between the two magnetic fields causes heating in the material of the core. Hysteresis may be lessened by adding a small amount of silicon to the alloy of the steel used in the core.

The core losses coupled with coil losses (copper losses) produce heat that raises the operating temperature of the electromagnetic device. The increased resistance of the copper wire in the coil due to the elevated temperature may be determined using Equation 9.8. Equation 9.8 is repeated here for convenience.

$$R = \frac{R_{AWG}\,(T + 234.5)}{254.5} \qquad \text{(repeated)} \qquad \textbf{(9.8)}$$

where

R = resistance at temperature T, in Ω

R_{AWG} = resistance of the Cu conductor at 20°C, in Ω

T = temperature of the operating coil (conductor with current flowing through it), in °C

Example
10.6

The coil resistance of a 24.0-V dc tubular solenoid coil is specified as 37.0 Ω at 20°C. The coil is constructed by winding 2710 turns of number 27 AWG solid copper wire onto the core. Determine

(a) The mmf, amp-turns, of the coil when it is cold (20°C).

(b) The mmf, amp-turns, of the coil when it is hot (operating at 85°C).

SOLUTION   (a) Find the amp-turns at 20°C using Equation 10.2.

Given   mmf = NI

Evaluate N = 2710 turns

$$I = \frac{E}{R} = \frac{24.0}{37.0} = 0.650 \text{ A}$$

Substitute mmf = 2710 × 0.650

Solve   mmf = 1.76 kA (amp-turns)

(b) Find the amp-turns (mmf) of the operating (85°C) solenoid using Equation 9.8 to determine the resistance of the coil and Equation 10.2 to determine the magnetomotive force.

Observation   First determine the resistance of the coil.

Given $R = \dfrac{R_{AWG}(T + 234.5)}{254.5}$

Evaluate $R_{AWG} = 37.0\ \Omega$

    $T = 85°C$

Substitute $R = \dfrac{37.0(85 + 234.5)}{254.5}$

Solve $R = 46.4\ \Omega$

Observation   Use the resistance of the operating (hot) coil to determine the current in the coil and the mmf.

Given $mmf = NI$

Evaluate $N = 2710$ turns

    $I = \dfrac{E}{R} = \dfrac{24.0}{46.5} = 0.516\ A$

Substitute $mmf = 2710 \times 0.516 = 1.4\ kA$ (amp-turns)

Observation   The mmf of the operating magnetic circuit (85°C) is reduced 20% by the increase in the coil resistance from 37 Ω (cold) to 46 Ω (hot).

> **As a Rule** When specifying the force of operation of a solenoid, the needed force is multiplied by a service factor (SF) of 1.3 to 1.5. For example, when 4.2 lb of force is needed then 5.5 to 6.3 lb is specified. This added margin of force compensates for the increasing resistance and the lowering of the magnetomotive force in the magnetic circuit due to an increase in temperature.

Applications

Because of their long life and very high reliability, solenoids of all types are widely applied as prime movers in mechanisms associated with computer peripheral equipment, business machines, electrical door locks, electrical contactors, flow valves, electrical brakes, etc.

10.3 MAGNETIC CONTACTOR

Introduction

The **magnetic contactor** is an electromagnetically operated switch that serves to open and close high-energy electric circuits. The contactor is designed to connect the source voltage to the load (lights, heaters, transformers, etc.) through sets of high-current–low-resistance contacts.

The solenoid principle is at the heart of the magnetic contactor. An armature (rather than a plunger) is electromagnetically pulled closed when voltage is applied to the coil circuit of the contactor. Figure 10.9 pictures some of the mechanisms that are attached to the armature to actuate sets of electrical contacts and to create an electromagnetically operated switch.

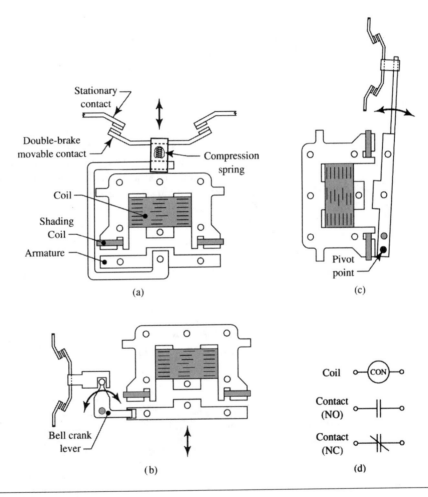

FIGURE 10.9

The magnetic force of attraction is used to close the armature of the solenoid to activate the double-brake contacts of the "main-line" electrical circuit. (a) The vertical action contactor uses gravity to open the contacts once the energy is removed from the coil. (b) The bell-crank contactor uses a bell-crank lever to transmit motion and force to the contacts once the coil is energized. (c) The clapper contactor uses the pivotal motion of the "clapper" (the armature) to activate the contacts. (d) In a control diagram, the contactor coil is represented by a circle with CON and the contacts are represented by two parallel lines. NO represents normally open contacts and NC represents normally closed contacts.

TABLE 10.2 Contactor Current Rating

NEMA Size	AC Current (A)	DC Current (A)
00	9	—
0	18	—
1	27	25
2	45	50
3	90	100
4	135	150
5	270	300
6	540	600
7	810	900
8	1215	1350
9	2250	2500

Magnetic contactors are rated by the National Electrical Manufacturers Association (NEMA) by a size number that corresponds to the attached load. The sizes, ranging from 00 to 9, are listed in Table 10.2, along with the load current carried by each set of contacts. Take, for example, the *three-pole ac contactor* used to control the three 20-A lighting circuits of Figure 10.10. The contactor must have three poles, one for each of the three lighting circuits. Based on the current in each load, a NEMA size 1 contactor is selected from Table 10.2. The 27-A rating for the size 1 contactor is for each individual contact set; thus, the three-pole contactor rated at 27 A is capable of switching three separate 27-A loads simultaneously.

Contact Design

Most ac contactors use *double-brake* contacts, as noted in Figure 10.9. The normally open (NO), movable double-brake contact is forced against the two stationary con-

FIGURE 10.10
An electrical diagram illustrating the use of a contactor to control three lighting circuits operated by a mechanical (automatic) switch in an electric timer. A lighting contactor enables many lighting circuits to be operated simultaneously.

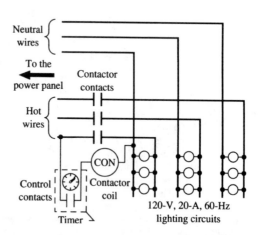

tacts, forming a bridge between the line and load terminals. The double-brake movable contact, called a **bifurcated** movable contact, completes the electrical circuit by creating a short circuit (very low resistance) between the fixed contact attached to the line terminal and the fixed contact attached to the load terminal.

The double-brake contact is mounted in a spring-loaded holder on an insulated T-bar, as noted in Figure 10.9. The compression spring is loaded (compressed) as the contacts come together during the closing of the armature. The spring serves to maintain a constant force between the closed contacts, and it also prevents contact bounce by cushioning the contacts as they close. The spring provides some compensation for the lack of *coplanarity* (alignment) in the stationary contacts.

Again, it should be understood that the normally open, bifurcated movable contact has no electrical connection to the external wiring as it is merely a shorting bar. The load and source wiring are connected to the stationary terminals through the *saddle clamp terminals*, as pictured in Figure 10.11.

Each set of contactor contacts (two stationary and one movable) is insulated from the other sets by an arc hood that serves to contain and quench the electrical arc that is present during the opening and closing of the contacts. The bifurcated movable contact (double-brake contact) used in most contactors provides for a higher current rating of the contactor for a smaller package size than would be possible if a single-brake contact device with the same current rating were used.

Double-brake contacts are made of a durable silver-cadmium alloy, which gives the contacts long life. The oxide that forms during normal operation (a dull, tarnished-looking substance) is very conductive and need not be removed. However, after years of service, the contacts will need replacing, since they become burned and pitted from the arcing during the opening and closing of the load circuit. Replacement kits containing new fixed and movable contacts and new compression springs are available to

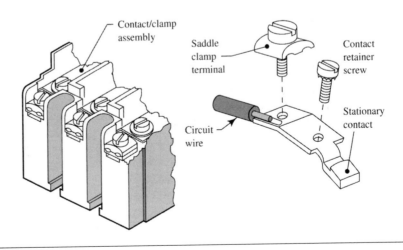

FIGURE 10.11
A solderless pressure clamp called a *saddle clamp* is used to make all electrical connections to the contactor.

repair worn or defective contacts in the contactor. Several recommendations for the maintenance of contactors may be found in Appendix C, Section C8.

Function

The contactor is made up of two sections, the switching section and the coil section, as illustrated in Figure 10.12. The terminals attached to the switching section for the three-pole contactor consist of the line terminals, labeled L1, L2, L3, and the load terminals, labeled T1, T2, T3. An additional, auxiliary contact set labeled 2, 3 is often present and is used in forming an interlock circuit with the coil circuit. The coil terminals that are attached to each end of the coil winding are used to energize the coil.

Because the purpose of the contactor is to control a very large load current in the switching section with a small control current in the coil section, it is important to keep in mind that the two currents are separate and independent of one another. That is, the load current does not flow through the coil, nor does the coil current flow through the load.

The idea that there are two electrically independent circuits that are functionally dependent adds an element of confusion for many students at this point in their development. The key is to wire the coil circuit as one circuit and the switching circuit as a second, independent circuit. However, when analyzing the operation of the contactor, think of the two sections of the device from a functional cause-and-effect point of view. That is, when the coil circuit is energized, the normally open switch contacts go closed and the load is energized. Note that this is a two-step, cause-and-effect function. First, the coil is energized; second, the load is connected to the line—electrically independent, functionally dependent!

Control Circuit Wiring

The circuit containing the coil is called the *contactor control circuit*, and it may be wired in one of two ways, either as a *two-wire control circuit* or as a *three-wire con-*

FIGURE 10.12
Pictorial of a magnetic contactor with the switch and coil sections identified. In the switch section, the source voltage (line voltage) is attached to the terminals labeled L1, L2, and L3. The load is attached to the terminals labeled T1, T2, and T3. An auxiliary set of contacts labeled 2, 3 may also be available. In the coil section, the coil is energized through the terminals attached to the ends of the coil winding.

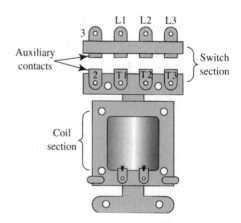

FIGURE 10.13
Wiring diagram for connections to the contactor. The switch circuit is symbolized by heavier lines, and the coil circuit is symbolized by lighter lines. In two-wire control, two wires lead from the pilot device to the contactor, where the wires are attached to terminals 1 and 3.

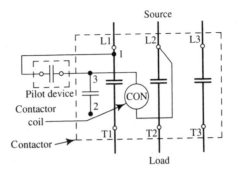

trol circuit. Each circuit has its unique applications, as is noted in the following discussion of each circuit.

Two-Wire Control Circuit   The two-wire control circuit, pictured in Figure 10.13, has the control circuit (coil circuit) noted with light, thin lines and the switch circuit indicated by heavy, broad lines. The operation of the coil circuit is through the contacts of a *maintained-contact pilot device* such as a toggle switch, a timer-controlled switch, a temperature switch (thermostat), a pressure switch, a float switch, etc.

In constructing the control circuit (coil circuit) of the contactor, two wires are needed for connection between the maintained-contact pilot device and the contactor. Figure 10.14 is an example of a two-wire control circuit using a thermostat (temperature switch) to turn on the contactor coil to energize a bank of three 2800-W, 240-V electric infrared heaters with fans connected in parallel. Notice that a three-pole contactor is being used in a two-pole application (240-V, 1-ϕ). Because three-pole contactors are more available than are two-pole contactors, their use in two-pole applications is fairly common.

As long as the contacts on the thermostat are closed and the source voltage is present, then the contactor coil (CON) is energized and the heat will be on.

FIGURE 10.14
A contactor controlled by a thermostat in a two-wire control circuit is used to turn on the high-wattage, high-current (35-A) infrared heaters. The line voltage thermostat cannot directly control the load current because of its low current rating (less than 15 A).

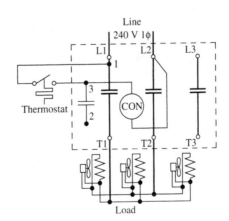

When the thermostat contacts open, the coil is de-energized, and the contacts in the switching section of the contactor then return to an open state. The contactor will operate automatically in response to the condition of the contacts (open or closed) of the pilot device (thermostat) without the intervention of an operator (person).

The two-wire control circuit does not provide for undervoltage protection. That is, when the *line voltage* (source voltage) is interrupted, the coil loses power and the contactor switch section opens, releasing the load from the line (undervoltage release). When power is restored to the line, the coil is re-energized, the contactor switch section once again closes (assuming that the pilot device is closed), and the load is again operational.

This situation can be dangerous if personnel are around equipment that might turn on unexpectedly or if an interrupted process is restarted without proper startup preparation. Two-wire control circuits are used only for remote or inaccessible applications or where an immediate return to service after a power failure is safe and desirable. These applications include systems that control lighting, heating, air conditioning, refrigeration, sewage treatment, irrigation, etc.

Three-Wire Control Circuit A three-wire control circuit, represented by the motor starter circuit of Figure 10.15, uses three wires to interconnect the start-stop pushbuttons (both *momentary-contact pilot devices*) to the contactor coil (*M*) portion of the motor starter to form a *holding interlock circuit*. Notice that the contactor coil is designated with an *M* when the contactor is serving as a motor starter.

The auxiliary contacts (2, 3) found on the contactor are used in the construction of the three-wire holding interlock circuit, which functions in the following manner:

1. The normally open *start* pushbutton is momentarily depressed, completing the coil circuit and pulling the contactor switch section closed, thus energizing the motor.

FIGURE 10.15
A motor starting circuit is activated by a three-wire control circuit. The pushbutton switches, used to start or stop the motor, are coupled with the auxiliary contacts (2, 3) to form a holding interlock circuit.

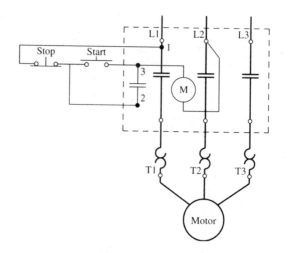

2. The auxiliary contacts (now closed) form an alternative, parallel pathway for the current in the coil circuit, thus allowing the contactor coil to remain energized after the *start* pushbutton is released. The coil circuit is *sealed on* (shunted) by the holding interlock circuit, which is made up of the auxiliary contacts.

3. Pressing the *stop* pushbutton opens the coil circuit and releases the contactor's switch section, thereby shutting off the power to the motor and resetting the holding interlock circuit. The motor will not restart until the *start* pushbutton is once again depressed.

Unlike the two-wire control circuit, the three-wire control circuit does provide undervoltage protection when the line voltage is interrupted and the coil loses power. Upon losing power, the contactor switch section opens, releasing the load from the line, as was the case with the two-wire control circuit. However, when power is restored to the line, the coil circuit is not automatically re-energized, and the contactor switch section remains open, with the load disconnected from the line. The three-wire control circuit can be restarted (once line voltage is restored) only by depressing the *start* pushbutton.

As a final observation, you might have noticed that in each of the control circuits (two-wire and three-wire), the voltage for the control circuits had the same magnitude as the voltage between lines L1 and L2 in the contactor. This arrangement works quite satisfactorily for most independent control circuits. However, in automated systems where many control circuits are used, it is customary to set the control voltage magnitude at a single, safe, low-voltage level (120 V, 24 V, etc.). A control transformer is used to supply a reduced voltage to the two-wire or three-wire control circuits through a separate fused circuit, as pictured in Figure 10.16.

FIGURE 10.16
A control transformer is used to supply a reduced voltage (24-V, 120-V) to the control circuits through a separate, fused circuit.

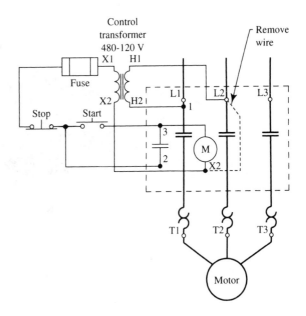

10.4 CONTROL RELAY

Introduction

The previous section introduced contactors, electromagnetic switches designed to switch large electric currents. This section deals with the **control relay,** an electromagnetic switch designed to switch small electric currents. The control relay is available in two classes—general-purpose and industrial (pictured in Figure 10.17).

The general-purpose control relay is usually designed to plug into a socket; when trouble develops, it is simply unplugged and replaced with a new unit. Because of its moderate cost and ease of replacement, the general-purpose control relay is the standard for commercial and industrial applications. The industrial control relay, on the other hand, is expensive but extremely versatile. Because of its high reliability and rugged, heavy-duty construction, the industrial control relay is widely used in industrial control circuit applications (machining centers, numerical control equipment, etc.). Besides a longer switching life, the industrial control relay has two additional advantages over the general-purpose control relay: a replaceable modular contact unit as well as a replaceable modular coil unit. Section C7 of Appendix C has information for relay contact maintenance.

Relay Structure and Terms

As was noted in Figure 9.9 (repeated here as Figure 10.18), the control relay is made up of three major sections: the coil and attendant frame; the armature, which serves as a movable lever; and the switch contacts attached to the armature.

The switch contacts move (open/close) when the armature is moved by the force of magnetic attraction resulting from electric current moving through the relay coil. Depending on the type and design of the relay, the loads switched by the

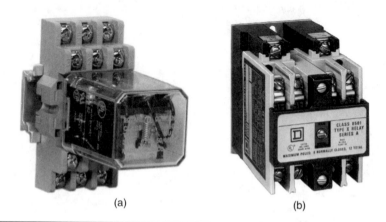

(a) (b)

FIGURE 10.17

The control relay is available in two classes: (a) The general-purpose control relay designed to plug into a socket; (b) The industrial control relay, a heavy-duty relay with replaceable contacts and coil. (*Courtesy of the Square D Company*)

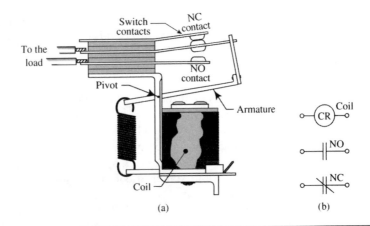

FIGURE 10.18
(a) The control relay (CR) is made up of the coil and frame, the armature (a movable lever), and the switch contacts. (b) In a control diagram, the relay coil is represented by a circle with the letters CR (control relay), and the contacts are represented by two parallel lines. NO represents normally open contacts, and NC represents normally closed contacts.

control relay may include pilot lights; alarms; solenoid coils; contactor coils; other relay coils; and small, light-duty motors. When compared to the loads switched by a contactor, all the loads switched by the control relay are small.

When selecting a control relay for an application, several terms must be understood in order to make an informed decision. The following terms are typical of those found in an industrial engineering catalog.

Coil voltage The nominal voltage needed to energize the relay coil.

Contact arrangement The poles, throws, and breaks in the relay switch section. Specified by the *form designation letter*. (See Figure 10.19.)

Contact voltage rating The maximum voltage that can be safely switched by the relay contacts.

Contact current rating The maximum current that can be safely switched by the relay contacts.

Power rating The nominal coil power rating (in watts for dc relays and volt-amperes for ac relays).

Operate time The typical time (in milliseconds) to activate the switch section once the coil voltage is applied to the coil.

Release time The typical time (in milliseconds) for the contacts in the switch section to return to their de-energized state once the coil voltage is removed.

Coil resistance The nominal ohmic resistance of the coil winding.

Maximum switching frequency The maximum number of operations (openings and closings) per minute at full rated load.

Mechanical life The minimum number of operations before failure of the armature/switch mechanism in millions of cycles.

Electrical life The minimum number of operations before failure at full rated load in hundred thousands of cycles.

Contact Arrangements

Relay contacts pictured in a schematic or a wiring diagram are always shown in the de-energized state unless specifically stated otherwise. The number of contacts and the arrangement of the contacts (called the *form*) are specified by a letter. The National Association of Relay Manufacturers (**NARM**)—as well as other standards organizations—sets standards for the *form* of the relay contacts. Figure 10.19 pictures some of the standard contact forms, along with the letter used to designate each form. A pictorial of the contact arrangements and a brief description of *poles*, *throws*, and *breaks* is also included. A complete listing of the relay and lever switch contact forms is included in Table A6 of Appendix A.

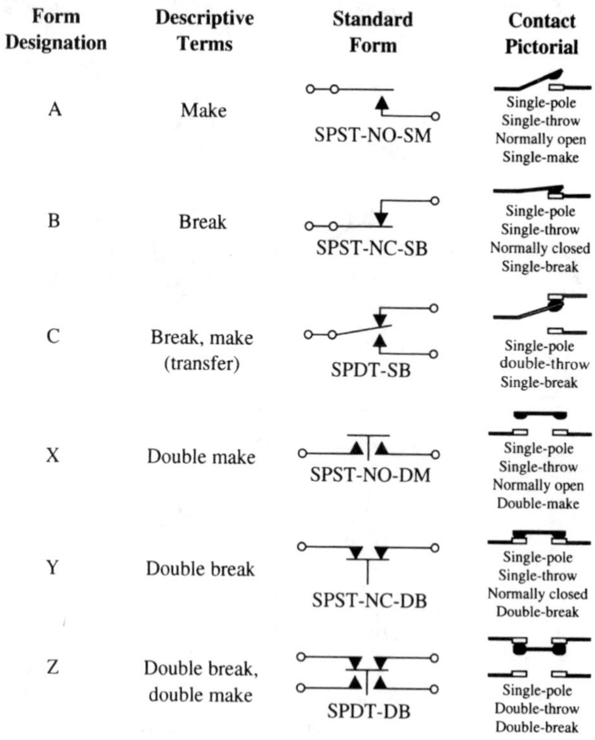

Form Designation	Descriptive Terms	Standard Form	Contact Pictorial
A	Make	SPST-NO-SM	Single-pole Single-throw Normally open Single-make
B	Break	SPST-NC-SB	Single-pole Single-throw Normally closed Single-break
C	Break, make (transfer)	SPDT-SB	Single-pole double-throw Single-break
X	Double make	SPST-NO-DM	Single-pole Single-throw Normally open Double-make
Y	Double break	SPST-NC-DB	Single-pole Single-throw Normally closed Double-break
Z	Double break, double make	SPDT-DB	Single-pole Double-throw Double-break

FIGURE 10.19

Some of the basic contact forms and terms used to describe *poles*, *throws*, and *breaks*:

SP	Single pole	ST	Single throw	SM	Single make
DP	Double pole	DT	Double throw	DM	Double make
NO	Normally open	SB	Single break		
NC	Normally closed	DB	Double break		

The contact forms pictured in Figure 10.19 are used to build up various combinations of contact configurations. For example, a switch section made up of two form C contacts would have two sets of single-pole, double-throw contacts or, stated another way, would be a two-pole, double-throw (DPDT) switch. Three form B contacts would provide a switch section with three sets of single-pole, single-throw contacts (3PST).

Contact and Coil Specifications

For general-purpose and industrial relays, contacts made of silver cadmium dioxide or silver cadmium dioxide flashed with gold are in wide use today. Relay contacts are rated in two ways: first, by current-handling capability and the voltage that may be safely switched (i.e., 10-A, 120/240-V ac resistive), and second, by the horsepower and voltage of the load (i.e., $\frac{1}{3}$-hp, 120/240-V ac).

The coil of the control relay is specified by voltage (ac or dc) and by power (watts for dc and volt-amperes for ac). The coil resistance is also specified. Coil voltages of 12, 24, 48, 120, and 240 V are common. Current in the coil is calculated using the specified coil voltage and power rating, as demonstrated in the following example.

Example *10.7*	Determine the operating current in the coil circuit of **(a)** A 24-V dc relay rated at 1.2 W. **(b)** A 24-V ac relay rated at 2.0 VA (volt-amperes).
SOLUTION	Find the current in each coil.
Given	$P = EI$; solving for I gives $$I = \frac{P}{E}$$
Evaluate	**(a)** $P = 1.2$ W $\qquad$ **(b)** $P = 2.0$ VA $E = 24$ V (dc) $\qquad\quad\;\; E = 24$ V (ac)
Substitute	$I = \dfrac{1.2}{24} \qquad\qquad\quad I = \dfrac{2.0}{24}$
Solve	$I = 50$ mA $\qquad\qquad\;\; I = 83$ mA

CHAPTER SUMMARY

- A series electromagnetic circuit is formed when a closed ferromagnetic pathway is wrapped with a coil of insulated wire, through which electric current is passed.
- A magnetic flux circulates in the core of the magnetic circuit when electric current passes through the coil.

- Like electrical conductors, the ferromagnetic material of the core is limited in its capacity to carry flux.
- The flux density (B) in the core of a magnetic circuit is measured in teslas (webers/square meter).
- Magnetomotive force (mmf) is the result of electric current circulating in the coil of wire wrapped around the ferromagnetic core.
- The magnetic field strength (H) takes into account the length of the path traveled by the magnetic flux.
- The permeability (μ) of a substance is an indication of its ability to carry magnetic flux when acted on by a magnetomotive force.
- An air gap alters the shape of the *B-H* magnetization curve for the ferromagnetic core material.
- An air gap in a magnetic circuit causes a tractive force to be developed that attempts to close the gap.
- The force resulting from an air gap is the basis of operation of the solenoid in an electromagnetic device.
- The solenoid transforms electric power into mechanical power in the form of force and motion (velocity).
- The ac operation of solenoids, contactors, and relays requires the use of a shading ring to prevent chatter.
- The tractive force of a solenoid is inversely proportional to the stroke length.
- The initial current is large in ac solenoids when the plunger is extended.
- An increase in operating temperature in the coil of a solenoid causes a decrease in the mmf and tractive force.
- The core losses in the magnetic circuit are due to hysteresis and eddy currents.
- The contactor is an electromagnetically operated switch that opens and closes high-energy electric circuits.
- The contactor, like the relay, is made up of two sections, the switching section and the coil section.
- Contactor and control relay coils and contacts are electrically independent but functionally dependent.
- The two-wire control circuit requires two wires for connection between the contactor and the pilot device.
- The three-wire control circuit uses three wires to interconnect the *start-stop* pushbuttons to the contactor to form a *holding interlock circuit.*
- Relay contacts are always shown in the de-energized state unless specifically stated otherwise.
- The coil of the control relay is specified by voltage (ac or dc) and by power (watts for dc and volt-amperes for ac).

SELECTED TECHNICAL TERMS

The following technical terms, abbreviations, and acronyms are defined in the glossary located after Chapter 16. You are encouraged to use the glossary to aid your understanding and to test your knowledge of these important terms.

bifurcated	ferromagnetic
control relay	flux
eddy current	flux density

hysteresis NARM
magnetic contactor permeability
magnetic flux rectilinear motion
mmf (magnetomotive force) solenoid

END-OF-CHAPTER QUESTIONS

Write T if the statement is true and F if the statement is false.

1. The magnetic field strength takes into consideration the length of the path taken by the magnetic flux as it circulates in the core.

2. A force is exerted on the poles on either side of an air gap in an effort to shorten the path by closing the gap.

3. The ferromagnetic material of the core that conducts magnetic flux has a limited capacity to carry flux.

4. The inrush current and the seated current in a direct current solenoid are the same.

5. An increase in resistance in the solenoid coil results in an increase in the magnetomotive force (mmf) and a decrease in the output force.

6. The three-wire control circuit uses a holding interlock circuit controlled by a maintained-contact pilot device.

7. A four-pole dc contactor rated at 50 A is capable of switching four separate 50-A loads simultaneously.

8. The toggle switch is an example of a momentary-contact device.

9. A double-brake movable contact is sometimes called a bifurcated movable contact.

10. The control circuit and the load circuit of a control relay are electrically independent of one another.

In the following, select the word or phrase that makes the statement true.

11. Flux density is measured in units of (webers, amp-turns, teslas).

12. A series electromagnetic circuit results when a closed ferromagnetic core is wrapped with insulated wire through which (magnetic flux, electric current, electromotive force) is passed.

13. Form Y relay contacts are (double-throw, double-pole, double-break) contacts.

14. The capacity of the magnetic circuit is nearly used up when the ferromagnetic material is in the (knee, linear, saturation) region of operation of the *B-H* curve.

15. A control transformer provides a reduced (current, power, voltage) to two- and three-wire control circuits.

Answer each of the following questions with a short answer in the form of a complete sentence. Include a restatement of the question in your answer.

16. What does *3 form C* mean when related to contact arrangements of a control relay?

17. What is the SI unit name of magnetomotive force, and what is the common unit name of magnetomotive force?

18. When would a three-wire control circuit be used instead of a two-wire control circuit?

19. Why is the inrush current five to ten times as great as the seated current in an ac solenoid?

20. Why is a shading coil used with an ac electromagnetic solenoid?

END-OF-CHAPTER PROBLEMS

Solve the following problems. Make sketches to aid in solving the problems and structure your work so it follows in an orderly progression and can easily be checked. Table 10.3 summarizes the formulas used in Chapter 10.

1. If the seated current in an ac solenoid is 280 mA, determine the range of expected inrush current (I_{inrush}) when the coil is initially energized with the plunger fully extended.

2. A solenoid is to be used to operate a mechanism requiring 3.8 lb to actuate it. Using the general range of service factor (SF) allowance for increased coil resistance due to heat, specify the range of operating force (F) needed for this application.

3. Determine the permeability (μ) of the cast iron core of the series magnetic circuit pictured in Figure 10.1(b) when the magnetic field strength (H) of the circuit is 600 A/m.

4. Determine the operating current (I) in the coil circuit of
   (a) A 48-V dc relay rated at 0.90 W.
   (b) A 120-V ac relay rated at 1.7 VA.

5. Determine the initial inrush current in the coil circuit of a 24-V ac relay rated at 2.4 VA when the inrush current is 1.8 times the operating current.

6. For the first 120-V continuous-duty ac solenoid listed in Table 10.1, determine the flux density (T) when the coil current is at its maximum with a stroke length of 1.0 in. The cross-sectional area of the air gap is 1.13 in^2.

7. The coil resistance of a 12.0-V dc rotary solenoid coil is specified as 7.72 Ω at 20°C. The coil is constructed by winding 550 turns of number 26 AWG onto the core. Determine

TABLE 10.3 Summary of Formulas Used in Chapter 10

Equation Number	Equation
10.1	$B = \dfrac{\phi}{A}$
10.2	$\text{mmf} = NI$
10.3	$H = \dfrac{NI}{\ell} = \dfrac{\text{mmf}}{\ell}$
10.4	$B = \mu H$
10.5	$F = \dfrac{B^2 A}{2\mu_0}$
9.8 (repeated)	$R = R_{\text{AWG}}\left(\dfrac{T + 234.5}{254.5}\right)$

    (a) The mmf (amp-turns) of the coil when it is cold (20°C).

    (b) The mmf (amp-turns) of the coil when it is hot (stabilized temperature of 105°C).

8. The armature, frame, and core of a 12.0-V dc control relay are made of mild steel. The average length of the magnetic circuit is 12 cm long when the relay is energized. The average cross-section of the magnetic circuit is 0.56 cm^2 and the coil, wound with 250 turns of number 35 AWG solid enameled copper wire, is energized with 60.0 mA of current. Assuming 20°C operating temperature, determine

    (a) The ohmic resistance (R) of the coil.

    (b) The flux density (B) in the magnetic circuit of the relay when the coil is energized.

    (c) The force (N) exerted on the armature to close it when the coil is energized.

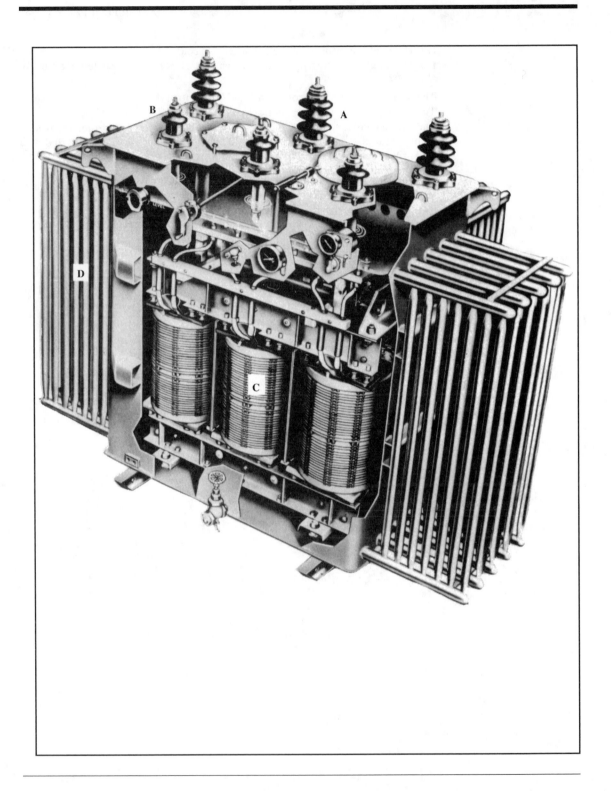

Chapter 11

Transformers and Power Distribution

About the Illustration. A substation transformer *is used to step down the three-phase (3-φ) high voltage provided by the utility company to a lower value of distribution voltage needed to power a plant or factory. The voltages provided by the utility company range from 2400 V through 69 kV, whereas the distribution voltages range from 480 V to 34.5 kV. This type of transformer is:* oil-immersed *and* convection-cooled *through the* radiation fins *(D). The utility lines are attached to the high-voltage bushings (A), and the lower-distribution voltage is taken from the low-voltage bushings (B). The cutaway view allows viewing of the transformer windings (C). Liquid level and temperature are indicated by gauges visible from the outside of the transformer. Because of its size, lifting hooks are provided on each of the top corners. (Courtesy of the Westinghouse Electric Company.)* ∎

INTRODUCTION

Over the years, the study of the ferromagnetic transformer has captured the imagination and focused the creative genius of such noted men as Michael Faraday, who, in 1831, first demonstrated the induction coil (transformer); Nikola Tesla, who did extensive experimentation with wireless power transmission; and Charles Steinmetz, who researched the relationship between hysteresis and flux density.

The modern-day transformer is virtually the same device used by Faraday to demonstrate electromagnetic induction. The principal function of the transformer remains the same to this very day—that of transforming voltage, current, and impedance from one level to another. Because of the simplicity in the construction of the ferromagnetic (iron-core) power transformer (copper wire, silicon-steel sheet metal, insulation), coupled with its longevity (no moving parts), you may fail to appreciate the importance of this machine to modern civilization. Also, you may not recognize the quality of design that allows the transformer to be analyzed with simple equations as an ideal linear device.

In this chapter, you learn about the construction, connection, and application of the iron-core transformer used in power-distribution systems and in control circuits.

CHAPTER CONTENTS

11.1 Power Generation, Transmission, and Distribution
11.2 Transformer Principles
11.3 Single-Phase Transformer
11.4 Three-Phase Transformer

PERFORMANCE OBJECTIVES

Once you have read and studied each section; worked through each example with pencil, paper, and calculator; worked through the end-of-chapter problems; and answered the end-of-chapter questions, then you should be able to

- Understand how power is generated and distributed.
- Describe the use of a switchboard, panelboard, and motor control center.
- Use the transformation ratio to determine transformer parameters.
- Explain the operation of the transformer.
- Differentiate among transformer types by construction and use.
- Identify the neutral in three-wire, single-phase connections and in three-phase wye connections.
- Understand the advantages of using three-phase power over single-phase power.
- Determine line voltage and phase voltage for delta- and wye-connected transformers and loads.
- Distinguish between delta- and wye-connected transformer primaries and secondaries.
- Know the conditions for a balanced three-phase load.

11.1 POWER GENERATION, TRANSMISSION, AND DISTRIBUTION

Electric Power Generation

Electric power is transmitted (moved) from the generating station through the switching station to the receiving substation by high-voltage transmission lines, as indicated in Figure 11.1. The synchronous generators (commonly called *alternators*)

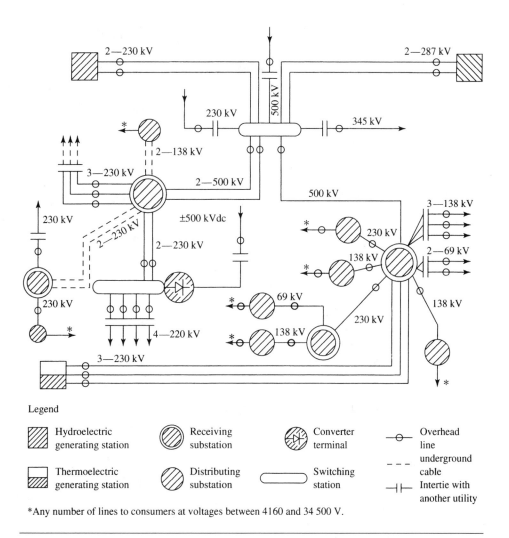

FIGURE 11.1

A three-phase power system showing generating stations, switching stations, receiving substations, distribution substations, and transmission line voltages.

used at the generating stations are **three-phase.** The alternators at thermoelectric generating stations are driven by steam turbines at speeds (angular velocities) up to 3600 rev/min. Many steam turbines have nameplate ratings of 400 MW or more, and an entire thermoelectric generating station might produce in excess of 1200 MW of electric power.

The steam turbine is powered from energy contained in coal, oil, natural gas, or uranium (nuclear energy). Figure 11.2 pictures a block diagram of a coal-fired thermoelectric generating system, where fuel is burned to heat water in a boiler to produce steam for the turbogenerator. The temperature and pressure of the steam from the boiler determines the output capacity of the electric generator. The higher

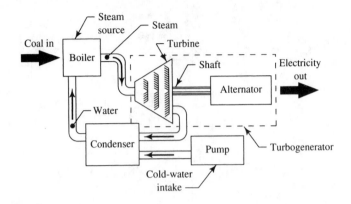

FIGURE 11.2
A block diagram of a coal-fueled, thermoelectric generating system.

the velocity of the steam passing through the turbine, the greater is the force on the turbine blades.

For efficient operation, the turbine is divided into a number of successive stages to get the most out of the energy contained in the steam. Each stage is connected through the same shaft to power the three-phase alternator at a constant speed. The steam enters the high-pressure side of the turbine at pressures up to 3700 lb/in^2, where it strikes the blades mounted in small-diameter wheels. As energy is taken from the steam to rotate the turbine, the pressure drops and the blades are mounted in proportionally larger wheels. Finally, at the low-pressure end of the turbine, the blades reach their maximum diameter. Low-pressure steam leaves the turbine and is sent on to the condenser, where it is changed (condensed) to water for delivery back to the boiler for reheating.

Transmission

The voltage from the three-phase alternator is raised by a transformer to a suitable level for transmission to the receiving substation. This concept is illustrated by the transmission-distribution system pictured in Figure 11.3. The transmission line voltage varies from 69 to 500 kV or more, depending on the distance over which the electric power is to be transmitted (about 1000 V/mi).

Distribution

As was noted in Figure 11.1, the electric power is sent from the receiving substation to the distribution substations within the city. The function of the distribution substation is to lower (step down) the high transmission voltage, using transformers, to a voltage level appropriate for distribution to industrial, commercial, and residential areas of the city. Like the transmission-line voltage levels, the distribution voltage

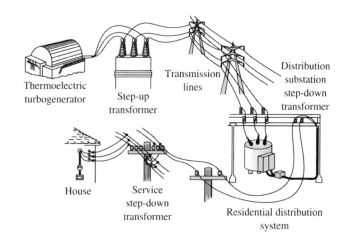

FIGURE 11.3
Pictorial of a generation, transmission, and distribution system.

levels depend upon transmission distance. Line voltages between 4160 and 24 000 V are common. Besides reducing the voltage by transformation, the distribution substation serves as a control point in the distribution system, where the power may be disconnected for maintenance or in case of trouble. The distribution substation has three main parts:

- The *input side* (high-voltage side), with the primary switch gear, including the protection system to monitor for fault conditions (an out-of-tolerance parameter). Included in this section are circuit breakers, meters, and an overcurrent limiting system.
- The *transformer section*, where three-phase **transformers** are connected for wye (Y) or delta (Δ) operation in the distribution systems. The voltage is stepped down (lowered) in this section of the substation to a level suitable for consumer use. The output level of the transformer voltage is also adjusted and regulated in this section of the substation.
- The *output side* (low-voltage side), with the secondary switch gear, including circuit breakers, load interrupter switches, and current limiters. Meters that check on the operation of the distribution system are also provided in this section.

As pictured in Figure 11.4, the distribution of electric power from the substation to the consumer is through high-voltage, three-phase lines to the final service transformer, where the three-phase voltage is lowered (stepped down) to an appropriate level used by the consumer. *Residential consumers* are provided with **single-phase** (1-ϕ) 240/120-V service. *Commercial businesses* receive both three-phase (3-ϕ), voltage (208 V) for motors and single-phase voltage (120 V) for lighting, appliances, and light-duty machine tools. *Industrial facilities* are fed with three-phase voltage (480 V) for powering large motors, heavy loads, and transformers. The transformers are located within the plant to provide single-phase voltage

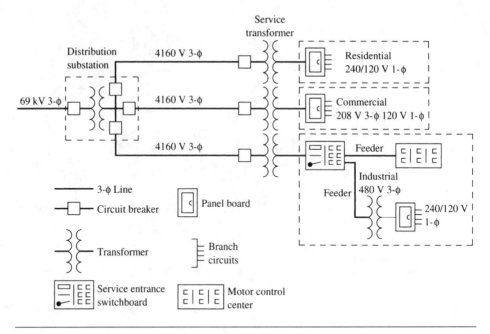

FIGURE 11.4

The voltage from the receiving substation is stepped down by the transformers located at the distribution substation to a lower voltage for safe and convenient delivery to the service transformers located near the residential, commercial, or industrial consumers of electric energy.

(240/120 V, 277 V) for lighting and light-duty machine tools, and reduced three-phase voltage (240 V) for smaller motors and loads.

Switchboard

The conductors from the utility company's distribution system are brought into the industrial plant through the service entrance switchboard equipment (see Figure 11.5). The power carried in the service entrance conductors from the utility company's transformer is broken down in the switchboard for distribution throughout the building's electrical system.

Besides dividing the incoming power and routing it throughout the plant, the switchboard contains all the necessary equipment for controlling, disconnecting (through the main switch), metering, protecting (through overcurrent devices), and recording the operation of the power system. Also, the connection to the building's grounding electrode is made at the switchboard.

Panelboard

An industrial panelboard, pictured in Figure 11.6, is a wall-mounted steel cabinet that receives power by way of a main feeder from the switchboard. Contained in the

FIGURE 11.5
The service conductors of the industrial plant are fed from the service transformer to the *switchboard*, which provides the main disconnect switch; the overcurrent devices for the feeder circuits; and the instrumentation and metering of the voltage and energy entering the facility. (*Courtesy of the Square D Company*)

panelboard are the overcurrent devices (circuit breakers or fused switches) for the branch circuits and/or feed circuits for additional panelboards.

Motor Control Center

Because three-phase motors are so common, the motor control center (pictured in Figure 11.7) is used to simplify the connection and control of industrial motors by providing a central location. The motor control center houses all the control circuits for two- and three-wire control of three-phase motors. The centers are designed to accept modular control units that are factory preassembled and prewired for start/stop, speed control, forward/reverse, and reduced-voltage starting. Because of the modularity of the control units, individual units can easily be replaced or their function modified with a minimum of lost time.

Because all the control functions for several motors are housed in one consolidated location and because all the control signals from pressure, limit, float, and similar switches are routed into the control center, the *interlocking* (interconnection) of machines used in the same assembly line is both simple and convenient. Also, all other control devices, including control relays, motor starters, control transformers, timers, counters, etc., are available in the motor control center for use in the interconnecting process.

FIGURE 11.6
The wall-mounted *panelboard* receives power from a feeder circuit brought in from the bottom of the panel. The branch circuits from the panelboard (that feed motors, transformers, and lighting systems) are protected with fusible switches or circuit breakers. (*Courtesy of the Square D Company*)

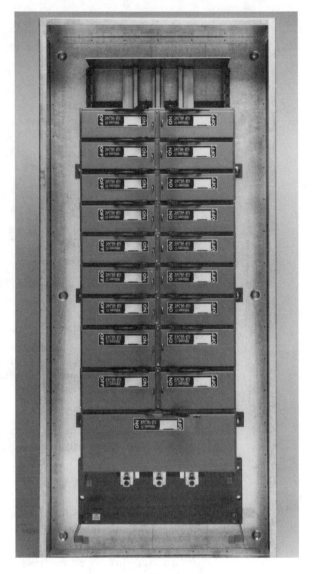

11.2 TRANSFORMER PRINCIPLES

Mutual Inductance

In the iron-core transformer of Figure 11.8(a), an alternating flux is set up in the core by the magnetomotive force (mmf) of the alternating current passing in the turns of wire of the coil on the left. The resulting flux (ϕ_m), which circulates in the magnetic core, links the turns of the coil on the right, inducing a voltage in this coil. The generation of a voltage in the second coil by a changing (ac) current in the first coil is called the *mutual induction*.

FIGURE 11.7
The *motor-control center*, powered by a feeder circuit from the switchboard, is used to simplify the connection and control of industrial motors by providing a central location for all the control equipment. (*Courtesy of the Westinghouse Electric Company*)

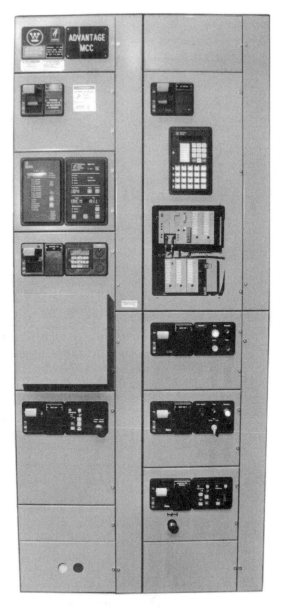

As pictured in Figure 11.8(b), the transformer consists of two coils positioned in close proximity to one another on a common, low-loss, laminated core so that the electromagnetic flux created by the alternating current in the first coil links the turns of the second coil. In this configuration, the coil connected to the source voltage is called the **primary winding** of the transformer, and the coil connected to the load is called the **secondary winding** of the transformer.

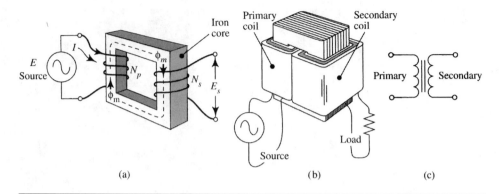

FIGURE 11.8
The iron-core transformer: (a) The mutual flux (ϕ_m) links the primary coil and the secondary coil inducing a voltage in both the primary and secondary coils. The induced secondary voltage is noted as (E_s); (b) The primary coil is connected to the source voltage, whereas the secondary coil is connected to the load; (c) Schematic symbol of the iron-core transformer.

The modern-day, high-power transformer is virtually 100% efficient (96 to 99%). This means that practically all the energy put into the transformer at the primary winding is transferred to the secondary winding with very little loss of energy. The cause-and-effect relationship between the two coils sharing the same ferromagnetic path is made possible by the changing *mutual flux* (ϕ_m) in the core of the transformer. It is the mutual flux that moves electrical energy from the primary winding (source) to the secondary winding (load).

Voltage Ratio

The transformer is quite versatile, and this versatility is seen when voltages are raised or lowered in ac distribution systems. Besides stepping up or stepping down voltage, the transformer will, by its nature, step up or step down the current. It should be understood, however, that the terms *step up* and *step down*, when used with transformers, indicate that the voltage and not the current is increased or decreased.

When the iron-core transformer is thought of as ideal, the following conditions may be assumed:

- The coils of the transformer will respond to variations in frequency and temperature in a linear fashion.
- The permeability of the core is constant for variations in frequency and temperature over the range of voltage applied to the primary winding.
- There is no loss in the mutual flux due to flux leakage and all the primary coil flux will link the secondary coil.

With the transformer operating as an ideal transformer, the voltages induced in the primary and secondary coils, as illustrated in Figure 11.9, may be expressed in terms of the effective (**rms**) values of voltage using the *general transformer equation*. The induced voltages depend on the number of turns of wire (N) in each of the coils, the frequency of the alternating current (f), and the amount of mutual flux in the core (ϕ_m). Thus,

$E_p = 4.44N_p f \phi_m$ is the general transformer equation for the voltage induced in the primary.

$E_s = 4.44N_s f \phi_m$ is the general transformer equation for the voltage induced in the secondary.

By dividing the secondary voltage equation into the primary voltage equation, a proportional equation is formed:

$$\frac{E_p}{E_s} = \frac{4.44 \, N_p f \phi_m}{4.44 \, N_s f \phi_m}$$

The following results when this equation is simplified:

$$\frac{E_p}{E_s} = \frac{N_p}{N_s} \qquad\qquad (11.1)$$

where

E_p = induced primary voltage, equal to the applied source voltage (E) in the ideal transformer, in V

E_s = induced secondary voltage, the voltage across the terminals of the secondary coil, in V

N_p = turns of wire in the primary coil, a unitless parameter

N_s = turns of wire in the secondary coil, a unitless parameter

Equation 11.1 indicates that, in the ideal transformer, the amount of voltage available at the secondary terminals of the transformer depends on the number of turns of wire on the primary coil compared to the number of turns of wire on the secondary coil. The ratio between the number of turns of wire in the primary coil and the number of turns of wire in the secondary coil is called the *transformation ratio* (N_p/N_s) of the transformer.

FIGURE 11.9
In the ideal transformer, the induced primary voltage (E_p) is equal in magnitude but opposite in polarity (a CEMF) to the applied source voltage (E). Thus, the sum of the voltages in the primary circuit is zero.

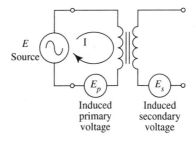

Example
11.1

The primary of a single-phase control transformer is constructed by winding 240 turns of wire to form the primary coil. Determine the voltage induced in the 48-turn secondary coil when the primary is attached to a 120-V, 60-Hz line.

SOLUTION

Find the secondary voltage by solving for E_s in Equation 11.1.

Given

$$E_s = \frac{E_p N_s}{N_p}$$

Evaluate

$E_p = 120$ V

$N_s = 48$ turns

$N_p = 240$ turns

Substitute

$$E_s = \frac{120 \times 48}{240}$$

Solve

$E_s = 24$ V

Transformer Action

The component of the total primary current that sets up the mutual flux in the transformer is called the *magnetizing current* (I_{mag}), and it is represented in Figure 11.10 as being 90° out of phase with the component of the primary current that provides the power for the load. The non-power-producing magnetizing current (I_{mag}) is small (less than 5% of the total current) when compared to the power-producing component of current.

    With a resistive load connected to the secondary of the ideal transformer, as pictured in Figure 11.11, the voltage induced in the secondary winding by the mutual flux will produce a current in the secondary circuit called the secondary current (I_s). The secondary current will, by Lenz's law, produce a magnetomotive force (mmf $= N_s I_s$) that will oppose and nullify (cancel) the primary magnetomotive force created by the current flow in the primary winding. The net result of this action is that additional current flows in the primary winding, developing additional primary magnetomotive force in response to the increase of the secondary current (due to the addition of a load) and the resulting increase in secondary magnetomotive force. Thus, the current in the primary of the transformer is set

FIGURE 11.10

With a resistive load attached to the secondary of the iron-core transformer, the currents in the primary coil consist of two components, I_{mag} (the magnetizing current) and I_p (the power-producing current). I (the total current) is the vector summation of $I_{mag} + I_p$. In the ideal transformer, $I_p = I$.

FIGURE 11.11
An ideal iron-core transformer with a resistive load attached to the secondary terminals will have a secondary current (I_s) in the load due to the induced secondary voltage, E_s. The current in the primary of the ideal transformer is I_p.

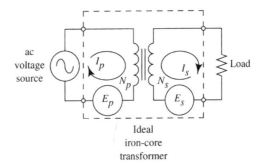

Ideal
iron-core
transformer

(controlled) by the amount of impedance (resistance, in this case) in the load attached to the secondary.

Current Ratio

Assuming an ideal transformer with a resistive load, then the power developed in the primary circuit (the source) is equal to the power transformed and delivered to the load attached to the secondary (the load):

$$P_{source} = P_{load}$$

With a lossless transformer and a resistive load, the power factor, $\cos(\theta)$, is unity, and it may be set equal to 1. Thus, the power in the source and load may be expressed in terms of voltage and current as

$$P_{source} = E_p I_p$$

$$P_{load} = E_s I_s$$

Since the source and load power are equal in an ideal transformer, the two equations may be set equal to each other:

$$E_p I_p = E_s I_s$$

Writing this equation as a proportion results in:

$$\frac{E_p}{E_s} = \frac{I_s}{I_p}$$

From Equation 11.1, $E_p/E_s = N_p/N_s$. Substituting N_p/N_s for E_p/E_s yields the following expression for the current ratio as it relates to the transformation ratio.

$$\frac{I_s}{I_p} = \frac{N_p}{N_s} \qquad\qquad \textbf{(11.2)}$$

where

   I_p = primary current, equal to the total current in the primary in the ideal transformer, in A

   I_s = secondary current, resulting from the induced secondary voltage, in A

N_p = turns of wire in the primary coil, a unitless parameter
N_s = turns of wire in the secondary coil, a unitless parameter

Equation 11.2 indicates that the current in the coils of the transformer is inversely related to the transformation ratio of the transformer.

Example *11.2*	Neglecting losses and assuming a unity power factor, determine the current taken by a transformer that is attached to a 240-V, 60-Hz, 1-ϕ line when the 48-V resistive load dissipates 62 W.
SOLUTION	First determine the power in the primary, then compute the current in the primary.
Given	Since the power into an ideal transformer is equal to the power out of the transformer, P_{source} is 62 W and $P = EI \cos \theta$. Solving for I gives $I = P/E$, since $\cos \theta$ is 1.
Evaluate	$P = 62$ W $E = 240$ V
Substitute	$I = \dfrac{62}{240}$
Solve	$I = 0.26$ A

11.3 SINGLE-PHASE TRANSFORMER

Transformer Types

Depending on voltage and power rating, transformers are either liquid-cooled or air-cooled. Air-cooled transformers are called *dry-type* transformers and are commonly used in low-power applications as control transformers or distribution transformers. Transformers are constructed so that the primary and secondary coils are electrically isolated from each other but are physically on the same core. The core design may be either a *shell-type* core, where the coils are surrounded by the core, or a *core-type* core, where the core is inside the coils, as shown in Figure 11.12.

The *control transformer*, as pictured in Figure 11.13(a), is commonly used to supply single-phase power at reduced voltages to the two- and three-wire industrial *control circuits* used in the control of contactors, relays, and motor starters. The primary line voltages may be as high as 480 V, whereas the secondary control voltage is typically 24, 48, or 120 V. These lower secondary voltages are much safer to work with in control circuit applications. In addition to stepping down the line voltage, the control transformer (for added safety) electrically isolates the control voltage circuit (secondary winding) from the line voltage circuit (primary winding).

The control transformer may be operated from either one phase of a three-phase line or from a single-phase line. For protection, the transformer, which is

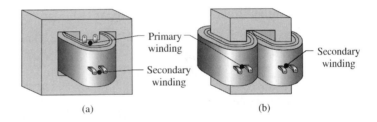

(a)   (b)

FIGURE 11.12

(a) The shell type of core construction has the primary and secondary coils wound around each other and the central part of the core with the remainder of the core surrounding the coils on the outside. (b) The core type of core construction has the primary coil wound around one side of the core and the secondary coil wound around the other side of the core. The core passes inside each of the coils.

physically small (rated under 1000 VA), is often equipped with a fuse in a fuse-holder connected into the secondary of the transformer, as noted in the schematic diagram of Figure 11.13(b). The control transformer has a characteristic large core to ensure nearly constant voltage even when a contactor solenoid is closing and requires a momentary overload current. Because of the added flux capacity of the iron core, the control transformer can withstand short-duration overloads with only small changes in output voltage.

In addition to the control transformer used to supply industrial control voltages, another common transformer found in an industrial power system is the single-phase *distribution transformer*. The single-phase, dry-type distribution transformer is used with three-phase industrial electrical systems to provide single-phase power to panelboards for the operation of fractional-horsepower motors, lights, and appliances.

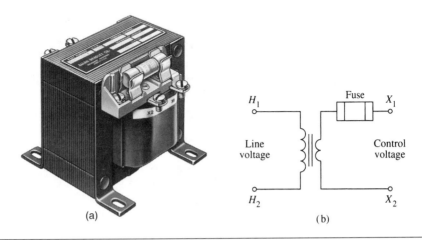

(a)   (b)

FIGURE 11.13

A control transformer: (a) With a built-in fuse holder. (*Courtesy of the Allen Bradley Company, Inc.*); (b) Schematic diagram with the line voltage attached to the H1 and H2 terminals and the control voltage for the load taken across terminals X1 and X2. The fuse is shown in the secondary circuit.

Single-Phase Electrical System

The standard three-wire, 120/240-V, single-phase voltages are developed in the secondary of a single-phase (1-ϕ) transformer, as pictured in Figure 11.14. The three-wire, two-voltage, single-phase system illustrated in Figure 11.14 uses a conductor connected from the midpoint, or *neutral*, of the secondary winding to the grounded *neutral bus* inside the lighting panelboard.

In the three-wire, 120/240-V, single-phase system, the neutral wire is the grounded conductor, whereas the other two wires are called the *hot wires*. The neutral conductor is identified with the white color coding of its insulation and, unlike the hot conductors, the neutral conductor must **never** be opened by switching or fusing.

When the two 120-V circuits are nearly balanced (an ideal condition)—that is, when approximately half the load is connected between each of the two ungrounded hot wires and the grounded neutral wire—the current in the neutral conductor will be small (near zero). The current in the neutral is the difference between the currents in the two ungrounded wires. This concept is pictured in Figure 11.15, where the neutral is shown carrying the unbalanced current (difference) of the two 120-V loads. The center-tapped secondary of the 120/240-V transformer is phased so that the two currents in the hot conductors move in the same direction at the same time. When the current flows in a clockwise direction through load A, the current through load B is also in a clockwise direction. Although both currents in the hot wires are moving in the same direction, this is not the case in the neutral conductor. The currents in the neutral are moving in

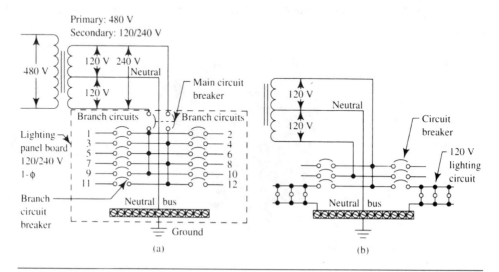

(a)   (b)

FIGURE 11.14

(a) The 480-V primary voltage is stepped down in the secondary to form a three-wire, two-voltage, single-phase electrical system where the neutral and the neutral bus are grounded. (b) The three-wire circuit is used to power a lighting panelboard, where the high (or hot) side of each circuit is connected to the light circuit through a circuit breaker.

FIGURE 11.15
The neutral carries the unbalanced current of 14 A, the difference between the currents in the two 120-V loads.

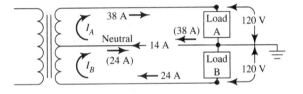

opposite directions to one another, and the net current is the difference between them ($38 - 24 = 14$ A).

Single-Phase Transformer Ratings

Transformers are rated by their apparent power (P_s) in volt-amperes (VA) or kilovolt-amperes (kVA). This information is usually listed on the transformer nameplate. In addition to the volt-ampere rating of the transformer, the primary and secondary voltage ratings are also listed on the nameplate. However, the primary full-load current may not be listed, and the secondary full-load current is only occasionally listed. When the current or currents are not listed, they can be determined from the nameplate voltage and apparent power (volt-ampere) ratings of the single-phase transformer by solving for current, as in Equation 9.18 ($P_s = EI$):

$$I = \frac{P_s}{E} \tag{11.3}$$

where

I = primary/secondary full-load current, in A
P_s = transformer apparent power rating, in VA
E = primary/secondary voltage, in V

The reason for rating transformers in volt-amperes (apparent power) instead of average power (watts) is to ensure that excessive current is not taken from the transformer when it is connected to loads that have less than a unity power factor (reactive load). It takes more current to produce a specified power (watts) in a reactive load than in a purely resistive load, since some of the current must produce wattless power (reactive power). From the equation for power, $P = EI \cos \theta$, if E is fixed and P is to remain constant, then I must increase to compensate for a decrease in the power factor, $\cos \theta$. Power factor–correcting capacitors may be placed across the load to null out the reactive current, thereby reducing the out-of-phase component of the line current and reducing the current demand through the transformer. The kVAR ratings of the capacitors for improving the power factor can be determined by using the information found in Table A8 of Appendix A.

The volt-ampere rating controls the maximum current that the transformer can safely carry in either the primary or secondary winding without overheating and exceeding the specified operating temperature of the transformer. If the transformer is rated in watts and if the load is operated at a power factor that is less than unity, then more and more current would be required to operate the load at the specified wattage as the power factor decreased. By rating the transformer in volt-amperes, definite voltage and current limits are established for all loads, independent of the load power factor.

Example 11.3

Determine the maximum primary and secondary currents permitted in a single-phase 37.5-kVA distribution transformer that is attached to a 480-V line when the transformer is supplying 240 V to a reactive load with a power factor of 0.78.

SOLUTION   Find both the primary and secondary current using Equation 11.3.

Given

$$I_p = \frac{P_s}{E_p} \qquad\qquad I_s = \frac{P_s}{E_s}$$

Evaluate

$P_s = 37\ 500$ VA $P_s = 37\ 500$ VA

$E_p = 480$ V $E_s = 240$ V

Substitute

$$I_p = \frac{37\ 500}{480} \qquad\qquad I_s = \frac{37\ 500}{240}$$

Solve

$I_p = 78$ A $I_s = 156$ A

Observation   Because the apparent power (volt-amperes) and not the average power (watts) is used to determine the maximum currents in the primary and secondary, the power factor is not used in the calculation.

Thus, the volt-ampere rating is intended to indicate the safe operating current (at a specified voltage) in the secondary winding of the transformer. Besides the secondary circuit, the transformer volt-ampere rating may also apply to the primary winding, since it is this winding that must take power from the line if the secondary winding is to deliver power to the load.

Single-Phase Nameplate Markings

The information provided on the transformer nameplate includes the frequency of operation (60 Hz), number of phases (1-ϕ), the volt-ampere rating (25 kVA), and the voltage rating (240-120/240). Certain markings associated with the transformer voltage specification need further explanation.

Bar (/) Voltages are from the same coil (120/240).
Dash (-) Voltages are from separate windings (480-120).
Cross (×) Voltages are from a dual-voltage winding (120×240).

The following examples will help you to understand the meaning of these markings.

120/240 240-V winding with a continuous winding that has been center-tapped (Figure 11.16(a))

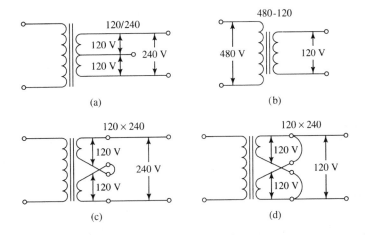

FIGURE 11.16
(a) A 240-V secondary winding with a center-tap connection to provide 120/240 V secondary. (b) A 480-V primary winding and a 120-V secondary winding to provide a 480-120 V transformer. (c) A dual-voltage 120 × 240-V winding connected in series to provide a 240-V secondary. (d) A dual-voltage 120 × 240-V winding connected in parallel to provide a 120-V secondary.

480-120 Two separate windings, one 480 V and the other 120 V (Figure 11.16(b))

120×240 Two separate but equal voltage windings (dual-voltage winding); both are 120 V and may be connected in series to provide 240 V, or in parallel to provide 120 V (Figure 11.16(c) and (d))

In general, specified voltages of 120 V, 208 V, 240 V, and 480 V are called *nominal voltages* (named voltage). These voltages are used when calculating the wire size and line drop. The actual voltages (measured voltages) usually vary from the nominal voltage by a few percentage points. For example, 120-V nominal may have actual voltages between 115 V and 125 V.

Transformer Connections

In addition to the nameplate information previously described, the high- and low-side terminals of the transformer are identified on the nameplate wiring diagram, as illustrated in Figure 11.17. As shown, the dual-voltage transformer has its high-side terminals indicated as H1, H2, H3, H4 and its low-side terminals as X1, X2, X3, X4. The even-numbered terminals (H2, H4, X2, X4) have the same instantaneous voltage polarity. Odd-numbered terminals (H1, H3, X1, X3) have the opposite polarity. That is, terminals marked H1 and H3 may have positive-going voltages at the same instant of time that H2 and H4 have negative-going voltages.

Because the terminal markings serve to identify the terminals with the same instantaneous polarity, the coils can be simply and correctly connected in series to operate at higher voltages or in parallel to operate at lower voltages by carefully

Acme Electric Company
Flagstaff, Arizona

Single-phase, dual-voltage, insulated transformer
kVA [15]   H.V. [240×480]   L.V. [120×240]   Hz [60]

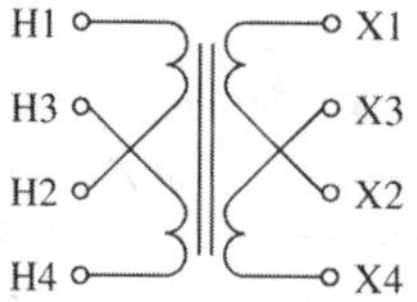

Primary Volts	Connect Primary Lines to	Inter-connect	Connect Secondary Lines to
480	H1 and H4	H2 to H3	
240	H1 - 3 & H2 - 4	H1 to H3 and H2 to H4	
Secondary Volts			
240		X2 to X3	X1 and X4
120/240		X2 to X3	X1 and X2-3 and X4
120		X1 to X3 X2 to X4	X1-3 and X2-4

FIGURE 11.17
The dual-voltage transformer has its high-side terminals indicated as H1, H2, H3, H4 and its low-side terminals indicated as X1, X2, X3, X4.

following the nameplate connection diagram. However, failure to follow correctly the system of terminal markings can result in short circuits that may damage the transformer and/or cause serious injury to the technician. When connecting transformers, these general rules need to be followed:

- When connecting transformer windings, always follow the recommendations of the manufacturer.
- Control and distribution transformers may be used to either step up or step down voltage. The high-voltage winding(s) are marked with an *H*, whereas the low voltage winding(s) are marked with an *X*.
- When connecting distribution and control transformers as step-down transformers, note that the terminals marked with an *H* are the primary terminals (high-voltage terminals); those marked with an *X* are the secondary terminals (low-voltage terminals).
- When connecting a dual-voltage transformer, as pictured in Figure 11.17, take special care in following the nameplate wiring diagram so that each terminal is correctly connected to produce the desired primary and secondary voltages. *For series operation*, to increase voltage, connect terminals 2 and 3 together, then attach input/output wiring to terminals 1 and 4. *For parallel operation*, to increase current, connect terminals 1 and 3 together and connect terminals 2 and 4 together, then attach input/output wiring to terminals 1,3 and 2,4.

Example
11.4

Demonstrate how the single-phase, 15-kVA, dual-voltage distribution transformer of Figure 11.17, with a nameplate voltage rating of 240×480 V on the high side and 120×240 V on the low side, can be connected to provide each of the following.

    (a) A 480-V primary and a 240-V secondary.
    (b) A 480-V primary and a 120-V secondary.
    (c) A 240-V primary and a 120/240-V secondary.

SOLUTION

Sketch a wiring diagram for each of the desired circuit conditions.

Given

The connection diagram, Figure 11.17

Observation

The two 240-V primary coils are connected in *series* to accommodate a line voltage of 480 V. The two 120-V secondary coils are connected in *series* to provide a load voltage of 240 V.

    (a) On the primary side of the transformer, interconnect H2 to H3 and attach the power lines to H1 and H4. On the secondary side of the transformer, interconnect X2 to X3 and attach the secondary lines to X1 and X4. Figure 11.18(a) shows the connections made.

Observation

The two 240-V primary coils are connected in *series* to accommodate a line voltage of 480 V. The two 120-V secondary coils are connected in *parallel* to provide a load voltage of 120 V.

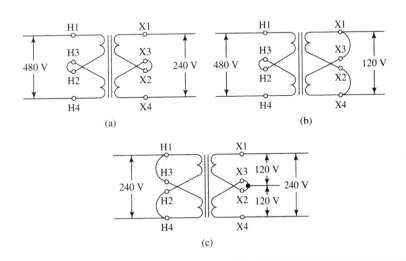

FIGURE 11.18
(a) 480-240 V transformer in which the primary is series connected (240 + 240 = 480 V), and the secondary is series connected (120 + 120 = 240 V). (b) 480-120 V transformer in which the primary is series connected (240 + 240 = 480 V), and the secondary is parallel connected (120 ∥ 120 = 120 V). (c) 240-120/240 V transformer in which the primary is parallel connected (240 ∥ 240 = 240 V), and the secondary is series connected with center tap (120 + 120 = 240/120 V).

(b) On the primary side of the transformer, interconnect H2 to H3 and attach the power lines to H1 and H4. On the secondary side of the transformer, interconnect X1 to X3 and interconnect X2 to X4, then attach the secondary lines to X1-3 and X2-4. Figure 11.18(b) shows the connections made.

Observation

The two 240-V primary coils are connected in *parallel* to accommodate a line voltage of 240 V. The two 120-V secondary coils are connected in *series* to provide a load voltage of 240 V; with the addition of a neutral at the center of the series-connected coils, 120 V is also provided.

(c) On the primary side of the transformer, interconnect H1 to H3 and interconnect H2 to H4, then attach the power lines to H1-3 and H2-4. On the secondary side of the transformer, interconnect X2 to X3 and attach the secondary lines to X1, X2-3, and X4. Figure 11.18(c) shows the connections made.

     The following example is designed to demonstrate how to determine the current in the primary and secondary windings for each of the three circuits used in the previous example.

Example 11.5

Determine the maximum current in each winding of the dual-voltage, 15-kVA, single-phase transformer of Figure 11.17 when it is connected to provide the following voltages.
    (a) A 480-V primary and a 240-V secondary.
    (b) A 480-V primary and a 120-V secondary.
    (c) A 240-V primary and a 120/240-V secondary.

SOLUTION

(a) Assuming an ideal transformer, find the current in the primary (I_p) and the current in the secondary (I_s) using the 15-kVA rating of the dual-voltage transformer.

Given

The 15-kVA rating of the transformer, the primary and secondary voltages, and Equation 11.3.

$$I_p = \frac{P_s}{E_p} \qquad\qquad I_s = \frac{P_s}{E_s}$$

Evaluate

$$P_s = 15 \text{ kVA} \qquad\qquad P_s = 15 \text{ kVA}$$

$$E_p = 480 \text{ V} \qquad\qquad E_s = 240 \text{ V}$$

Substitute

$$I_p = \frac{15\,000}{480} \qquad\qquad I_s = \frac{15\,000}{240}$$

Solve

$$I_p = 31.3 \Rightarrow 31 \text{ A} \qquad I_s = 62.5 \Rightarrow 63 \text{ A}$$

(b) Assuming an ideal transformer, find the current in the primary (I_p) and the current in the secondary (I_s) using the 15-kVA rating of the dual-voltage transformer.

Given   The 15-kVA rating of the transformer, the primary and secondary voltages, and Equation 11.3.

$$I_p = \frac{P_s}{E_p} \qquad\qquad I_s = \frac{P_s}{E_s}$$

Evaluate $P_s = 15\,\text{kVA}$ $P_s = 15\,\text{kVA}$

$E_p = 480\,\text{V}$ $E_s = 120\,\text{V}$

Substitute $I_p = \dfrac{15\,000}{480}$ $I_s = \dfrac{15\,000}{120}$

Solve $I_p = 31.3 \Rightarrow 31\,\text{A}$ $I_s = 125 \Rightarrow 130\,\text{A}$

(c) Assuming an ideal transformer, find the current in the primary (I_p) and the current in the secondary (I_s) using the 15-kVA rating of the dual-voltage transformer.

Given   The 15-kVA rating of the transformer, the primary and secondary voltages, and Equation 11.3.

$$I_p = \frac{P_s}{E_p} \qquad\qquad I_s = \frac{P_s}{E_s}$$

Evaluate $P_s = 15\,\text{kVA}$ $P_s = 15\,\text{kVA}$

$E_p = 240\,\text{V}$ $E_s = 240\,\text{V}$

Substitute $I_p = \dfrac{15\,000}{240}$ $I_s = \dfrac{15\,000}{240}$

Solve $I_p = 62.5 \Rightarrow 63\,\text{A}$ $I_s = 62.5 \Rightarrow 63\,\text{A}$

Observation   Since the two 120-V secondary coils are connected in series, the current in each is the same, and it is equal to 62.5 A, or 7500 VA, divided by 120 V (7500/120 = 62.5 A).

11.4 THREE-PHASE TRANSFORMER

Characteristics of a Three-Phase System

The voltage in the "electric mains" provided by the utility company is three-phase (3-ϕ) voltage, which is generated by a three-phase generator (alternator) consisting of three separate coils spaced 120° apart, as pictured in Figure 11.19. When the three coils of the elementary generator of Figure 11.19(a) are rotated, three alternating voltages are produced for each cycle of rotation, as illustrated in Figure 11.19(b). As pictured, the three sinusoidal voltages starting from zero, one after the other and 120° apart from each other, are called *three-phase voltage* (3-ϕ).

Three-phase power-distribution systems are the standard of the world, with 50 and 60 Hz being the preferred frequencies of the alternating current and voltage.

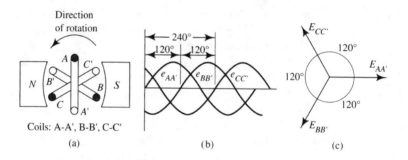

Direction of rotation

Coils: A-A', B-B', C-C'

(a)

(b)

(c)

FIGURE 11.19

Three-phase voltage is generated in a three-phase alternator (generator): (a) The three coils in the generator are 120° apart, and when rotated they generate three sine wave voltages; (b) The sinusoidal voltages generated are 120° apart; (c) The generated voltages can be represented in a phasor voltage diagram.

Three-phase power systems have several advantages over single-phase power systems, which make three-phase power preferable to single-phase power. Among the advantages are the following:

Constant power to the load As pictured in Figure 11.20(a), three-phase power is constant in a balanced load (one in which the same power and power factor are present in each of the three phases). As pictured, the instantaneous power (p_A, p_B, p_C) produced in each of the three coils is equal (the same); the summation of these three powers is indicated by the straight line labeled p_T, the total instantaneous power. Because the power curve is constant (a straight line), a constant unvarying torque will be produced when a three-phase motor is attached to the three-phase line.

Since the three-phase power is not pulsating, there is little or no mechanical vibration associated with the operation of a three-phase motor. However, this is not the case with the operation of a single-phase motor operating from a single-phase line. The pulsating nature of single-phase

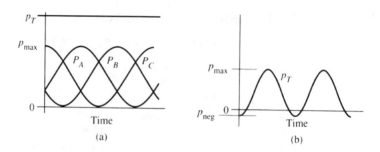

FIGURE 11.20

Power from: (a) A three-phase power line with a balanced load; (b) A single-phase power line with an inductive load. *Note:* A balanced three-phase load has the same power and the same power factor in each of the three loads.

power, as illustrated in Figure 11.20(b), produces mechanical vibration during operation of a single-phase motor. As shown in Figure 11.20(b), single-phase inductive power passes through zero four times in each power cycle.

A rotating magnetic field With the positioning of the coils in the three-phase alternator, the generated voltages are 120° apart, as noted in Figure 11.19(c). Each of these voltages produces currents, which, collectively, set up a magnetic field that has a constant flux density. The magnetic field produced by the voltages is *rotating*; that is, it moves from pole to pole. It is this rotating field that makes the self-starting of three-phase motors possible. Single-phase motors are not self-starting because the magnetic field in the single-phase motor does not rotate.

Smaller size, less weight, and less cost Three-phase motors are considerably smaller in size and weight when compared to a single-phase motor with the same nameplate power rating. Because of their lower production cost, three-phase motors are less expensive than comparably rated single-phase motors.

Smaller, lighter, and less expensive dc power supplies When three-phase power is rectified and filtered for direct current (dc) power, smaller-size (both physically and electrically) components can be used than would otherwise be possible in a similarly rated single-phase power supply. The reduction in size and weight of the three-phase power supply results in a considerable savings in both cost and space.

Three-Phase Power Distribution

The primary side of the three-phase distribution transformer is connected to the utility power lines in one of two possible configurations, delta or wye, depending on the level of the voltage and the nature of the transformer being used. Table 11.1 lists the common primary voltages for both three-phase and single-phase distribution transformers, and Table 11.2 gives some of the available kVA ratings.

The three-phase transformer primary represents a three-phase load to the incoming power line; as such, it may be connected to the three-phase line as a **wye (Y) connection** or as a **delta (Δ) connection**, as illustrated in Figure 11.21. In either

TABLE 11.1 Typical Voltage Ratings of Distribution Transformers

Single-Phase (V)		Three-Phase, Line-to-Line (V)	
120	480	208	600
208	600	240	2400
240	2400	480	4160

TABLE 11.2 Typical Kilovolt-Ampere Ratings of Distribution Transformers

Single-Phase (kVA)		Three-Phase (kVA)	
1	50	3	75
1.5	75	6	112.5
2	100	9	150
3	125	15	225
5	167.5	22.5	300
7.5	200	25	500
10	250	30	600
15	333	37.5	750
25	500	45	1000
37.5		50	

connection, the number of lines and the primary voltage rating of the transformer must be compatible with the number of wires and the voltage rating of the incoming three-phase line. That is, a four-wire, 4160-V three-phase line can be connected only to a transformer primary that can be configured to operate safely with four wires and 4160 V. Furthermore, the transformer must be rated to operate at that voltage.

Line and Phase Parameters

When connecting a three-phase transformer, an understanding of the difference between phase voltage and line voltage is necessary. The primary and secondary voltages listed on the nameplate of a three-phase transformer are the *phase voltages* of

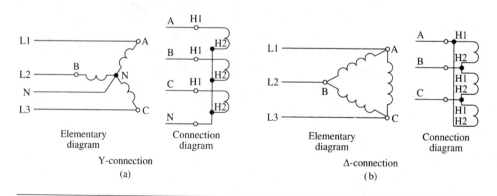

FIGURE 11.21

Primary connection of a three-phase transformer with the secondary winding not shown: (a) Wye-connection using terminals A, B, C, and N. This is a four-wire, wye-connected three-phase system; (b) Delta-connection using terminals A, B, and C. This is a three-wire, delta-connected three-phase system. *Note*: N = neutral.

the primary or secondary, unless the voltage is followed by the letter Y, which indicates that the voltage is the *line voltage.*

The relationship between line voltage and phase voltage is illustrated in Figure 11.22, where the primary of a 2400-208Y/120-V transformer is delta-connected and the secondary is wye-connected. As illustrated, each 2400-V phase of the three-wire, delta-connected primary is wired to the 2400-V line voltage. Additionally, the four-wire, wye-connected secondary provides (sources) a line voltage of 208 V (line-to-line) along with a phase voltage of 120 V (line-to-neutral) to the load.

Delta-Connection

➢ **As a Rule** In a balanced three-phase system, because *each winding* (phase) is connected between two lines in *delta*-connected primaries and/or *delta*-connected secondaries,

1. The voltage between the lines is equal to the voltage across the phase:

$$E_{\text{line}} = E_{\text{phase}} \qquad \textbf{(11.4)}$$

2. The line current is the vector addition of two-phase currents, and the line current is equal to the square root of 3 (written as $\sqrt{3}$) times the phase current.

$$I_{\text{line}} = \sqrt{3} \times I_{\text{phase}} \qquad \textbf{(11.5)}$$

where, for the delta-connection,

E_{line} = voltage between any two lines (L1-L2, L2-L3, L3-L1 of the primary of Figure 11.22), in V

E_{phase} = voltage across any one of the three phase (windings) of the transformer, in V

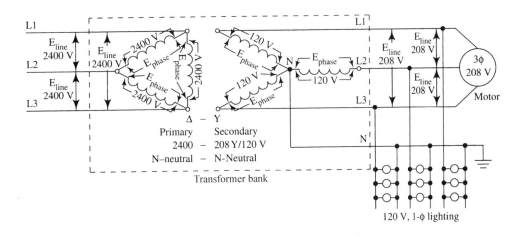

FIGURE 11.22
Line and phase voltage of the delta-connected primary and the wye-connected secondary.
Primary: delta-connected, 2400-V line and phase.
Secondary: wye-connected, 208-V line and 120-V phase.
Nameplate: 2400-208Y/120 V.

I_{line} = current in any one of the three lines (L1,2,3), in A

I_{phase} = current in any one of the three phases (windings) of the transformer, in A.

$\sqrt{3} = 1.73$

Wye-Connection

➢ **As a Rule** In a balanced three-phase system, because *two windings* (phases) are connected between two lines in *wye*-connected primaries and/or *wye*-connected secondaries,

1. The line voltage is the vector addition of two-phase voltages, and the line voltage is equal to the square root of 3, or $\sqrt{3}$ times the phase voltage:

$$E_{line} = \sqrt{3} \times E_{phase} \tag{11.6}$$

2. The current in the lines is equal to the current in the phases:

$$I_{line} = I_{phase} \tag{11.7}$$

where, for the wye-connection,

E_{line} = voltage between any two lines (L1-L2, L2-L3, L3-L1 of the secondary of Figure 11.22), in V

E_{phase} = voltage across any of the three phases (windings) of the transformer (line-to-neutral), in V

I_{line} = current in any one of the three lines (L1,2,3), in A

I_{phase} = current in any one of the three phases (windings) of the transformer, in A

$\sqrt{3} = 1.73$

Example 11.6

Determine the line voltage in the following wye-connected, three-phase transformers when

    (a) The phase voltage is 120 V.
    (b) The phase voltage is 277 V.
    (c) The phase voltage is 2400 V.

SOLUTION   **(a)** Find the line voltages using Equation 11.6.

Given $E_{line} = \sqrt{3} \times E_{phase}$

Evaluate $E_{phase} = 120$ V

Substitute $E_{line} = \sqrt{3} \times 120$

Solve $E_{line} = 208$ V

    (b) Find the line voltages using Equation 11.6.

Given $E_{line} = \sqrt{3} \times E_{phase}$

Evaluate	$E_{\text{phase}} = 277 \text{ V}$
Substitute	$E_{\text{line}} = \sqrt{3} \times 277$
Solve	$E_{\text{line}} = 480 \text{ V}$

(c) Find the line voltages using Equation 11.6.

Given	$E_{\text{line}} = \sqrt{3} \times E_{\text{phase}}$
Evaluate	$E_{\text{phase}} = 2400 \text{ V}$
Substitute	$E_{\text{line}} = \sqrt{3} \times 2400$
Solve	$E_{\text{line}} = 4160 \text{ V}$
Observation	Nameplate secondary voltage ratings for wye-connected (Y) transformers for the preceding voltage ratings are noted as 208Y/120 V, 480Y/277 V, 4160Y/2400 V. In this notation, the line voltage (with the letter Y) is stated first, followed by the phase voltage (line/phase).

Three-Phase Transformer Connections

Some three-phase transformers, as indicated in Figure 11.23, are not prewired into a delta- or wye-connection. When this is the case, the same transformer primary winding may be configured in either a delta- or a wye-connection (depending on the incoming power line voltage). The secondary winding may also be configured in a delta- or wye-connection, depending on the voltage needs of the load. Also note that the three transformer coils in the secondary are the voltage source for the connected load.

Figure 11.24(a) pictures the primary of a 2400-V, three-phase transformer connected to a three-phase, 4160-V line using a wye-connection, and Figure 11.24(b) shows the same transformer connected to a three-phase, 2400-V line using a delta-connection. Each connection uses a 208-V, wye-connected, four-wire, three-phase secondary with a grounded neutral.

FIGURE 11.23
The three-phase transformer may be constructed with three windings on a single core as a *consolidated, three-phase transformer*, as pictured, or with three separate, identical, single-phase transformers called a *three-phase transformer bank*. Three-phase transformers are designed, constructed, and wired for delta or wye operation with specific voltages.

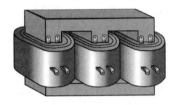

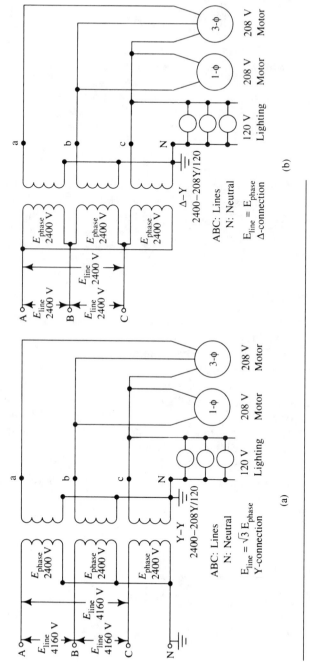

FIGURE 11.24

By reconnecting the primary of the three-phase transformer, both a 4160-V line and a 2400-V line can be accommodated. (a) Because two primary coils are between each pair of line wires in a wye-connected transformer, the line voltage needed to energize the primary is $\sqrt{3}$ times as great as the phase voltage (4160 V = 2400 V × $\sqrt{3}$). (b) Since the lines are connected directly across the coils in the primary of the delta-connected transformer, the line voltage is equal to the phase voltage (2400 V = 2400 V).

➢ **As a Rule**

- Delta and wye are the two principal ways to connect either primary or secondary three-phase transformer coils.
- The primary and secondary may be connected in four basic combinations plus the open delta (V-connection), as illustrated in Figure 11.25.
- If three separate, identical, single-phase transformers are used to form a delta-delta- (Δ-Δ) connected three-phase transformer bank and if one of the three single-phase transformers fails or is taken out of service, then the other two unaffected transformers may be operated in an open-delta (V-V) configuration at 57.7% (not 66.7%) of the original three-phase transformer bank rating.
- The wye-wye connection must have a metallic conductor connecting the neutral of the wye-connected primary to the neutral of the wye source, which is grounded. Without this connection, harmonic currents are produced, resulting in distortion in the voltage waveform, induced hum in telephone lines, and possible damage to the transformer from high voltage caused by *series resonance*.
- When wiring transformers into a circuit, it is very important that all transformer connections be made as shown on the manufacturer's nameplate.

Example 11.7

(a) Connect the three separate single-phase transformers pictured in Figure 11.26(a) into a three-phase, delta-to-wye transformer bank to supply load power for single-phase lighting and three-phase motor operation.

(b) Connect the three separate single-phase transformers pictured in Figure 11.26(a) into a three-phase, delta-to-delta transformer bank to supply load power for three-phase motor operation.

SOLUTION

(a) Sketch the connection diagram for a three-wire, delta-connected primary and a four-wire (neutral grounded), wye-connected secondary.

Given Figure 11.25(c)

Solve The three-phase, delta-to-wye connection is illustrated in Figure 11.26(b).

Observation The neutral of the wye-connected secondary is provided for connection of the single-phase lighting load between each of the three phases and the grounded neutral. The three-phase motor load is connected across the three-phase lines (a, b, and c). The lighting loads are evenly distributed among the three phases to maintain a balanced secondary load.

(b) Sketch the connection diagram for a three-wire, delta-connected primary and a three-wire, delta-connected secondary.

Given Figure 11.25(c)

Solve The three-phase delta-to-delta connection is illustrated in Figure 11.26(c).

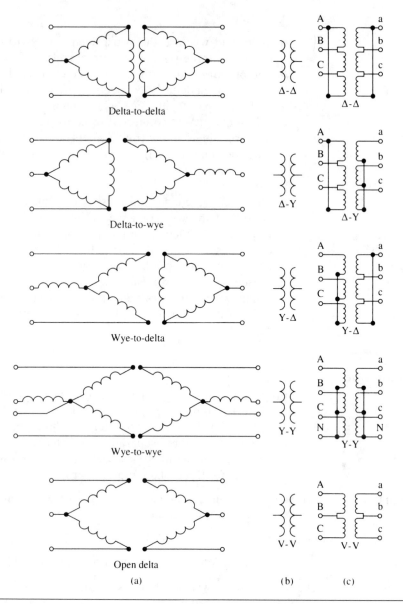

Delta-to-delta

Delta-to-wye

Wye-to-delta

Wye-to-wye

Open delta

(a)   (b)   (c)

FIGURE 11.25
Three-phase connections of primary and secondary windings: (a) Elementary connection diagrams;
(b) Single-line diagrams; (c) Connection diagrams.

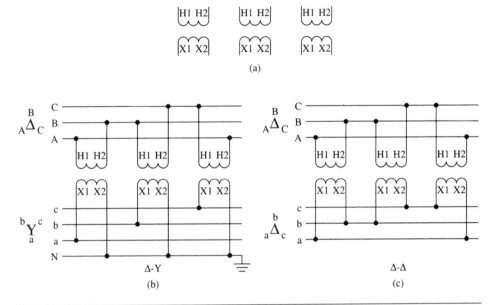

(a)

Δ-Y

(b)

Δ-Δ

(c)

FIGURE 11.26

Illustration for Example 11.7: (a) Three separate, single-phase transformers with the high-side (H) and low-side (X) terminals noted; (b) Transformers connected into a delta-to-wye, three-phase transformer bank; (c) Transformers connected into a delta-to-delta, three-phase transformer bank.

Power in the Balanced Three-Phase Load

The load attached to a three-phase line may be balanced, unbalanced, delta-connected, or wye-connected. Figure 11.27 pictures a three-phase motor (a balanced load) connected to a three-phase line through terminals T1, T2, and T3. The motor may be a delta-connected load, as shown in Figure 11.27(a), or a wye-connected

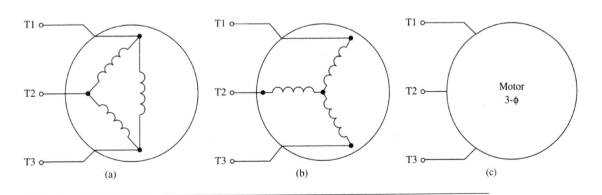

FIGURE 11.27

Three-phase motor connections: (a) Delta-connection; (b) Wye-connection; (c) Schematic symbol for a three-phase motor connected either delta or wye.

load, as in Figure 11.27(b). Figure 11.27(c) pictures the general schematic symbol for a three-phase motor.

The term *balanced load* means that the three branches or phases of the load are each equal; that is, the impedance of each branch is the same. Because the currents are the same in each phase of a balanced load, there is no net current at the neutral in the wye-connected motor and no need for a neutral wire.

The power supplied from a transformer secondary to a balanced three-phase load (delta- or wye-connected) is three times the power supplied to one of the three phases of the load:

$$P_{3\text{-}\phi} = 3P_{\text{phase}} = 3(E_{\text{phase}})(I_{\text{phase}})\cos\phi \qquad \textbf{(11.8)}$$

where

$P_{3\text{-}\phi}$ = power in the three-phase balanced load, in W
P_{phase} = power in one of the phases of the balanced load, in W
E_{phase} = voltage across one of the phases (see Figure 11.22, delta-connected primary) of the load, in V
I_{phase} = current in one of the phases of the load, in A
$\cos\phi$ = power factor of the three-phase load, where the power factor angle, ϕ, is the angle between the phase voltage and the phase current and not the angle between the line voltage and the line current

Example 11.8

Determine the power dissipated in a three-wire, wye-connected, balanced, three-phase load when E_{line} is 480 V, I_{phase} is 24.0 A, and the power factor angle (ϕ) is 28.4°.

SOLUTION   First determine the phase voltage with Equation 11.6, and then find the power using Equation 11.8.

Given $P_{3\text{-}\phi} = 3(E_{\text{phase}})(I_{\text{phase}})\cos\phi$

Evaluate $E_{\text{phase}} = \dfrac{E_{\text{line}}}{\sqrt{3}} = \dfrac{480}{\sqrt{3}} = 277\text{ V}$

$I_{\text{phase}} = 24.0\text{ A}$

$\cos\phi = \cos 28.4° = 0.8796$

Substitute $P_{3\text{-}\phi} = 3 \times 277 \times 24.0 \times 0.8796$

Solve $P_{3\text{-}\phi} = 17.5\text{ kW}$

Although Equation 11.8 seems straightforward in its application, the line voltage and line current, not the phase voltage and current, are easily measured. Thus, the power equation would be in its most useful form when expressed in terms of line voltage and current.

For the delta-connected balanced load, where the phase voltage is equal to the line voltage ($E_{\text{phase}} = E_{\text{line}}$) and the phase current is equal to the line current divided by $\sqrt{3}$ ($I_{phase} = I_{line}/\sqrt{3}$), Equation 11.8 may be stated in terms of E_{line} and I_{line} as

$$P_{3\text{-}\phi} = 3(E_{line})\left(\frac{I_{line}}{\sqrt{3}}\right)\cos\phi$$

For the wye-connected balanced load, where the phase current is equal to the line current ($I_{phase} = I_{line}$) and the phase voltage is equal to the line voltage divided by $\sqrt{3}$ ($E_{phase} = E_{line}/\sqrt{3}$, Equation 11.8 may be stated in terms of E_{line} and I_{line} as

$$P_{3\text{-}\phi} = 3\left(\frac{E_{line}}{\sqrt{3}}\right)(I_{line})\cos\phi$$

Since the power equations for both delta- and wye-connected balanced loads are the same, the equation for power in a three-phase balanced load in terms of line voltage and current is

$$P_{3\text{-}\phi} = \left(\frac{3}{\sqrt{3}}\right)(E_{line})(I_{line})\cos\phi$$

Simplifying $3/\sqrt{3}$ by rationalizing the denominator yields

$$\frac{3}{\sqrt{3}} \times \frac{\sqrt{3}}{\sqrt{3}} = \frac{3\sqrt{3}}{3} = \sqrt{3}$$

The equation for power in the three-phase, wye- or delta-connected load in terms of line voltage and current is

$$P_{3\text{-}\phi} = \sqrt{3}(E_{line})(I_{line})\cos\phi \tag{11.9}$$

where

$P_{3\text{-}\phi}$ = power in the three-phase balanced load, in W
E_{line} = voltage between any two terminals (T1-T2, T2-T3, T3-T1) connected to the balanced load (Figure 11.27(c)), in V
I_{line} = current in any line connected to the balanced load, in A
$\cos(\phi)$ = power factor of the three-phase load, where the power factor angle, ϕ, is the *angle between the phase voltage and the phase current* and not the angle between the line voltage and the line current

Example 11.9

Determine

(a) The power taken from the line to operate a 230-V, three-phase, 5-hp induction motor with a line current of 15.2 A and a power factor of 0.83.

(b) The efficiency of the 5-hp motor when it is producing 4.5 hp from the power of part (a).

SOLUTION   (a) Use Equation 11.9 to find the power needed to operate the motor.

Given	$P_{3-\phi} = \sqrt{3}(E_{line})(I_{line})\cos\phi$
Evaluate	$E_{line} = 230$ V
	$I_{line} = 15.2$ A
	$\cos\phi = 0.83$
Substitute	$P_{3-\phi} = \sqrt{3} \times 230 \times 15.2 \times 0.83$
Solve	$P_{3-\phi} = 5.0$ kW

(b) Find the efficiency of the motor when it takes in 5.0 kW and produces 4.5 hp.

Given	$\eta = \dfrac{P_{out}}{P_{in}}$
Evaluate	$P_{out} = 4.5 \text{ hp} \times 746 \dfrac{\text{W}}{\text{hp}} = 3.36$ kW
	$P_{in} = 5.0$ kW
Substitute	$\eta = \dfrac{3360}{5000}$
Solve	$\eta = 0.67$

Three-Phase Transformer Ratings

Like single-phase transformers, three-phase transformers are rated by their apparent power (P_s) in volt-amperes (VA) or kilovolt-amperes (kVA). When the primary and secondary full-load currents are not listed on the nameplate, the nameplate voltage and apparent power specifications may be used to determine them. Since the transformer is ideal, the power factor ($\cos\phi$) is equal to 1, and the apparent power in the three-phase transformer is

$$P_{s3-\phi} = \sqrt{3}E_{line}I_{line} \qquad (11.10)$$

Solving for I_{line} results in

$$I_{line} = \frac{P_{s3-\phi}}{E_{line}\sqrt{3}} \qquad (11.11)$$

where

I_{line} = primary/secondary full-load line current, in A
$P_{s3-\phi}$ = 3-ϕ transformer apparent power, in VA
E_{line} = primary/secondary line voltage, in V

Example
11.10

Determine the full-load secondary current in a three-phase transformer rated at 30 kVA when the 208-V secondary is wye-connected.

SOLUTION   Find the current in the secondary using Equation 11.11.

Given	$I_{\text{line}} = \dfrac{P_{s3\text{-}\phi}}{E_{\text{line}}\sqrt{3}}$
Evaluate	$P_{s3\text{-}\phi} = 30\ \text{kVA}$
	$E_{\text{line}} = 208\ \text{V}$
Substitute	$I_{\text{line}} = \dfrac{30\ 000}{208\sqrt{3}}$
Solve	$I_{\text{line}} = 83\ \text{A}$

Transformers, like other ferromagnetic devices, have core losses as well as winding losses (copper loss) that cause them to operate at less than 100% efficiency and at elevated temperatures. Even with losses, transformers are among the most efficient machines, with efficiencies greater than 96% for high-power units.

As was noted in Chapter 10, ferromagnetic devices (transformers, solenoids, relays, motors, etc.) all experience an increase in temperature when operating due to I^2R loss (winding losses) and to core losses (hysteresis and eddy current). Because losses serve to raise the operating temperature of the transformer, the transformer can't be operated at full power above a specified rise in the operating temperature.

CHAPTER SUMMARY

- Electric power is moved from the generating station through the switching station to the receiving substations by high-voltage, three-phase transmission lines.
- Electric power enters an industrial plant through the service entrance switchboard equipment.
- Overcurrent devices are part of a panelboard.
- The terms *step up* and *step down* are used with transformers to indicate that voltage is increased or decreased.
- The transformation ratio is the ratio between the number of turns of wire in the primary and secondary coils.
- The amount of current in the primary winding of an ideal transformer depends on the demand of the load connected to the secondary winding of the transformer.
- The control transformer is used to step down voltage for use with two- and three-wire control circuits.
- In a single-phase, three-wire system, the neutral wire is identified by the white color coding of its insulation.
- Transformers are rated in volt-amperes (apparent power) rather than in watts (average power) to ensure that their maximum current rating is the same for all loads.
- Specified voltages of 120, 208, 240, 480, 600, etc., are nominal voltages used to calculate wire size, etc.
- The letters H and X are used to identify the high (H) and low (X) sides of the transformer.
- Single-phase power pulsates, whereas three-phase power is constant.

- When wye-connected, the three-phase line voltage is $\sqrt{3}$ times the phase voltage; however, when delta-connected, the line and phase voltages are equal.
- In a balanced, wye-connected three-phase load, the neutral carries no current.
- In a balanced three-phase load, power (watts) is equal to 3 times the product of the phase voltage, phase current, and power factor ($3E_{phase}I_{phase} \cos \phi$), or power is equal to $\sqrt{3}$ times the line voltage, line current, and power factor ($\sqrt{3} \, E_{line} \, I_{line} \cos \phi$).

SELECTED TECHNICAL TERMS

The following technical terms, abbreviations, and acronyms are defined in the glossary located after Chapter 16. You are encouraged to use the glossary to aid your understanding and to test your knowledge of these important terms.

delta connection	single-phase
primary winding	three-phase
rms (root mean square)	transformer
secondary winding	wye connection

END-OF-CHAPTER QUESTIONS

Write T if the statement is true and F if the statement is false.

1. Besides distributing power throughout an industrial building, the switchboard is the point of connection to the building's grounding electrode.
2. When used with transformers, the terms *step up* and *step down* indicate an increase or decrease in the current.
3. The transformation ratio is a ratio between the turns of wire in the secondary and the turns of wire in the primary (N_s/N_p).
4. The neutral, connected to ground in the three-wire, single-phase, 120/240-V supply is the center of the secondary winding.
5. The phase voltage is equal to 1.73 times the line voltage in a wye-connected load.
6. Control voltages of 24 and 120 V are available from distribution transformers.
7. An industrial panelboard, a wall-mounted steel cabinet, receives power from a feeder in the motor control center.
8. A three-phase transformer bank is constructed by connecting three separate single-phase transformers into either a delta or wye configuration.
9. The line and phase voltage are equal in a wye-connected, three-phase transformer.
10. The high-side terminals of a transformer are indicated by the letter H.

In the following, select the word or phrase that makes the statement true.

11. Transformer operation is due to the principle of (self-induction, electromagnetic induction, mutual induction), in which a voltage is generated in one coil by a current in another coil.
12. An industrial building is supplied by a three-wire, 480-V, three-phase electrical system. Assuming a transformer with separate primary and secondary windings is

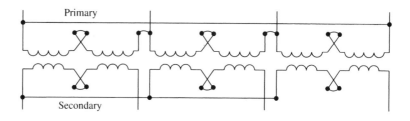

FIGURE 11.28
Transformer for Question 14.

used, the transformer voltage specifications for a three-wire, 120/240-V feeder to a single-phase panelboard would be (480Y/120/240, 480-120/240, 480×120/240).

13. If a transformer with a primary rated at 240 V has a transformation ratio of 5:1, then the secondary voltage is (24 V, 36 V, 48 V).

14. The three-phase transformer shown in Figure 11.28 is a (wye-delta, wye-wye, delta-wye, delta-delta) connection.

15. The transformer's mutual flux moves electrical energy from the (load, primary, secondary) winding to the (source, primary, secondary) winding.

Answer each of the following questions with a short answer in the form of a complete sentence. Include a restatement of the question in your answer.

16. Besides reducing the voltage by transformation, what other functions does a distribution substation provide?

17. What are three advantages of using a three-phase power system over a single-phase power system?

18. What are some of the reasons to use a motor control center to control three-phase motors?

19. What are the characteristics of a balanced three-phase load?

20. Explain why transformers are specified in volt-amperes rather than watts.

END-OF-CHAPTER PROBLEMS

Solve the following problems. Make sketches to aid in solving the problems and structure your work so it follows in an orderly progression and can easily be checked. Table 11.3 summarizes the formulas used in Chapter 11.

1. A 240-24-V control transformer is constructed by winding 865 turns of wire to form the secondary coil of the transformer. Determine the number of turns in the primary coil.

2. Determine the transformation ratio of the control transformer of Problem 1.

3. Sketch the connection of the single-phase, dual-voltage transformer of Figure 11.29 for 240-24-V operation.

4. Connect the transformer bank pictured in Figure 11.26(a) for three-phase, three-wire, wye-to-delta operation. Sketch the connection diagram and show both the primary connection to lines A, B, C and the secondary connection to lines a, b, c.

TABLE 11.3 Summary of Formulas Used in Chapter 11

Equation Number	Equation
11.1	$\dfrac{E_p}{E_s} = \dfrac{N_p}{N_s}$
11.2	$\dfrac{I_s}{I_p} = \dfrac{N_p}{N_s}$
11.3	$I = \dfrac{P_s}{E}$
11.4 ⎫ Delta-connection 11.5 ⎭	$E_{\text{line}} = E_{\text{phase}}$ $I_{\text{line}} = \sqrt{3} \times I_{\text{phase}}$
11.6 ⎫ Wye-connection 11.7 ⎭	$E_{\text{line}} = \sqrt{3} \times E_{\text{phase}}$ $I_{\text{line}} = I_{\text{phase}}$
11.8	$P_{3\text{-}\phi} = 3P_{\text{phase}} = 3(E_{\text{phase}})(I_{\text{phase}})\cos\phi$
11.9	$P_{3\text{-}\phi} = \sqrt{3}(E_{\text{line}})(I_{\text{line}})\cos\phi$
11.10	$P_{s3\text{-}\phi} = \sqrt{3}E_{\text{line}}I_{\text{line}}$
11.11	$I_{\text{line}} = \dfrac{P_{s3\phi}}{E_{\text{line}}\sqrt{3}}$

5. Determine the primary full-load line current of an ideal, delta-connected, three-phase, 15-kVA transformer attached to a 240-V line.

6. A three-phase delta transformer bank made up of three 25-kVA transformers is rated at 75 kVA. Determine the rating (kVA) of the transformer bank when one transformer is removed for maintenance and the remaining transformers are operated in an open-delta configuration.

7. The substation transformer pictured at the start of this chapter has a rating of 600 kVA, 2.40 kV/480 V. Determine the currents in the primary and secondary lines when the transformer is connected delta-delta.

8. Determine the maximum current in each winding of the single-phase dual-voltage 5.0-kVA control transformer of Figure 11.29 when it is connected to provide 240-48 V.

FIGURE 11.29
Transformer for Problem 3.

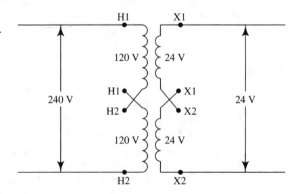

9. Determine the power dissipated in a three-wire, wye-connected, three-phase balanced load when E_{phase} is 277 V, I_{phase} is 24 A, and the power factor angle (ϕ) is 35°.

10. Determine the secondary current in a single-phase distribution transformer that is attached to a 480-V line when the transformer is supplying 240 V to a 16-kW reactive load with a power factor of 0.78.

11. Determine the line current needed to operate a 460-V, 15-hp induction motor from a three-phase, 60-Hz line when the motor has a rated efficiency of 0.89 and a power factor of 0.85.

12. Three single-phase distribution transformers rated at 333 kVA, 2400 V-600 V, 60 Hz are connected wye-delta to a 4160 V, three-phase line. If the load is 680 kVA, determine

    (a) The incoming and outgoing line currents.

    (b) The primary and secondary phase currents.

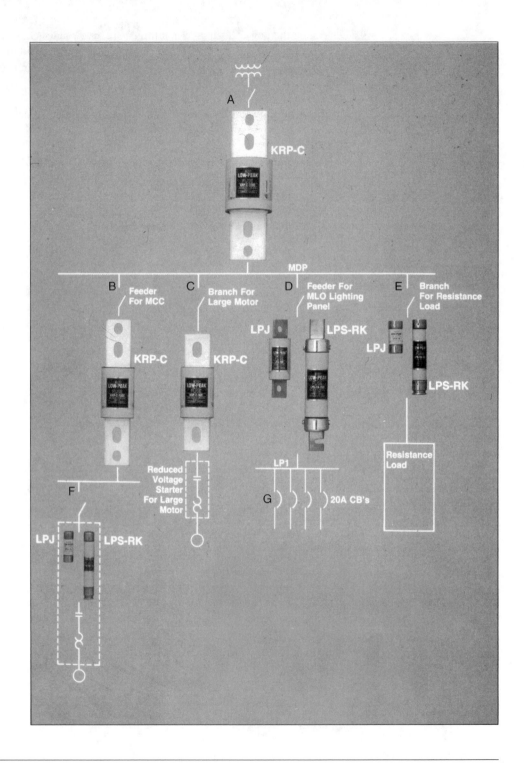

Chapter 12

Overcurrent Protection

About the Illustration. *The overcurrent protection points in an industrial plant's electrical system are emphasized in this electrical diagram. The fused disconnect main switch (A) provides power from the* service transformer *to the* switchboard's *overcurrent devices for the* feeder circuits. *Pictured are the feeder circuits for a motor-control center (MCC) (B), large motor fused disconnect (C), fused disconnect for a lighting panel (D), and a fused disconnect for a resistive heater (E). Additional overcurrent protection is provided to each machine motor by a fused disconnect (F) and at the lighting panel to each light circuit by a 20-A circuit breaker (G). (Courtesy of Bussmann, Cooper Industries, Inc.)* ∎

INTRODUCTION

Electrical circuits are protected by overcurrent devices (fuses and circuit breakers) and overload devices so that the ampacity (current rating) of the conductors is not exceeded. When the current rating of the circuit is exceeded, then the insulation on the wiring may overheat, melt, and possibly catch fire. Several devices, including fuses, circuit breakers, and overload relays (both thermal and magnetic), are introduced in this chapter.

Additionally, you will learn about overcurrent conditions and how overcurrent and overload devices are used to prevent damage to the electrical circuit and its wiring and equipment. Furthermore, the dangers to people (fire, electrocution, etc.) associated with fault and overload conditions are addressed.

CHAPTER CONTENTS

PERFORMANCE OBJECTIVES

Once you have read and studied each section; worked through each example with pencil, paper, and calculator; worked through the end-of-chapter problems; and answered the end-of-chapter questions, then you should be able to

- Distinguish among overcurrent conditions resulting from short circuit, ground fault, and overload.
- Explain the difference between a high-impedance ground fault and a low-impedance ground fault.
- Differentiate between a fault and an overload condition.
- List the two major categories of overcurrent devices.
- Relate the specific heat of a material to the fusing process.
- Understand what is meant by the inverse time characteristic of an overcurrent or overload device.
- Know the four basic parameters used to specify overcurrent devices.
- Describe the operation of both thermal and magnetic overload relays.

12.1 OVERCURRENT CONDITIONS

Introduction

An **overcurrent** condition is any current level above the rated continuous current of a circuit. Thus, a current level even slightly over the rated continuous circuit current is an overcurrent condition. Protection from overcurrent conditions is provided by *overcurrent devices* and *overload devices*. These two types of devices provide protection for a variety of overcurrent conditions, ranging from levels slightly over the rated continuous circuit current (an **overload**) to levels that are many times the rated continuous circuit current.

The three major circumstances leading to an overcurrent condition in an electrical power distribution system are a short circuit, a ground fault, and an overload. Of these three conditions, a short circuit and a ground fault represent a *fault* in an electrical system. A fault may occur in any part of the electrical distribution system, including transformers, conductors, and loads. A fault condition is any abnormal condition that provides an unintentional path for current outside the conductor, such as two conductors coming in contact with one another, the breaking down of the insulation in a motor, or one or more phases in the power system shorting to ground.

Short Circuit Condition

A **short circuit** results when the circuit conductors (wires) touch each other, causing excess current in the electrical system. The excess current is called the **fault current.** When conductors touch, the current in them quickly rises to several times their rated safe capacity due to the lessening of the impedance in the circuit. The rating of conductors by the amount of current they can safely carry is called the **ampacity** of the conductor. The ampacity of conductors in a metal raceway for typical operating conditions of 30°C (86°F) or less is given in Appendix A, Table A7.

The increased current (fault current) results in overheating of the conductors, which—if left unchecked—would result in damage or destruction to the source (transformer), conductors, and/or load. The heat generated in the conductive path of the electric system must be kept within the design limit of the system, or the wiring will overheat, causing a serious fire hazard. The energy converted to heat in an electrical circuit depends directly on the square of the current, the resistance, and the time. Thus,

$$W = Pt = I^2Rt \qquad (12.1)$$

where

W = work, in J
P = power, in W
I = current, in A
R = resistance, in Ω
t = time, in s

In order for the current to increase during a short circuit, the impedance of the circuit must proportionally decrease. Thus, when wires short-circuit, the impedance limiting the current flow is small, and the resulting current is large.

The impedance of the electrical system that limits the fault current is a combination of the wiring resistance and the equivalent impedance of the transformer supplying power to the short circuit. The following example explores these concepts.

Example 12.1

Determine the quantity of heat produced (joules) for 10 ms (the time needed by the overcurrent device to clear the fault) in the three-wire, 120/240-V, single-phase circuit pictured in Figure 12.1 under

(a) Normal operation, pictured in Figure 12.1(a).
(b) Short-circuit operation, pictured in Figure 12.1(b).

SOLUTION

Find the work (heat) produced in each circuit during the duration of the fault. Use Equation 12.1.

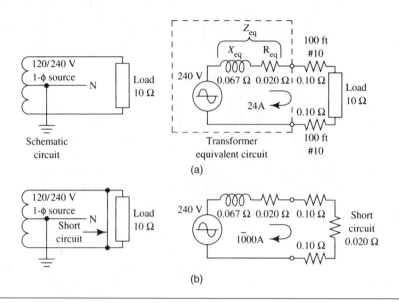

FIGURE 12.1

Heat is produced when current flows in an electrical circuit. (a) Under normal operation, circuit current produces a manageable amount of heat in the conductors and load. (b) Under a short-circuit condition, circuit current produces large amounts of heat in the conductors, which is manageable only if the over-current device quickly interrupts the fault current.

Given	**(a)** $W = I^2 Rt$	**(b)** $W = I^2 Rt$
Evaluate	$I = 24$ A	$I = 1000$ A
	$R = 10\ \Omega$	$R = 0.24\ \Omega$
	$t = 10$ ms	$t = 10$ ms
Substitute	$W = 24^2 \times 10 \times 0.010$	$W = 1000^2 \times 0.24 \times 0.010$
Solve	$W = 58$ J	$W = 2.4$ kJ

As noted in the preceding example, the purpose of the overcurrent device is to limit the duration (time) of the fault, thereby keeping the production of heat and the increase of temperature in the electrical system (transformers, wires, motors, etc.) to a minimum. However, if the short circuit is allowed to remain and the fault current is not interrupted, then insulation will burn, wires will melt, and a fire could very likely result.

Ground Fault Condition

A *ground fault condition* is created when a "hot" wire comes in contact with an object at or near ground potential. The resulting current is usually high; however, it is possible for it to be low. The magnitude of the fault current will depend on the impedance in the path between the hot wire and ground.

When the impedance to ground is high, the fault current is small, and it may not be sufficient to be sensed by the overcurrent device. Hence, the circuit remains in operation. Since the circuit remains active, any object in the path between the hot wire and ground is at the same potential (voltage) as the hot wire (120 to 480 V). This condition represents a very serious shock hazard when touched by a person, because the person touching the electrified object may represent a lower impedance path to ground than the electrified object. When this is the case, a marked increase in the fault current will result, with most of the current flowing through the person's body, resulting in electrocution.

To guard against small-current ground fault conditions, a *ground fault inter-rupter* (**GFI**) is used to sense low values of current (typically 5 mA) passing to ground and safely disconnect the circuit. The ground fault interrupter protects against the hazards of current passing to ground through a person's body (electric shock). By quickly (in less than 25 ms) disconnecting the circuit from the line, the GFI keeps a person who is part of a ground fault path from being harmed. Once the GFI has opened the line, the circuit breaker–like device must be manually reset.

Figure 12.2 pictures two ground fault conditions. The first condition, pictured in Figure 12.2(a), is a hot wire touching a metal box (a low impedance path to ground). This condition creates a high-current ground fault. The fault current is limited only by the resistance of the conductor (0.10 Ω) and the impedance of the transformer ($R = 0.02\ \Omega$, $X_L = 0.07\ \Omega$). The resulting fault current is 120 V/0.14 Ω, or 860 A. This is sufficient current to open the overcurrent device quickly and interrupt the ground fault current.

The second condition is illustrated in Figure 12.2(b). Here the hot wire has contacted a damp wooden fence, which has an impedance of 85 Ω to ground. The resulting fault current is 120 V/85 Ω, or 1.4 A. This is not sufficient current to open the standard overcurrent devices. However, if a GFI has been installed in the circuit, the ground fault current would be quickly interrupted.

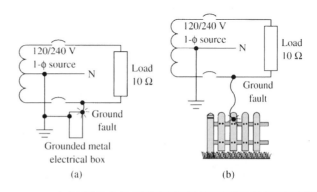

(a)   (b)

FIGURE 12.2
(a) A hot wire contacts a grounded metal electrical box (very low impedance to ground), creating a high-current ground fault of 860 A. (b) A hot wire contacts a damp wooden fence (a moderate impedance to ground), creating a low-current ground fault of 1.4 A.

In summary, when a ground fault occurs and the object that the hot wire contacts is not properly grounded, the object becomes electrified. This condition can cause serious injury or death by electric shock. The solution to this kind of ground fault is to install a ground fault interrupter in the circuit.

Overload Condition

An overload results when the connected load requires more current than the wiring has been rated to carry. This condition may be caused by a motor being overloaded due to

- Too much load placed on the machine it is driving.
- Excessive friction in the drive system of the machine it is driving due to bearing failure or drive-train wear.
- A jam in a machine driven by the motor.

An overload can also result when an appliance is used on a circuit with too little ampacity for its demand. Overloads can result from the gradual deterioration (aging) of dc power supply components in electronic equipment or from the breakdown of insulation in transformers and motors.

Unlike a short circuit or a ground fault, an overload does not occur due to wires coming in contact with one another or with grounded objects. The excess current produced by an overload remains within the conductors, causing overheating and deterioration to the conductor insulation. If uninterrupted, an overload can lead to the failure of

- The bearings in motors, due to the grease being melted out by the excess heat produced by the overcurrent condition.
- The insulation of the windings in the motor, due to overheating.
- The insulation of the windings in the transformer, due to overheating.

An uncontrolled overload can, in its extreme case, lead to fire and destruction of life and property. If the overcurrent device is sized too large, as illustrated in Figure 12.3, then an overcurrent condition may persist and not be interrupted in a timely manner.

FIGURE 12.3
An overload can persist when the wrong-size overcurrent device is installed. The 30-A fuse is not compatible with the ampacity of the conductor or the normal continuous current of the load. With 28 A in the circuit, the load is 87% overloaded, and the fuse will never open.

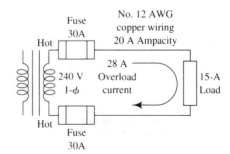

12.2 OVERCURRENT PROTECTION DEVICES

Introduction

The purpose of overcurrent protection is to ensure that excess circuit current resulting from a fault condition or an overload will be interrupted before it can cause damage and/or destruction to

- The transformer that is the source of circuit current.
- The conductors (wiring) that carry circuit current.
- The electrical load being supplied with energy from the circuit current.

In addition to the overcurrent condition associated with a fault, **arcing** is also usually present when a short circuit or a ground fault occurs. Arcing results from conductors coming together and then separating. Once the arc is struck and the air is **ionized** (made conductive), then considerable heat is generated in the arc zone, even though the current is passing through the air from one conductor to the other. If the distance between the wires remains small and the current is not interrupted quickly by the overcurrent device, the wires may melt.

When fault currents are large, so are the magnetic fields associated with the currents. The force created by the magnetic fields is sufficient to bend and move conductors within a motor or a transformer, causing mechanical fatigue and, in some cases, breaking the wire. The movement of the wire may be sufficient to cause wires to touch and arc. Each of these conditions may lead to the failure of the equipment.

> ➤ **As a Rule** Overcurrent protection is required by the *National Electrical Code* (**NEC**) at every point in the electrical distribution system where the size (ampacity) of the conductors is reduced.

In the industrial electrical system, as noted in Figure 12.4, this includes the service entrance (the switchboard), panelboards, and motor-control centers. Additionally, *overload protection* is usually provided as part of the motor starter for motors and as part of the control transformer for the control circuit.

Overcurrent Devices

Fuses and circuit breakers, as pictured in Figure 12.5, are the two types of overcurrent devices in common use to interrupt and clear short-circuit, ground fault, and most overload conditions. However, fuses and most circuit breakers are not designed to provide overload protection to all motors—they will, of course, interrupt short-circuit and high-current ground fault conditions. Protection from low-current overloads and ground faults in motors usually requires the use of a separate overload device.

Of the two overcurrent devices shown in Figure 12.5, the fuse, which is thermally operated, must be replaced after the overcurrent condition is *cleared* (removed). The circuit breaker, which may be either thermally or magnetically operated, is simply reset (usually manually) after the overcurrent condition is cleared.

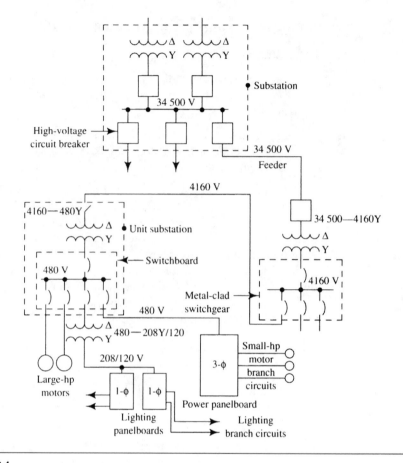

FIGURE 12.4

An electrical distribution system used to provide power to an industrial site. Overcurrent devices are provided at each point in the system where the conductor size (ampacity) is reduced. *Note*: The breakers in the panelboards are not shown.

Fuses

Fuses are thermally operated devices that destroy themselves by melting a fusible metal link when excess current passes through the link. When the fuse melts, the fault or overload currents are interrupted and the load, which is in series with the fuse, is disconnected from the line. The simplicity of the *fusing action* (melt and clear) is predictable, since the quantity (amount) of heat is easily determined. The quantity of heat (Q) produced in the link is the same as the work (W) done by the electric current, which is equal to the product of the current squared (I^2), the resistance (R) of the link, and the time (t) the fault or overload current flows:

$$Q = W = I^2Rt \tag{12.2}$$

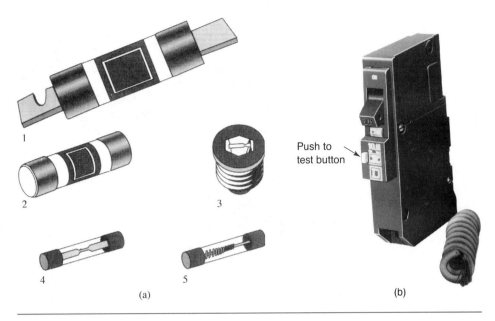

FIGURE 12.5

Fuses and circuit breakers are used to interrupt and clear short-circuit, ground fault, and overload conditions. (a) Cartridge fuses (1, 2, 4, 5) and plug fuse: (1) Dual-element cartridge fuse with knife blades designed for use in class R rejection fuse clips. (2) Common cartridge fuse with ferrules designed for use with fuse clips. (3) Common plug fuse with Edison base designed for use in a screw socket and recommended for replacement applications only. (4) Fast-acting glass fuse with ferrules designed for use with fuse clips or fuse holders of electronic equipment. (5) Time-delay glass fuse with ferrules designed for use with fuse clips or fuse holders of instruments and appliances. (b) A molded-case circuit breaker with UL *Class A* ground fault protection (GFI). *(Courtesy of the Square D Company)*

where

Q = quantity of heat equal to the work, in J
I = fault or overload current, in A
R = resistance of the fuse link, in Ω
t = time the fault or overload current flows, in s

The fusible metal link (Figure 12.6), which is designed to melt at a given current level in a predictable period of time, is housed in a fiber, ceramic, or glass tube that is fitted with metal end caps. The *fuse link*, or *element*, may be covered with fine silica sand to minimize and contain arcing once the fuse melts. The metal link is connected, by soldering, either to the copper end caps or copper blades. When installed in the fuse clips housed within a metal enclosure, the fuse completes the electric circuit. Some types of fuses have removable end caps that permit replacement of the "blown" (melted-open) fuse link; however, most fuses are simply removed from their clips and replaced once they have opened.

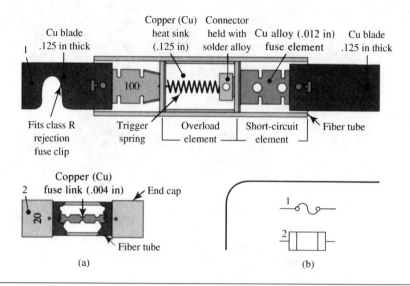

FIGURE 12.6

(a) The fuse design includes: (1) The dual-element, time-delay fuse with two series-connected elements. The *overload element* is attached to heavy copper heat sinks to provide time delay for motor starting while protecting against overloads. The *short-circuit element* melts under short-circuit conditions. This type of fuse (class RK5) can handle five times its rated current for 10 s without fusing. Rating: 100 A, 250 V, 200 000 A interrupting rating, class RK5. (2) Single-element, fast-acting fuse. The link has several sections that melt simultaneously, helping to lessen arcing. Rating: 20 A, 250 V, 50 000 A interrupting rating, class K5. (b) Schematic fuse symbols for use in (1) electronic schematic diagrams and (2) power or control circuit diagrams.

Example *12.2*	Determine the quantity of heat produced in each dual-element, 100-A, time-delay fuse (pictured in Figure 12.6) during the 3.0-s start-up of the three-phase, 230-V, 25-hp induction motor shown in the circuit of Figure 12.7. The starting current is five times the 62-A running current. The average resistance of the fuse elements (links) is 56 mΩ during the starting period.
SOLUTION	Find the work done (joules) when five times the running current of 62 A flows through the 56-mΩ link resistance.
Given	$Q = W = I^2 Rt$
Evaluate	$I = 5 \times 62 \text{ A} = 310 \text{ A}$
	$R = 0.056 \ \Omega$
	$t = 3.0 \text{ s}$
Substitute	$Q = 310^2 \times 0.056 \times 3.0$
Solve	$Q = 16 \text{ kJ}$

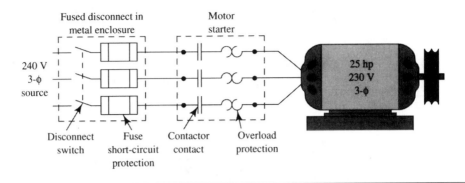

240 V
3-φ
source

Fused disconnect in
metal enclosure

Motor
starter

25 hp
230 V
3-φ

Disconnect
switch

Fuse
short-circuit
protection

Contactor
contact

Overload
protection

FIGURE 12.7
Motor branch circuit of Example 12.2.

The quantity of heat produced during start-up of the motor is equal to the work of 16 kJ done by the electric current.

Specific Heat of Fuse Materials

Once the quantity of heat produced in the fuse is known, this information can be used to approximate the increase in temperature of the fuse link. As you may know, the production of heat in the material of the fuse link increases the temperature (a measure of the energy in the material) of both the fuse link and the support material. The change in the temperature of the fuse link is an important part of the fusing process. Each fusible material, including copper, copper alloys, silver, and zinc, requires a certain quantity of heat to raise a unit mass of the material 1°C. This property of the material is called the **specific heat** of the material. The specific heat of a material is defined as the quantity of heat (joules) needed to change the temperature of a unit mass (a specified mass) of the material 1°C. For example, it takes 4190 J to raise 1 kg of water 1°C, but only 390 J to raise 1 kg of copper 1°C. A temperature conversion table is located in Section A2 of Appendix A.

As indicated in the following equation, the approximate change in the fuse-link temperature (due to electrical current passing through the link) is related to the quantity of heat, the mass of the fuse structure, and the specific heat (see Table 12.1) of the fuse link and its support structure.

$$\Delta T = \frac{Q}{mc} \qquad (12.3)$$

where

ΔT = change in the temperature of the fuse material, in °C;
i.e., $\Delta T = T_{hot} - T_{cold}$
Q = quantity of heat, in J
m = mass of the fuse structure, in kg
c = specific heat of the fuse material, in J/kg·°C

TABLE 12.1   Properties of Selected Fusible Metals and Alloys

Material Name	Specific Heat (J/kg·°C)	Melting Point (°C)
Brass (60 Cu/40 Zn)	390	840
Bronze (80 Cu/20 Sn)	360	870
Copper	390	1080
Silver	230	960
Zinc	390	420

Example 12.3

Determine the change in the temperature of the 96-g copper fuse structure of Figure 12.6 and Example 12.2 after the start-up of the motor is completed and 16 kJ of work has been done to produce 16 kJ of heat.

SOLUTION

Find the temperature change in the copper fuse structure due to the 16 kJ of heat produced.

Given

$$\Delta T = \frac{Q}{mc}$$

Evaluate

$$Q = 16 \text{ kJ}$$

$$m = 96 \text{ g} = 0.096 \text{ kg}$$

$$c = 390 \text{ J/kg·°C} \qquad \text{from Table 12.1}$$

Substitute

$$\Delta T = \frac{16\,000}{0.096 \times 390}$$

Solve

$$\Delta T = 427 \Rightarrow 430°C$$

If the temperature of the fuse in this example started at an **ambient temperature** of 20°C (room temperature), then the final temperature of the fuse structure after the 3-s start-up period of the motor would be 450°C. Since the melting temperature of copper is 1080°C, the fusible element (link) would not have melted. However, had the fusible element been made of zinc instead of copper, then the fuse link would have melted, because the melting point of zinc is 420°C.

Inverse Time Characteristic

An inverse relationship between the time for fusing (melt and clear) and the amount of overcurrent exists for all fuses. Small-overcurrent conditions take longer to clear than do large-overcurrent conditions. This characteristic, illustrated in Figure 12.8, is referred to as the *inverse time characteristic* of fuses.

As pictured in Figure 12.8, a fuse can carry 100% of its *rated continuous current* (**RCC**) for an indefinite period of time. However, at 110% RCC (10% over-

FIGURE 12.8
The semilog, time-current curve of a quick-acting glass instrument fuse shows the inverse time characteristic found with all types of fuses.

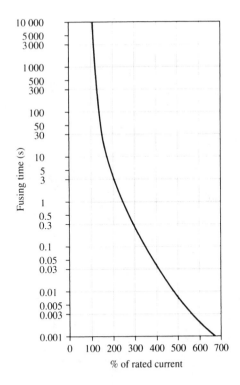

load), the fuse will melt and clear (open) in approximately 10 000 s (about 3 h) and at 125% RCC (25% overload), it will open in about 300 s (5 min). At 100% overload (200% of RCC), the fuse will open in roughly 3 s. An overload condition of seven times the rated continuous current (700% RCC) will cause the fuse to open in less than 0.001 s (<1 ms).

Fuse Types

Various types of fuses are available to protect the electrical distribution circuits that bring power to the load and to protect the load itself. As an example, the electrical circuits in your home are protected from overcurrent by fuses or circuit breakers; at the same time your television set (a load attached to the electrical circuit) may be protected by a fuse or circuit breaker within the set. Because of the need for flexibility in selecting overcurrent devices, several types of fuses (as illustrated in Figure 12.9) are available. The fuses discussed here are among those available.

Nonrenewable Cartridge Fuses Nonrenewable cartridge fuses (Figure 12.9(a)) are used in general-purpose applications (lighting, heating, etc.) where the fuse is loaded to no more than 80% of its RCC amperage. These common, inexpensive fuses are not recommended for motor circuits, since they must be sized at 400% of their amperage rating, which would require an expensive, large-amperage disconnect switch. Ratings fall in a range of 1/8–600 A in two voltages, 250 V and 600 V, and a 10 000-A (typical) interrupting rating (higher ratings are available and are marked on the fuse

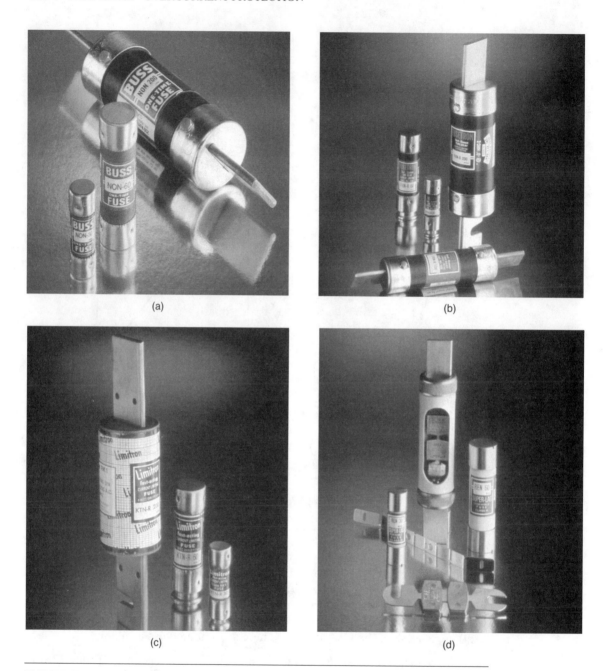

(a)

(b)

(c)

(d)

FIGURE 12.9

Several types of cartridge fuses (catalog symbol): (a) Nonrenewable cartridge fuses (NON and NOS); (b) Dual-element (time delay) cartridge fuses (FRN-R and FRS-R); (c) Fast-acting (current-limiting) cartridge fuses (KTN-R and KTS-R); (d) Renewable link cartridge fuses (REN and RES). *(Courtesy of Bussmann, Cooper Industries, Inc.)*

body). These fuses have very little time delay and cannot be used in class R rejection clips, which accept only high-interrupting rated fuses with a cutout in the blade, as illustrated in Figure 12.9(b).

Dual-Element (Time-Delay) Cartridge Fuses The dual-element cartridge fuse (Figure 12.9(b)) is a very popular fuse for applications such as mains, feeders, branch circuits, motors, welders, and transformers. They will load to 80% of RCC amperage for general applications and to 125% (typical) for good starting and overcurrent protection in motors. Ratings include $\frac{1}{10}$-600 A in two voltages, 250 V and 600 V, and a 200 000-A interrupting rating. These fuses are designed to handle five times their amperage rating for a minimum of 10 s, and they fit into all styles of fuse clips, including the class R rejection clips.

Fast-Acting (Current-Limiting) Cartridge Fuses The current-limiting cartridge fuse (Figure 12.9(c)) is a very fast-acting fuse that may be used to protect against overload and fault conditions and is best used with nonvarying loads or loads with little inrush current. It is susceptible to *nuisance fusing* (opening) when used in circuits with momentary overloads (motors), power spikes (lighting), or any other cause of fluctuating current. Ratings include 1–600 A in two voltages, 250 V and 600 V, and a 200 000-A interrupting rating. This fuse fits into all styles of fuse clips, including the class R rejection clips.

Renewable Link Cartridge Fuses Renewable link cartridge fuses (Figure 12.9(d)) are used for general-purpose applications where the fuse is loaded to no more than 80% of its RCC amperage. These expensive (initial cost) fuses are not usually used in motor circuits, since they must be sized at 300% of their amperage rating; this requires an expensive, large-amperage disconnect switch. Ratings include 1–600 A in two voltages, 250 V and 600 V, and a 10 000-A interrupting rating. These fuses have very little time delay and cannot be used in class R rejection clips, which accept only high-interrupting rated fuses. Their main advantage is that the end caps unscrew, permitting the removal of the spent fuse link and the installation of a new (inexpensive) fuse link.

Miscellaneous Cartridge Fuses These fuses (not shown) are physically small in size and are usually associated with protection of the load. Typical applications include installation in electronic equipment (instrument fuses) and electrical apparatus (indicating fuses). These fuses are characterized by their small size ($\frac{1}{4}$- and $\frac{13}{32}$-in diameter) and their glass, ceramic, or fiber tubes. Ratings include $\frac{1}{16}$-30 A; voltages of 32, 125, 250, and 600 V; and an interrupting rating that varies from type to type.

Molded-Case Circuit Breakers

The molded-case circuit breaker, as pictured in Figure 12.10, is a single integrated unit housed in an insulated case that is suitable for use on circuits rated at or below 600 V. The circuit-breaker unit is made up of a switching section and an overcurrent sensing/actuation section. Molded-case circuit breakers are installed in preassembled cabinets and, unlike most other wiring devices, cannot be interchanged among different manufacturers.

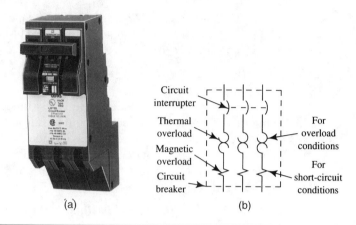

(a)

Circuit interrupter

Thermal overload

Magnetic overload

Circuit breaker

For overload conditions

For short-circuit conditions

(b)

FIGURE 12.10

(a) A *thermal-magnetic*, molded-case, three-pole, three-phase circuit breaker with the three switch levers "ganged" together to ensure that all three phases of the load are disconnected when one phase is overloaded or faulted. *(Courtesy of the Square D Company)* (b) Schematic symbol for a three-phase circuit breaker with both magnetic and thermal overloads. The dashed lines indicate that the three circuit interrupters (switches) are activated together.

A molded-case circuit breaker, unlike a fuse, will interrupt a fault current or an overload without destroying itself. Once the fault is cleared or the overload is reduced, then the circuit breaker is simply reset by moving the external lever through the manufacturer's recommended reset procedure. This same lever may be manually operated as a disconnect switch to isolate the electrical load from the source of power by moving it to the *on* or *off* position.

The switch lever is *trip-free*, so it cannot be held closed during a fault or overload. Also, for safety, the circuit breaker switch levers are "ganged" together when multiple units are assembled into two- or three-pole circuit breakers for use with 240-V, single-phase circuits and 208-, 240-, 480-, and 600-V, three-phase circuits. The purpose of mechanically connecting the switch levers together in multipole circuit breakers is to ensure that all lines (except neutral) are disconnected from the power source when just one pole experiences an overload or a fault condition.

Thermal and Magnetic Overload Operation

Although manual operation of the circuit breaker is an important feature, it is the thermal or magnetic operation of the circuit breaker (in response to overload and fault current conditions) that is of paramount importance. Thus, the molded-case circuit breaker has three distinct modes for operation—manual, thermal, and magnetic. Not all circuit breakers are manufactured with all three modes of operation. However, all circuit breakers can be manually and thermally operated; some, with large current ratings, have magnetic operation.

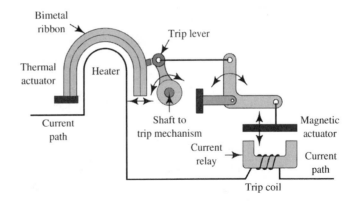

FIGURE 12.11

An artist's representation of the mechanics of the trip mechanism in the circuit breaker. The representation shows how the heat and magnetism produced by the overload or fault current is converted to motion to activate the mechanism.

Thermal operation of the circuit breaker requires that a **bimetal** ribbon (two metals bonded together) be heated by the circuit current passing through a series-connected heater, as pictured in Figure 12.11. Because the two metals making up the bimetal ribbon expand at different rates as they heat up, the ribbon is forced to move in one direction in response to the heat produced by the current flow in the heater. When an overcurrent condition exists, the force of movement from the bimetal ribbon is sufficient to activate the circuit breaker by "tripping" the trip mechanism and opening the spring-loaded switch contacts. The *trip time* (time to open the circuit) of the *thermal actuation* of the circuit breaker, like the fuse, depends on the inverse time characteristic of the heating of the bimetal ribbon. That is, larger current flow produces a shorter trip time.

By adding a current relay to the circuit breaker, as illustrated in Figure 12.11, the large magnetic field associated with the short circuit current can be used to trip the circuit breaker. Because *magnetic actuation* of the circuit breaker is very fast, the trip time is very small. Quick activation of the trip mechanism ensures that the short circuit will be interrupted within the time it takes for just a few cycles of the short-circuit current to flow.

In summary, the time-delay overload feature of a circuit breaker comes from the design characteristics of the bimetal used in the thermally operated portion of the trip mechanism, whereas the instantaneous operation of the circuit breaker is provided by the magnetically operated portion of the trip mechanism. Circuit breakers that have a combination of thermal and magnetic actuation have an inverse time tripping characteristic for overloads up to about ten times the rated continuous current of the circuit breaker. Above this level of current (short circuits), the tripping time is virtually instantaneous.

Fusible Disconnect

When a switch is used to isolate parts of an electric circuit, it is called a *disconnect switch*, or simply a *disconnect*. The spring-actuated, knife-blade switch mechanism is mounted in a metal case with an external operating handle, as pictured in Figure 12.12(a). Since the switch is not designed to interrupt short-circuit current, fuses are incorporated with the switch to produce a single device called the *fusible disconnect*, as illustrated in Figure 12.12(b).

Enclosed switches are used as disconnects for the main service into a building and for feeder and branch circuit operation. Disconnects, using either circuit breakers or a combination of fuses and switches (fusible disconnect), are used with motors for short-circuit protection and to isolate the motor from the line for maintenance, as pictured in Figure 12.7.

Modern disconnect switches rely on a spring-actuated mechanism for quick-make and quick-break motion. This mechanism provides a constant speed for uniform opening and closing of the switch that is independent of the speed of movement of the externally operated handle. Additionally, a door interlock is provided to prevent the enclosure door from being opened when the switch is closed and energized. Fusible disconnects are available with voltage ratings of 250 and 600 V and various current ratings and configurations for use with single-phase, 120-V, two-wire circuits; single-phase, 240-V, three-wire circuits; and three-phase, 208-, 240-, 480-, and 600-V circuits (as pictured in Figure 12.13).

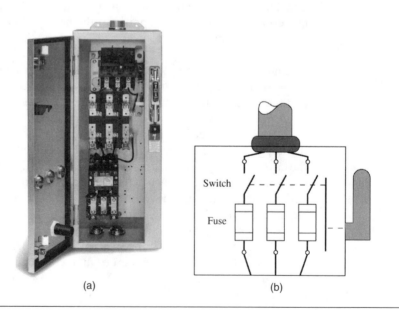

(a)   (b)

FIGURE 12.12

(a) A three-pole fused disconnect shown with the door open and the fuses removed. *(Courtesy of the Square D Company)* (b) Pictorial of the three-pole fused disconnect shown with the switch and fuse symbols identified.

FIGURE 12.13
Fused disconnects. When interrupting: (a) A single-phase, 120-V circuit uses a one-pole fused disconnect. The *hot wire* is switched and fused, whereas the neutral is never switched or fused. (b) A single-phase, 240-V circuit uses a two-pole fused disconnect. Both hot wires are switched and fused. (c) Three-phase, 208-, 240-, 480-, or 600-V circuits use a three-pole fused disconnect. All three hot wires are switched and fused.

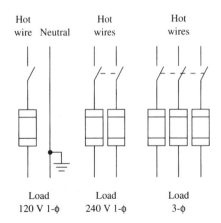

12.3 PARAMETERS OF OVERCURRENT PROTECTION DEVICES

Requirements for Fuses and Circuit Breakers

There are four basic parameters that must be considered when selecting an overcurrent device for installation in the electrical power-distribution system. These important parameters are the *voltage rating*, the *rated continuous current*, the *interrupting rating*, and the *response time* of the fuse or circuit breaker. Each of these parameters must be compatible with the characteristics of the load, conductors, and transformer in order to protect the electric circuit when overcurrent conditions occur.

Voltage Rating The voltage rating marked on the fuse or circuit breaker is an indication that the overcurrent device can safely interrupt fault conditions in an electrical circuit when the circuit voltage is equal to or less than the rated voltage. This voltage requirement, which is covered by the NEC, must be adhered to if the fuse or circuit breaker is to interrupt the fault current safely.

The standard voltage ratings of 32, 125, and 250 V (for small-dimension miscellaneous cartridge fuses) and the voltage ratings of 250 and 600 V (for cartridge fuses and molded-case circuit breakers) have sufficient range to cover the needs of automotive, electronic, and low-voltage power-distribution circuits.

Because fuses are sensitive to changes in current but not to changes in voltage, a fuse may be operated at voltages below the maximum fuse rating without change to its fusing characteristics. However, if fuses rated below the circuit voltage are used in a circuit, they may not interrupt the arcing associated with the melting and clearing of the fuse. The arc, which is sustained due to the use of an undervoltage fuse, may be maintained indefinitely, as is the overcurrent condition. Thus, a fuse or circuit breaker rated at 250 V may safely be used in a 240-V, 120-V, or 24-V circuit, but it must not be used in a 480-V circuit.

Rated Continuous Current In general, the rated continuous current (RCC) of an overcurrent device (a fuse or circuit breaker) is selected to be equal to or greater than the ampacity of the circuit conductors and electrical equipment being protected. For example, as pictured in Figure 12.14, either a switch and 30-A fuse or

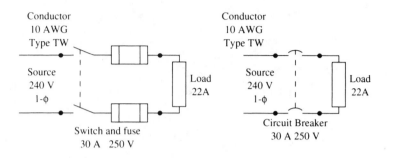

FIGURE 12.14
A fuse or a circuit breaker may be used to provide overcurrent protection to the electrical circuit.

a 30-A circuit breaker may be used to disconnect and protect the load and the conductors of the circuit. The current rating of the overcurrent device (30 A) is determined by first confirming the load current (22 A) and then selecting a conductor from Table A7 (Appendix A) with sufficient ampacity to handle the load current with some capacity left over. In the case of the illustrated circuit, a number 10 AWG, type TW copper conductor is selected.

The maximum load current is generally held to 80% of the rated current of the overcurrent device when the load is operated continuously for 3 h or more (a continuous load) or when the load is a single piece of equipment or a single appliance. However, if the load is an electric motor, then a fuse or circuit breaker may have a rated continuous current several times greater than the full-load motor current or the ampacity of the conductors supplying current to the motor. This is permitted because the motor and the circuit conductors are protected against overload by an additional overload device in the motor starter or in the motor itself that will disconnect the motor from the circuit. The maximum ratings of fuses and circuit breakers, as percentages of the full-load motor current, are listed in Table 12.2 by the type of motor.

Cartridge fuses with voltage ratings of 250 and 600 V have current ratings divided into six current series for each voltage, as indicated in Table 12.3. The six current series listed in Table 12.3 start with current values up through 30 A and finish with values of 450 through 600 A. Each series is exclusively sized so that fuses in a given series will fit only into the fuse clips for that series. Thus, the design of the fuse clips prevents an undervoltage or an overcurrent fuse from being installed in the clips. Fuses rated at 250 or 600 V are available with the following current ratings:

30-A series	15, 20, 25, 30 A
60-A series	35, 40, 50, 60 A
100-A series	70, 80, 90, 100 A
200-A series	110, 125, 150, 175, 200 A
400-A series	225, 250, 300, 350, 400 A
600-A series	450, 500, 600 A

Circuit breakers are available in six current series based on the frame size of the circuit breaker. The currents for the six frame sizes are 50, 100, 225, 400, 600, and 800 A. Each frame size has several available currents and several operating voltages

TABLE 12.2 Maximum Rating or Setting of Motor Branch-Circuit, Short-Circuit, and Ground Fault Protective Devices

	Percent of Full-Load Current			
Type of Motor	Nontime Delay Fuse	Dual Element (Time-Delay) Fuse	Instantaneous Trip Breaker	Inverse Time Breaker*
Single-phase, all types				
No code letter	300	175	700	250
All ac single-phase and polyphase squirrel-cage and synchronous motors† with full-voltage, resistor or reactor starting:				
No code letter	300	175	700	250
Code letters F to V	300	175	700	250
Code letters B to E	250	175	700	200
Code letter A	150	150	700	150
All ac squirrel-cage and synchronous motors† with autotransformer starting:				
Not more than 30 amps				
No code letter	250	175	700	200
More than 30 amps				
No code letter	200	175	700	200
Code letters F to V	250	175	700	200
Code letters B to E	200	175	700	200
Code letter A	150	150	700	150
High reactance squirrel-cage				
Not more than 30 amps				
No code letter	250	175	700	250
More than 30 amps				
No code letter	200	175	700	200
Wound-rotor —				
No code letter	150	150	700	150
Direct-current (constant voltage)				
No more than 50 hp				
No code letter	150	150	250	150
More than 50 hp				
No code letter	150	150	175	150

For explanation of code letter marking, see Table 430-7(b).

For certain exceptions to the values specified, see Sections 430-52 through 430-54.

*The values given in the last column also cover the ratings of nonadjustable inverse time types of circuit breakers that may be modified as in Section 430-52.

†Synchronous motors of the low-torque, low-speed type (usually 450 rpm or lower), such as are used to drive reciprocating compressors, pumps, etc., that start unloaded, do not require a fuse rating or circuit-breaker setting in excess of 200 percent of full-load current.

Reprinted with permission from NFPA 70-1993, the National Electrical Code®, *Copyright © 1992, National Fire Protection Association, Quincy, MA 00269. This reprinted material is not the complete and official position of the National Fire Protection Association, on the referenced subject which is represented only by the standard in its entirety.*

(ranging from 120 V to 600 V) for each current. All breakers within a given frame size have the same physical dimensions, contact rating, and interrupting rating. When selecting fuses or circuit breakers for an application, always consult the *National Electric Code* for guidance.

Interrupting Rating The interrupting rating of a fuse or circuit breaker is an indication of the device's ability to open and clear the very large short-circuit

TABLE 12.3 Cartridge Fuse Classification

Current Series* (A)	Fuse Clip Accepts	Nominal Size Diameter × Length (in)	
		250 V	600 V
Up through 30	Ferrule	$\frac{9}{16} \times 2$	$\frac{13}{16} \times 5$
35 through 60	Ferrule	$\frac{13}{16} \times 3$	$\frac{11}{16} \times 5\frac{1}{2}$
70 through 100	Knife blade	$1 \times 5\frac{7}{8}$	$1\frac{1}{4} \times 7\frac{7}{8}$
110 through 200	Knife blade	$1\frac{1}{2} \times 7\frac{1}{8}$	$1\frac{3}{4} \times 9\frac{5}{8}$
225 through 400	Knife blade	$2 \times 8\frac{5}{8}$	$2\frac{1}{2} \times 11\frac{5}{8}$
450 through 600	Knife blade	$2\frac{1}{2} \times 10\frac{3}{8}$	$3 \times 13\frac{3}{8}$

*Each series has several sizes; for example, the 100-A series is available with RCCs of 70, 80, 90, and 100 A.

current or the ground fault current passing in the device when a fault condition occurs within the electrical system. The interrupting rating (also called the interrupting capacity) must be sufficient to handle safely all available fault current without causing extensive damage to the electrical components in the system.

The interrupting ratings of standard size 250- and 600-V fuses and molded-case circuit breakers are found on the label or body of the device. When the rating is not printed on the device, it may be assumed that the interrupting rating is 10 000 A for fuses and 5000 A for circuit breakers. Since smaller-size fuses and circuit breakers may have ratings less than these, it is best (and safest) to contact the manufacturer for the interrupting rating.

Fuses with a notch at one end of the blade or an undercut ferrule at one end are designed to fit class R rejection fuse clips. These fuses are modern, whereas fuses without the notch or undercut are of an older design. These newer-design fuses, classified as RK1 and RK5, have a 200 000-A interrupting rating, whereas those without the notch or undercut have a 10 000-A interrupting rating unless specifically marked with a higher rating.

It is extremely important that fuses and circuit breakers not be used in circuits where the fault currents can exceed their interrupting rating. When the available fault current does exceed the overcurrent protection device's interrupting rating, the fault current will not be interrupted, and serious damage to equipment and injury to personnel may result. Without sufficient impedance to limit fault current and without sufficient interrupting capacity, a circuit will have no way to disconnect the fault current and, as a result, it will literally destroy itself attempting to interrupt the fault current. When replacing overcurrent devices, it is critical that the replacement fuse or circuit breaker have the same or higher interrupting ratings.

➢ **As a Rule** Always replace a blown fuse or a failed circuit breaker with one that has the same voltage, current, and interrupting ratings.

Violation of this rule may result in melting of the conductors, an explosion in the electrical equipment, fire, destruction to equipment, and/or injury to personnel.

Response Time Characteristics The time of response of a fuse or circuit breaker falls into one of two general categories: quick-acting or slow-acting. The names associated with quick-acting fuses include *fast-acting*, *medium-acting*, and *non-time delay*. The names for slow-acting fuses include *slow blow*, *time delay*, and *dual element*, whereas *instant trip* and *time limited* are names associated with circuit breakers.

Slow-acting overcurrent devices (fuses and circuit breakers with time delay) are used in circuits where short-duration, large transient currents may be present as when starting motors or turning on dc power supplies. Since these short-duration overcurrent conditions are expected, they are viewed as having little consequence. However, when fast-acting fuses or instant-trip circuit breakers are used in this application, *nuisance fusing*, or *tripping*, may result.

The *time-current curves* of Figure 12.15 picture the typical response-time characteristic of a fuse and a circuit breaker. Both of these inverse time curves, provided by the manufacturer of the overcurrent devices, represent the thermal response of metal to the passage of electric current. In the fuse, it is the heating of the metal fuse element (link); in the circuit breaker, it is the heating of the bimetal ribbon.

As noted in Figure 12.15, both time-current curves have the same relative actuation time for a given current up to about 700 A. However, with magnetic actuation

FIGURE 12.15
A typical time-current curve for a fuse (solid line) and circuit breaker (dashed line). The two curves have similar response times up to about seven times the rated current (700 A). Above this current, the circuit breaker trips due to its magnetic overload feature.

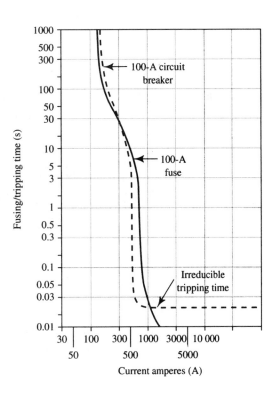

added to the circuit breaker, the time limitation is not dependent on the size of the overcurrent condition above the preset activation level (7 to 10 times RCC) of the current relay. Instead, once the fault current reaches the preset activation level, then the inertia of the moving parts of the trip mechanism during the unlatching of the breaker contacts is the limiting factor. Thus, magnetic actuation of the circuit breaker provides a much faster trip time than does thermal actuation for overcurrent conditions above the activation level of the current relay.

The time limitation that is inherent in all circuit breakers, with or without magnetic activation, is called the *irreducible tripping time*, as noted in Figure 12.15. Because the irreducible tripping time of a circuit breaker depends on the mass of all the associated elements of the trip mechanism and how fast the elements are accelerated, the irreducible tripping time is not constant for all circuit breakers. However, the tripping time is quite small, usually within the time it takes for just a few cycles of the 60-Hz fault current to flow (typically within 20 to 40 ms).

Magnitude of the Fault Current

As was stated earlier, it is the impedance of the electrical system that limits the amount (magnitude) of the fault current. This impedance is a combination of the resistance of the conductors and the equivalent impedance of the transformer supplying power to the short circuit. Since the transformer impedance greatly affects the amount of available fault current during a short-circuit or a ground fault condition, then the transformer impedance may be used to give an approximation of the magnitude of the fault current. An appropriate interrupting rating for the overcurrent devices protecting the circuit can be selected once an approximate value for the magnitude of the fault current is known.

The impedance of the transformer is usually stated on the nameplate as percent impedance (% impedance) or per-unit impedance (PU impedance). The transformer percent impedance is an indicator of the amount (percent) of primary voltage needed to produce full-load current in the secondary of the transformer when the secondary is short-circuited. For example, when the transformer impedance of a 30-kVA, three-phase, 480-V (secondary) transformer is 2%, the approximate available short-circuit current is 50 times the rated secondary current of 36 A. That is, the short-circuit current is 50×36 A, or 1800 A. The approximate available short-circuit current may be determined by applying the following equation(s).

For 3-ϕ transformers: For 1-ϕ transformers:

$$I_{sc} = \frac{P_s}{\sqrt{3} \times E \times \%Z} \times 100 \qquad I_{sc} = \frac{P_s}{E \times \%Z} \times 100 \qquad \textbf{(12.4)}$$

where

I_{sc} = approximate available short-circuit current from the transformer's secondary, in A

P_s = transformer apparent power rating, in VA

E = transformer secondary line voltage, in V

$\%Z$ = nameplate percent impedance of the transformer

Note: In Equation 12.4, when the impedance is given as a per-unit value, multiply the per-unit impedance by 100 to get percent impedance.

Example 12.4

Determine the approximate short-circuit current available from a 75-kVA transformer secondary with a per-unit transformer impedance of 0.034 when the transformer is single-phase with a 240-V secondary and three-phase with a 240-V secondary.

SOLUTION

Find the approximate short-circuit current available from each secondary.

single-phase transformer   *three-phase transformer*

Given

$$I_{sc} = \frac{P_s}{E \times \%Z} \times 100 \qquad I_{sc} = \frac{P_s}{\sqrt{3} \times E \times \%Z} \times 100$$

Evaluate

$$P_s = 75\,000 \text{ VA}$$

$$E = 240 \text{ V}$$

$$\%Z = \text{PU} \times 100$$

$$\%Z = 0.034 \times 100 = 3.4$$

Substitute

$$I_{sc} = \frac{75\,000}{240 \times 3.4} \times 100 \qquad I_{sc} = \frac{75\,000}{\sqrt{3} \times 240 \times 3.4} \times 100$$

Solve

$$I_{sc} = 9.2 \text{ kA} \qquad I_{sc} = 5.3 \text{ kA}$$

Observation

The minimum interrupting rating of the 250-V fuses or circuit breakers protecting the secondary of either of the transformers of this example is 10 000 A.

Conductor Temperature Rise

As observed in the previous example, the magnitude of the fault current (I_{sc}) is many times greater than the normal conductor current. Although this large current only flows for a brief period of time before the overcurrent device opens, the quantity of heat produced ($Q = I_{sc}^2 Rt$) is substantial resulting in a rapid rise in the conductor's temperature. Because of the conductor's thermal properties, the heat doesn't instantly dissipate from the conductor; instead, it may cause the wire and its insulation to overheat. Overheated insulation is a major cause of insulation deterioration resulting in electrical equipment failure.

It is the I^2t rating of the conductor that defines the temperature rise in the wiring and subsequently the final temperature of the conductor's insulation. The I^2t rating comes from substituting equation 12.2 ($Q = W = I^2Rt$) into equation 12.3 ($\Delta T = Q/(mc)$) resulting in

$$\Delta T = \frac{I^2Rt}{mc} = \frac{R}{mc}(I^2t)$$

If the factor R/mc is assumed to be constant, then the sudden rise in conductor temperature (ΔT) during the circuit fault is directly related to the I^2t factor and

$\Delta T \propto I^2 t$. The $I^2 t$ rating of a conductor depends on the cross-sectional area, the permitted temperature rise of the insulation, and the thermal capacity (specific heat) of the wiring. The following equation is used to evaluate the $I^2 t$ rating of a conductor.

$$I^2 t = kA^2 \log\left(\frac{x + T_{fin}}{x + T_{int}}\right) \tag{12.5}$$

where

k = thermal constant of the conductor material, 115×10^3 for copper wiring and 52.2×10^3 for aluminum wiring

A = conductor cross-sectional area, in mm^2

x = inferred zero point of the conductor material, 234°C for copper and 228°C for aluminum

$\log$ = calculator's common log keystroke

T_{fin} = final temperature, in °C

T_{int} = initial temperature, in °C

$I^2 t$ = surge current rating, in A$^2\cdot$s

I = fault current (I_{sc}), in A

t = duration of fault, in s

Example 12.5

The wiring between the load and the secondary of the three-phase, 240-V, 75-kVA transformer of Example 12.4 is copper AWG No. 4/0 with a diameter of 460 mils (.460 in/11.68 mm) selected to handle up to 180 A of line current under normal operation. Determine

(a) The $I^2 t$ rating of the conductor during a short-circuit of 5.3 kA if the conductor temperature at the time of the short-circuit is 32.0°C and the wire is limited to a maximum temperature of 200°C for brief time periods.

(b) The period of time (t) the short-circuit current (I_{sc} = 5.3 kA) can flow before the conductor's 200°C temperature rating is exceeded.

SOLUTION

(a) Using Equation 12.5, find $I^2 t$ for the AWG 4/0 conductor.

Given

$$I^2 t = kA^2 \log\left(\frac{x + T_{fin}}{x + T_{int}}\right)$$

Evaluate

$k = 115 \times 10^3$ for copper

$A = \pi d^2/4 = \pi(11.68)^2/4 = 107.2$ mm^2

$x = 234$°C for copper

$T_{fin} = 200$°C

$T_{int} = 32.0$°C

Substitute

$$I^2 t = 115 \times 10^3 (107.2)^2 \log\left(\frac{234 + 200}{234 + 32.0}\right)$$

Solve

$$I^2 t = 281 \times 10^6 \text{ A}^2\cdot\text{s}$$

SOLUTION

Given

(b) Using $I^2t = 281 \times 10^6 \text{ A}^2 \cdot \text{s}$, solve for t.

$I^2t = 281 \times 10^6 \text{ A}^2 \cdot \text{s}$

$t = 281 \times 10^6 / I^2$

Evaluate

$I = I_{sc} = 5.3 \times 10^3 \text{ A}$

Substitute

$t = 281 \times 10^6 / (5.3 \times 10^3)^2$

Solve

$t = 10 \text{ s}$

In summary, by knowing the circuit's parameters, then the *voltage rating*, the *rated continuous current*, the *interrupting rating*, and the *response time* of the fuse or circuit breaker may be selected to be compatible with the characteristics of the load, conductors, and the transformer.

➤ **As a Rule** It is essential that the *National Electrical Code*® be complied with when selecting and installing overcurrent protection devices.

12.4 MOTOR OVERLOAD PROTECTION

Introduction

An overload in a motor will produce overheating in the windings, lubricant, and bearings of the electric motor. The overload in the motor may be due to an excessive load on the power-transmission elements, which could occur from a misaligned drive belt coming off the pulley and wedging between the pulley and shaft, from a jam in a conveyor system, or from an excessively large load on a machine tool.

The heat produced by the overload condition results from the motor's operating current being elevated to a value above its rated *full-load current* (**FLC**). If left unchecked, the heat will produce a *winding failure*. In this case, the winding fails because the insulation on the wire has deteriorated due to the excess heat. Once the motor fails, then it must be taken out of service, torn down, and rewound—an expensive repair.

As you now know, fuses and circuit breakers used in motor circuits are oversized (125% or more) to allow starting of the motor. The motor-starting current may be five to seven times the normal running current. Since the usual overcurrent devices (fuses and circuit breakers) are overrated and time-delayed to facilitate starting, they can't always sense the presence of an overload current, which might range from 110% to 250% over the FLC. Thus, the job of the overload protector in the motor circuit is to sense an overload condition and then safely disconnect the motor from the line before excessive overheating occurs.

Running Overload Protection

Overload protection is needed while the motor is in operation (running) to protect it from overheating. Running overload protection is provided by several devices that either sense the motor current or the heat produced by the current. Devices

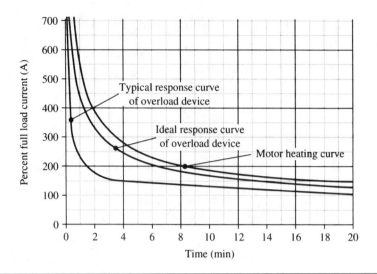

FIGURE 12.16

Thermal switches and motor-overload relays are typical overload devices used to prevent overheating in the motor. An ideal overload device would track just below the motor-heating curves, where it can sense when the FLC is exceeded—interrupting the overcurrent condition before the recommended operating temperature of the motor is exceeded.

that sense motor conditions include fuses, circuit breakers, thermal switches, and overload relays. The ideal overload device will track the *motor heating curve*, as pictured in Figure 12.16, and provide protection to the motor by disconnecting it from the line before the overload can cause overheating. Thermal switches and motor overload relays are designed for this purpose.

Thermal Switches

In fractional-horsepower and small-horsepower motors, overload protection may be provided by a thermal sensor placed in the motor when it is manufactured. The sensor responds to an increase in temperature and disconnects the motor from the electric circuit. The simplest of these sensors is a *temperature-sensitive bimetal switch* (illustrated in Figure 12.17) embedded in the motor frame. When heated to a predetermined temperature by the current passing through the metal disk, the bimetal disk switch suddenly changes (snaps) from a *concave* to a *convex* shape—releasing the switch contacts and disconnecting the motor from the line. Once the motor cools, the bimetal switch is reset. Depending on design, the resetting of the switch may be carried out either *manually* by depressing a reset button (as pictured) or *automatically*, when the bimetal disk cools and changes (snaps) back to a concave shape, closing the switch contacts once again.

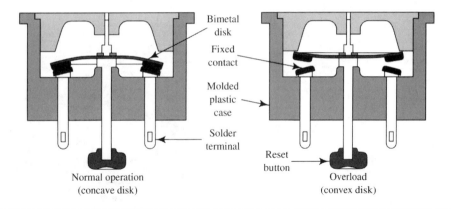

Normal operation
(concave disk)

Overload
(convex disk)

FIGURE 12.17
When heated, the bimetal thermal overload switch snaps from a concave curve to a convex curve at a pre-determined temperature. This action is used to open a set of overload contacts in the motor power circuit.

Motor Overload Relays

Larger-horsepower motors are protected from overloads by using a magnetic motor starter (Figure 12.18), which, in its simplest form, is nothing more than a magnetic contactor (Section 10.3) combined with an overload device. The overload device is called an *overload relay*.

The overload relay is made up of a current-sensing unit connected in series with the line current and combined with a mechanism that is actuated by the sensing unit. The overload relay mechanism removes the current from the motor directly, by tripping a trip mechanism and releasing the spring-loaded power contacts on the starter, or indirectly, by opening a set of *overload* (OL) *contacts* in the three-wire control circuit of the contactor, as pictured in Figure 12.19. Motor overload relays are activated by means of thermal (heat) or magnetic (electromagnetic flux) energy.

Two types of *thermal overload relays* are commonly used in motor starters to indirectly open the power contacts. The first type (the *solder-pot overload relay*) uses a heater to melt a low-temperature metal alloy, which releases a ratchet-and-pawl mechanism, allowing a spring-loaded overload contact to open. The second type (*the bimetal overload relay*) uses a heater to activate (move) a bimetal ribbon, which is linked to a snap-action switch that opens the overload contacts.

The solder-pot overload relay uses a low-melting-point **eutectic** (u-tek´-tik) **alloy** that solders the shaft of the ratchet to the journal of the heat-sink structure attached to the heater, as illustrated in Figure 12.20(a). Because of the thermal inertia and the inverse time-current response of the heater structure, normal starting of the motor will not melt the eutectic alloy. However, once the FLC of the motor is exceeded, then it is only a matter of time before the temperature rises to the melting point of the eutectic alloy. When the eutectic alloy melts, the ratchet and pawl (Figure 12.20(b)) are free to move and release (open) the spring-loaded,

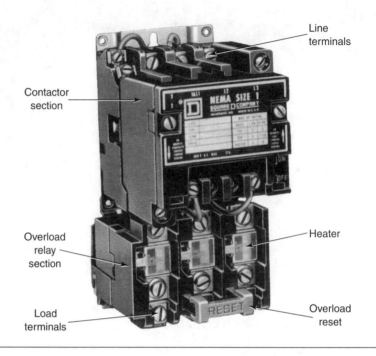

FIGURE 12.18
The magnetic motor starter is essentially a magnetic contactor coupled with a thermal overload relay. *(Courtesy of the Square D Company)*

normally closed, overload contacts. The interchangeable heater element (heater-ratchet assembly) is selected to match the motor's full-load current and service factor ratings.

When the normally closed (NC) overload (OL) contacts are opened in a *magnetic motor starter*, the contactor three-wire control circuit and the contactor magnetic coil are de-energized, allowing the power contacts to open and disconnect the motor from the line. In a *manual starter*, the solder-pot overload relay directly opens the power contacts on the starter, disconnecting the motor from the line.

The eutectic alloy used in the solder-pot overload relay (as with all eutectic alloys) has the unique property of changing immediately from a solid to a liquid at one particular temperature (called the **eutectic point**). The eutectic point is the lowest temperature at which an alloy melts. Above the eutectic point, the alloy is liquid; below the eutectic point, it is solid. A eutectic alloy has no plastic state.

Once the solder-pot overload relay cools, then the ratchet shaft is once again held immobile by the solidified metal alloy. The overload contacts and the thermal relay must be manually reset by depressing the reset button. When the reset button is depressed, the pawl is relatched by forcing it over the stationary ratchet wheel. The activation spring or springs are compressed and the overload contacts are closed. With the ratchet and pawl held stationary, the *thermally activated, manually reset, eutectic alloy, overload relay* is once again operational.

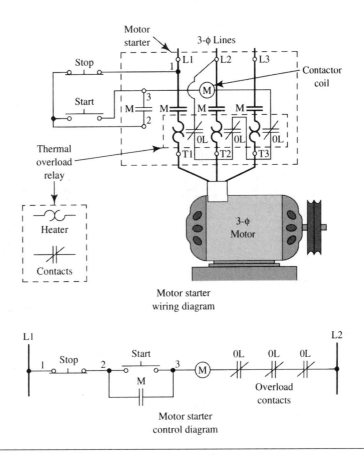

FIGURE 12.19
The overload (OL) contacts are placed in the control circuit of the contactor portion of the motor starter. When an overload condition is sensed, the OL contacts are opened, causing the contactor coil (M) to open, releasing the contactor contacts and disconnecting the motor from the line.

The bimetal overload relay utilizes the motion of a bimetal U-shaped ribbon acting through a mechanical linkage to open the overload contacts of a *snap action* switch. The bimetal ribbon, pictured in Figure 12.21, is heated by the motor current passing through the heater. The overload contacts, which are connected into the three-wire control circuit of the contactor in the motor starter, are opened when the temperature reaches a preset level. Since most bimetal overload relays have a means of adjusting their operating temperature from roughly 85% to 115% of the specified nominal value, the bimetal overload relay's operating temperature can be altered to compensate for variations in the ambient temperature surrounding the motor. The bimetal overload relay has a wide selection of heaters, which enable the overload relay operating characteristics to be tailored to both the motor's operating characteristics (FLC, SF, etc.) and operating conditions. When initially selecting the heater for the bimetal overload relay, you will find manufacturer's *heater selection*

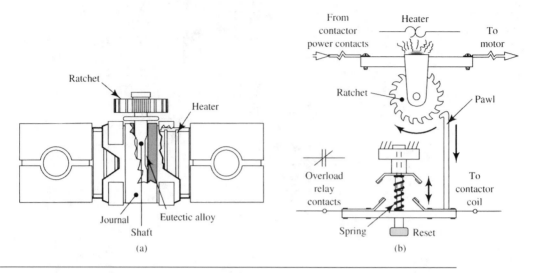

FIGURE 12.20
The solder-pot overload relay operates on the eutectic alloy-and-ratchet principle. (a) The heater section of the solder-pot overload relay. (b) Once the metal alloy melts, the ratchet is allowed to spin, releasing the pawl and opening the overload contacts.

charts very helpful. The *thermally activated*, *indirectly heated*, *bimetal overload relay* may be selected to reset manually or automatically.

The magnetic overload relay (pictured in Figure 12.22) uses the *solenoid principle* for the driving force needed to trip a set of overload contacts. Motor current passes through the coil of the *current relay*, and when a preset level of overload current is reached, the plunger in the center of the current coil is pulled up into the coil, striking a trip pin, which, in turn, opens a normally closed (NC) set of overload contacts. Because the NC overload contacts are connected in series with the contactor's

FIGURE 12.21
The bimetal overload relay uses a U-shaped bimetal ribbon heated by a U-shaped heater to trip a snap action switch. (*Courtesy of the Square D Company*)

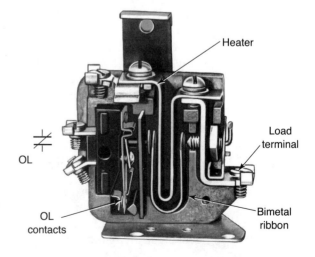

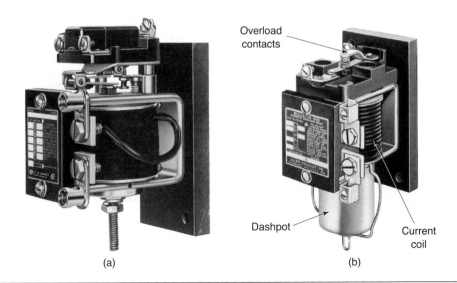

Overload
contacts

Dashpot

Current
coil

(a)   (b)

FIGURE 12.22
The magnetic overload relay uses a current coil with a plunger (a solenoid) to actuate the normally
closed overload contacts: (a) Instantaneous trip-current relay; (b) Inverse time-current relay. *(Courtesy
of the Allen Bradley Company, Inc.)*

three-wire control circuit, current to the motor is interrupted. The actuation of
the plunger and the resulting opening of the overload contacts may be realized
either instantaneously or not instantaneously, depending on whether an *instantaneous trip-current relay* or an *inverse time-current relay* is used.

Since the electromagnetically operated, instantaneous, trip-current relay has
an adjustable current trip point, it may be preset to an overload current level. Once
this current level is reached, the overload contacts are opened in the three-wire
motor starter circuit, disconnecting the motor from the power line. The adjustable-
current feature of the magnetic overload relay allows a machine motor to be shut
down at a predetermined current level when a jam in a conveyor system or an overload in the drive system occurs. When the overload is sensed by the instantaneous
trip-current relay, the system is quickly shut down. To prevent nuisance tripping, the
overload contacts are shorted during the start-up and acceleration period of the motor. The reset of the instantaneous trip-current relay is automatic, and it occurs
quickly, since cooling of the trip mechanism is not needed.

The inverse time-current relay is a magnetically operated overload relay that
can have its trip current and trip time adjusted to match the motor's operation and
load characteristics. Like thermal overload relays, this type of magnetic overload relay has an inverse time-tripping characteristic. However, unlike thermal relays, the
inverse time-current relay's current trip point does not vary with a change in the
ambient air temperature surrounding the motor. Although quite a bit more expensive than thermal overloads, the inverse time-current relay has unmatched features,
which allow it to be used in special applications where devices with normal time-
current curves will not function suitably.

The flexibility of the inverse time-current relay comes from its adjustable current trip point and the use of a **dashpot** (a fluid-filled, shock-absorber-like device) that mounts below the current coil. During operation, the dashpot damps (slows) the motion of the solenoid plunger and sets the time-current characteristic. By adjusting the opening of the port that controls the transfer (flow) of fluid from the reservoir to the piston assembly of the dashpot, the time-current characteristic of the inverse time-current relay can be custom-tailored to handle any number of applications. The inverse time-current relay is available with either automatic or manual reset.

CHAPTER SUMMARY

- A short circuit and a ground fault represent a fault in an electrical system.
- The electrical wiring can overheat, causing a serious fire hazard, when the heat in an electric system is not kept within the design limit of the system.
- The fault current is limited by a combination of the wiring resistance and the equivalent impedance of the transformer supplying power to the short circuit.
- A ground fault condition is created when a "hot" wire comes in contact with an object at or near ground potential.
- *Winding failure* in motors and transformers is due to excess heat created by overloads—an overcurrent condition.
- The ground fault interrupter (GFI) is used to sense low values of ground fault current and safely disconnect the circuit.
- An uncontrolled motor overload can lead to failure of the bearings and the insulation of the windings.
- Fuses and circuit breakers are the predominant types of overcurrent devices used to interrupt and clear faults and most overloads.
- Fuses are thermally operated devices that destroy themselves by melting a fusible metal link when excess current passes through the link.
- The specific heat of a material is the quantity of heat needed to change the temperature of a unit mass of the material 1°C.
- Most overcurrent protection devices have an inverse relationship between the time for opening and clearing and the amount of overcurrent.
- A *molded-case circuit breaker* will interrupt a fault current or an overload without destroying itself.
- Circuit breakers are manually and thermally operated, although some, with large current ratings, are magnetically operated.
- A circuit breaker or a fusible disconnect is used with a motor to protect it from a fault condition and to isolate it from the line when maintenance is done.
- Four basic parameters used to select an overcurrent device are the voltage rating, rated continuous current, interrupting rating, and the response time.
- The amount (magnitude) of fault current may be approximated with the aid of the transformer nameplate percent impedance.
- *Motor-running overload protection* is provided by thermal switches and by overload relays that sense motor-circuit conditions and disconnect the motor from the line.

SELECTED TECHNICAL TERMS

The following technical terms, abbreviations, and acronyms are defined in the glossary located after Chapter 16. You are encouraged to use the glossary to aid your understanding and to test your knowledge of these important terms.

ambient temperature	GFI
ampacity	ionize
arcing	NEC
bimetal	overcurrent
dashpot	overload
eutectic alloy	RCC
eutectic point	short circuit
fault current	specific heat
FLC	

END-OF-CHAPTER QUESTIONS

Write T if the statement is true and F if the statement is false.

1. A ground fault condition is created when the neutral wire touches an object at or near ground potential.
2. Arcing is usually associated with a short circuit or a ground fault condition.
3. A circuit breaker must be replaced after an overcurrent condition is cleared.
4. A GFI protects against the hazards of electric shock.
5. A fuse rated at 125 V can be used in a 24-V circuit.
6. In order for the current to increase during a short circuit, the impedance of the circuit must proportionally decrease.
7. A current level even slightly over the rated continuous circuit current is an overcurrent condition.
8. The current caused by an overload remains within the confines of the current-carrying conductor.
9. Overload protection is needed while a motor is running to protect the motor from overheating.
10. When a fault occurs, one purpose of an overcurrent device is to keep the production of heat and the increase of temperature in the electrical system to a minimum.

In the following, select the word or phrase that makes the statement true.

11. A fault condition is any abnormal condition that provides a path for overcurrent (within, outside) the conductor.
12. The production of heat in a fuse link (increases, decreases, does not change) the energy in the material.
13. The purpose of the overload device in the motor starter circuit is to sense a(n) (ground fault, short circuit, overload) condition and disconnect the motor from the line before excessive overheating occurs.
14. When fault currents are large, the (circuit impedance, magnetic field, reluctance) is also large.

15. The noninstantaneous trip current relay used in a motor starter has an adjustable (time, current, time and current) feature to match the motor's operation and load characteristics.

Answer each of the following questions with a short answer in the form of a complete sentence. Include a restatement of the question in your answer.

16. What is the difference between an overload condition and a fault (short-circuit or ground fault) condition?
17. From where does the flexibility in the inverse time-current relay come?
18. What does the specific heat of a material indicate?
19. What is the unique property of a eutectic alloy?
20. What is the operating principle of the temperature-sensitive, thermal bimetal switch used in some motors?

END-OF-CHAPTER PROBLEMS

Solve the following problems. Make sketches to aid in solving the problems and structure your work so it follows in an orderly progression and can easily be checked. Table 12.4 summarizes the formulas used in Chapter 12.

1. Using the conversion factors found in Table A1 of Appendix A, convert 15 kWh of energy to joules of energy.
2. Determine the amount of energy (in joules) that is changed to heat and light when a 60-W lamp is operated for 30 min.
3. Determine the energy converted to heat in (a) joules and (b) kilowatt-hours when a 480-V load is short-circuited and it takes 65 ms to open the breaker. The combined resistance of the wiring, transformer, and short circuit is 580 mΩ.

TABLE 12.4 Summary of Formulas Used in Chapter 12

Equation Number	Equation
12.1	$W = Pt = I^2Rt$
12.2	$Q = W = I^2Rt$
12.3	$\Delta T = \dfrac{Q}{mc}$
12.4	For 1-ϕ transformers: $I_{sc} = \dfrac{P_s}{E \times \%Z} \times 100$ For 3-ϕ transformers: $I_{sc} = \dfrac{P_s}{\sqrt{3} \times E \times \%Z} \times 100$
12.5	$I^2t = kA^2 \log\left(\dfrac{x + T_{fin}}{x + T_{int}}\right)$

4. Determine the change in the temperature of a 45-g bimetal ribbon with a specific heat of 420 J/kg·°C that is used in the trip circuit of a circuit breaker. The heater produces 2200 J of heat during the 2.5 s it takes the breaker to open.

5. Determine the approximate short-circuit current available from a 208-V, 45-kVA, three-phase transformer secondary with a transformer impedance of 1.85%.

6. Determine the time it will take for a 12-g zinc fuse element to melt (fuse) when the cold temperature of the fuse link (ambient) is 22°C. The average power dissipated during the fusing of the zinc fuse link is 360 W. Use the data for zinc found in Table 12.1 and the identity $Q = W = Pt$ (joules).

7. An AWG No. 4 aluminum conductor normally carries 52 A of current to an industrial heater. Determine the period of time that an 880-A fault current can flow before the conductor's 150°C temperature rating is exceeded. The conductor temperature at the time of the fault was 82.4°F.

8. A small piece of AWG No. 32 copper wire is to be used as a fusible link in an inexpensive consumer product. If the ambient temperature is 28.0°C, determine the period of time (s) for the fuse link to open (melt) and interrupt the 42.0-A fault current.

Stepper
motor

Programmable
motion control

Motor
driver

C

A

B

System Configuration

Chapter 13

Sequential Process Control

About the Illustration. *Sequential process control may incorporate a motion-control system to drive stepper motors in repetitive operations. Pictured here is Superior Electric's SLO-SYN® motion-control equipment. Motion commands for the operation and control of the stepper motor (C) are entered into the motor driver (B) through the programmable motion control (A). Programming of the programmable motion control unit is done by temporarily connecting a personal computer to the unit. (Courtesy of Superior Electric.)* ■

INTRODUCTION

A sequential process may be used to assemble manufactured parts into a finished product on an assembly line, to machine parts (drill, mill, turn, etc.) in an automated machining center, or to combine ingredients into food products (baked goods, canned goods, etc.) in an automated food-processing plant. The control of a sequential process may be described in several ways, including flowcharts, function charts, timing diagrams, symbolic logic (Boolean algebra), and ladder diagrams.

If the technician is to be successful in maintaining an automated system, both the control of the sequence of events in the process and the nature of the processes must be understood. The ladder diagram aids in understanding the sequence of operations, since it both diagrams the events of the discrete state process control and symbolizes the control circuit hardware.

This chapter studies the terms, symbols, and constructs of the ladder diagram and reviews various control devices, including the control relay. Additionally, the terms, symbols, and concepts of symbolic logic are introduced.

CHAPTER CONTENTS

PERFORMANCE OBJECTIVES

Once you have read and studied each section; worked through the end-of-chapter problems; and answered the end-of-chapter questions, then you should be able to

- Understand the structure of the ladder diagram.
- Recognize and name the graphic symbols that represent industrial electronic equipment in the ladder diagram.
- Identify and use alpha designators for devices.
- Explain the operation of the solid-state relay.
- Know the guidelines used for interpreting a ladder diagram.
- Identify basic motor-control circuits and understand their operation.
- Identify the state of the contacts in both an ON-delay and an OFF-delay timer.
- Relate switch and relay logic to ladder diagrams for the control of sequential processes.
- Interpret the truth tables of common logic functions.
- Read a ladder diagram and understand the sequence of events needed to control the process.

13.1 LADDER DIAGRAM CONVENTIONS

Introduction

The *ladder diagram*, or elementary diagram as it is sometimes called, is a common method of showing how the steps of a process that is controlled by a sequence of discrete state events (ON/OFF, OPEN/CLOSED, HIGH/LOW, ONE/ZERO, START/STOP) are related to the hardware used to implement the control circuit. A process of this nature is called a **sequential** process. A sequential process has a beginning, an end, and a series of one or more steps in between.

The means of initiating (starting) or terminating (stopping) the process steps is based on either the passage of an interval of time (*time-driven*) from a timer *within* the control circuit or on an *external* event (*event-driven*), such as the pushing of a START button, the closing of a limit switch, or the activation of a relay coil.

Most sequential processes consist of a combination of event-driven and time-driven steps.

Since the control relay was the first device used to implement hard-wired control circuits, the control diagram was called a relay ladder diagram. The name has been shortened to *ladder diagram*. The ladder diagram has become the standard way of representing the logical sequence of events needed to implement the control of a sequential process.

A popular control language called *ladder logic* is formed when the orderly sequence of events of discrete state process control is combined with the logic and structure of the ladder diagram. Ladder logic is the preferred language for programming the modern-day *programmable logic controller* (**PLC**).

Use and Structure of a Ladder Diagram

A ladder diagram representing the control system of a sequential process provides a record for the technical personnel so that they can install the system and do routine maintenance on the system. Additionally, the ladder diagram of a system is used to make modifications to the system and to aid in the troubleshooting and repair of the system when components fail.

The ladder diagram is created from a set of standard electrical (not electronic) *graphic symbols* that represent industrial electric control equipment (control relays, contactors, motor starters, solenoids, time-delay relays, limit switches, pushbutton switches, etc.). Each graphic symbol (Table 13.1) used in the ladder diagram is given a unique alphanumeric *device designator* to differentiate it from all other graphic symbols representing the functions and components of the ladder diagram. Table 13.2 lists many of the standardized abbreviations used as device designators. *Connecting lines* are used to interconnect the graphic symbols between the vertical *rails* to form the horizontal *rungs* of the ladder.

The various devices that are symbolized in the ladder diagram are divided into two general classes, *input devices* and *output devices*. The input devices (control or pilot devices) are placed on the *left* of the ladder diagram, whereas the output device(s) are shown on the *right* of the diagram.

Figure 13.1 pictures a simple ladder diagram for the control of a pilot light. Notice the use of standard electrical graphic symbols to represent the input and output devices. Also note the use of an alphanumeric device designator to give each device a unique designation. Finally, look at how the input devices on the left are interconnected to the output device on the right to form the desired control function.

Input Devices

The input devices, located on the left side of the ladder diagram of Figure 13.1, are the two limit switches (1LS and 2LS) and the normally open pushbutton switch (1PBS). The output device is the pilot light (1LT). The switches (input devices) are connected in series and parallel with one another to form the control logic of the diagram. The presence or absence of light from the pilot light (1LT) depends on the

TABLE 13.1 Graphic Symbols for Industrial Electrical Diagrams

Switches										
Liquid Level		Vacuum and Pressure		Temperature		Flow (Air, Water etc.)		Foot		Toggle
Normally Open	Normally Closed	Normally Open	Normally Closed	Normally Open	Normally Closed	Normally Open	Normally Closed	Normally Open	Normally Closed	
FS	FS	PS	PS	TAS	TAS	FLS	FLS	FTS	FTS	TGS

Limit							Selector	
Normally Open	Normally Closed	Neutral Position		Maintained Position	Proximity Switch		2 – Position	3 – Position
					Closed	Open	SS	SS
LS	LS		Actuated	LS	PRS PES	PRS PES		1 2 3
		LS	LS					
Held Closed	Held Open	NP	NP					
LS	LS							

Pushbutton			Contacts							
Single circuit	Double circuit		Solid State Relay	Time Delay After Coil				Relay, etc		Thermal Overload
		Mushroom Head		Energized		De-energized				
Normally Open	PBS	PBS	Normally Open	Normally Open	Normally Closed	Normally Open	Normally Closed	Normally Open	Normally Closed	
PBS			SSR	TR	TR	TR	TR	CR M CON	CR M CON	OL
Normally Closed										
PBS										

Resistors, Capacitors, etc.							
Resistor	Heating Element	Fixed	Polarized Electrolytic	Horn, Siren, etc.	Buzzer	Bell	Fuses (All Types)
RES	HTR	CAP	CAP	AH	ABU	ABE	FU

Coils						Connections, etc.			
Solenoids, Brakes, etc.				Relays, Timers, etc.	Solid State Relay, etc.	Conductors		Ground	Chassis or Frame Not Necessarily Grounded
General	2-Position Hydraulic	3-Position Pneumatic	2-Position Lubrication			Not Connected	Connected		
SOL	SOL 2-H	SOL 3-P	SOL 2-L	CR TR M CON PRS	SSR PES			GND	CH

TABLE 13.2 Standardized Abbreviations Used as Device Designators

Abbreviation	Term	Abbreviation	Term
ABE	Alarm, bell	M	Motor starter
ABU	Alarm, buzzer	MB	Magnetic brake
AH	Alarm, horn	MTR	Motor
CAP	Capacitor	NLT	Neon light
CB	Circuit breaker	OL	Overload relay
CH	Chassis connection	PBS	Pushbutton switch
CON	Contactor	PES	Photoelectric switch
CR	Control relay	PRS	Proximity switch
CRM	Control relay, master	PS	Pressure switch
D	Diode	RES	Resistor
DISC	Disconnect switch	RH	Rheostat
DZ	Diode, zener	SOL	Solenoid
FLS	Flow switch	SS	Selector switch
FS	Float switch	SSR	Solid state relay
FTS	Foot switch	T	Transformer
FU	Fuse	TAS	Temperature switch
GND	Ground connection	TB	Terminal block
HTR	Heating element	TCS	Thermocouple switch
LS	Limit switch	TGS	Toggle switch
LT	Light, pilot	TR	Time-delay relay

condition (OPEN or CLOSED) of the switches. That is, when the events of the process cause 1PBS to CLOSE at the same time 1LS is CLOSED, the pilot lamp is ON. The lamp is also ON when conditions are such that both 1LS and 2LS are CLOSED at the same time.

As observed in Figure 13.1, the input devices govern the flow of electric current to the output devices by enabling the completion of the electric circuit. All the control (pilot) devices previously discussed in Section 9.4 may be used as input devices in the control circuit of a sequential process. As was previously noted, control devices (pilot devices) may be operated either as *momentary-contact devices* or as *maintained-contact devices*.

Momentary-contact devices are characterized by a spring-loaded mechanism that returns the device to its nonactivated position once the force of activation is

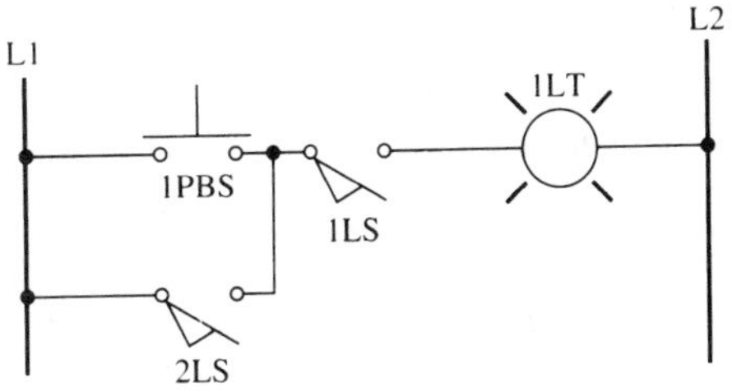

FIGURE 13.1
A ladder diagram for the control of a pilot light. The input devices (pilot devices) are shown on the left, whereas the output devices are pictured on the right. L1 and L2 identify the *rails*.

removed. The pushbutton switch and the *held-open* or *held-closed* limit switch of Table 13.1 are examples of momentary-contact devices. Maintained-contact devices are characterized by the fact that they change state each time they are activated. The device does not revert back to its previous state after activation. The selector switch and the toggle switch of Table 13.1 are examples of maintained-contact devices. The following switch and relay contacts (control devices) included in Table 13.1 serve as input devices in ladder diagrams:

Limit switch	LS	Proximinity switch	PRS
Float switch	FS	Pressure switch	PS
Temperature switch	TAS	Flow switch	FLS
Foot switch	FTS	Toggle switch	TGS
Selector switch	SS	Pushbutton switch	PBS
Control relay	CR	Motor starter	M
Contactor	CON	Time-delay relay	TR

Output Devices

Output devices include all the devices that represent a resistive or inductive load, such as heating elements, relay and contactor coils, alarms, lights, solenoid coils for activating clutches, valves, etc. The heating elements, coils, lights, alarms, and solenoid coils included in Table 13.1 that are found in the output section of the ladder diagram are as follows:

Control relay coil	CR	Motor starter coil	M
Contactor coil	CON	Timer coil	TR
Overload contact*	OL	Pilot light	LT
Neon light	NLT	Solenoid coil	SOL
Heating element	HTR	Horn or siren	AH
Buzzer	ABU	Bell	ABE

Control Relay

As you know, the control relay (pictured in Figure 13.2) is a low-power device that changes state when voltage is applied to the coil. With the application of voltage to the relay coil, current flows and magnetomotive force (mmf) is created to energize the magnetic circuit. Once mmf is present in the magnetic circuit, flux flows and the air gap between the steel armature and the steel core is forced closed, changing the state of the set(s) of contacts.

By convention, the relay contacts are pictured in ladder diagrams in their de-energized state. Thus, when the relay coil is energized, the set(s) of contacts

---

*Overload contact(s) may appear in the input side of the diagram.

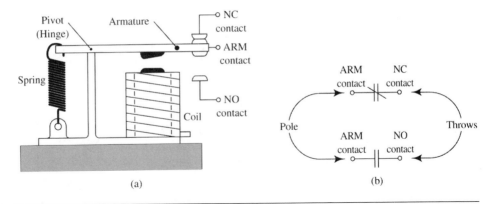

FIGURE 13.2
The control relay is shown in the de-energized state. (a) The armature (ARM) contacts change state when the relay coil is energized. (b) The pole in an SPDT set of contacts is the armature (ARM) contacts while the throws are the NC and NO contacts.

change state. The normally open (NO) contacts are closed and the normally closed (NC) contacts are opened.

As illustrated in Figure 13.2, the armature contact is the **pole,** or moving contact, in the set of contacts. The stationary contacts, called the **throws,** are either normally open or normally closed. When the relay coil is energized, the armature contact is moved from making contact with the NC contact to making contact with the NO contact.

The control relay can have any number of sets of contacts made up of an armature contact, an NO contact, and an NC contact. Control relays are commonly available with one, two, three, or more (up to several dozen) sets of contacts. The contacts of the relay are identified in the ladder diagram by using an alphanumeric designator. The contacts of the control relay are designated by the letters CR, and the motor starter is denoted by the letter M. Table 13.2 summarizes many of the letter designators associated with the contacts of relays.

Contacts in the Ladder Diagram

Figure 13.3 pictures a ladder diagram with the auxiliary set of the motor starter (M) contacts shown in rung 2 and rung 3. In rung 2, the left contact is the armature contact, whereas the right contact is the normally open contact. In rung 3, the left contact is the armature contact, and the right contact is the normally closed contact. As illustrated, one set of relay contacts can be used in a control circuit to serve as both a normally open and a normally closed contact.

Since the ladder diagram is not a wiring diagram, the actual connection of the armature contact would involve connecting it to the *high side* of the control circuit. The NO contact would be connected in parallel with the START pushbutton switch, whereas the NC contact would be connected in series with the pilot lamp.

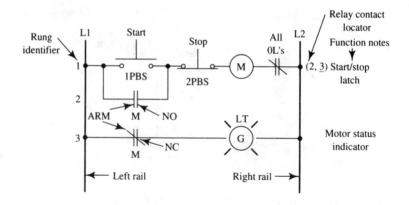

FIGURE 13.3
The contacts in rung 2 and rung 3 are from the same set of contacts. The armature contact is on the left in rung 2, whereas the NO contact is on the right. The armature contact is on the left of the contact in rung 3, and the NC contact is on the right.

To aid in identifying the rungs of the ladder diagram, the rungs are numbered along the left rail; the numbers in parentheses along the right rail are used to designate the location of the relay contacts of the coil in that rung. A line beneath a number in parentheses indicates a normally closed contact.

Solid-State Relay (SSR)

The solid-state relay (SSR), pictured in Figure 13.4, is used to replace the electromagnetic relay (EMR) if compatibility with solid-state digital circuitry is needed or if the physical operating conditions warrant its use. Because it has no moving contacts, the solid-state relay is more reliable than the electromagnetic relay, since the

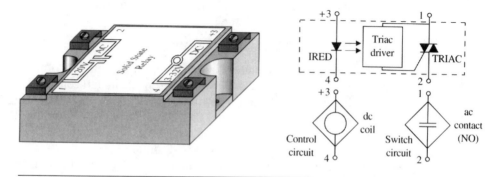

FIGURE 13.4
Solid-state relay (SSR). The coil circuit (infrared emitting diode, IRED) is optically isolated from the ac contacts (triac switch circuit), creating a dc control circuit (terminals 3 and 4) electrically independent of the ac power contacts (terminals 1 and 2).

SSR is not susceptible to contact bounce due to mechanical shock and vibration. Like the EMR, the SSR isolates the low-voltage, low-power control circuit from the high-voltage, high-power load circuit. However, unlike the electromagnetic relay, the solid-state relay has a faster switching time and a longer lifetime because it uses solid-state components and an **optocoupler/isolator.**

The solid-state relay is available for general-purpose applications as a form A (SPST, NO) relay with dc (3 to 32 V) or ac control circuits (90 to 280 V ac) and either a dc or an ac switching circuit of up to 500 V.

The solid-state relay circuit pictured in Figure 13.4 illustrates how the control circuit turns on the switch circuit. The dc control circuit (3 to 32 V) is used to turn on an infrared light-emitting diode (IRED). The light energy from the IRED is *optically* coupled to the solid-state triac driver, a triggering circuit. The trigger circuit (driver) is used to change the state of the triac in the switch circuit from OFF to ON, thereby completing the ac switch circuit. The load current passes through the low-impedance path provided by the ON (conducting) triac. The activation of the IRED in the SSR control circuit is similar in function to the activation of the EMR coil, whereas turning ON the SSR triac switching circuit is equivalent to closing the NO relay contacts in the EMR.

Because of the semiconductor components, the solid-state relay is susceptible to transient "spikes" (both voltage and current), which can damage it. To guard against the presence of spikes from inductive loads (motors, solenoids, relays, etc.), surge-suppression circuits are added across (in parallel with) the inductive output devices in the load circuit and across the ac output switching circuit of the solid-state relay.

As illustrated in Figure 13.5(a), a surge suppressor may be placed in parallel with an ac-powered inductive load. This device may be a metal-oxide varistor

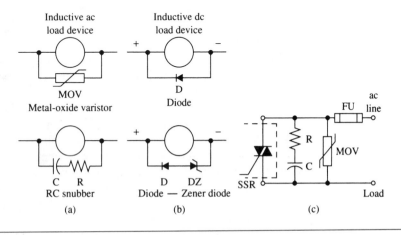

FIGURE 13.5
Surge suppression is placed in parallel with: (a) An ac powered inductive load; (b) A dc powered inductive load; (c) The ac switch of the SSR. The RC snubber is often accompanied by a metal-oxide varistor (MOV) to limit the peak voltage and to suppress electrical noise.

(**MOV**) or a resistor-capacitor (RC) *snubber* circuit. The suppressor device placed in parallel with a dc powered inductive load is usually a diode or a back-to-back diode-zener diode combination, as shown in Figure 13.5(b). The ac switch section (triac) of the solid-state relay is typically protected by an RC snubber circuit placed in parallel with the switch. Additionally, as pictured in Figure 13.5(c), a metal-oxide varistor may be added in parallel with the RC snubber to suppress electromagnetic interference (**EMI**), which can affect the operation of the solid-state relay. A series-connected fuse may be added to the switching circuit to prevent the load from taking excessive current and damaging the ac switch.

Interpreting the Ladder Diagram

Several conventions and practices are followed when the ladder diagram is read. The following statements contain many of the important points needed to interpret and understand ladder diagrams.

- Power connections to the rails are not usually shown but are, instead, simply implied.
- Inputs (switch contacts, relay contacts, etc.) are placed in the ladder diagram on the left-hand side.
- Outputs (coils, lights, solenoids, etc.) are placed in the ladder diagram on the right-hand side.
- Regardless of the actual number of contacts, all the thermal and magnetic overload relay contacts may be shown as one contact, with the label *All OLs*.
- Inputs can be connected to control more than one output. When this is the case, the outputs are connected in parallel.
- The rungs of the ladder are numbered from top to bottom on the left-hand side of the diagram.
- The sequences of events pictured in the ladder diagram is read from top to bottom.
- Control relays, timing relays, contactors, and motor starters usually have more than one set of auxiliary contacts, which may appear anywhere in the ladder diagram.
- Relay contacts are identified with the relay coil alphanumeric device designator.
- A locating system consisting of rung numbers placed within parentheses may be used on the right side of the ladder diagram to identify the location and type of relay contact(s) controlled by a given relay coil.
- When numbers are placed within the body of the ladder diagram, they represent nodes (intersections) where devices are connected together. These numbers correspond to *wire labels* (numbers) attached to the actual circuit wiring, which are used to identify the wiring during installation and maintenance of the control system.
- Notes may be added to the right of the diagram to indicate function or components in that particular rung.

13.2 LADDER DIAGRAMS OF BASIC MOTOR-CONTROL CIRCUITS

Introduction

In this section you will study the operation of several conventional ladder diagrams used to control the operation of motors. These circuits will allow you to begin to master and understand ladder diagrams and, at the same time, will provide you with an opportunity to familiarize yourself with several basic ladder configurations that are standard to the industry.

Although there are several ways a control circuit might be shown, the electrical industry has established a set of conventions that all technical personnel must adopt and use in order to have a compatible set of knowledge with other technicians and electricians. It is very important for your understanding of control circuits that you master the techniques of reading, interpreting, and applying the rules and conventions of ladder diagrams. Use the information of the previous section to assist you in understanding the following ladder diagrams.

START/STOP Circuit

The START/STOP circuit pictured in Figure 13.6 is a three-wire control circuit that provides *undervoltage protection*. The auxiliary contacts, found on the contactor, are used in the construction of the three-wire holding circuit, which *seals*, or *latches*, the contactor coil ON. The three-wire START/STOP circuit functions in the following manner:

1. The NO START pushbutton (2 PBS) is momentarily depressed, completing the coil circuit and pulling the contactor switch section closed, energizing the motor.
2. At the same time, the auxiliary contacts (M) close and form an alternative parallel pathway around the START pushbutton switch (2 PBS). This alternative circuit allows the contactor coil to remain energized after the START pushbutton is released. The contactor coil of the motor starter is *latched*, or *sealed*, ON.
3. Pressing the STOP pushbutton opens the sealed circuit, de-energizing the contactor coil (M) and releasing the contactor's switch section. This action shuts off the power to the motor and resets the holding contact to normally open. The motor will not restart until the START pushbutton is once again depressed by the machine operator.

FIGURE 13.6

Ladder diagram for a START/STOP three-wire control circuit for a magnetic motor starter.

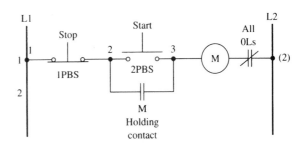

In the START/STOP three-wire control circuit, should the control voltage be interrupted due to either a power failure, a low coil voltage, or an overload relay trip (OLs in rung 1 open), the motor starter will be de-energized and the contacts of the contactor power section will drop open. The holding contact (M in rung 2) will open and unlatch the contactor coil, thus preventing the circuit from restarting once power is restored or the overload relay contacts are reset. Once the undervoltage condition is cleared, depressing the START pushbutton will once again energize the motor contactor and start the motor.

The START/STOP three-wire circuit may have additional START and STOP pushbutton switches added to the control circuit, as illustrated in Figure 13.7. It is important to note that STOP buttons are always wired in series with the contactor coil so that any one of them can open the circuit and de-energize the contactor coil. START buttons are always wired in parallel with the holding contact, so any one of them can close the circuit and energize the contactor coil. The horsepower ratings for NEMA starters used in full-voltage starting are given in Appendix A, Table A9.

Jog Circuit

The three-wire jog circuit is a variation of the START/STOP three-wire circuit. In the circuit pictured in Figure 13.8, a jog-run selector switch (SS) is used to select either the run mode or the jog mode. In the run mode, as illustrated in Figure 13.8(a), the circuit functions as a standard three-wire START/STOP circuit, with the holding contact (M) in parallel with the START pushbutton switch. When the selector switch is moved to the jog position (Figure 13.8(b)), then the holding contact (M) is no longer in parallel with the START pushbutton switch, and the START/STOP circuit cannot be sealed. In this state, the motor can be intermittently operated (jogged) by depressing and releasing the start button.

The jog feature is used with machine tools and automated systems to momentarily start the machine in order to move it to a desired location for installing tooling or checking clearances before the system is put into operation.

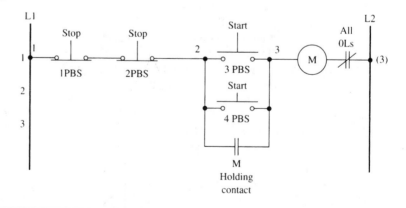

FIGURE 13.7
Ladder diagram for a multiple START/STOP three-wire control circuit for a magnetic motor starter.

FIGURE 13.8
Ladder diagram for a three-wire control circuit for jog or run using a START/STOP pushbutton switch and a JOG/RUN selector switch: (a) Selector switch in the RUN position; (b) Selector switch in the JOG position.

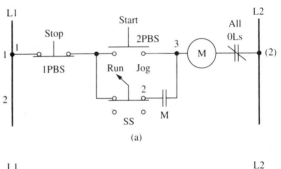

(a)

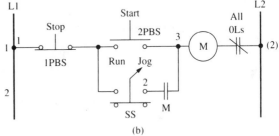

(b)

Interlock Circuits

The control circuits of two electric motors may be interlocked, as illustrated in Figure 13.9, to prevent the first motor (1M) from operating unless the second motor (2M) is running. Conveyor belt systems are interlocked in this manner to prevent goods from piling up when they are transferred from one belt to another. In this circuit,

1. The NO contact in rung 1 remains open until the START pushbutton (4PBS) in rung 3 is depressed, energizing motor starter coil 2M;
2. Once coil 2M is energized, then coil 1M can be energized by depressing the START pushbutton (2PBS) in rung 1.

By interlocking the motors in this manner, the desired starting sequence is ensured.

FIGURE 13.9
Ladder diagram for an interlock circuit in which the first motor will not start until the second motor is running.

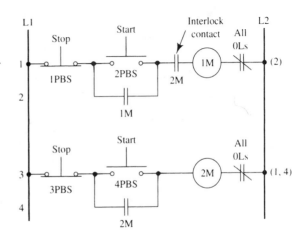

As pictured in Figure 13.10, two motors may be interlocked to prevent them from operating at the same time and putting an excessive demand on the power line. In each of the following cases, assume that both motor starters are de-energized.

Case 1 When the START pushbutton switch (2PBS) is closed in rung 1, the coil of starter 1M is energized; at the same time, the contacts (1M) in series with contactor coil 2M are opened, thus disabling rung 3 and motor starter 2M.

Case 2 When the START pushbutton switch (4PBS) is closed in rung 3, the coil of starter 2M is energized; at the same time, the contacts (2M) in series with contactor coil 1M are opened, thus disabling rung 1 and motor starter 1M.

Compelling Circuit

A circuit controlling the operation of a machine that requires that the drive motors be brought to a stop before changing the direction of travel is called a *compelling control circuit*. Figure 13.11 pictures such a circuit. Here two motors are interlocked to prevent damage by inadvertent operation of both the UP and DOWN motors at the same time.

In this circuit, the operator is "compelled" (required) to open either the UP or DOWN circuit by depressing the STOP pushbutton (1PBS) before changing the direction of travel. In each case, the motor is brought to a stop before the hoist is operated in the opposite direction. The following is the sequence of events controlled by the ladder diagram of Figure 13.11.

1. Depressing the UP pushbutton (2PBS) in rung 1 energizes the up contactor coil, 1M, causing the hoist to move in an upward direction.
2. At the same time, the normally closed contact (1M) in rung 3 is opened and the normally open contact (1M) in rung 2 is closed, *sealing in* the circuit around the UP pushbutton switch (2PBS).

As long as the up contactor in rung 1 is energized, depressing the DOWN pushbutton switch in rung 3 will have no effect. This is, of course, due to the open

FIGURE 13.10
Ladder diagram for an interlock circuit in which the interlocks are placed in series with each of the two motor-control circuits to prevent both motors from being operated at the same time.

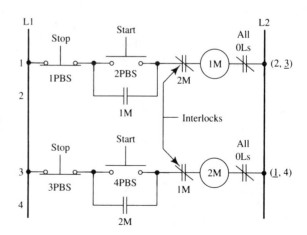

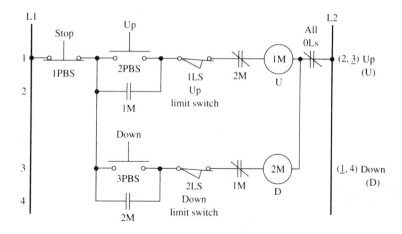

FIGURE 13.11
Ladder diagram of the control circuit for an overhead hoist in which the operator is compelled to stop the motor before changing direction.

contact (1M) in series with the DOWN contactor coil 2M in rung 3. In order for the hoist to move down, one of two events must take place in rung 1: Either the limit switch (1LS) must open, or the STOP pushbutton (1PBS) must be depressed, opening the circuit and releasing the UP contactor coil 1M. Once the UP contactor coil is de-energized, then the down pushbutton may be depressed, energizing the DOWN contactor coil. Thus, the overhead hoist is protected by an interlock in each of the two (UP and DOWN) motor-control circuits.

In addition to the electrical interlocks, the travel of the hoist is limited by the two limit switches installed in the hoist and wired in series into the hoist control circuit. The limit switches prevent the hoist cable from binding when it reaches the end of its travel. When the hoist reaches its upper limit, the UP limit switch (1LS) is opened, which opens the UP contactor coil (1M). When the DOWN limit is reached, the DOWN limit switch (2LS) is opened, which opens the DOWN contactor coil (2M).

Reverse Circuit

Electric motors are often used with machines that operate in both forward and reverse. Both fractional-horsepower, single- and three-phase motors can be operated in forward and reverse with a manually operated drum switch. However, if the motor is under three-wire control (magnetic starter) or if it has a large horsepower rating, then a *reversing magnetic motor starter* (pictured in Figure 13.12) is used to control forward and reverse operation.

To reverse a three-phase motor, two of the three line wires to the motor are interchanged. This is carried out by the reversing motor starter, which is an assembly made up of two three-phase contactors (one for forward and one for reverse), a set of dual interlocks (mechanical and electrical), and a three-phase thermal overload relay.

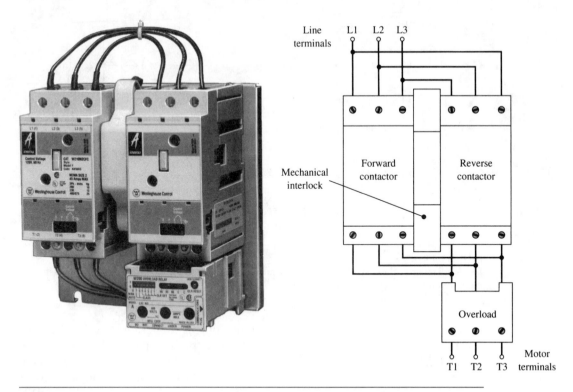

FIGURE 13.12

A reversing magnetic motor starter is used to control forward and reverse operation of the three-phase motor. The reversing starter consists of two contactors, an interlock module, and an overload relay. Notice that the wiring sequence of the reverse contactor is L3, L2, L1, whereas the forward contactor is wired L1, L2, L3. L1 and L3 are interchanged between the contactors. (*Courtesy of the Westinghouse Electric Company.*)

The reversing motor starter, as illustrated in the ladder diagram of Figure 13.13, is wired to a three-function (FORWARD, STOP, and REVERSE) pushbutton switch station (three pushbutton switches in the same enclosure). To prevent both contactors from being energized at the same time, mechanical and electrical interlocks are used in each contactor circuit. As pictured in the ladder diagram of Figure 13.13, the two contactor circuits are *cross interlocked*. That is, a normally closed contact from the forward contactor (1M) is placed in series with the reverse contactor coil, and a normally closed contact from the reverse contactor (2M) is placed in series with the forward contactor coil.

The mechanical interlock is achieved by using a pushbutton switch with both a normally open and a normally closed contact. These pushbutton switches break the circuit before they make the circuit (break before make). The dashed lines in Figure 13.13 indicate that the two contacts are *mechanically connected* and activated by the same pushbutton. The normally closed contact of the forward pushbutton switch is connected in series with the reverse contactor coil circuit in rung 3, whereas the nor-

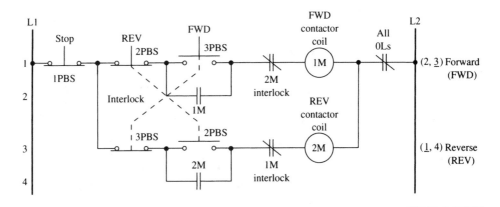

FIGURE 13.13
The ladder diagram of the control circuit of the reversing magnetic motor starter with the electrical and mechanical interlocks noted.

mally closed contact of the reverse pushbutton switch is connected in series with the forward contactor coil circuit in rung 1.

With the motor in forward operation, depressing the reverse pushbutton will close the normally open pushbutton switch contact in rung 3 and will open the forward contactor circuit (rung 1), causing it to drop out and close the interlock contact (1M) in rung 3. The motor will be instantly connected to the reverse contactor for reverse rotation. The reverse circuit will be sealed and will continue to operate in reverse until either the STOP pushbutton or the FWD pushbutton is depressed.

This is not a compelling circuit, since the STOP pushbutton does not need to be depressed when changing directions. Because the motor is not stopped before changing directions, the motor and the machine must be designed to withstand the sudden changes in torque associated with going directly from full forward angular velocity to full reverse angular velocity without first stopping.

13.3 TIMING RELAYS

In addition to contacts, switches, and relay coils, the *timing relay* is commonly found in sequential process-control circuits. Timing relays, or timers, as they are often called, are available in two general types—the ON-delay timing relay and the OFF-delay timing relay. Each type is used as an integral component of the ladder diagram. Figure 13.14 details the graphic symbol for each of the two types of timing relays.

ON-Delay Timing Relay

The *ON-delay timing relay* (TR) provides the time delay after the coil is energized. The ON-delay timer contact symbol has the arrow on the contact pointing up, as noted in Figure 13.14(a). The tail of the arrow indicates that the time delay starts

FIGURE 13.14
(a) The relay contacts of the ON-delay timer change state after the coil is energized and the timer times out.
(b) The relay contacts of the OFF-delay timer change state after the coil is de-energized and the timer times out.
Note: NO = normally open
TO = timed open
NC = normally closed
TC = timed closed

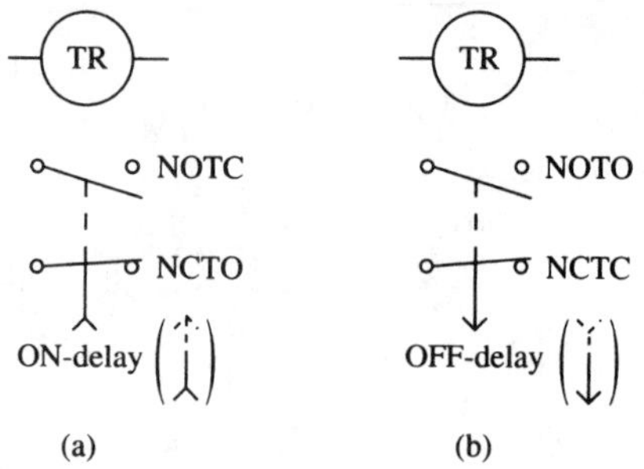

(a)

(b)

when the coil of the ON-delay TR is energized. Even though the timer coil is energized, the relay contacts do not immediately change state; instead, they remain in the de-energized state, with the normally open contacts open and the normally closed contacts closed. The change in the state of the contacts occurs only after the timer times out. That is, with the ON-delay timer, the normally open contacts are "timed closed" (NOTC) and the normally closed contacts are "timed open" (NCTO) after the relay coil is energized. The interval of time between the energizing of the ON-delay timer coil and the change in state of the contacts can be adjusted over a fairly wide range of values.

OFF-Delay Timing Relay

The *OFF-delay timing relay* provides the time delay after the coil is de-energized. The OFF-delay timer contact symbol has the arrow on the contact pointing down, as noted in Figure 13.14(b). The tip of the arrow indicates that the time delay starts when the coil of the timing relay is de-energized (opened). The operation of the OFF-delay timer is such that the contacts immediately change state when the timer coil is energized. Thus, the normally open contacts close and the normally closed contacts open. Once energized, the change in the state of the OFF-delay contacts occurs only after the coil of the OFF-delay timer is de-energized and the timer times out. That is, once the coil of the OFF-delay timer is de-energized, the normally open contacts remain in the closed state until they are timed open (NOTO) and the normally closed contacts remain in the open state until they are timed closed (NCTC). Like the ON-delay TR, the interval of time delay of the OFF-delay TR can be adjusted over a wide range of values. Table 13.3 summarizes the operating sequence for the ON-delay, OFF-delay, and instantaneous relays.

The ladder diagram of Figure 13.15 is a partial diagram of a cleaning system control circuit. The system consists of a material handler (not shown), a heater for heating the cleaning solution, and a pump to move the cleaning solution through the spray heads.

When the pump motor is initially turned on, its operation is delayed for 2.0 min while the cleaning solution in the holding tank is heated to above room

TABLE 13.3 Operating Sequence for Timers and Instantaneous Relays

Symbol	Function and Description of Sequence
	ON-delay timed closed Contact normally open: contact timed closed once relay is energized; contact opens instantly when relay is de-energized. (NOTC)
	ON-delay timed open Contact normally closed: contact timed open once relay is energized; contact closes instantly when relay is de-energized. (NCTO)
	OFF-delay timed closed Contact normally closed: contact opens instantly when relay is energized; contact timed closed when relay is de-energized. (NCTC)
	OFF-delay timed open Contact normally open: contact closes instantly when relay is energized; contact times open when relay is de-energized (NOTO).
	Instantaneous Contact normally closed: contact opens instantly when relay is energized; contact closes instantly when is relay is de-energized. (NC)
	Instantaneous Contact normally open: contact closely instantly when relay is energized; contact opens instantly when relay is de-energized. (NO)

temperature. With the START pushbutton depressed, the operation of the heater and pump, as pictured in Figure 13.15, is as follows:

1. The tank heater (HTR) in rung 1 is energized, since the thermostat switch (TAS) is closed at room temperature (25°C). The thermostat is set at an upper limit of 60°C to prevent the cleaning solution from overheating.
2. The heater ON light is lighted (green).
3. The control relay (1CR) is energized, causing the normally open contact in rung 2 to close and seal the system on.
4. The ON-delay timer is started.
5. The motor OFF light is lighted (red).

After a 2.0-min delay:

1. The contacts in rungs 5 and 6 will change state.
2. The *normally open timed closed* (NOTC) contact in rung 5 will close, energizing the motor starter turning on the pump.
3. The *normally closed timed open* (NCTO) contact in rung 6 will open, turning off the motor OFF light.
4. The contact (1M) in rung 7 will close, turning on the motor ON light.
5. The contact in rung 8 will close, energizing the remainder of the control circuit.

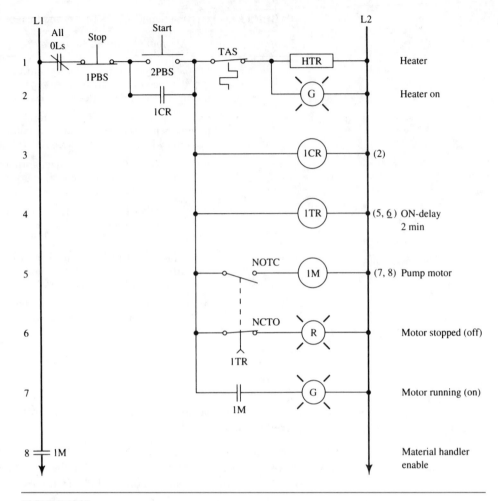

FIGURE 13.15
Ladder diagram of a cleaning system that uses an ON-delay timer to delay the start of the pump motor.

13.4 SWITCH AND RELAY LOGIC

Introduction

In order for you to understand fully the operation of industrial control circuits, you must understand symbolic logic and the logic functions. Wherever switches, relays, or programmable logic controllers (PLC) are used for control, a knowledge of logic will help you to understand the operation of the control circuit. Control circuits are made up of basic logic functions that are used by themselves or in combination with one another to perform the desired control function.

Symbolic logic was developed in 1849 by George Boole, who developed the rules for what is now called **Boolean algebra.** His ideas were utilized by Claude Shannon in 1938 to analyze relay and switching circuits by writing equations with symbols to represent the state (open or closed) of the relays and switch contacts. The terms that make up the equations correspond to the various relay and switch contacts in the circuit.

Ladder diagrams, which are made up of switch and relay contacts to form the logic for process control, are ideal for the study of symbolic logic. Because the input devices (switches, relays, etc.) have two discrete states, each device is either OPEN or CLOSED and it can be nothing else. Since relays and switches have only two states, it is natural that the binary numbers one (1) and zero (0) be used to describe these two conditions. The CLOSED state is indicated by a one (1), whereas the OPEN state is indicated by a zero (0).

A set of SPDT relay contacts from a control relay (CR) will be used to represent a two-state logic condition. As illustrated in Figure 13.16, each of the two relay contacts is labeled with a different logic variable, which is used to represent the discrete state of each contact when the relay coil is turned ON and OFF.

The state of the normally open (NO) contact is indicated by the variable A, and the state of the normally closed (NC) contact is indicated by the variable $\overline{A}$, read "not A." Because of the two-state nature of the SPDT relay contacts, the following will happen when the pushbutton in rung 1 of Figure 13.16 is depressed or released:

> When the pushbutton is *depressed*, the NO contact must close and the NC contact must open.
> When the pushbutton is *released*, the NO contact must open and the NC contact must close.

Since the state of an open contact is indicated by 0 and the state of a closed contact is indicated by 1, when the NO contact is open, $A = 0$, and when the NO contact is closed, $A = 1$. Likewise, when the NC contact is open, $\overline{A} = 0$; when it is closed, $\overline{A} = 1$.

The table next to the ladder diagram in Figure 13.16 shows the state of the two relay contacts (A and $\overline{A}$) in relation to the control signal applied to the relay coil. As

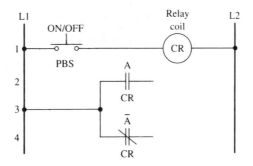

FIGURE 13.16
The contacts of the control relay are each labeled with a variable that is used to represent the state of each contact when the relay coil is turned ON and OFF. *Note:* The NO contact is open when $A = 0$ and closed when $A = 1$. The NC contact is open when $\overline{A} = 0$ and closed when $\overline{A} = 1$.

illustrated in the table, when the control signal to the relay coil is not present (relay OFF), then the NO contact is open ($A = 0$) and the NC contact is closed ($\overline{A} = 1$). However, when there is a control signal to the relay coil (relay ON), then the NO contact is closed ($A = 1$) and the NC contact is open ($\overline{A} = 0$).

Logic Expressions

The ladder diagram of Figure 13.17(a) pictures an NO contact from the control relay in series with a pilot light. The presence or absence of light depends on the state of the relay contact A. The state of the output device is represented by f_o, which stands for the output function. In this case, the output function (f_o) represents the condition of the light (the output device), which is represented as light lighted = 1, light dark = 0. That is, when the light is lighted, $f_o = 1$; when the light is dark, $f_o = 0$.

The conditions in the table of Figure 13.17(b) define the operation of the circuit. This type of table is called a *truth table*. The truth table contains information on the state of

- The input devices (switch and relay contacts);
- The output devices (lights, contactors, etc.);
- The control devices. Control devices include the mechanisms that actuate the input devices, which may include relay coils, switch handles and pushbuttons, or thermostat bimetal strips.

Truth tables usually don't include the state of the relay coil, because the state of the relay coil is mirrored by the NO contact, which is always included. However, for emphasis and clarity, the state of the relay coil will be included in the truth tables in this section.

From the truth table of Figure 13.17(b), we learn that when contact A is open ($A = 0$) the light is dark ($f_o = 0$), and when contact A is closed ($A = 1$), the light is

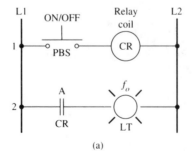

Control	Input	Output
Relay Coil State	Contact State	Light State
	A	f_o
OFF (0)	0	0
ON (1)	1	1

(a)   (b)

FIGURE 13.17

A ladder diagram and truth table for the control of a pilot light. (a) The presence or absence of light depends on the state of the relay and its relay contact (A), which is controlled by the ON/OFF pushbutton switch. (b) The truth table is a tabulation of the operation of the control circuit. Notice that when the relay coil is ON, the contact is closed ($A = 1$) and the light is lighted ($f_o = 1$).

lighted ($f_o = 1$). A logical equation that represents the direct relationship that exists between the state of the relay contact and the condition of the light in the diagram is

$$f_o = A \qquad\qquad (13.1)$$

where

f_o = output function, where 1 = lighted and 0 = dark
A = state of the contact, where 1 = closed and 0 = open

Logic Operators

Only three basic logic operators are needed to develop all the logical decisions needed to control a sequential process. The three logical operations as pictured in Figure 13.18 are as follows:

The *AND operator* is indicated by any of the usual signs of multiplication, as in $A \cdot B, A \times B$, or AB, which is read "A AND B."
The *OR operator* is indicated by a plus sign ($+$), as in $A + B$, which is read "A OR B."
The *NOT operator* is indicated by a bar over the letter (called an overbar), as in $\overline{A}$, which is read "NOT A."

AND Function

The *AND function* depends upon two or more events occurring at the same time in the control circuit. When events occur in combination with one another, the desired output function will result. For example, a microwave oven will start when certain conditions are present in conjunction with one another. Let f_o = microwave oven

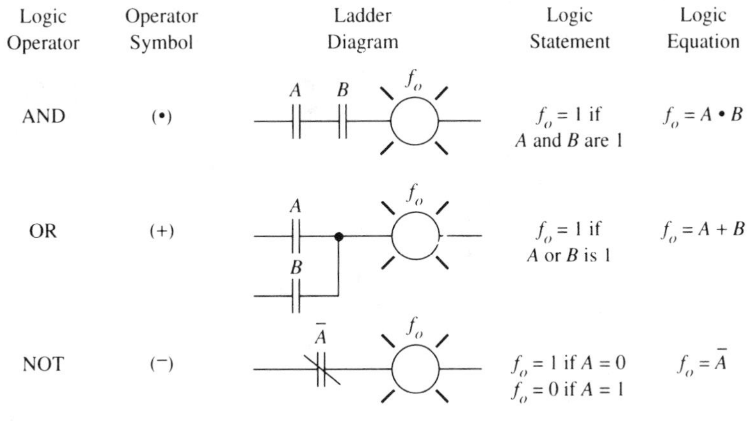

Logic Operator	Operator Symbol	Ladder Diagram	Logic Statement	Logic Equation
AND	($\cdot$)		$f_o = 1$ if A and B are 1	$f_o = A \cdot B$
OR	($+$)		$f_o = 1$ if A or B is 1	$f_o = A + B$
NOT	($^-$)		$f_o = 1$ if $A = 0$ $f_o = 0$ if $A = 1$	$f_o = \overline{A}$

FIGURE 13.18
The three basic logic operators (AND, OR, NOT) with their operator symbols, ladder diagrams, logic statements, and logic equations.

starts, D = door closed, T = timer set, P = power level selected, and S = start button depressed. Then

$$f_o = D \text{ AND } T \text{ AND } P \text{ AND } S$$

$$f_o = D \cdot T \cdot P \cdot S$$

That is, the microwave oven will start *if the door is closed and the timer is set and the power level is selected and the start button is depressed.*

In ladder diagrams, the AND function is seen as two or more open switch or relay contacts connected in *series* with each other. In Figure 13.19(a), the pilot light will light only when both sets of relay contacts are closed at the same time. The truth table of Figure 13.19(b) confirms this fact, since it also shows that the light will light only if both sets of contacts are closed. When $A = 1$ (closed) *and* $B = 1$ (closed), then $f_o = 1$ (lighted). Observe that when either one or both of the contacts are open, then the light is dark ($f_o = 0$). The logical equation that represents the condition for lighting the light is

$$f_o = A \text{ AND } B$$

$$f_o = A \cdot B \tag{13.2}$$

The output function from an AND circuit is ON (1) only when all the input device's logic states are 1 (closed) at the same time. If one or more input states are open (0), then the output function of the AND circuit is OFF (0).

OR Function

The *OR function* allows for a choice between two or more events occurring at the same time. For example, the courtesy light in a two-door automobile will light when either the left door or the right door is opened (causing the door switch to be closed) or when the interior light switch is activated (closed). Let f_o = courtesy light

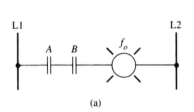

Control		Input		Output
Relay Coil State		Contact State		Light State
A	B	A	B	f_o
Off	Off	0	0	0
Off	On	0	1	0
On	Off	1	0	0
On	On	1	1	1

(a)   (b)

FIGURE 13.19
The AND function: (a) Two or more normally open contacts connected in series with one another represent the AND function in a ladder diagram. (b) The truth table shows that the light will light only if both contacts are closed ($A = 1$ and $B = 1$). Observe that when either one or both of the contacts is open, then the light is dark ($f_o = 0$).

lighted, L = left door open, R = right door open, and S = interior switch closed. Then,

$$f_o = L \text{ OR } R \text{ OR } S$$

$$f_o = L + R + S$$

That is, *the courtesy light will be lighted if the left door is opened or the right door is opened or the interior switch is closed.*

In a ladder diagram, the OR function is seen as two or more open switch or relay contacts connected in *parallel* with each other. In Figure 13.20(a), the pilot light will light when either one of the sets of relay contacts is closed. The truth table of Figure 13.20(b) shows that the OR function includes the possibility of both sets of contacts being closed (1). This type of OR operator is *inclusive* in that it includes the possibility of one or both of the sets of contacts being closed. The term OR means an inclusive OR. The logical equation that represents the conditions for lighting the light is

$$f_o = A \text{ OR } B \text{ OR } A \text{ AND } B$$

$$f_o = A + B + A \cdot B \Rightarrow A + B \tag{13.3}$$

The output function from an OR circuit is OFF (0) only when all the input device's logic states are 0 (open) at the same time. If one or more input states are closed (1), then the output function of the OR circuit is ON (1).

Table 13.4 has been included to give a quick reference to the rules for the AND function (series-connected switch or relay contacts) and the OR function (parallel-connected switch or relay contacts).

XOR Function

The *exclusive OR (XOR) function* provides for one or the other of the input devices to be closed (1) in order for the desired output function to be energized ($f_o = 1$). If

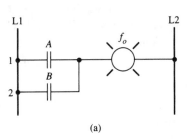

Control		Input		Output
Relay Coil State		Contact State		Light State
A	B	A	B	f_o
Off	Off	0	0	0
Off	On	0	1	1
On	Off	1	0	1
On	On	1	1	1

(a) (b)

FIGURE 13.20
The OR function: (a) Two or more normally open contacts connected in parallel with one another represent the OR function in a ladder diagram. (b) The truth table indicates that the light will light when one or both contacts are closed ($A = 1$ or $B = 1$). The pilot light is dark ($f_o = 0$) only when both contacts are open.

TABLE 13.4 Logic Rules for Contacts and Switches

Logic Statement	Meaning of the Logic Statement	Equivalent Contact Circuit
AND	SERIES CONNECTIONS	
$0 \cdot 0 = 0$	An *open* in series with an *open* is *open*.	
$0 \cdot 1 = 0$	An *open* in series with a *closed* is *open*.	
$1 \cdot 1 = 1$	A *closed* in series with a *closed* is *closed*.	
$A \cdot \overline{A} = 0$	A contact in series with its NOT is *open*.	
OR	PARALLEL CONNECTIONS	
$0 + 0 = 0$	An *open* in parallel with an *open* is *open*.	
$0 + 1 = 1$	An *open* in parallel with a *closed* is *closed*.	
$1 + 1 = 1$	A *closed* in parallel with a *closed* is *closed*.	
$A + \overline{A} = 1$	A contact in parallel with its NOT is *closed*.	

both inputs are closed at the same time, then the XOR function *excludes* the output function from being ON.

In ladder diagrams, the XOR function is seen as two sets of SPDT switch or relay contacts connected in *series-parallel* with each other, as illustrated in Figure 13.21(a). The truth table of Figure 13.21(b) shows that the XOR function will permit the light to be lighted when contacts A AND NOT B are both closed OR when contacts NOT A AND B are both closed. A logical equation that represents this statement is

$$f_o = A \cdot \overline{B} + \overline{A} \cdot B \qquad (13.4)$$

The exclusive OR function may be used as a *cross interlock* between two circuits when one or the other (but not both) output devices must be active. An example for the use of the XOR function is in a standby power system, where the *standby power* comes from a diesel-electric alternator, which is started when city power is lost. As illustrated in Figure 13.22, each ladder diagram contains one of the two logic conditions that produces an output from the XOR circuit. In the *city power control circuit*, it is the function A AND NOT B, whereas in the *standby power control circuit*, it is the function NOT A AND B.

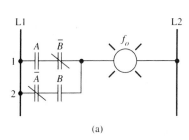

Control		Input		Output
Relay Coil State		Contact State		Light State
A	B	$\bar{A}\ \bar{B}$	$A\ B$	f_o
Off	Off	1 1	0 0	0
Off	On	1 0	0 1	1
On	Off	0 1	1 0	1
On	On	0 0	1 1	0

(a)   (b)

FIGURE 13.21
The exclusive OR (XOR) function: (a) The light will be lighted when contact A is closed and contact NOT B is closed (rung 1) OR it will be lighted when contact NOT A is closed AND contact B is closed (rung 2). However, the light will not be lighted if both contacts A and B are closed or if both contacts A and B are open. (b) The truth table for the XOR function.

FIGURE 13.22
A standby power system with *cross interlocks* between (a) the city power control circuit and (b) the standby power control circuit. The XOR function is used to form the interlock.

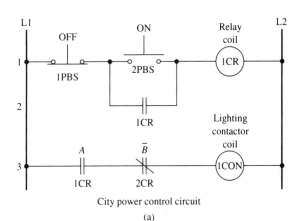

City power control circuit

(a)

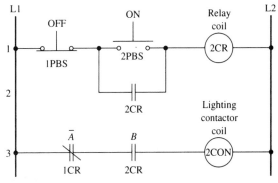

Standby power control circuit

(b)

When the city power is on, the lighting system operates from the contactor (1CON) in the city power circuit and the standby power contactor (2CON) is locked out, since the NC contact (1CR) in rung 2 of the standby power circuit is open. When city power is lost, the standby power is switched on by depressing the ON pushbutton in rung 1 of the standby power circuit, thus energizing the standby contactor (2CON). The city power is locked out, since the NC contact (2CR) in rung 2 of the city power circuit is open. If for some reason both power systems were active at the same time and both control circuits were energized, then both systems would be locked out.

NOT Function

The *NOT operator* is also called the *inversion operator*, because the logic is inverted. That is, the AND and OR function examines (looks at) the NO input devices (switch contacts and relay contacts) to determine their states. However, when the logic is inverted, the NC contacts are examined instead of the NO contacts. This results in a complementary logic called NAND and NOR logic. NAND is a contraction of NOT-AND and NOR is a contraction of NOT-OR.

➤ **As a Rule** With NAND and NOR logic, the NC contacts are examined.

Figure 13.23(a) is a ladder diagram illustrating the NOT function. The diagram consists of an NC relay contact in series with a pilot light. The truth table of Figure 13.23(b) illustrates the inverse nature of the NOT function.

The truth table shows that the pilot light is lighted ($f_o = 1$) when the relay coil is de-energized, the NO contact is open ($A = 0$), and the NC contact is closed ($\overline{A} = 1$). The pilot light is dark ($f_o = 0$) when the relay coil is energized, the NO contact is closed ($A = 1$), and the NC contact is open ($\overline{A} = 0$). The logical equation that represents the condition for lighting the light is

$$f_o = \overline{A} \tag{13.5}$$

Thus, if the output device is to be energized ($f_o = 1$), then *the relay coil must not be energized.*

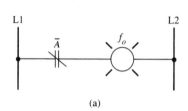

Control	Input	Output
Relay Coil State	Contact State	Light State
	$\overline{A}$	A	f_o
Off	1	0	1
On	0	1	0

(a)   (b)

FIGURE 13.23
The NOT function: (a) The NOT function is represented as a normally closed contact. (b) The inverse nature of the NOT function is seen in the truth table. The light is lighted when the relay coil is OFF and the normally open contact (A) is open.

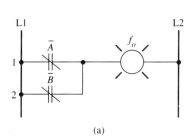

Control		Input		Output
Relay Coil State		Contact State		Light State
A	B	$\overline{A}\ \overline{B}$	$A\ B$	f_o
Off	Off	1 1	0 0	1
Off	On	1 0	0 1	1
On	Off	0 1	1 0	1
On	On	0 0	1 1	0

(a)   (b)

FIGURE 13.24
The NAND function: (a) Two or more normally closed contacts connected in parallel with one another represent the NAND function in a ladder diagram. (b) The truth table indicates that the light will light when one or both normally closed contacts are closed ($\overline{A} = 1$ OR $\overline{B} = 1$ OR $\overline{A}$ AND $\overline{B} = 1$; then $f_o = 1$). The pilot light is dark ($f_o = 0$) only when both normally closed contacts are open.

NAND Function

In a ladder diagram, the *NAND function* appears as two or more NC switch or relay contacts connected in *parallel* with each other. As noted in Figure 13.24(a), the pilot light will go dark only when both contacts are opened. If either one or both of the sets of NC relay contacts remains closed ($\overline{A} = 1$ or $\overline{B} = 1$), then the light will remain lit. The truth table of Figure 13.24(b) confirms this observation. When reading the truth table, remember that $A = 0$ when $\overline{A} = 1$ and vice versa. The logical equation that represents the conditions for lighting the light is

$$f_o = \overline{A} \text{ OR } \overline{B} \text{ OR } \overline{A} \text{ AND } \overline{B}$$
$$f_o = \overline{A} + \overline{B} + \overline{A} \cdot \overline{B} \Rightarrow \overline{A} + \overline{B} \qquad \textbf{(13.6)}$$

Figure 13.25 pictures a ladder diagram that incorporates two NC float switches (FS) connected in parallel to form a NAND function. The NAND function is used to

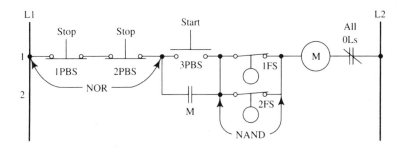

FIGURE 13.25
A ladder diagram for controlling a pump motor that is used to fill two tanks. The two normally closed float switches connected in parallel form a NAND function. The two normally closed pushbutton switches form a NOR function.

control the filling of two reagent tanks (the main and the auxiliary tank) that hold product for a plastic-making process. The two tanks have interconnecting pipes that are arranged in such a manner that the main tank fills first. Once the main tank is full, then an overflow pipe carries product to the auxiliary tank. The pump is controlled by a float switch in each tank, and each tank has a separate stop pushbutton switch, which may be used by an operator to stop the pump manually if necessary. Once the pump is started, the tanks will continue to fill until both the float switches are opened, signaling that each tank is full. With the two float switches open, the circuit controlling the motor starter of the pump will also be open, and the pump will be de-energized.

NOR Function

In a ladder diagram, the *NOR function* is pictured as two or more NC switch or relay contacts connected in *series* with each other. In Figure 13.26(a), the pilot light will light only when both sets of NC relay contacts are closed at the same time. The truth table of Figure 13.26(b) shows that when either one or both of the normally closed contacts are open, then the light is dark ($f_o = 0$). Once again, remember when reading the truth table that $A = 0$ when $\overline{A} = 1$ and vice versa. The logical equation that represents the condition for lighting the light is

$$f_o = \overline{A} \text{ AND } \overline{B}$$
$$f_o = \overline{A} \cdot \overline{B} \tag{13.7}$$

Besides incorporating a NAND function in its operation, the pump motor control circuit of Figure 13.25 also uses two NC pushbuttons to form a NOR function to open the pump motor starter circuit. If either of the NC series-connected STOP pushbutton switches is depressed while the pump is operating, then the pump motor starter will be de-energized and the pump motor will stop.

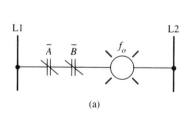

Control		Input		Output
Relay Coil State		Contact State		Light State
A	B	$\overline{A}\,\overline{B}$	$A\,B$	f_o
OFF	OFF	1 1	0 0	1
OFF	ON	1 0	0 1	0
ON	OFF	0 1	1 0	0
ON	ON	0 0	1 1	0

(a)   (b)

FIGURE 13.26
The NOR function: (a) Two or more normally closed contacts connected in series with one another represent the NOR function in a ladder diagram. (b) The truth table shows that the light will light only if both normally closed contacts are closed ($\overline{A} = 1$ AND $\overline{B} = 1$; then $f_o = 1$). Observe that when either one or both of the normally closed contacts are open, then the light is dark ($f_o = 0$).

TABLE 13.5 Logic Functions

Logic Function	Ladder Diagram	Truth Table	Logic Gate Symbol
AND	$f_o = A \cdot B$	$\begin{array}{cc\|c} A & B & f_o \\ \hline 0 & 0 & 0 \\ 0 & 1 & 0 \\ 1 & 0 & 0 \\ 1 & 1 & 1 \end{array}$	$A \cdot B$
OR	$f_o = A + B$	$\begin{array}{cc\|c} A & B & f_o \\ \hline 0 & 0 & 0 \\ 0 & 1 & 1 \\ 1 & 0 & 1 \\ 1 & 1 & 1 \end{array}$	$A + B$
XOR	$f_o = \bar{A} \cdot B + A \cdot \bar{B}$	$\begin{array}{cc\|c} A & B & f_o \\ \hline 0 & 0 & 0 \\ 0 & 1 & 1 \\ 1 & 0 & 1 \\ 1 & 1 & 0 \end{array}$	$\bar{A} \cdot B + A \cdot \bar{B}$
NOR	$f_o = \bar{A} \cdot \bar{B}$	$\begin{array}{cc\|c} A & B & f_o \\ \hline 0 & 0 & 1 \\ 0 & 1 & 0 \\ 1 & 0 & 0 \\ 1 & 1 & 0 \end{array}$	$\bar{A} \cdot \bar{B}$
NAND	$f_o = \bar{A} + \bar{B}$	$\begin{array}{cc\|c} A & B & f_o \\ \hline 0 & 0 & 1 \\ 0 & 1 & 1 \\ 1 & 0 & 1 \\ 1 & 1 & 0 \end{array}$	$\bar{A} + \bar{B}$
NOT	$f_o = \bar{A}$	$\begin{array}{c\|c} A & f_o \\ \hline 0 & 1 \\ 1 & 0 \end{array}$	$\bar{A}$

To summarize, relay and switch logic has been in use for many years, and it is still the choice for simple process control systems. However, in recent years, the *programmable logic controller* (PLC) has replaced relays as the principal form for process control. Table 13.5 is designed to help you understand the logic functions. Included in this table are the various logic functions, along with their ladder diagrams, truth tables, and logic gate symbols.

13.5 PROCESS-CONTROL CIRCUITS

As you now know, an input device initiates the sequence of events in the control of a process by the change in state of a switch or relay contact. Once the process is started, then the sequence of each event in the process is controlled by the logic functions of the *decision* part of the control circuit, which in turn enables the output device(s) to carry out the desired action.

When troubleshooting a control circuit, you, as a technician, must be aware of the three-part structure of the ladder diagram. In each of the following ladder diagrams, first identify the individual events that make up the sequence of events in the process control, then follow each event through the input, decision, and output segments of the circuit.

Hot-Air Dryer

The ladder diagram of Figure 13.27 is the heater part of the control circuit for an industrial hot-air dryer. The distinguishing characteristic of this part of the control circuit is the use of an ON-delay relay to bypass the low-temperature thermostat during the initial heating cycle at the start-up of the system.

Once operational, the temperature of the hot air out of the system centers around the setting of the high-temperature thermostat that controls the heater contactor. The overload contacts of the blower motor thermal relay are placed in series with the control relay in rung 1.

Liquid-Heating System

Before liquid is added to other ingredients in a food manufacturing process, a measured amount of the liquid is heated in a vat. The metering of the liquid is achieved through the use of the float switch illustrated in the ladder diagram of Figure 13.28.

Once the system is energized, the liquid is allowed to flow into the vat (through the solenoid-controlled inflow valve in rung 3) until the float switch in rung 4 is closed and the float switch in rung 3 is opened. The closing of the float switch in rung 4 signals that the vat has the desired amount of liquid and the liquid is ready to be heated.

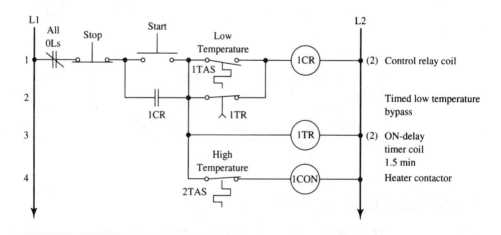

FIGURE 13.27
The ladder diagram for the control of the heater portion of a hot-air dryer.

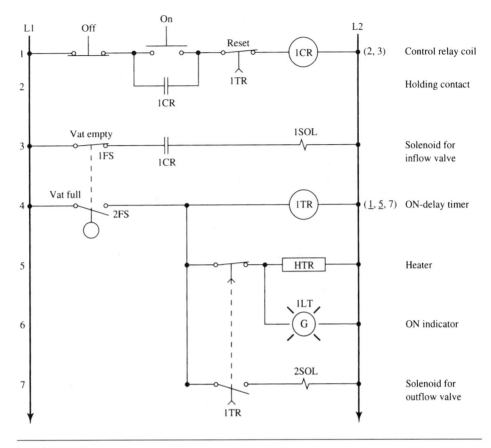

FIGURE 13.28
The ladder diagram for the control of a heating system to heat liquid. The inflow and outflow valves are solenoid-controlled.

The heating cycle is timed by the ON-delay timer, which controls the heater through the NCTO contact in rung 5. Once the timer times out, the contacts in rungs 5, 7, and 1 change state. With the change in state, the heater is off; the outflow solenoid in rung 7 is energized, allowing the liquid to exit the tank; and the system is reset (shut down) by the opening of the timed contact in rung 1, which releases the holding contact in rung 2. Once the liquid level in the vat drops, the float switch in rung 4 opens, resetting the ON-delay timer to its de-energized state.

Material-Handling System

When an automatic material-handling system is made up of two or more powered elements, the system should be interlocked to sequence the start-up of the system. A method for shutting down the system when jams or backups occur is also an important consideration in this type of material-handling system. Possible damage to the product being transported can be avoided by quickly detecting obstructions in the material-handling system.

The three-belt conveyor system of Figure 13.29, used in conjunction with an automated assembly line, employs ON-delay timers and interlocks to sequence the starting of the system. Also, photoelectric control is used to detect backups on the conveyor.

The *photoelectric switch* (PES) incorporates a visible, or IR, light-source transmitter, a receiver to detect remotely (without touching) the presence or absence of product on the conveyor, and a set of switchable contacts. A control signal, created by the solid-state electronic circuit of the PES, is used to trigger the set of solid-state contacts open or closed. The photoelectric switch has a number of standard features, including *instantaneous closing contacts* and ON- and OFF-delay *timed closing contacts*.

The conveyor system of Figure 13.29 has a photoelectric transmitter/receiver physically located along each conveyor to detect blockages, backups, or jams in the system. To prevent nuisance stopping of the conveyor, the ON-delay feature in each of the photoelectric control circuits is set for a 7-s delay. As long as products move along the assembly line in their usual 5-s intervals, the photoelectric switch does not change state.

With the PES active (1TGS closed in rung 6, turning the light beams ON), the PES contacts in rung 1 go closed. When a jam occurs, the light beam is interrupted (broken) for more than 7 s, causing the PES in rung 6, 7, or 8 to change state and open the PES contacts in rung 1. When an obstruction occurs, one of the *conveyor-jammed annunciator lights* will light and a horn will sound, signaling that the conveyor system is blocked. As a final reminder, all relay contacts, including those of the photoelectric switch, are shown in the de-energized state in a ladder diagram.

Machine-Control System

Figure 13.30 illustrates a machine tool that has its spindle controlled with a solenoid-activated clutch. The table of the machine is moved by hydraulic power provided by the hydraulic pump. The table travel and the direction of travel are controlled by a three-position hydraulic valve. The direction of travel is determined by which of the two solenoids is energized.

The valve symbol pictures three position blocks between the two solenoids. The valve is shown in the OFF position (table stopped), with the fluid bypassed around the hydraulic cylinder. By energizing one or the other of the solenoids, the valve ports are moved either left or right. Depending on which solenoid is active, the valve ports are aligned to permit fluid either to enter the cylinder on the left and exit on the right or to enter on the right and exit on the left. Thus, the cylinder ram may be moved either left or right. The stops on the table limit the travel by tripping limit switches. The ladder diagram for the control of the table travel and the operation of the spindle is pictured in Figure 13.31.

Study both Figures 13.30 and 13.31 to determine the operation of the system. Remember first to identify the individual events that make up the sequence of events being controlled and then to follow each event through the input, decision,

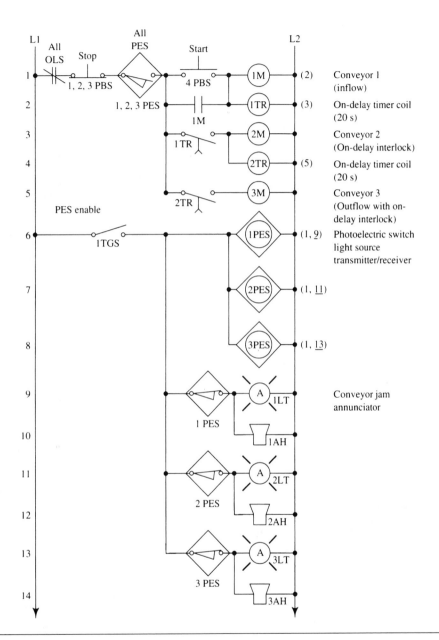

FIGURE 13.29
A conveyor system with interlocks and ON-delay for sequential start and photoelectric switches to detect a blockage in the system.

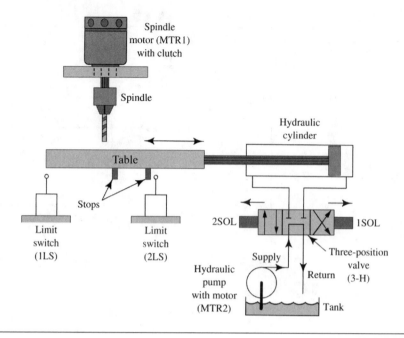

FIGURE 13.30
A pictorial of a machine tool that has its table moved by hydraulic power.

and output parts of the control circuit. The rungs of the ladder are read as logic statements. For example, to energize motors 1 (MTR1) and 2 (MTR2), the master start must be closed AND the master stop must be closed AND both overloads must be closed ($f_o = 2PBS \cdot 1PBS \cdot OLs1 \cdot OLs2$).

CHAPTER SUMMARY

- A ladder diagram is used to represent the sequence of discrete state events in a sequential process.
- Graphic symbols and device designators are used to represent hardware in the ladder diagram.
- An understanding of conventions and practices is needed when reading ladder diagrams.
- The auxiliary contact on the contactor is used in three-wire control circuits to latch the contactor coil on.
- Two powered circuits may be cross interlocked to prevent both motors from being operated at the same time.
- The ON-delay relay provides the time delay after the coil is energized.
- The OFF-delay relay provides the time delay after the coil is de-energized.
- The three basic logic operators are the AND, OR, and NOT; the NAND and NOR result when the NOT is applied to the AND and OR.
- The OR function is inclusive, whereas the XOR is exclusive.

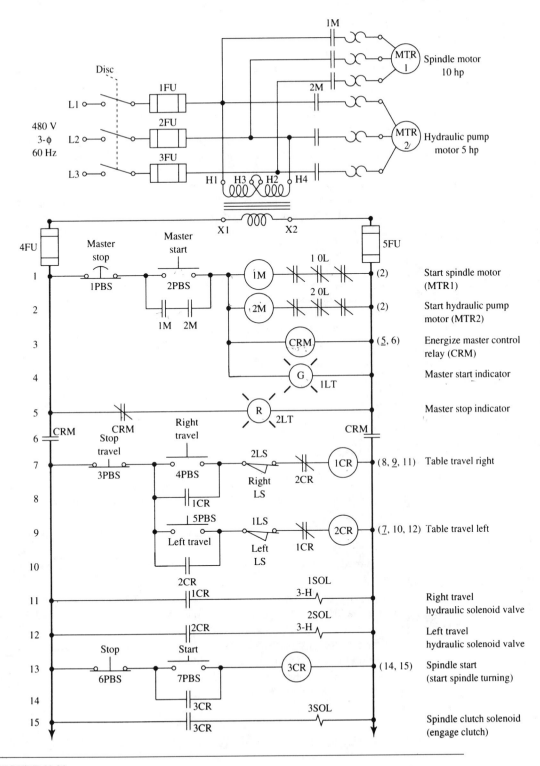

FIGURE 13.31
The ladder diagram for the control of the table travel and the operation of the spindle of the machine tool pictured in Figure 13.30.

- The AND function is seen in the ladder diagram as two or more series-connected switch or relay contacts; the OR function is seen as two or more parallel-connected switch or relay contacts.
- The NC contacts are *examined* with NAND and NOR logic. The NO contacts are *examined* with AND and OR logic.
- The *photoelectric switch* (PES) uses light to detect remotely the presence or absence of an object and to produce a control signal using solid-state circuitry to control a set of contacts.
- When troubleshooting a control circuit, first identify the individual events that make up the sequence of events in the process control and then follow each event through the input, decision, and output segments of the circuit.

SELECTED TECHNICAL TERMS

The following technical terms, abbreviations, and acronyms are defined in the glossary located after Chapter 16. You are encouraged to use the glossary to aid your understanding and to test your knowledge of these important terms.

Boolean algebra	PLC
EMI	pole
MOV	sequential
optocoupler	throw

END-OF-CHAPTER QUESTIONS

Write T if the statement is true and F if the statement is false.

1. The state of the relay contacts is shown in the ladder diagram with the relay coil de-energized.
2. The steps in a sequential process are started and stopped by either an internal event or an external timer.
3. A metal-oxide varistor is often used with a solid-state relay to speed up its operation.
4. The contactor coil is designated by CON on a ladder diagram.
5. The overload (OL) relay contacts are shown on the right side of the ladder diagram, connected in parallel with the output device.
6. With the START pushbutton depressed, the conveyor system of Figure 13.29 will start without first closing the toggle switch (1TGS) in rung 6.
7. A line beneath a number in parentheses on the right side of the ladder diagram is an indication of where the NC relay contact is located.
8. A normally open contact in parallel with its own normally closed contact in a ladder diagram will always result in a closed circuit.
9. In three-wire control circuits, the STOP pushbutton switches are always wired in series with the output device.
10. A NOT function is seen in a ladder diagram as a normally open contact in series with the output device.

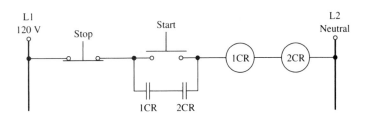

FIGURE 13.32
Ladder diagram for end-of-chapter question 19.

In the following, select the word or words that make the statement true.

11. The SSR is available as a form A relay with contacts that are (SPST, SPDT, DPST).
12. An OFF-delay timing relay is used to start a motor 30 s after the timer is turned off by placing a set of (NOTC, NCTO, NOTO, NCTC) contacts in series with the motor starter.
13. A closed switch or relay contact is noted in a truth table as a (0, 1).
14. The SPDT relay contact set is made up of one NC contact, one NO contact, and (two, three or more, one) armature(s) (ARM) with two contacts.
15. The NAND function is seen in a ladder diagram as two or more normally (open, closed) contacts in (parallel, series) with each other.

Answer each of the following questions with a short answer in the form of a complete sentence. Include a restatement of the question in your answer.

16. What is compelled to be done by the operator when a compelling control circuit is in use?
17. In Figure 13.27, why are the blower motor thermal overload contacts placed in series with the control relay coil?
18. How does the exclusive OR (XOR) differ from the inclusive OR?
19. Why won't the two identical 120-V relay coils function properly when they are connected in series, as pictured in Figure 13.32?
20. In a reversing circuit, how can the forward and reverse contactors be prevented from both being energized at the same time?

END-OF-CHAPTER PROBLEMS

Solve the following problems. Structure your work so it follows in an orderly progression and can easily be checked. Tables 13.4 and 13.5 summarize the rules and logic functions used in Chapter 13.

1. For the ladder diagram of Figure 13.33,
   (a) Complete the truth table.
   (b) Does $f_o = A \cdot B$ represent the condition for energizing the relay coil?
2. The ladder diagram of Figure 13.34 pictures a light that is controlled by two switches from two positions.
   (a) Complete the truth table for the output function.
   (b) Does $f_o = A \cdot B + \overline{A} \cdot \overline{B}$ represent the conditions for the light to be lighted?

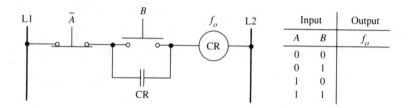

Input		Output
A	B	f_o
0	0	
0	1	
1	0	
1	1	

FIGURE 13.33
Ladder diagram and truth table for end-of-chapter problem 1.

3. For the circuit of Figure 13.29, how long after the first conveyor is started does the third conveyor start? Each timer has 20 s of ON-delay.

4. Complete the three-wire control wiring diagram of Figure 13.35 so it seals the single-phase, 230-V magnetic motor starter when the START pushbutton switch is depressed.

5. Draw a ladder diagram for the three-wire, motor-control circuit of Problem 4 and add a pneumatic pressure switch that will prevent the motor from starting until there is sufficient air pressure in the system. Also include an additional STOP push-button switch.

6. Develop a ladder diagram to provide an ON-delay of 2 min using two 1-min, fixed, ON-delay timers to control a solenoid. Use a toggle switch (ON/OFF) to initiate the timing sequence.

7. Develop a ladder diagram to control the dispensing of one of three kinds of canned soft drinks from a vending machine. A dispensing solenoid is energized when coins have been inserted (which activates a proximity switch) and one of the three selection pushbuttons has been depressed. Make provisions in the circuit to prevent more than one can from being dispensed when more than one selection pushbutton is depressed.

8. For the circuit in Figure 13.36, the motor of the cooling pump is kept operating for 2 min after the ON/OFF toggle switch is turned OFF and the temperature activated switch (TAS) has opened. The ladder diagram of Figure 13.36 illustrates how the TAS, sensing the coolant temperature, is used to signal a timing relay (TR) to start timing and turn off the pump motor starter.

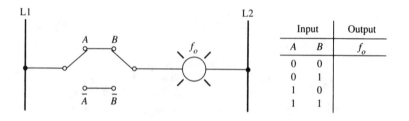

Input		Output
A	B	f_o
0	0	
0	1	
1	0	
1	1	

FIGURE 13.34
Ladder diagram and truth table for end-of-chapter problem 2.

FIGURE 13.35
Wiring diagram for end-of-chapter problem 4.

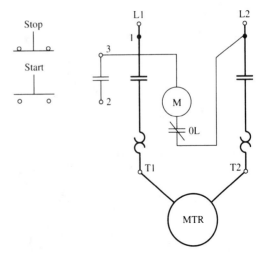

Using either an outline form or a paragraph form, explain the sequence of events *from* the initial closing of the ON/OFF toggle switch (and coolant heating) *to* when the ON/OFF toggle switch is opened and the pump motor finally shuts down. You may assume that the coolant is cool enough at start-up and that the TAS is open. The TAS will close once the coolant reaches 145°F, and it will open when the coolant reaches 125°F.

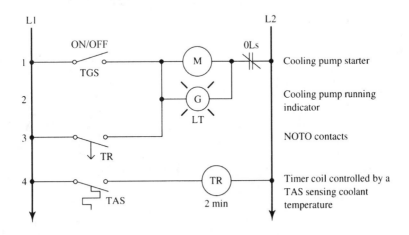

FIGURE 13.36
Ladder diagram for end-of-chapter problem 8.

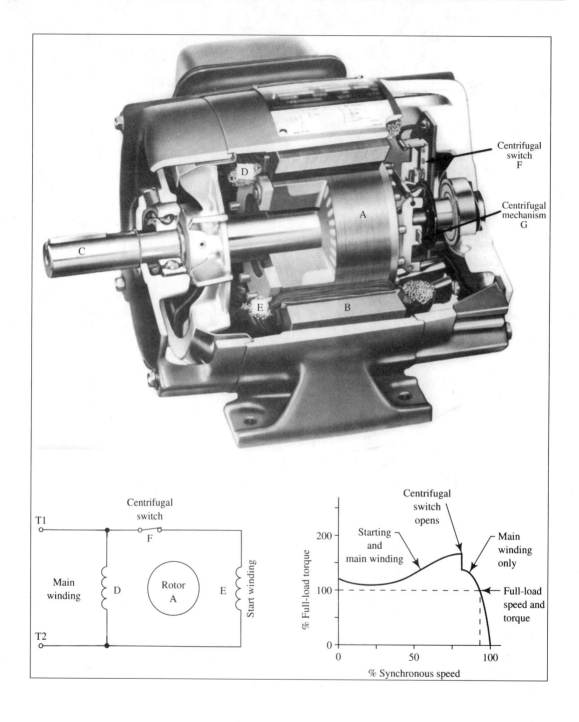

Centrifugal
switch
F

Centrifugal
mechanism
G

A

C

D

E

B

Centrifugal
switch

T1

F

Main
winding

D

Rotor
A

E

Start winding

T2

Centrifugal
switch
opens

Starting
and
main winding

Main
winding
only

200

100

0

% Full-load torque

Full-load
speed and
torque

0   50   100

% Synchronous speed

Chapter 14

Alternating-Current Electric Motors

About the Illustration. *As noted in the schematic diagram, the* split-phase squirrel-cage induction motor *is designed to start and operate on single-phase current by providing two windings to create a rotating magnetic field. The two windings (DE) wound on stator core (B) are the* main winding *and the* start winding. *The start winding, which has more resistance than the main winding, is switched out of the motor circuit by the* centrifugal switch *(F) once the motor reaches about 80% of its full-load speed (see the speed-torque diagram). The centrifugal switch is activated by the centrifugal force created by the rotating rotor shaft (C). The centrifugal switch (F) and its mechanism (G) are attached to the keyed rotor shaft (C) along with the rotor (A). (Courtesy of Magnetek-Century Electric Company.)* ■

INTRODUCTION

The fundamental operating principle of all electric motors is virtually the same. The rotation of the rotor is due to the interaction of the flux fields of the stator and the rotor, producing a net difference of force on the rotor conductors that tends to move them at right angles to the flux. Once a rotating magnetic field moves around the stator, the flux in the rotor will follow it, creating rotary motion. The rotating magnetic field needed to produce accelerating torque is produced by the resultant flux from the polyphase (two- and three-phase) currents in the stator (stationary) windings in combination with the electromagnetically induced currents and flux in the rotor conductors of ac single-phase and three-phase squirrel-cage induction motors. Table 14.1 lists many of the commonly used types of electric motors. Those in bold are studied in this chapter.

In addition to the motor principles, the nameplate parameters will also be investigated in conjunction with the meaning of key terms and concepts. The speed-torque characteristics of various induction motors (ac three-phase and ac single-phase) will be looked at, along with the electrical circuits and components needed for their control. Additionally, the principles related to the gearmotor will be covered.

CHAPTER CONTENTS

PERFORMANCE OBJECTIVES

Once you have read and studied each section; worked through each example with pencil, paper, and calculator; worked through the end-of-chapter problems; and answered the end-of-chapter questions, then you should be able to

- Distinguish between the wound-rotor and the squirrel-cage induction motor.
- Describe how the rotating magnetic field is produced in the three-phase and the single-phase induction motor.
- Determine the synchronous speed of the squirrel-cage induction motor.
- Understand the concept of slip.
- Differentiate among starting torque, breakdown torque, full-load torque, and no-load torque.
- Interpret and understand motor nameplate parameters.
- Identify and relate the speed-torque motor characteristic to the NEMA design type.
- Know why a reduced-voltage controller is used to start an induction motor.
- Explain the use of slip rings and the rotor rheostat to control the operating speed of the wound-rotor motor.
- Name the methods used to start single-phase squirrel-cage motors.
- Describe how a gearmotor is constructed and understand its application.
- Calculate the locked rotor current from the nameplate code letter.
- Specify the type of built-in thermal switch needed to protect a motor from overheating.
- List the types of motor enclosures.

14.1 THREE-PHASE INDUCTION MOTORS

Introduction

The *induction motor* is the most frequently used type of ac motor. It is both simple and hardy in its construction and when combined with its reliable operating characteristics, it requires very little maintenance. The induction motor consists of two major parts: the **stator** (stationary part) and the **rotor** (rotating part). The rotor is fastened to the motor's output shaft, which is supported by bearings.

Both single-phase and three-phase induction motors are widely applied in manufacturing plants, where they are used to power almost all the plant's machine tools and process machinery. Because of their reduced cost and smooth, noise-free operation, three-phase motors are the usual choice for industrial applications. Information on the inspection and maintenance of three-phase squirrel-cage motors is given in Appendix C, Section C10.

The two types of three-phase induction motors discussed in this section are the *squirrel-cage* and the *wound-rotor motors*. Both motors operate on the same basic principle and have the same stator construction; however, they differ in their rotor construction.

When an induction motor is connected to a three-phase ac line, it is the stator coils that are powered by the 60-Hz ac line. The rotor is not electrically connected to

TABLE 14.1 Classification of Selected Electric Motors

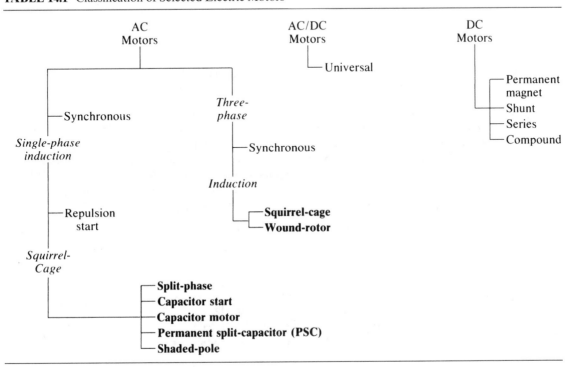

the ac line; it receives its electric current through *electromagnetic induction* from the transformer formed between the stator (primary) and the rotor (secondary), which leads to the name *induction motor*.

In an induction motor, it is the *transformer action* between the stator windings and either the rotor bars in the squirrel-cage type of rotor or the rotor windings in the wound rotor that sets up the current and flux in the rotor. The torque needed to spin the rotor results from the interaction between the stator's rotating magnetic field and the induced current and the resulting magnetic field of the rotor.

Rotating Magnetic Field

The operation of the induction motor (single-phase or three-phase) depends on the presence of a rotating magnetic field. In the three-phase motor, this field is established by the three-phase current in each of the phase windings in the stator. Because the phase currents are 120° apart, the magnetic fields associated with each phase are also 120° apart.

The simplified drawing of a stator winding of a three-phase, two-pole induction motor is pictured in Figure 14.1(a). With the stator attached to the three-phase line, the current in each winding changes with time (as illustrated in Figure 14.1(b)), creating a varying magnetic flux at each set of poles. Because of the phase difference in the current of the stator windings, as well as the connection of the windings into phase groups, the individual flux in each pair of poles is combined into a *resultant flux* (the total flux), which moves around the stator at a speed equal to the 60-Hz frequency of the current (3600 rev/min). Figure 14.1(c) pictures the movement of the resultant flux as it rotates around the surface of the stator poles. The resultant flux is the *rotating magnetic field* of the three-phase induction motor.

Synchronous Speed

The *synchronous speed* of the motor is the speed at which the rotating magnetic flux moves around the stator. In the two-pole stator winding (two poles per phase), represented by Figure 14.1(a), the rotating field makes one trip around the stator for each (one) cycle of current in the stator windings. The four-pole (per phase) stator winding has two sets of coils connected in series, which results in a rotating field that makes one trip around the stator for every two cycles of current in the stator windings. In general, the synchronous speed of the ac motor is

$$n_s = \frac{f \times 60}{p} \tag{14.1}$$

where

n_s = synchronous speed, in rev/min
f = frequency of the line current, in Hz
p = pairs of poles per phase; 2 poles = 1 pair
60 = conversion factor from seconds to minutes

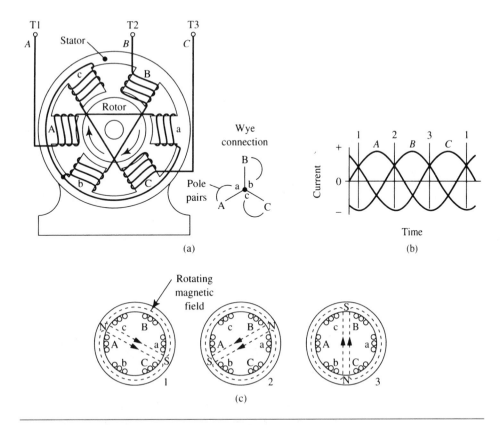

FIGURE 14.1
(a) A simplified representation of the stator winding (wye-connected) of a three-phase, two-pole induction motor. At any instant of time, a two-pole motor has one north and one south pole resulting from the rotating field. (b) The current in each phase changes with time, creating a varying magnetic flux at each set of poles. Each current waveform is 120° apart. (c) The individual flux at each pair of poles is combined into a resultant flux (the rotating magnetic field), which moves around the stator as a north and south pole. The numbers correspond to the numbered points along the time axis in (b).

Example
14.1

Determine the synchronous speed of a three-phase, four-pole motor operating from a 240-V, 60.0-Hz line.

SOLUTION

Find the synchronous speed when there are two pairs of poles per phase (four-pole motor = 2 pairs of poles).

Given

$$n_s = \frac{f \times 60}{p}$$

Evaluate

$f = 60.0$ Hz

$p = 2$

Substitute $n_s = \dfrac{60.0 \times 60}{2}$

Solve $n_s = 18\overline{0}0$ rev/min

Since the direction of rotation of the rotor is the same as the direction of the rotating magnetic field, the following is true.

> ➤ **As a Rule** The direction of the three-phase motor may be changed by interchanging any two of the three line wires, thereby changing the direction of the rotating magnetic field.

Torque

The *torque* needed to spin the rotor and move the load attached to the rotor shaft is developed when the rotating magnetic field of the stator *induces* current in the conductors in the rotor. The current flow in the rotor conductors produces a flux around the conductor, which interacts with the resultant flux from the stator's rotating magnetic field.

As illustrated in Figure 14.2, the magnetic field from the stator poles combines with the flux from the current-carrying rotor conductor to produce a *net right angle force* on the rotor conductors. It is this force that produces torque and rotates the rotor shaft and the load.

From the illustration of Figure 14.2(b), the *flux density* above and below the conductors is not constant. Because the field directions are aiding, the flux density is greater above the conductor on the left. For the same conductor, the flux density is less below the conductor, since the directions of the fields are opposing. As pictured, the force (*F*) is directed away from the region of *increased flux density* and toward the region of *lessened flux density*. The loop is rotated from the horizontal position to the vertical position.

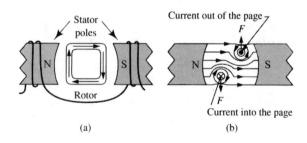

(a)   (b)

FIGURE 14.2
A pictorial representation of one pair of poles in an induction motor. (a) The stator poles with the induced current circulating in one of the rotor conductors as viewed from the top. (b) The current-carrying rotor conductor viewed head-on. Notice the conductor with a dot indicating the tip of the current arrow and the conductor with a cross indicating the tail of the current arrow. The stator flux is indicated by lines between the poles, whereas the circular lines represent the rotor flux.

> **As a Rule** A current-carrying conductor in a magnetic field will move at right angles to the field.

Slip

The rotor of the induction motor cannot operate at the same speed as the rotating magnetic field (the synchronous speed) because there would be no relative motion. Without relative motion between the stator and the rotor, there would be no transformer action and no induced current in the rotor conductors. Without current flow in the rotor, there would be no rotor magnetic field, no torque, and no shaft rotation.

> **As a Rule** To develop torque in any type of induction motor (even with no load), the rotor speed must always be less than the synchronous speed.

The difference between the rotor speed (shaft speed) and the synchronous speed is called **slip.** Slip is usually expressed as a percentage of the synchronous speed, as noted in the following equation.

$$\%\text{slip} = \frac{n_s - n_r}{n_s} \times 100\% \tag{14.2}$$

where

$\%\text{slip}$ = slip expressed as a percentage of the synchronous speed
n_s = synchronous speed, in rev/min
n_r = rotor speed, in rev/min

Example 14.2

Determine the slip in a two-pole, three-phase, 60.0-Hz, squirrel-cage induction motor that has a shaft speed of 3490 rev/min.

SOLUTION   First find the synchronous speed and then solve for the slip. *Note*: 2 poles = 1 pair.

Given
$$n_s = \frac{f \times 60}{p} = \frac{60.0 \times 60}{1} = 3600 \text{ rev/min}$$

$$\%\text{slip} = \frac{n_s - n_r}{n_s} \times 100\%$$

Evaluate $n_s = 3600 \text{ rev/min}$

$n_r = 3490 \text{ rev/min}$

Substitute $\%\text{slip} = \frac{3600 - 3490}{3600} \times 100$

Solve $\%\text{slip} = 3.1\%$

Squirrel-Cage Motor

The main parts of the *squirrel-cage induction motor* are the stator and rotor, as shown in Figure 14.3. The stator is made up of *laminations* (thin pieces) of sheet

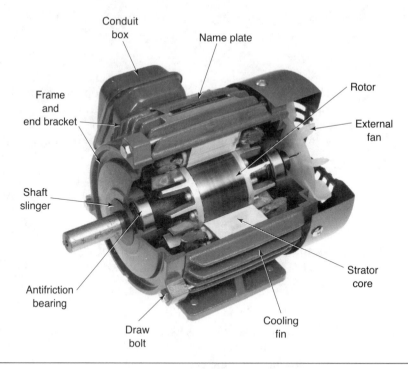

Conduit box

Name plate

Frame and end bracket

Rotor

External fan

Shaft slinger

Antifriction bearing

Strator core

Draw bolt

Cooling fin

FIGURE 14.3

A squirrel-cage induction motor with the stator and rotor noted. This motor has a totally enclosed fan-cooled (TEFC) enclosure. *(Courtesy of TECO American, Inc.)*

steel, which are stacked together to form the stator core. The core has slots to receive the windings of insulated wire, which form the *stator coils*. The coils are connected together (series or parallel) into *phase groups*, which make up the three phases of the stator. The phases may be connected to the line in either delta or wye.

The squirrel-cage *rotor core* is made of round, die-stamped pieces of sheet steel. The thin pieces of steel are stacked to form the laminated core, which has equally spaced slots in the perimeter running parallel to the shaft. Molten metal (aluminum, copper, etc.) is cast into the slots in the outside edge of the core to produce the conductors of the squirrel-cage rotor. The cast connecting *end rings*, which short-circuit the parallel conductors together, are also formed at the same time as the conductors. The shape of the rotor's conductive path, as pictured in Figure 14.4, is similar to a squirrel's (or hamster's) exercise wheel, which gives the name *squirrel cage*.

The conductors in the squirrel cage are not insulated, since they are much more conductive than the surrounding laminated steel. Thus, the induced rotor current passes through the parallel conductors and the end rings but not through the steel core. The slots in the rotor core are **skewed** to create parallel conductors that are shifted relative to the axis of the shaft. Skewing the rotor conductors improves both the smoothness (produces more uniform torque) and quietness (reduces hum) during the motor's operation.

FIGURE 14.4
An artist's representation of the
squirrel-cage rotor with the laminated
core removed. The round parallel
conductors are interconnected by the
end rings.

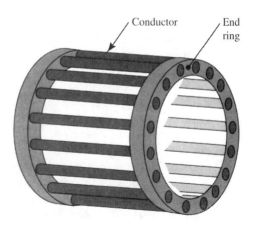

Squirrel-Cage Motor Speed-Torque Characteristics

The National Electrical Manufacturers' Association (NEMA) has adopted several
motor designs to include a wide variety in the *speed-torque characteristics* of the
squirrel-cage induction motor. Of these, designs B, C, and D are the most common.
The speed-torque characteristic for each of these three designs is pictured in Figure
14.5(a), whereas the key characteristics are indicated on the design B speed-torque
curve of Figure 14.5(b).

 The motor speed-torque curves of Figure 14.5 give a visual indication of how a
motor of a given design will behave as it accelerates up to speed while overcoming
the *driven weight* and *inertia* of the load. The maximum speed and torque in the normal operating range of the motor that produces the rated output power (hp) is
called the *full-load speed* (rev/min) and the *full-load torque* (lb-ft).

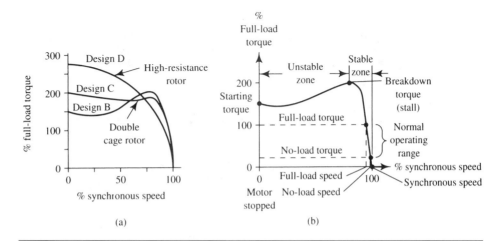

FIGURE 14.5
Speed-torque characteristics of the three-phase, squirrel-cage induction motor: (a) For the NEMA
designs B, C, and D; (b) With the key features noted; design B curve.

When viewing the curve of Figure 14.5(b), you might have noticed that the full-load speed is less than the synchronous speed. As was previously noted, this situation always occurs in induction motors, since a current (and the resulting flux and torque) is induced in the rotor only when there is a relative difference in speed between the stator-current frequency and the rotor-current frequency. **Full-load slip** is the term given to the difference between the synchronous and full-load speeds. Table 14.2 lists typical full-load speeds for two-, four-, six-, and eight-pole three-phase induction motors.

As previously shown, slip is an indication of how well the motor speed is regulated as well as how the motor might behave when driving cyclic (varying torque) loads. Slip is one of the parameters that varies among the NEMA design types.

The shapes of the curves of Figure 14.5(a) are not the same due to the variation of resistance (R) and inductance (L) in the design of the motor, which results in different starting torques and breakdown torques. The torque and current characteristics of typical three-phase, squirrel-cage induction motors are listed in Table 14.3 by NEMA design letters. Some of the terms used in conjunction with the squirrel-cage motor speed-torque characteristics include the following:

Starting torque The maximum torque produced by the motor at the instant the motor starts. The motor can start the load if the starting torque is greater than the load torque.

Breakdown torque The maximum torque that the motor can develop at rated voltage and frequency without stalling and dropping in speed. The breakdown torque may be used as an indicator of how the motor will respond to a sudden, momentary increase in load or a sudden drop in the terminal voltage. The motor will *stall* if a load greater than the breakdown torque is applied to the motor.

Rated torque The torque produced by the motor when it is continuously producing the rated output power while operating at the rated terminal voltage and rated line frequency.

Squirrel-Cage Motor NEMA Design Types

The difference in the characteristic curves of the three motor designs (B, C, and D) of Figure 14.5(a) is due to variations in the design of their rotors.

TABLE 14.2 Synchronous and Typical Full-Load Speeds of 60-Hz, Three-Phase Induction Motors

Poles	Full-Load Speed (rev/min)	Synchronous Speed (rev/min)
2	3500	3600
4	1750	1800
6	1150	1200
8	875	900

TABLE 14.3   Torque and Current Characteristics of Common 60-Hz, 1800-rev/min, 25-hp, Three-Phase, Squirrel-Cage Induction Motors

NEMA Design	Starting Torque (%)	Breakdown Torque (%)	Starting Current (%)	% Slip
B[a]	150	200	600	2–5
C	200	190	600	2–5
D	280	[b]	600[c]	12[d]

[a] NEMA design A is a variation of design B, not commonly used.
[b] Design D motors don't stall when overloaded; the torque curve has negative slope. Used for very high inertia and loaded starts.
[c] Smaller motors may have larger (higher %) starting current.
[d] Motors with higher % slip values.

Design B motors have low resistance (R) and low inductance (L) in the rotor, which gives them their characteristic moderate starting torque and their high breakdown torque. Because of these characteristics, design B motors are *general-purpose* motors, which are widely used with machine tools, fans, blowers, rotary pumps, and lightly loaded conveyor and material-handling systems.

Design C motors have high starting torque with medium breakdown torque. These properties are due to the *double-cage* construction of the rotor, as illustrated in Figure 14.6, where both rotor bars are placed in the same slot, one above the other. When compared to the design B motor of Table 14.3, this rotor design produces higher starting torque without an increase in starting current.

With double-cage construction, the design C motor operates with the advantages of a *high-resistance* outer rotor circuit during starting (high torque), and a *low-resistance, high-inductance* inner rotor circuit while running. In the design C rotor, the outer winding produces the high starting and accelerating torque, whereas the inner winding provides the running torque at good efficiency. Because of the variation in impedance due to the changing frequency in the rotor (f_r) between starting $(f_r = 60$ Hz) and running conditions $(f_r < 2$ Hz), the outer winding is the principal winding when starting, and the inner winding plays a lesser role. The roles are reversed when the motor is running.

Since the running characteristics of the design C motor are similar to those of the design B motor, the percent slip of less than 5% is the same, as is the breakdown torque of about 200%. With higher starting torque, the design C motors are used to power conveyors, reciprocating pumps, compressors, and machines with high inertial loads (flywheels, etc.).

FIGURE 14.6
Double-cage rotor used in NEMA design C squirrel-cage induction motors.

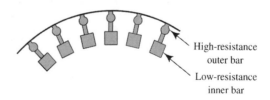

High-resistance outer bar

Low-resistance inner bar

Design D motors have very high starting torque and no breakdown torque, since the characteristic curve has *negative slope* due to the *very high resistance* in the rotor. This characteristic allows design D motors to start very heavy loads with high inertial loads. Once in operation, the design D motor can handle momentary overloads, since it has a high slip characteristic (>10%). Although less efficient than the other designs in operation, the design D motor is applied where loads vary widely, as with punch presses, shears, elevators, and oil-well pumps.

Starting Squirrel-Cage Induction Motors

Depending on the size (hp, kW) of the motor, the induction motor may be started by connecting the motor directly across the three-phase line using a *manual starter* (for smaller size motors of 10 hp or less, <7.5 kW) or with a *magnetic motor starter* for two-wire and three-wire control (remote starting) and/or when the load is greater than 7.5 kW (>10 hp).

Because some industrial loads (conveyors, printing presses, etc.) need to be started gradually or because the starting current would be too large, induction motors use *reduced-voltage starting*. By starting large motors with a reduced voltage, the starting current is lowered and the line voltage drop associated with starting large motors is minimized. For example, a 230-V, 50-hp, three-phase squirrel cage induction motor has a maximum *locked-rotor current* (starting current) of 725 A and a full-load current (maximum running current) of 130 A. Without reduced-voltage starting, this large motor would require oversize conductors, disconnects, and overcurrent devices (at increased expense). When starting larger motors, the severity of the resulting drop in the voltage of the feeder circuit(s) depends on the capacity of the motor circuit (wire size, transformer kVA, etc.).

Reduced-voltage starting is designed to minimize the disruption to the power system, to lessen mechanical shock to the drive train, and to provide a smooth start for the load. In addition to reducing the voltage and starting current, the starter also lowers the starting torque. The greatest starting torque is developed by the motor when it is started on full-line voltage.

> **As a Rule** The starting torque is proportional to the square of the terminal voltage. That is to say, one-half voltage results in one-quarter torque.

$$\%\tau = V_{pu}^{2} \times 100\% \qquad \textbf{(14.3)}$$

where

    $\%\tau$ = torque expressed as a percentage of the rated starting torque
    V_{pu} = reduced motor terminal voltage expressed as a per-unit value of the rated terminal voltage

As an example, when the motor's terminal voltage is reduced to 70% of its rated value (0.70 as a per-unit value), the starting torque is about 50% ($0.70^2 \times 100\% = 49\%$) of its full-voltage amount.

Reduced-Voltage Starting

One of several electrical circuits may be used to reduce the voltage. Among these are the *primary resistor starter*, the *autotransformer starter*, and the *wye-delta starter*.

Wye-delta Starters   Wye-delta starters use a specially constructed three-phase delta-wye squirrel-cage induction motor to reduce the voltage. When wye-connected at start-up, the starting voltage is 58% ($1/\sqrt{3}$). Once the motor is up to operating speed, the starter reconnects the windings into a delta-connection for full-voltage operation. Since the voltage is reduced to 58% of its rated value at starting, the starting torque is reduced to 33% $(1/\sqrt{3})^2$ of its rated starting torque; because the motor is *inductive* in nature at start-up, the starting current is also reduced to 33% of its value.

Example *14.3*	For a three-phase, 230-V, delta-wye motor started with a reduced voltage wye-delta starter, determine (a) The starting voltage. (b) The percent of available starting torque.
SOLUTION	**(a)** Find the phase voltage of the wye-connected motor when the line voltage is 230 V.
Given	From Equation 11.6, the voltage of the wye-connected motor windings is
	$$E_{\text{phase}} = \frac{E_{\text{line}}}{\sqrt{3}}$$
Evaluate	$E_{\text{line}} = 230 \text{ V}$
Substitute	$$E_{\text{phase}} = \frac{230}{\sqrt{3}}$$
Solve	$E_{\text{phase}} = 133 \text{ V}$
Observation	The voltage across the motor winding (the phase voltage) when wye-connected is 133 V.
	(b) Determine the % torque.
Given	$\%\tau = V_{\text{pu}}^2 \times 100\%$
Evaluate	$V_{\text{pu}} = \dfrac{133}{230} = 0.578$
Substitute	$\%\tau = 0.578^2 \times 100$
Solve	$\%\tau = 33\%$
Observation	The wye-delta, reduced-voltage controller switches the six wires of the delta-wye motor so that the wye connection is present at starting. This starts the motor at 58% of its rated terminal voltage and 33% of its normal starting torque and current.

Primary Resistance Starter The primary resistance starter is made up of a *magnetic motor starter* (contactor with overload relay) combined with one or more additional *contactors* coupled with the necessary *timing relay(s)* and *power resistors* (in series with the motor during start-up).

Figure 14.7 illustrates the power and control circuit for a *primary resistance reduced-voltage starter* used to electromagnetically start a three-phase squirrel-cage motor. Due to the *resistive* nature of this type of starter (unlike the inductive wye-delta starter), the current is reduced in direct proportion to the reduction in the voltage. Therefore, a design B motor started with 80% of the normal line voltage would have 80% of its starting current and 64% of its starting torque ($\%\tau = 0.80^2 \times 100\% = 64\%$).

In conclusion, *reduced-voltage starting* is used to start large squirrel-cage induction motors to lessen the *loading effect* of the motor. By reducing the voltage, the torque at starting is also reduced by the square of the voltage reduction. Because the starting current and starting torque are reduced, the time to accelerate the motor to full-load speed is increased. The transition time from START to RUN is typically 2 to 3 s per resistor section in the primary resistor starter, 8 to 10 s in the wye-delta starter, and 6 to 7 s in the autotransformer starter.

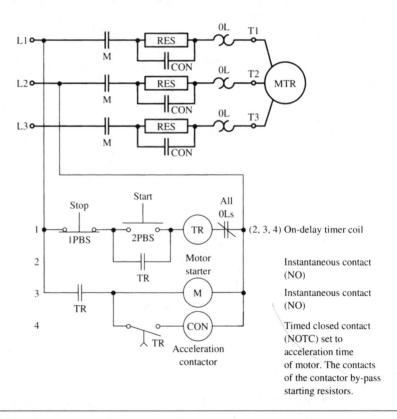

FIGURE 14.7
A primary resistance-type reduced-voltage starter that incorporates a resistor connected in series with each motor winding to produce a voltage drop and reduce the terminal voltage at the motor.

Wound-Rotor Motor

The wound-rotor, three-phase induction motor (also called a *slip-ring motor*) differs from the squirrel-cage induction motor in the design of the rotor, which includes *slip rings*, through which external resistance is introduced into the rotor circuit. The stator construction is the same in each motor. By using a wound-rotor rather than a squirrel-cage design, the induction motor becomes more flexible in its operation. Low rotor resistance for high efficiency under normal running conditions is combined with high resistance, high torque, and low starting current at start-up.

Both these optimum operating conditions are possible because of the design and operation of the wound-rotor. The rotor is wound with an insulated three-phase, wye-connected winding, with the end of each phase connected to a slip ring, as illustrated in Figure 14.8. Each slip ring has a *brush* (usually made of carbon) riding on it, which facilitates the connection of an external *motor resistor* in series with each of the three rotor windings. As pictured, the stator is delta-connected, whereas the rotor is wye-connected. The external variable resistance, called a **rheostat,** is also wye-connected. Details for the replacement of slip ring brushes and the maintenance of the slip rings and rheostat are given in Appendix C, Section C11.

Operation of the Wound-Rotor Motor

Under manual control, the rotor rheostat provides a means of varying the rotor-circuit resistance during the starting current and producing a variable starting torque. As the motor gains speed during acceleration, the resistance of the rotor rheostat (Figure 14.8) is reduced, thereby controlling the torque of the motor so that optimum torque/current is provided over the period of acceleration. Once the motor reaches full speed, the brushes are *short-circuited*, and the motor operates similar to a squirrel-cage motor. Because high torque can be maintained by adjusting the rotor resistance, the wound-rotor motor is used to start very high inertial loads.

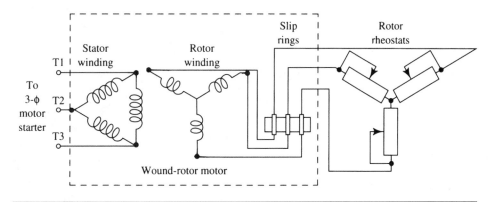

FIGURE 14.8

Circuit of the wound-rotor motor. The stator windings are delta-connected, but the rotor windings are wye-connected. The open end of each rotor phase winding is connected to a slip ring. The external rheostat is wye-connected to the rotor windings through the brushes and slip rings.

Speed Control of the Wound-Rotor Motor

Since the resistance in the rotor circuit can be varied during operation, a method is provided for controlling the speed of the motor. With an increase in rotor resistance under loaded operation, a given torque is created and the slip is increased, thus reducing the operating speed. This concept is illustrated in the speed-torque curves of Figure 14.9, which pictures three speed-torque curves for the wound-rotor motor, along with the load's speed-torque curve. The load is a blower (fanlike device), which, as pictured, has an increasing torque as it increases in speed. The increasing torque is due to the rise in work needed to move a larger volume of air through the blower as the speed increases.

Operating points are established at the intersections of the motor's curves with the blower's curve. Each operating point is indicated by a labeled dot. When the motor is operated with high resistance in series with the rotor, the point of operation is at the low-speed point. The speed of operation is approximately 75% of the synchronous speed. Likewise, when the motor is operated with a medium amount of series rotor resistance, then the speed is roughly 85% of the synchronous speed. Finally, when the motor is operated with the brushes shorted (no resistance in the rotor), the operating speed is about 95% of the synchronous speed. Thus, the blower may be operated at any one of three speeds.

A motor circuit for the three-speed *manual operation* of the blower motor is pictured in Figure 14.10. Here, two contactors (CON) are used to switch external resistance in and out of the rotor circuit. When the motor starter (M) is initially energized under the load of the blower, the rotor has all the resistance (high resistance) in series with it, and the motor operates at the low-speed operating point of Figure 14.9. As noted in the speed-torque curves, the operating torque produces a lowered speed at reduced torque with an increased slip.

When medium-speed operation is selected by shorting out one set of rotor resistors with the contacts of the first contactor (1CON), the motor speeds up and the slip declines. At high-speed operation, the remaining set of rotor resistors is shorted by the second contactor's contacts (2CON). This also shorts the brushes that contact the slip rings, causing the wound-rotor motor to operate like a general-purpose, design B, squirrel-cage motor.

FIGURE 14.9
Speed-torque characteristic curves of the wound-rotor motor combined with the load (blower) speed-torque curve. The operating points are indicated for each relative rotor resistance.

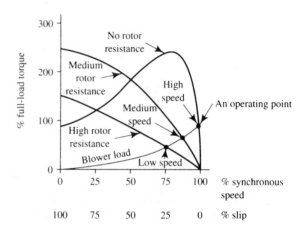

FIGURE 14.10
Wiring diagram of the wound-rotor blower motor with speed control provided by the external motor resistors.

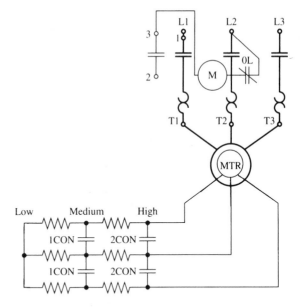

In recent years, the use of wound-rotor motors has been diminishing due to the increased availability and cost-competitiveness of *adjustable-frequency drives* for the control of ac squirrel-cage induction motors. These drive systems are smaller and have *soft-starting* features that eliminate the inrush of starting current and allow the motor to be operated over a wide range of speeds.

14.2 SINGLE-PHASE INDUCTION MOTORS

Introduction

Single-phase, squirrel-cage induction motors are designed to start and operate on single-phase current by providing two windings to create the rotating magnetic field. The two windings consist of the *main winding* (also called the run winding) and the *start winding* (also called the auxiliary winding). Like the three-phase, squirrel-cage induction motors, the *locked-rotor* (starting) current is roughly four to six times the full-load operating current for the single-phase induction motor. *Note*: Section C9 of Appendix C contains information for the general maintenance and troubleshooting of the single-phase, squirrel-cage motor.

As illustrated in Figure 14.11, the main winding is placed directly across the single-phase line, and the start winding—along with its *phase-shifting component*—is placed in parallel with it. As illustrated, the start winding (an inductor) is placed in series with a capacitor (Figure 14.11(a)) or in series with an *equivalent resistance* (Figure 14.11(b)) to produce a shift in the phase (ϕ) between the current in the main winding (I_m) and the current in the start winding (I_s).

By creating a phase difference with a phase-shift element between the main and start current (Figure 14.12), the motor operates as a *two-phase* device with a

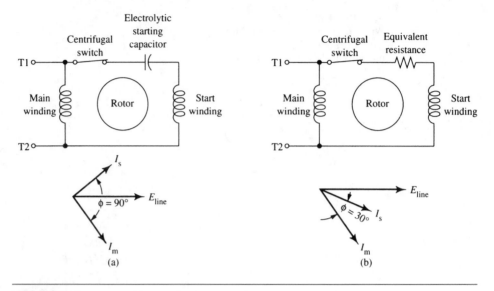

FIGURE 14.11
Schematic diagrams of the single-phase, squirrel-cage induction motors: (a) Capacitor-start motor along with the phase diagram representing the start and main stator currents; (b) Split-phase motor along with the phase diagram representing start and main stator currents.

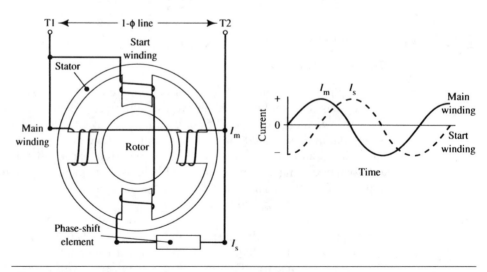

FIGURE 14.12
A simplified representation of a two-pole, single-phase, squirrel-cage induction motor with the phase shifting element in series with the start winding. The out-of-phase current in each phase changes with time, creating a varying flux at each set of poles. The combination of the two fluxes produces the rotating magnetic field. As pictured, the ac waveforms are out of phase with each other by 90°.

TABLE 14.4 Synchronous and Typical Full-Load
Speeds of 60-Hz, Single-Phase Motors

Poles	Full-Load Speeds (rev/min)			Synchronous Speeds (rev/min)
	Shaded-Pole	PSC	Other Types	
2	3000	—	3450	3600
4	1550	1625	1725	1800
6	1050	1075	1140	1200
8	—	825	850	900

rotating magnetic field. The synchronous speed, as noted in Table 14.4, depends on the number of poles in the stator.

Split-Phase Motor

Figure 14.11(b) pictures the schematic diagram of the resistance-start, induction-run motor, or *split-phase motor*, as it is commonly called. The equivalent resistance in the start winding comes from the resistance of the small-diameter wire used to wind the starting winding (not from an external resistor). The start winding, which has fewer turns than the main winding (to minimize its inductive properties), is switched out of the motor circuit by the *centrifugal switch* once the motor reaches about 80% of its full-load speed. The centrifugal switch, attached to the rotor shaft through a mechanism, is opened by the centrifugal force resulting from the rotation of the shaft.

During the starting of the split-phase motor, both stator windings are connected across the line to produce the starting torque. Since the start winding has much more resistance and less inductance than the main winding, the current in the start winding is *out of phase* with the current in the main winding. The main current (I_m in Figure 14.11(b)) is shifted about 50° from the line voltage. The phase difference between the currents is about 30°. Once the rotor is sufficiently accelerated, the centrifugal switch mechanism is actuated, opening the switch and removing the current from the start winding. With the start winding out of the circuit, the single-phase induction motor is operated with only the main winding connected.

Once started (rotating), and with only the main winding connected, the motor still develops sufficient torque to operate because of the flux produced by the induced current in the rotor. The necessary rotating magnetic field is produced by the *resultant flux* from the main stator winding and the *quadrature* (at 90°) *flux* from the rotor. Figure 14.13 pictures a typical speed-torque curve of a split-phase, squirrel-cage induction motor. The following are the typical characteristics of the split-phase induction motor:

- The motor has small horsepower, usually not more than $\frac{1}{2}$ hp.
- Starting torque is small, about 125% of full-load torque.

FIGURE 14.13
Speed-torque characteristics of a split-phase, squirrel-cage induction motor.

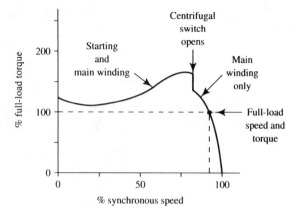

- Most motors can be reversed. *Note*: Always check the nameplate for connection information.
- Applications include the operation of washing machines, belt-driven power tools, fans, and blowers.

Capacitor-Start Motor

Figure 14.11(a) pictures the schematic diagram of the capacitor-start, induction-run motor, or *capacitor-start motor*, as it is commonly called. When compared to the split-phase motor, this type of single-phase, squirrel-cage induction motor has an improved starting torque. By placing an *ac electrolytic capacitor* in series with the start winding, the phase shift (ϕ) between the main winding current and the start winding current is nearly 90°. This more nearly approximates the starting action of a two-phase rotating magnetic field than does the split-phase induction motor (90° versus 30° phase shift between currents when starting).

Since a capacitor is used to shift the phase, the start winding is made with larger wire (lower resistance), which lowers the I^2R loss and the associated heating during acceleration. The mechanics of the starting and operation of the motor are similar to those of the split-phase motor in that a centrifugal switch or, in some models, a *starting relay* is used to open the start winding once the motor has accelerated to sufficient speed (70% to 80% of full-load speed). Once the switch opens, the motor operates as an induction-run motor with similar running properties to that of the split-phase motor with about the same full-load torque. The following are the usual characteristics of the capacitor-start induction motor:

- The motor has moderate horsepower, usually not more than 2 hp.
- Starting torque is high, about 250% of full-load torque.
- Because the ac electrolytic starting capacitor is designed for intermittent service (typically 20 starts per hour), the motor cannot be used where repeated starting and stopping is required.
- Most motors can be reversed. *Note*: Always check the nameplate for connection information.

■ Applications include the operation of refrigeration equipment, compressors, conveyors, pumps, and commercial machinery.

Capacitor Motor

Figure 14.14(a) pictures the schematic diagram of the capacitor-start, capacitor-run motor, or *capacitor motor*, as it is commonly called. This type of single-phase induction motor is instantly recognizable from the two capacitor housings mounted opposite each other on the outside of the motor. Because of this unique feature, the capacitor motor is sometimes referred to as a *two-value capacitor motor*.

As illustrated in Figure 14.14(a), a high-value (100 to 330 μF) ac electrolytic capacitor is used for starting, whereas a lower value (3.3 to 15 μF) *oil-filled capacitor* is used for running. During starting, the electrolytic capacitor is placed in parallel with the oil-filled capacitor, since the centrifugal switch is closed when the motor is started. Once the rotor is accelerated to about 75% of the full-load speed, the switch opens, removing the electrolytic starting capacitor from the circuit. Since the low-loss, continuous-duty, oil-filled run capacitor remains in the start winding, the motor operates as a true two-phase motor (90° phase shift) with constant torque, unlike the split-phase motor with its pulsating torque and vibration. The capacitor motor is used in applications requiring high starting torque combined with smooth, quiet operation. Because the power factor is improved by the presence of the run capacitor, the efficiency is improved and the line current is reduced. The capacitor motor combines the high starting torque of the capacitor-start motor with the *quiet operation* of the permanent split-capacitor motor. The following are the usual characteristics of the capacitor motor:

■ The motor has moderate horsepower, usually not more than 3 hp.
■ Starting torque is high, about 300% of full-load torque.

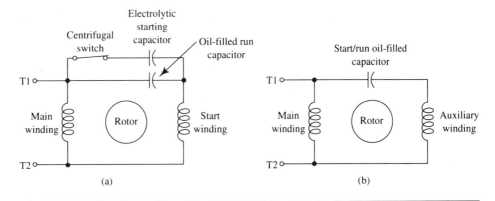

FIGURE 14.14
Schematic diagrams of the single-phase, squirrel-cage induction motors: (a) Capacitor motor; (b) Permanent split-capacitor motor (PSC).

- Operation is smooth and quiet.
- Because the ac electrolytic starting capacitor is designed for intermittent service, the motor cannot be used where repeated starting and stopping is required.
- Most motors can be reversed. *Note*: Always check the nameplate for connection information.
- Applications include high-efficiency operation of refrigeration equipment, air compressors, conveyors, pumps, blowers, and machine tools.

Permanent Split-Capacitor (PSC) Motor

Figure 14.14(b) pictures the schematic diagram of the permanent split-capacitor motor, or *PSC motor*, as it is commonly called. The PSC motor has two windings, the *main winding* and the *auxiliary winding*, that are identical in construction. The auxiliary winding is not switched out of the circuit but instead stays in the circuit connected to the START/RUN oil-filled capacitor. Because the size of the low loss, continuous duty, oil-filled capacitor is selected to optimize the operating characteristics, the motor has starting characteristics approximating those of the split-phase motor, with slightly higher starting torque. The smooth- and quiet-running characteristics of the PSC are similar to those of the capacitor motor.

Since no centrifugal switch is needed for its starting, the PSC motor can be started and stopped any number of times, and it can be reversed without first bringing the motor to a complete stop, as must be done in other motors so the centrifugal switch may close.

The reversing of the motor is accomplished by switching the capacitor out of the auxiliary circuit and placing it into the main winding circuit, as illustrated in Figure 14.15. The speed of the PSC motor can be controlled by controlling the line voltage with a solid-state triac circuit. The following are the usual characteristics of the permanent split-capacitor motor:

- The motor has moderate horsepower, usually not more than $\frac{1}{2}$ hp.
- Starting torque is moderate, about 150% of full-load torque.
- Operation is smooth and quiet, with no pulsation or hum.
- It can be quickly reversed with repeated starts and stops.
- Speed can be controlled.
- Maintenance costs are reduced since a centrifugal switch is not used.
- Applications include operation of commercial and industrial coolers, furnace blowers, air conditioner condensers, heat pumps, and other shaft-mounted, air-over fan and blower equipment. The PSC motor is also used in small- to medium-torque gearmotors.

Shaded Pole Motor

Although the *shaded pole motor* is a single-phase, squirrel-cage rotor motor, its principle of operation differs from that of the other types of squirrel-cage motors previously covered. The magnetic field does not rotate but is instead swept across the face of the stator pole pieces, creating a small torque on the rotor.

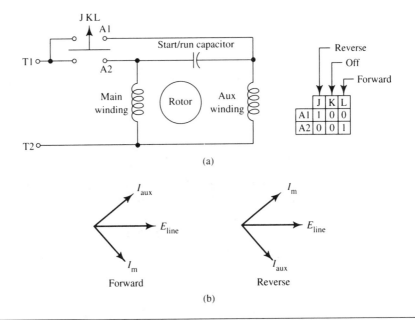

(a)

(b)

FIGURE 14.15
(a) The PSC motor is reversed using a three-position selector switch. With the selector in the L position, the capacitor is connected in series with the auxiliary winding and the motor moves in the forward direction. In the K position, the motor is off. By moving the selector to J, the capacitor is connected in series with the main winding and the motor moves in the reverse direction. (b) The phase relations between I_m and I_{aux} in the PSC in the forward and reverse directions.

A *shading coil* (low-resistance, single-turn copper winding) is placed around one-quarter of the width of the pole on one side of each pole, as pictured in Figure 14.16. When the stator winding is energized, a current is induced in the shading coil, which, in turn, produces an opposing flux and causes the field flux to move across the pole face in a nonuniform manner. The sweeping action of the flux moving across the pole face produces sufficient torque to start the rotor moving. Once the rotor

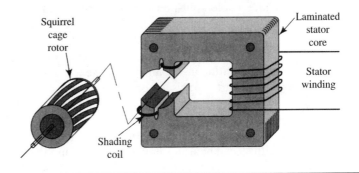

FIGURE 14.16
Open-frame type of a two-pole, shaded-pole motor.

starts rotating, the induced field in the rotor takes over from the shading coil's interaction with the stator field to produce a rotating magnetic field. Once up to full-load speed, the motor maintains a fairly uniform, full-load speed and torque, although the full-load slip is typically 10% or more.

Although the shaded pole motor is very inefficient (5% to 30%), the power demand of the least expensive open-frame type of shaded pole motor is small ($\frac{1}{250}$ to $\frac{1}{20}$ hp); thus the cost of operation is not a consideration. Because the shaded pole motor is the simplest and least costly motor to produce, it is found in many household products, including range hood vent fans, hair dryers, kitchen appliances, etc. The following are the usual characteristics of the shaded pole motor:

- The motor has small horsepower, usually not more than $\frac{1}{6}$ hp.
- Starting torque is low, about 50% of full-load torque.
- For its size, the full-load current is high and efficiency is low.
- It cannot be reversed without dismantling the motor.
- It can be speed-controlled.
- It has good reliability.
- Commercial applications include operation of refrigeration condensers, central air-conditioning outdoor condensers, heater blowers, and ventilators. The shaded pole motor is also used in small-torque gearmotors.

Gearmotors

By combining a motor with a *gearhead* (a speed reducer), as pictured in Figure 14.17, reduction in speed and increased torque are realized. However, this realization is accompanied by a loss of energy from the friction in the gear set. The **gearmotor** provides a simple and compact way of lowering motor speeds below the lowest standard operating speed of 900 rev/min. Depending on the motor type and the gear ratio, the gearmotor is capable of speeds ranging from less than 1 rev/min to greater than 300 rev/min with torques ranging from under 10 lb-in to more than 4000 lb-in.

The types of motors used to drive the gearhead include dc permanent magnet, universal, stepper, servo, three-phase induction, single-phase induction (PSC, shaded pole, split-phase, capacitor-start), and synchronous motors.

The gearmotor is available in two general shaft types, *parallel shaft* and *right-angle shaft* (offset shaft). The parallel-shaft type uses either a combination of helical gears (for low noise) and spur gears (for economy) or *planetary gears* (three spur gears rotating inside an internal gear) to produce a compact drive system. The right-angle-shaft type uses either worm gears or helical bevel gears. Although higher in torque, heavier in construction, and more able to handle shock loads, the right-angle gearhead is less efficient than a comparably-sized parallel-shaft gearhead. The service life of the gearhead of the gearmotor is determined by the life of the bearings, with sleeve bearings having a shorter life than ball bearings.

The gearmotor is used in motion-control applications where reduction in angular velocity (speed) is needed to improve positioning resolution or when a high-speed motor (900 to 3600 rev/min) is used to drive an extremely low-speed

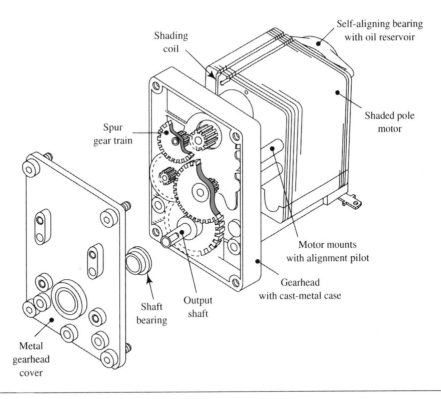

FIGURE 14.17

A gearmotor consists of a motor integrated with a gearhead. The angular velocity (speed) is reduced, but the torque is increased. The gearmotor pictured is for light-to-moderate duty. *(Courtesy of Merkle-Korff Industries, Inc.)*

machine. Rather than deal with a bulky belt and pulley drive for speed reduction and torque increase, a small-envelope gearmotor can be connected directly to the machine to amplify torque and decrease speed.

Gearmotors are usually specified by the full-load torque and speed at the output shaft. The horsepower may not be specified. When horsepower is needed, it can be computed from the torque and speed. The following formula, which yields horsepower, has been normalized for the common parameters of speed (rev/min) and torque (lb-in) found in engineering catalogs:

$$P = \frac{\tau\omega}{63{,}000} \qquad \textbf{(14.4)}$$

where

P = power at the shaft, in hp

τ = torque at the output shaft, in lb-in

ω = angular velocity (speed) of the output shaft, in rev/min

Note: 63,000 results from 5250 being multiplied by 12, the conversion factor between feet and inches.

Example
14.4

Determine the output power at the shaft of a gearmotor when the torque is 120 lb-in and the speed is 5.0 rev/min.

SOLUTION   Find the power in hp.

Given
$$P = \frac{\tau\omega}{63{,}000}$$

Evaluate
$$\tau = 120 \text{ lb-in}$$
$$\omega = 5.0 \text{ rev/min}$$

Substitute
$$P = \frac{120 \times 5.0}{63{,}000}$$

Solve
$$P = .010 \text{ hp}$$

The gearmotor is used in the drive trains of many industrial, commercial, and residential machines and appliances. The following are a few of the applications that utilize the gearmotor: home appliances, small pumps, business machines, copy machines, conveyors, vending machines, ribbon drives, printers, security cameras, valve actuators, pick-and-place robots, and automated tellers.

14.3   MOTOR NAMEPLATE PARAMETERS

The National Electrical Manufacturers' Association (NEMA) has developed standards for the design and rating of electric motors. When motors are standardized, it is possible to interchange them from one manufacturer to another, thereby making motor replacement possible even in older machines. The electrical and mechanical parameters needed to compare the application to the performance of the motor are contained on the *nameplate* fastened to the motor frame. The nameplate of an induction motor is pictured in Figure 14.18. In addition to the manufacturer's name and the type of motor, the nameplate usually contains most of the following categories of information.

Voltage (VOLTS, or V)   For normal operation, the line voltage must be within ±10% of the *nominal* (named) voltage listed on the nameplate. If the line voltage is lower than 10% of the nominal nameplate voltage, then the torque will be lowered and the motor current will be increased, causing overheating. When the line voltage is greater than 10% of the nominal voltage, then ac motors will have excessive magnetizing current in the windings, causing overheating.

The specified nameplate terminal voltage is typically 4% less than the nominal line voltage in order to compensate for wiring loss in the power-distribution system. Thus, a three-phase motor designed to operate on a 480-V line is specified (nameplate) as a 460-V motor (0.96 × 480 V = 460 V). Some of the nameplate voltages for ac motors include 115/230 V for single-phase motors operated on 120/240-V lines, 200 V for a 208-V line, 230 V for a 240-V line, and 460 V for a 480-V line. Because

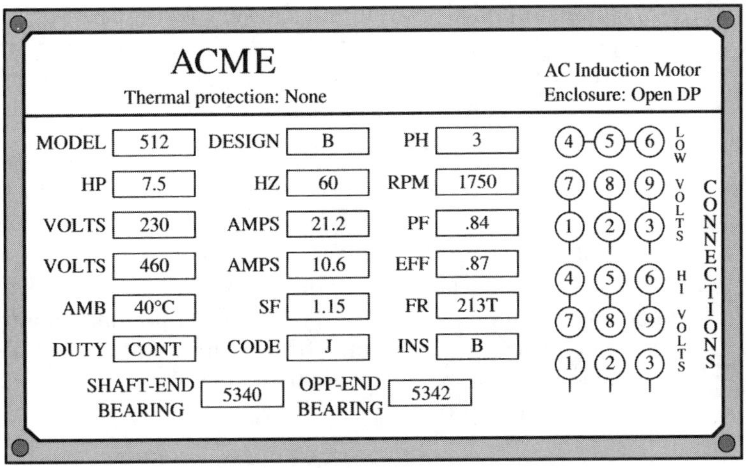

FIGURE 14.18
The nameplate of an induction motor.

many of the motor's parameters depend on the terminal voltage, this parameter must be an exact match when replacing a motor.

Current (AMPS, or A)   The current listed on the nameplate is the full-load current at the rated terminal voltage and power (horsepower, kilowatts). The current parameter is used to size the conductors, determine the rating of the fuses or circuit breakers, and select an appropriate heater for the thermal overload relay in the motor starter.

A single-voltage motor will have one current listed, whereas a dual-voltage motor will have two currents listed, as illustrated in the nameplate of Figure 14.18. As you know, the starting current is typically 400% to 600% of the full-load running current. The no-load operating current is always less than the nameplate current. By comparing the measured line current (with the motor operating under load) to the nameplate current, a technician can make a good evaluation of the motor's operating conditions.

Power (hp or kW)   The power rating is typically the maximum permitted continuous power output by the motor under load without exceeding the specified temperature rise. The BES unit for motor power is the horsepower; the SI unit is the kilowatt. When replacing a motor, select a motor with a power rating equal to or greater than that of the motor being replaced.

Phase (PHASE, PH, or ϕ)   The rating is commonly single- or three-phase. The phase must be matched exactly when the motor is replaced.

Frequency (Hz)   Nameplate frequency is specified in hertz, with 60 Hz being the typical frequency in the United States. Most ac motors have been designed to accommodate a ±5% variation in the line frequency. Variation in the frequency of power supplied from utilities is very rare. However, power supplied in remote regions (oil exploration, etc.) from a motor-generator may experience variations in frequency with changes in the load. An increase in the line frequency will result in an increase in the revolutions per minute of the motor. Always use the same nameplate frequency when motors are changed.

Speed (rev/min)   The speed at the shaft with the motor under full load is called the full-load speed. The full-load speed of an ac induction motor is always less than the synchronous speed. The synchronous speed varies, depending on the characteristics of a particular type of motor. For example, a four-pole induction motor has a synchronous speed of 1800 rev/min, a nominal nameplate speed of 1750 rev/min, and an actual speed range from 1740 to 1770 rev/min. When replacing motors, the speed should be matched to within 5% of the nameplate of the motor being replaced.

Insulation Class (INS CLASS, or INS)   The insulation class letters (A, B, F, H, etc.) are used to establish the maximum allowable operating temperature of the motor winding, thereby ensuring a substantial service life for the winding insulation. Most T-frame motors are class B insulation, whereas older motors are usually class A. Class B limits the maximum insulation temperature to 130°C (266°F), class A is limited to 105°C (221°F), class F is limited to 155°C (311°F), and class H is limited to 180°C (356°F).

Design (DESIGN)   The NEMA design letters (A, B, C, and D) designate the motor's speed-torque characteristics, including the amount of starting torque, starting current, breakdown torque, etc. Design B motors, which are very common, are used for general application requiring medium starting torque. Design C motors are used for loads with high starting inertia, since the design provides a high starting torque. Design D motors have extra large starting torques to ensure the starting of very high inertia loads. Design A motors are not too common, since they are just a variation of the design B motor with a higher locked rotor current (starting current).

Code (CODE)   The NEMA code letters, in conjunction with a table of values of kilovolt-amps per horsepower (kVA/hp), are used to calculate the motor's locked rotor current (current into the stator at starting). The code letters, which range from A through V, are used to specify the maximum rating of the motor's branch circuit overcurrent device.

The tabular kVA/hp values represented by the code letters are used in conjunction with the following equations (single-phase and three-phase) to determine the approximate locked rotor current.

For single-phase motors,

$$I_{LR} = \frac{\text{kVA/hp} \times P \times 1000}{E} \qquad \textbf{(14.5)}$$

For three-phase motors,

$$I_{LR} = \frac{\text{kVA/hp} \times P \times 1000}{\sqrt{3}E} \qquad \textbf{(14.6)}$$

where

I_{LR} = locked rotor current, in A

kVA/hp = kilovolt-amps per horsepower associated with the NEMA code letter

P = power rating of motor, in hp

E = motor nameplate voltage rating, in V

Example 14.5

Determine the minimum and the maximum value of locked rotor current for a three-phase, 460-V, 7.5-hp motor with a code J on the nameplate.

SOLUTION

Using the published value of 7.1 to 8.0 kVA/hp (the NEMA range of values for code J), compute the minimum and the maximum values of locked rotor current for the 7.5-hp motor.

Given

$$I_{LR} = \frac{\text{kVA/hp} \times P \times 1000}{\sqrt{3}E}$$

Evaluate

kVA/hp = 7.1 to 8.0

P = 7.5 hp

E = 460 V

Substitute

$$I_{min} = \frac{7.1 \times 7.5 \times 1000}{\sqrt{3} \times 460}$$

$$I_{max} = \frac{8.0 \times 7.5 \times 1000}{\sqrt{3} \times 460}$$

Solve

I_{min} = 67 A I_{max} = 75 A

Frame (FR) Since the frame number refers to the NEMA system of motor dimensions, it is vital that the frame size be matched exactly when replacing a motor so that the shaft will be at the correct height and the mounting holes will line up. If the equipment is older, you need to be aware that the frame-size numbering system was changed in 1952 and 1964. When the frame designator on the motor to be replaced cannot be found, consult a transformation chart for the modern frame-size number. The suffix T is a modern designator, whereas the suffix U is an older, discontinued NEMA frame designator.

Ambient Temperature (AMB) The ambient temperature rating specifies the maximum temperature of the air around the motor. A nameplate ambient temperature of 40°C (104°F) means that the motor can be operated at full load as long as the surrounding air temperature does not exceed 104°F (40°C). Sometimes the nameplate specifies temperature rise rather than ambient temperature; these are not the same. That is, a rating of 40°C ambient is not the same as a rating of 40°C rise.

Service Factor (SF) The quantity of reserve overload capacity built into the motor is indicated by the service factor. Motors with service factors greater than

1.0 are used in applications where momentary overloads might occur. A service factor of 1.15 is an indication that an overload of 15% can be tolerated by the motor. When replacing the motor, always select a service factor that is equal to or greater than the nameplate value being replaced.

Power Factor (PF) The motor's power factor can be used with induction motors to aid in selecting capacitors for power factor correction. The installation of the capacitors across the motor will reduce current demand in the feeder circuit to the motor. The reduction in line current is due to the capacitors, which supply the *magnetizing*, or kilo-var, current instead of the line. See Table A8.

Efficiency (EFF) The efficiency of the motor is the ratio between the output and input power, usually marked on the nameplate as a decimal fraction (i.e., .78, .86, etc.). In general, a motor's efficiency increases as the horsepower increases. Very large motors have efficiencies in excess of 95%, whereas fractional-horsepower motors typically have efficiencies less than 80%. Most motors are at their maximum efficiency when operated above 80% of their full-load rating. Since motor efficiency is significantly diminished when the motor is underloaded, the power rating (hp, kW) of the motor must be matched to the load in order to get as much efficiency from the motor as possible.

Electrical, magnetic, and frictional losses contribute to the efficiency of the motor. These losses include winding losses (I^2R), magnetic losses (eddy currents, hysteresis, air gap, etc.), and mechanical losses (bearing friction, gear friction, etc.).

Duty (DUTY) Most motors are continuous-duty motors. However, when designed for intermittent duty, the motor will be marked with the permitted amount of operating time, which should not be exceeded as the motor will overheat. When an intermittent duty motor is replaced, it may be upgraded to a continuous-duty motor.

Enclosure (ENCL) Frequently, motors are open to receive air over and around the windings. These motors have an *open* construction. The ventilation openings on all sides of the motor provide adequate ventilation. However, the motor must be used indoors where clean, dry air is available. In addition to the open-construction enclosure, motors are available with the following motor housings:

Drip-proof (DP) Drip-proof motor housings have openings in the end shields that are shaped so that moisture falling vertically (between 75° and 90°) to the ventilation openings will not enter the interior of the motor. The drip-proof enclosure is usually operated indoors in moderately clean locations.

Total enclosed (TE) Totally enclosed housings have no openings for ventilation in the enclosure. Although not airtight or waterproof, the motors with TE ratings are used in dirty, damp, and oily locations. These are:

■ *Totally enclosed, fan-cooled (TEFC) enclosure.* The TEFC motor housing has an external fan within a shroud that encompasses the totally enclosed motor so that air is moved over and around the motor housing.

■ *Totally enclosed nonventilated (TENV) enclosure.* The TENV motor housing, which has no external fan or shroud, depends on the natural convection (movement) of the air for cooling the motor housing.

- *Totally enclosed air-over (TEAO) enclosure.* The TEAO motor housing uses air from the driven device (as in a forced-air heating and cooling system) to provide a cooling airflow over and around the motor housing.
- *Explosion-proof (EX PRF) housing.* The EX PRF housing totally encloses the motor windings in a strong, fairly thick case designed to withstand the force of an internal explosion without allowing the flames or the products of the explosion to escape. Because these motors are designed for use in hazardous environments, where explosive materials and atmospheres are present, they must be used in accordance with the National Electric Code for hazardous locations.

Bearing Type (BGR) The antifriction ball bearing and the sleeve journal bearing are the two basic types of bearings most commonly found in motors. Of the two, the ball bearing type is used with high loads and where frequent lubrication is impractical, and the sleeve bearing type is used where quiet operation and cost are factors. Since most sleeve bearing motors can be operated in any position, it is important when replacing a motor that the same type of bearing be used.

Thermal Protection Some motors have a built-in thermal switch to disconnect the motor from the line when the preset temperature becomes excessive due to an overload condition. Built-in thermal protection is used with motors that are automatically started, are unattended in remote locations, or are fractional horsepower. The three basic types of built-in protection are as follows:

Automatic (Auto) An automatic thermal switch resets automatically after the motor cools. This type of thermal protection should never be used if unexpected restart could pose a safety hazard.

Manual (Man) A manual thermal switch is reset when an external button is depressed. This type of thermal protection is used instead of automatic reset to ensure operator safety.

Impedance (Imp) An impedance thermal switch or—as it is frequently called—*impedance protected*, is designed so that the windings have sufficient impedance to protect the motor from burning out with the rotor locked (stalled) for up to 15 days.

When replacing a thermally protected motor, it is very important that the same type (automatic, manual, or impedance) be selected. Never use motors with automatic reset where unexpected (automatic) starting presents a hazard. When this is the case, always use a manual-reset, thermally protected motor.

CHAPTER SUMMARY

- The induction motor is made up of two major parts: the stator (stationary part) and the rotor (rotating part).
- The two types of three-phase induction motors are the squirrel-cage and the wound-rotor motors.
- In the induction motor, the rotor is not electrically connected to the ac line; it receives its electric current through electromagnetic induction.

- The operation of the induction motor depends on the presence of a rotating magnetic field.
- The synchronous speed of the motor is the speed at which the rotating magnetic flux moves around the stator.
- The direction of rotation of the three-phase induction motor is changed by interchanging any two of the three line connections.
- Torque to spin the rotor and rotate the load is developed when the rotating magnetic field of the stator induces current in the conductors in the rotor, producing a net right angle force on the rotor conductors.
- Slip, the difference between the rotor speed and the synchronous speed, must be present in the induction motor in order to develop torque.
- NEMA has adopted several motor designs to accommodate a variety of speed-torque characteristics of the squirrel-cage induction motor.
- Reduced-voltage controllers are used to start large motors by reducing the motor terminal voltage and lowering the starting current, thereby lessening the drop in line voltage associated with starting large motors.
- The wound-rotor induction motor uses slip rings to connect external resistance into the rotor winding to control the speed-torque characteristic of this type of induction motor.
- The single-phase, squirrel-cage induction motor operates as a two-phase motor with a rotating magnetic field because a phase shift is created between the currents in the start and main windings of the stator.
- The gearmotor (a gearhead combined with a motor) provides a simple and compact way of lowering motor speeds below the lowest standard operating speeds and increasing output torque.
- The electrical and mechanical parameters needed to compare an application to motor performance are contained on the nameplate fastened to the motor frame.

SELECTED TECHNICAL TERMS

The following technical terms, abbreviations, and acronyms are defined in the glossary located after Chapter 16. You are encouraged to use the glossary to aid your understanding and to test your knowledge of these important terms.

breakdown torque	rotor
full-load slip	skewed
gearmotor	slip
rated torque	starting torque
rheostat	stator

END-OF-CHAPTER QUESTIONS

Write T if the statement is true and F if the statement is false.

1. Under load, the rotor of the induction motor operates at the synchronous speed.
2. The starting torque of the design B motor is the maximum torque that the motor can develop.
3. Motors designed to operate on a 480-V line are rated as 460-V motors.
4. Squirrel-cage and wound-rotor motors are both types of synchronous motors.
5. The rotating field in a single-phase induction motor results from the two-phase flux developed by the currents in the stator windings.

6. The rotor conductors in the squirrel cage are skewed to improve the starting current and the starting torque.

7. The conductors of a squirrel-cage rotor are formed from insulated copper wire.

8. A four-pole (1800-rev/min synchronous speed), three-phase, wound-rotor induction motor can be operated at a shaft speed of 1400 rev/min.

9. The two types of motor bearings in common use are the ball thrust and the sleeve bearing.

10. The single-phase motor is started when the phase is shifted between the currents in the start and the main windings.

In the following, select the word or phrase that makes the statement true.

11. The difference between the angular velocity of the motor shaft and the synchronous speed of the rotating field of the stator is called (slide, creep, slip).

12. The motor in a home air conditioner is usually protected from excessive heating by a(n) (manual, automatic, impedance) type of built-in thermal protector.

13. The type of motor housing having no external fan or shroud is the (TEAO, TENV, TEFC).

14. The single-phase induction motor with the smallest starting torque is the (split-phase, PSC, shaded pole, capacitor) motor.

15. The difference in the speed-torque characteristic curves of the NEMA motor designs is due to variations in the design of their rotors. The (design B, design C, design D) motor has a double squirrel-cage rotor.

Answer each of the following questions with a short answer in the form of a complete sentence. Include a restatement of the question in your answer.

16. Based on a comparison of their speed-torque characteristics, how does the operation of the design D motor differ from the design B motor?

17. In terms of their definition, what is the difference between the rated torque and the breakdown torque of an induction motor?

18. How does the wound-rotor induction motor differ in construction and operation from the three-phase, squirrel-cage induction motor?

19. When replacing an induction motor, which of the nameplate parameters must be matched exactly and which parameters may be increased?

20. Why are reduced-voltage starters used to start large squirrel-cage induction motors?

END-OF-CHAPTER PROBLEMS

Solve the following problems. Make sketches to aid in solving the problems and structure your work so it follows in an orderly progression and can easily be checked. Table 14.5 summarizes the formulas used in Chapter 14.

1. A 460/230-V, three-phase, 15-hp motor has a full-load current of 21 A when connected to a 480-V line. When this motor is connected to a 240-V line, this motor will
   (a) Develop 30 hp.
   (b) Take 42 A.
   (c) Develop 7.5 hp.
   (d) Take 10.5 A.

TABLE 14.5 Summary of Formulas
Used in Chapter 14

Equation Number	Equation
14.1	$n_s = \dfrac{f \times 60}{p}$
14.2	$\%\text{slip} = \dfrac{n_s - n_r}{n_s} \times 100\%$
14.3	$\%\tau = V_{pu}^{\,2} \times 100\%$
14.4	$P = \dfrac{\tau\omega}{63{,}000}$
14.5 (1ϕ)	$I_{LR} = \dfrac{\text{kVA/hp} \times P \times 1000}{E}$
14.6 (3ϕ)	$I_{LR.} = \dfrac{\text{kVA/hp} \times P \times 1000}{\sqrt{3}E}$

2. A $\frac{1}{3}$-hp split-phase induction motor with a nameplate speed of 1725 rev/min is connected to a 6.0-in-diameter fan pulley through a V-belt drive. When the diameter of the motor pulley is 4.0 in, the speed of the fan is
   (a) 1150 rev/min
   (b) 1160 rev/min
   (c) 1170 rev/min

3. Compute the synchronous speed of a single-phase, six-pole PSC motor operating from a 120-V, 60.0-Hz line.

4. Determine the %slip in a $\frac{1}{2}$-hp, single-phase, 115-V, 60.0-Hz, four-pole, capacitor-start induction motor with a measured shaft speed of 1725 rev/min.

5. A single-phase, 115-V, $\frac{1}{3}$-hp motor with a Code L on the nameplate is started as a capacitor start motor. Determine the minimum and the maximum value of locked rotor current. Code L is equivalent to 9.0 to 10.0 kVA/hp.

6. The primary resistor type of reduced-voltage starter is used to start a 30-hp, 460-V, three-phase induction motor. If the terminal voltage of the motor is reduced from 460 V to 345 V by the starting resistors in each phase of the motor, determine
   (a) The percent of available starting torque (%τ) to start the motor at the reduced voltage.
   (b) The locked-rotor starting current at the reduced voltage when the full-voltage, locked-rotor starting current for the 30-hp motor is specified as 217 A.

7. If the power from the motor into the gearhead of a gearmotor is .330 hp, determine
   (a) The power (hp) at the output shaft of the gearhead when the torque is 294 lb-in and the speed is 45 rev/min.
   (b) The efficiency of the gearhead from input to output.

8. Develop a ladder diagram for the control of the wound-rotor blower motor used to drive the three-speed blower of Figure 14.10. Use a three-wire control circuit

with the motor starter. The speed is switched from LOW/START to MEDIUM to HIGH by manually depressing one of four pushbutton switches (STOP, LOW/START, MEDIUM, HIGH). Some of the constraints in the control circuit include the following:

- All the external resistance must be in the rotor circuit when started.
- The motor must be switched from low speed to medium speed before being switched to high speed.
- The motor must never be started by depressing either the MEDIUM pushbutton or the HIGH pushbutton.
- The motor must be stopped by depressing the STOP pushbutton.

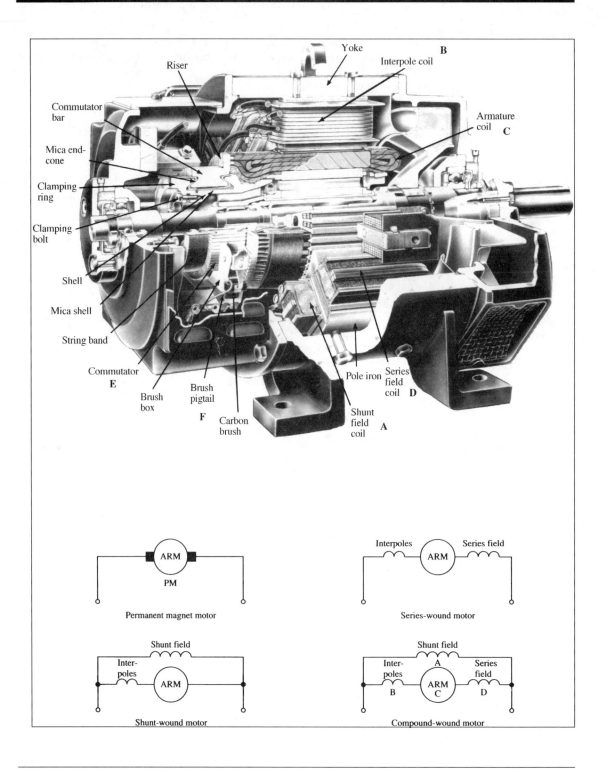

Riser

Yoke

Interpole coil

B

Commutator bar

Armature coil **C**

Mica end-cone

Clamping ring

Clamping bolt

Shell

Mica shell

String band

Commutator **E**

Brush box

Brush pigtail

F

Carbon brush

Pole iron

Series field coil **D**

Shunt field coil **A**

ARM

PM

Permanent magnet motor

Interpoles

ARM

Series field

Series-wound motor

Shunt field

Inter-poles

ARM

Shunt-wound motor

Shunt field

Inter-poles

A

ARM C

B

Series field

D

Compound-wound motor

Chapter 15

Direct-Current Motors and Alternating-Current Drives

About the Illustration. *The rotating element in the dc motor is called the* armature. *It is made up of a laminated iron core with parallel slots in it to receive the current-carrying conductors to create the* armature coil *(C). The* commutator *(E) along with the armature is integrated into the motor shaft. The conductors from the armature coil are connected to the* segments *of the commutator, which are energized by the* carbon brushes *(F) riding on the commutator. An external magnetic field is produced by either a* permanent magnet *(PM),* shunt field coil *(A),* series field coil *(D), or compound winding (shunt and series coils).* Interpoles *(B) are located between the main field poles to minimize sparking between the brushes and commutator. (Courtesy of the Reliance Electric Company.)* ■

INTRODUCTION

The fundamental operating principle of all electric motors is virtually the same. The rotation of the armature is due to the interaction of two flux fields, one from the field (stator) and the other from the armature (rotor). These two flux fields produce a net difference of force on the armature conductors, which tends to move them at right angles to the resultant of the flux fields. In the dc motor, the flux between the north and south field poles (stator) is unidirectional, as is the flux created in the armature (rotor). *Brushes* and a *commutator* are used to switch the polarity of the current in the armature (*commutation*), thereby reversing the direction of the flux in the armature structure. Table 15.1 lists many of the commonly used types of electric motors. Those motors in bold, along with the stepper motor, are studied in this chapter.

In addition to the principles of operation, both the dc motor and the stepper motor speed-torque characteristics and terminology are studied. Also, the electrical circuit and starting techniques are examined for the control and operation of each type of dc motor (permanent magnet, shunt, series, compound, and universal) as well as the stepper motor. The operation of ac induction motor drives is also studied.

CHAPTER CONTENTS

PERFORMANCE OBJECTIVES

Once you have read and studied each section; worked through each example with pencil, paper, and calculator; worked through the end-of-chapter problems; and answered the end-of-chapter questions, then you should be able to

- Identify the four types of dc motors.
- Describe applications for dc motors.
- Understand the use of interpoles in dc motors.
- Explain why starting resistors are needed when starting dc motors.
- Know the methods for controlling the speed of dc motors.
- Interpret the speed-torque characteristic curves for the four types of dc motors.
- Describe the operating characteristics of the universal motor.
- Understand why ac motors with solid-state, variable-frequency drives are replacing dc motors.
- Name and know the function of each element in the drive system of a stepper motor.
- Define the basic parameters used to specify the operating characteristics of the stepper motor.
- Explain the operating principle of the stepper motor.
- Calculate the resolution of the stepper motor.
- Explain the difference between the holding torque and the detent torque using the speed-torque curve of a stepper motor.

15.1 DC MOTORS

Introduction

Direct-current motors (dc motors) are used in industrial applications where control of speed and torque is an important consideration. When high starting torque, a wide range of speeds, quick stopping (high deceleration torque) and reversal, and/or smooth operation at low speeds are needed in an industrial process, the dc motor has been the traditional choice. However, in recent years, advancements in *variable-*

TABLE 15.1 Classification of Selected Electric Motors

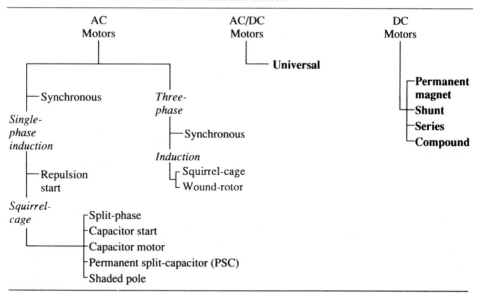

speed, *adjustable-frequency* drives for ac induction motors have made ac motors competitive in price with dc motors for many of these same applications, without the added expense associated with maintaining dc motors.

In industrial manufacturing applications, dc motors and dc gearmotors are commonly used with commercial solid-state *adjustable voltage controllers* to provide speed and torque control. The adjustable dc controller furnishes overcurrent protection along with ac operation from 120/240 single-phase lines to supply power to small (up to 2 hp) 90- or 180-V dc motors so that they can operate with loads that demand *constant* or *variable torque* over a wide range of adjustable speeds. Some examples of the types of torque loads commonly controlled using dc motors in combination with commercial adjustable dc motor controllers are:

- Constant-torque loads, including conveyors, positive-displacement pumps, and variable-speed processing equipment
- Variable-torque loads, including fans, blowers, ventilators, and centrifugal pumps

DC Motor Operation

The rotating element in the dc motor is called the **armature** (rotor). It is made up of a laminated core (to lessen eddy-current loss) with parallel slots in it to receive the current-carrying conductors, as illustrated in Figure 15.1. The **commutator** and the armature are attached to the motor shaft, with the commutator located on one end of the shaft.

FIGURE 15.1
An elementary representation of a two-pole dc motor with the field windings, armature, armature conductors, commutator (two segments), and brushes pictured.

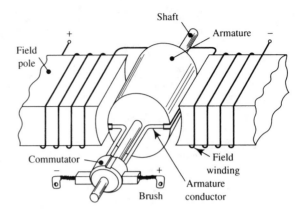

The conductors from the armature are connected to the *segments* of the commutator, which are energized by the carbon brushes riding on the commutator. The current supplied by the brushes creates a magnetic field in the armature windings. This magnetic field interacts with the external magnetic field created by the windings around the *field poles* (the stator).

The external magnetic field is produced by one of four methods: permanent magnet (PM), shunt winding, series winding, or compound winding. In many small-horsepower motors ($\frac{1}{1000}$ to $\frac{1}{4}$ hp), the field is produced by a permanent magnet. In all other motors, the field poles are wound with a field winding through which current is passed, creating a dc electromagnet that produces the necessary magnetic field.

Once the field and the armature circuit are energized, the fields in each interact to produce a torque that causes the armature to rotate. In the elementary dc motor of Figure 15.1, the armature conductor is rotated to a vertical position, causing the commutator to move under the brushes. Here the conductors are reconnected so that the current moves through the armature coil in the opposite direction. Thus, the loop continues to rotate in the same direction, gaining speed as it moves, since the commutator continues to **commutate** (periodically reverse) the armature current. To ensure smooth operation, most commercial motors have four or more poles (Figure 15.2), and the armature is wound with many conductors, each connected to opposite segments of the commutator, forming a closed loop.

The torque produced by the motor is directly proportional to the product of the field flux (ϕ), the armature current (I_A), and a proportionality constant (k) representing the number of poles and conductors in the armature. This relationship is noted in the following equation:

$$\tau = kI_A\phi \tag{15.1}$$

Interpoles

The brushes are located along a *neutral plane* (illustrated in Figure 15.2) that is formed in the resultant magnetic field of the armature and the field coils. The *neutral*

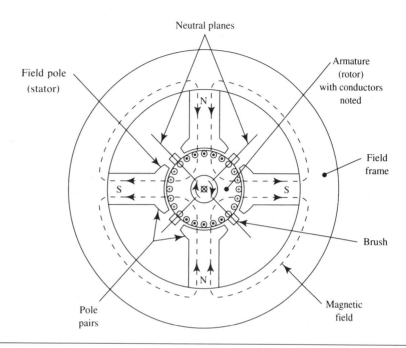

FIGURE 15.2
Pictorial of a four-pole motor.

planes pass through the radial center of the armature at right angles to the field flux. The brushes are located along the neutral plane, since the countervoltage (CEMF) at each segment of the commutator is about equal and the voltage is at its minimum. With minimal voltage present when the commutator segments are shorted together by the brushes, minimal arcing is produced, which extends the life of the brushes and allows the commutator to run at a cooler temperature.

When the speed of the motor increases, the intersection of the fields shifts position, and the neutral plane is twisted (moved) from its low-speed position to a new position. To compensate for this movement, the brush holders must be manually shifted in the same direction as the movement in the neutral plane so that arcing remains at a minimum. To avoid the need for manual adjustment of the brushes in dc motors, **interpoles** wound with *compensating windings* are commonly added to the design of dc motors to null out the effect of the shift in the neutral plane. The magnetomotive force (mmf) created by the heavy interpole windings, placed in series with the armature, as shown in Figure 15.3(a), interacts with the armature field and straightens the resultant flux so that the neutral plane does not shift. As illustrated in Figure 15.3(b), the narrow interpoles are located halfway between the main poles, and their use eliminates the need physically to move the brush holders.

Direct-current motors are classified by the method used to produce the field flux. Thus, a dc motor is designated as a permanent magnet motor, a shunt-wound motor, a series-wound motor, or a compound-wound motor. The term *wound* is

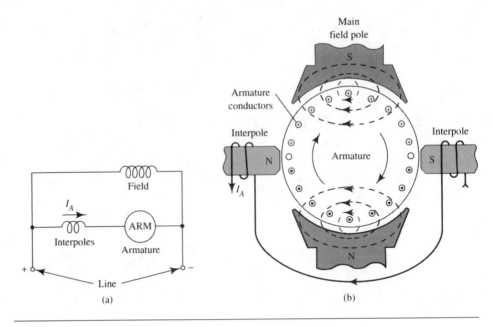

FIGURE 15.3

The interpoles are narrow poles that are placed between the main poles. (a) The interpole windings are placed in series with the armature. (b) Interpoles are used to align the field flux around the neutral plane, so arcing between the brushes and the commutator is minimized even when the armature current is varying.

often omitted from the motor name. Thus, dc motors are referred to as PM motors, shunt motors, series motors, or compound motors.

Characteristics of DC Motors

When initially energized, the armature current (I_A) can be very high in the permanent magnet, series, and compound types of dc motors. Once acceleration of the armature starts, the generated *counterelectromotive force* (CEMF) will oppose the armature current, reducing it to its rated full-load value at rated full-load speed. Since torque is directly proportional to armature current, large values of armature current are accompanied by large values of torque. Thus, when the starting current is ten times its full-load current value, the torque is also ten times its full-load torque.

Except for small-horsepower (under $\frac{1}{2}$ hp) dc motors of all types, the initial starting current is limited to protect the brushes and commutator by adding resistance to the armature circuit. When resistors are added to the armature circuit, the starting current can be limited to a safe value (200% to 300% of full-load current), which will not overheat the brushes and the commutator, yet it will allow the dc motor to accelerate high inertial loads.

Figure 15.4 pictures the power and control circuit of an adjustable-speed, dc motor-starter circuit (two-stage), in which a tapped resistor is placed in series with the armature to limit starting current. Since there is no CEMF to limit the armature current as the motor is started, the starting current without starting resistors may be

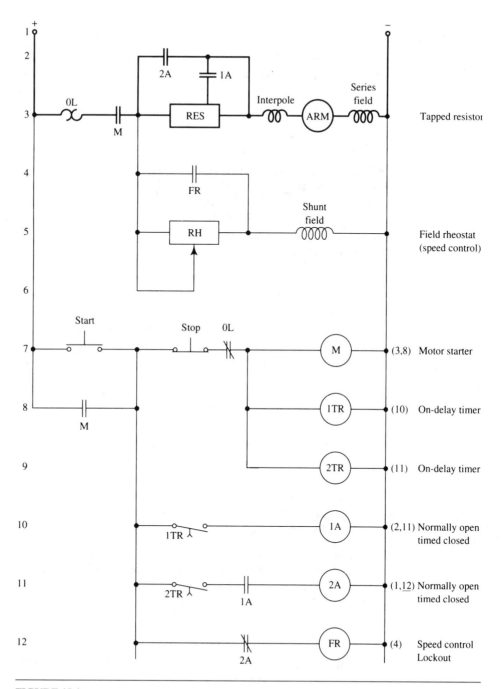

FIGURE 15.4
The power and control circuit of an adjustable-speed, dc motor-starter circuit for a compound motor.

10 to 15 times the full-load current, or—in the case of a 1-hp motor—several hundred amperes.

Needless to say, direct connection of a dc motor across the line is not done, since the overcurrent devices would open before the motor got started. Instead, a *dc motor starter* is used with all but the smallest dc motors. The starter provides for the removal of the starting resistance, in steps, as the armature increases in speed. With the armature rotating near *base speed* (full-load speed), the countervoltage is sufficient to limit current to its rated value without the aid of an external resistance.

The *speed-control rheostat* in the shunt field circuit of the compound motor circuit, pictured in Figure 15.4 (rung 5), is not part of the start circuit, since it is bypassed by the contacts of the *field rheostat control relay* (FR) when the START pushbutton is depressed. Once the starting resistances have been bypassed and the ON-delay timing relay (2TR) has timed out, then the field-rheostat-control relay coil (FR) is de-energized, opening the contacts across the field rheostat and fixing the motor to the preset speed directed by the position of the speed-control rheostat.

Speed Regulation

The response of a dc motor's speed to a change in load is an important consideration when selecting a motor to do a particular job. The *speed regulation* of the motor is a good indication of how it will behave when the load is increased or decreased. In general, dc motors will increase in speed when the mechanical load is removed and decrease in speed when a load is applied. The variation in speed resulting from a change in load depends on the type of motor. The speed regulation (expressed as a percent) for a dc motor is determined by applying the following formula.

$$\text{Speed regulation} = \frac{n_{nl} - n_{fl}}{n_{fl}} \times 100\% \qquad \textbf{(15.2)}$$

where

Speed regulation, expressed as a percent (no units)

n_{nl} = no-load speed, in rev/min or rad/s

n_{fl} = full-load speed, in rev/min or rad/s

Note: Shunt and permanent-magnet motors are called *constant-speed* motors because their speed regulation is between 5% and 15%. Series and compound motors are called *variable-speed* motors because their speed regulation is much greater than 15%.

Example 15.1

Determine the speed regulation of a $\frac{1}{7}$-hp permanent magnet motor operating under load at 1650 rev/min when the no-load speed is specified as 1800 rev/min.

SOLUTION

Find the percent speed regulation.

Given

$$\text{Speed regulation} = \frac{n_{nl} - n_{fl}}{n_{fl}} \times 100\%$$

Evaluate	n_{nl} = 1800 rev/min
	n_{fl} = 1650 rev/min

Substitute Speed regulation $= \dfrac{1800 - 1650}{1650} \times 100$

Solve Speed regulation = 9.1%

As illustrated in Figure 15.5, the speed changes 8 to 10% for the shunt and the permanent magnet motors, up to 25% for the compound motor (depending on the amount of compounding), and an indeterminate amount for the series motor. The speed regulation for the unloaded series motor cannot be determined because, as pictured, the speed continually gets higher as the torque demand of the load gets smaller. The series motor is subject to **runaway.** To prevent runaway, the series motor must be operated under load at all times, which is done by directly coupling the series motor to its load or by coupling the load through a speed reducer. Operating a dc series motor unloaded can lead to its destruction, since the armature will reach high speeds, thus creating large, destructive *centrifugal forces*.

Permanent-Magnet Motor

The permanent-magnet motor, or PM motor, as it is commonly called, uses permanent magnets for its field poles; it does not have a wound field. Because the field flux is fixed by the magnets, the starting torque is directly proportional to the armature current ($\tau \propto I_A$). In smaller, unrestricted PM motors, the starting torque may be as high as 600% of the full-load torque; of course, the armature current is also 600% of the full-load current.

The schematic representation of the PM motor and its speed-torque characteristics for variation in terminal voltage are pictured in Figure 15.6. With its high torque and small physical size, the PM motor has the highest horsepower-to-weight ratio of any dc motor. It also has good efficiency, since power is not needed to create the field in the motor. Because the armature's flux density is small and the field's flux

FIGURE 15.5
Speed-torque characteristics for the PM, shunt, series, and compound dc motors at rated voltage with no external resistance.

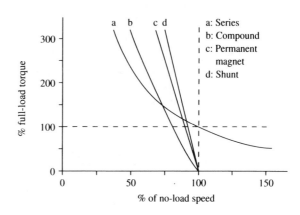

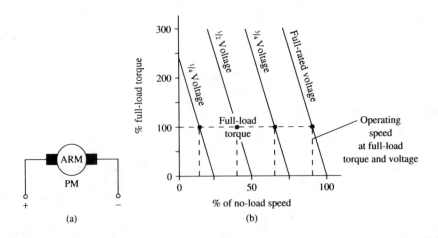

FIGURE 15.6

Permanent magnet motor: (a) The schematic representation of the PM motor depicts only the armature. (b) The motor speed-torque characteristic curves for a PM motor under terminal voltage control. The speed diminishes as the terminal voltage is reduced. However, the full-load torque remains constant.

density is much greater, the field is not twisted much. Thus, interpoles are not needed to maintain the position of the neutral plan in the PM motor.

By varying the armature's terminal voltage (as illustrated in Figure 15.6(b)), the speed can be varied. As pictured, the full-load torque remains constant at all speeds, as does the speed regulation, even though the speed is decreased. By using a variable-voltage supply (adjustable dc controller), the no-load motor speed can be made to follow the variation in voltage. Thus a lower voltage produces a proportionally lower motor speed.

Shunt-Wound Motor

The shunt-wound motor, or *shunt motor*, utilizes field windings wrapped around field poles to form an electromagnet to provide the magnetic field for the motor. As pictured in Figure 15.7(a), the field winding is connected in shunt (parallel) with the armature.

With the shunt field winding connected across the line, the field flux remains constant (like the PM motor). Thus, the shunt motor performs very much like the PM motor, although the starting torque is somewhat smaller. By using an adjustable controller, the speed of the shunt motor may be varied by changing the voltage of the armature. The variable-voltage, torque-speed characteristics are like those of the PM motor of Figure 15.6(b). Like the PM motor, the shunt motor is reversed by interchanging the line connection to the armature terminals.

By adding a *rheostat* (variable resistor) in series with the field winding, as illustrated in Figure 15.7(b), a simple and reliable means of changing the armature speed is possible. This inexpensive method, which is used to speed the motor up (not to slow it down), weakens the field flux and reduces the countervoltage. As shown in

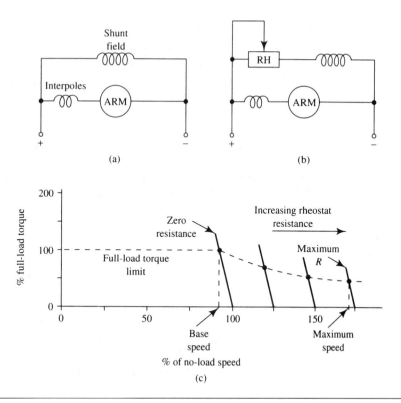

FIGURE 15.7

Shunt-wound motor: (a) The schematic representation of the shunt motor. (b) The speed of the motor may be increased by adding a rheostat in series with the shunt field winding. (c) The motor speed-torque characteristic of a shunt motor with a field rheostat installed in the field winding circuit. Below base speed, the full-load torque is constant; above base speed, the torque varies inversely with speed, producing a constant output power.

the speed-torque characteristic of Figure 15.7(c), the *base speed* of the motor is the full-load speed (at full-load torque and full terminal voltage), at which the motor operates with no additional field resistance (zero resistance).

As resistance is added to the shunt field winding by increasing the resistance of the rheostat, the full-load torque is lowered and the full-load speed is increased from the base speed to the maximum speed. The torque varies inversely with the change of speed, yielding a constant full-load output power ($P = \tau\omega$). Thus, when the rheostat resistance is increased, the speed increases and the torque decreases, thereby maintaining a constant output power from the motor.

> ➤ **As a Rule** The field must never be opened when the shunt motor is running with no load.

Operating an unloaded dc shunt motor and then opening its field winding will lead to the destruction of the armature, since the armature will reach high speeds (*runaway*), creating large, destructive centrifugal forces.

Series-Wound Motor

The field of the series-wound motor, or *series motor*, is connected in series with the armature, as shown in Figure 15.8(a). Since all the armature current passes through the field winding, the winding is made with just a few turns of heavy wire to lessen the resistance in the winding.

Because of the series connection, a change in load changes the armature current (I_A), the field flux, and the speed of rotation. Thus, a decrease in load current due to a lessening of the load on the shaft of the motor lowers the field flux density, lessening the countervoltage and increasing the shaft speed. As shown in the speed-torque curve of Figure 15.8(b), the series motor has poor speed regulation.

Because the flux in the field winding varies with load conditions, the series motor does not have a specified no-load speed. Instead, the speed varies inversely with the field flux and the load on the shaft. As the shaft is unloaded, the field flux diminishes, and the motor shaft speed increases.

> ➤ **As a Rule** The load must never be completely removed from the shaft of the series motor.

With the load completely removed, the speed will increase unimpeded to a point where the motor runs away, creating high centrifugal force and causing mechanical damage to the armature. To prevent runaway, the series motor must be operated under load at all times. This is done by directly coupling the series motor to its load or by coupling it to the load through a speed reducer.

Although the speed is not well regulated, the series motor produces very large amounts of torque for its size and weight. This is due to the starting torque being proportional to the square of the armature current($\tau \propto I^2$), which means that a starting current of three times the operating current will produce a starting torque of

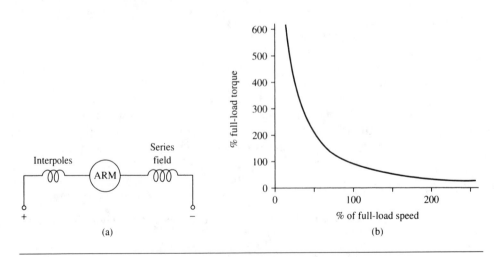

(a)   (b)

FIGURE 15.8

The series-wound motor: (a) The schematic representation of the series motor. (b) The motor speed-torque characteristic of a series motor. As the motor shaft is loaded, the speed drops and the torque rises.

nine times the full-load torque. Series motors are used as *traction motors* to provide the motive power for diesel-electric locomotives. A series motor works extremely well in applications requiring high torque and low speed for starting and high speed and low torque for running, as in the case of starting and moving trains, streetcars, and winches.

Compound-Wound Motor

The compound-wound motor, or *compound motor*, has both a shunt field winding and a series field winding wound on the field poles and connected as pictured in Figure 15.9(a). When the two field windings are connected to aid one another, the compound motor is classed as a *cumulative-compound motor*, and the motor has a specified full-load speed, which means that it can be safely operated unloaded.

The speed regulation of the cumulative-compound motor is not as good as the shunt motor, but because of the series field winding, it does respond better to variations in demand for torque than does the shunt motor. The speed-torque characteristic is shown in Figure 15.9(b). Like the shunt motor, the speed may be either decreased by varying the armature's terminal voltage or increased by adding resistance to the shunt field circuit. Cumulative-compound dc motors find application in machines that require some speed regulation when sudden loads are applied to the running motor as with punch presses, shears, etc.

Universal Motor

The universal motor is a type of series motor that derives its name from the fact that it is specifically designed to operate on both alternating and direct current. Since the field and the armature windings are connected in series with the armature current (I_A), the polarity of both fields changes in step with the change in the direction of

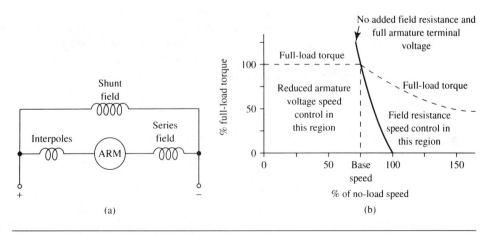

FIGURE 15.9

The compound-wound motor: (a) The schematic representation of the compound motor. (b) The motor speed-torque characteristics of a compound motor.

TABLE 15.2 Selected Starting and Operating Parameters of DC Motors

Motor Type	Starting		Speed Regulation	Speed Control	Reversible
	Torque	Current			
PM	High	High	Good	Yes	Yes
Shunt	Moderate	Moderate	Very good	Yes	Yes
Series	Very high	High	Varying	Yes	Yes
Compound	High	High	Moderate	Yes	Yes
Universal	High	High	Varying	Yes	Yes

the alternating current and the position of the commutator. Although the motor will operate satisfactorily on ac, the universal motor is usually specified as an intermittent-duty motor. This rating is due to the heating of the brushes caused by arcing between the brushes and the commutator, resulting from the lack of a neutral plane during ac operation. Because of the arcing, the universal motor is used only for light-duty, intermittent applications such as hand-held portable power tools (drill, router, hand saw, etc.) or household appliances and equipment (can opener, food processor, hand mixer, vacuum cleaner, etc.).

The speed-torque characteristics of the universal motor are virtually the same as those of the series-wound dc motor. Each has high starting torque and high no-load speed. Like the series motor, runaway is prevented in the universal motor by either directly coupling the load to the motor or by coupling the motor through a gear train to the load.

Having been specifically designed for high-speed operation, the universal motor typically operates at speeds up to and beyond 20,000 rev/min. To operate at these high speeds, the universal motor is made small and lightweight with a small laminated armature and laminated field poles, thus making it acceptable for use in hand-held portable tools and household appliances. The speed of the universal motor can be controlled by varying the voltage or by adding series resistance to the circuit. The direction of rotation is changed by interchanging the armature terminal connections. Table 15.2 summarizes some of the characteristics of the dc motors covered in this section. Information for the replacement of brushes and the maintenance of commutators is included in Appendix C, Section C12.

15.2 ADJUSTABLE-FREQUENCY AC DRIVES

Introduction

The use of adjustable-frequency, solid-state ac motor drives to precisely control the speed and to regulate the torque of ac squirrel-cage induction motors has rapidly grown. In the past, these features were obtainable only with dc motors and dc motor controls. By using a competitively priced, adjustable-frequency solid-state drive to control a squirrel-cage induction motor, the somewhat complicated, high maintenance dc motor can be replaced. This move results in a net savings in the initial cost of the motor, as well as the cost of maintenance, since the squirrel-cage induction motor is less expensive, more powerful, smaller, and has no brushes or commutator—both high maintenance items in dc motors.

The heart of the adjustable-frequency ac drive is the switching action of the power semiconductors, which may be insulated gate bipolar transistors (IGBTs), gate turn-off switches (GTOs), or thyristors (SCRs). These three-leaded, power solid-state devices are triggered ON by a *gate triggering processor* that controls the frequency of the gate pulse as well as the SCR's gate firing angle (turn-on point). When used in an appropriate electronic circuit (rectifier/inverter, converter), it is the semiconductor switching devices in conjunction with the trigger processor that provide an adjustable frequency at the drive's output to the three-phase induction motor. Adjusting the three-phase drive's frequency up or down causes the motor's synchronous speed to also move up or down, as demonstrated by Equation 14.1 ($n_s = f \times 60/p$). With the number of pairs of poles (p) constant, the synchronous speed (n_s) of the induction motor depends directly on the applied frequency (f).

Example 15.2

A 460-V, 60-Hz, three-phase, four-pole, squirrel-cage induction motor has a shaft speed of 1750 rev/min at full-load torque. If the motor is attached to an adjustable frequency ac drive that delivers a constant volts per hertz ratio, determine

    (a) The synchronous speed of the motor when the drive output frequency is 60.0 Hz, 37.5 Hz, and 15.0 Hz.

    (b) The slip speed (n_{sl}) at full-load torque for each of the three synchronous speeds.

    (c) The operating speed of the rotor (n_r) at full-load torque for each of the three synchronous speeds.

    (d) The percent slip at full-load torque for each of the three synchronous speeds.

    (e) The shape of the torque-speed curve for each of the three drive frequencies.

SOLUTION

    (a) Using Equation 14.1, calculate the synchronous speed (n_s) at 60.0 Hz, 37.5 Hz, and 15.0 Hz for the four-pole squirrel-cage induction motor.

Given $n_s = (f \times 60)/p$

Evaluate $f = 60$ Hz $f = 37.5$ Hz $f = 15.0$ Hz

             $p = 2$ pairs $p = 2$ pairs $p = 2$ pairs

Substitute $n_s = 60.0 \times 60/2$ $n_s = 37.5 \times 60/2$ $n_s = 15.0 \times 60/2$

Solve $n_s = 1800$ rev/min $n_s = 1125$ rev/min $n_s = 450$ rev/min

    (b) For a constant V/Hz drive, the slip speed is uniform at full-load torque regardless of the synchronous speed. It is calculated using the nameplate frequency.

Given $n_{sl} = n_s - n_r$

Evaluate $n_s = (60.0 \times 60)/2 = 1800$ rev/min   (Equation 14.1)

            $n_r = 1750$ rev/min

Substitute	$n_{sl} = 1800 - 1750$
Solve	$n_{sl} = 50$ rev/min (for all synchronous speeds)
Observation	Slip speed (n_{sl}) is the same at full-load torque regardless of the synchronous speed.

(c) The operating speed of the rotor (n_r) at full-load torque is equal to the difference between the synchronous speed (n_s) and the slip speed (n_{sl}).

Given	$n_r = n_s - n_{sl}$		
Evaluate	$n_s = 1800$ rev/min	$n_s = 1125$ rev/min	$n_s = 450$ rev/min
	$n_{sl} = 50$ rev/min	$n_{sl} = 50$ rev/min	$n_{sl} = 50$ rev/min
Substitute	$n_r = 1800 - 50$	$n_r = 1125 - 50$	$n_r = 450 - 50$
Solve	$n_r = 1750$ rev/min	$n_r = 1075$ rev/min	$n_r = 400$ rev/min

(d) Using Equation 14.2, solve for the percent slip for each synchronous speed (n_s) at 60.0 Hz, 37.5 Hz, and 15.0 Hz.

Given	%slip $= [(n_s - n_r)/n_s] \times 100\%$		
Evaluate	$n_s = 1800$ rev/min	$n_s = 1125$ rev/min	$n_s = 450$ rev/min
	$n_r = 1750$ rev/min	$n_r = 1075$ rev/min	$n_r = 400$ rev/min
Substitute	%slip $= [(1800 - 1750)/1800] \times 100$		
	%slip $= [(1125 - 1075)/1125] \times 100$		
	%slip $= [(450 - 400)/450] \times 100$		
Solve	%slip $= 2.78\%$ for a drive frequency of 60.0 Hz		
	%slip $= 4.44\%$ for a drive frequency of 37.5 Hz		
	%slip $= 11.1\%$ for a drive frequency of 15.0 Hz		

(e) See Figure 15.10 for the speed-torque curves for each of the three synchronous frequencies.

The shape of the torque-speed curve of the squirrel-cage induction motor (Figure 15.10) depends on the voltage and the frequency applied to the stator of the motor. Since torque is proportional to the square of the terminal voltage and the synchronous speed is directly related to the drive frequency, the drive system is designed to change voltage in direct proportion to the change in the drive frequency (stator voltage is halved when frequency is halved).

Volts per Hertz Ratio

The relationship between voltage and frequency is stated as a *volts per hertz ratio* (V/Hz). In Figure 15.10, as the curves shift position along the speed axis with each change in drive frequency, the V/Hz ratio produces a constant current in the stator

FIGURE 15.10

Family of speed-torque curves for a three-phase, design B squirrel-cage induction motor controlled by an adjustable frequency ac drive operating with volts per hertz torque control.

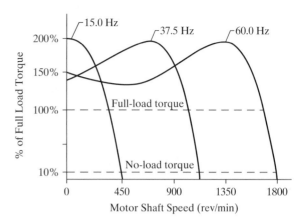

and a constant flux between the stator and the rotor, resulting in the shape of the torque-speed curve remaining the same, with the same amount of torque available at full load in each curve.

As the drive frequency is increased from zero, the voltage from the ac drive increases to maintain the V/Hz ratio at a constant value (resulting in a constant torque). An induction motor with a nameplate value of 460 V and 60 Hz has a V/Hz ratio of 7.6 V/Hz, while a 230 V, 60 Hz induction motor has a 3.8 V/Hz ratio. The voltage in the first motor increases 7.6 V for each 1 Hz increase in frequency. The increase in voltage is needed to compensate for a similar increase in the motor's equivalent inductive reactance (X_L) that increases in direct proportion with the increase in frequency $(X_L = 2\pi fL)$. As long as the voltage and frequency are held at the given V/Hz ratio, the motor will operate properly.

Advantages of an Adjustable Frequency AC Drive

As you know, when a three-phase induction motor is started at full line voltage, it takes at least six (6) times its nameplate current and produces between 150% and 200% of its full-load torque. At start-up, the current and torque do not track together as is the case once the motor clears the breakdown torque region of the speed-torque curve (see Figure 14.5). At start-up, the slip is very high, resulting in a substantial starting torque and a very large starting current.

By utilizing an adjustable frequency ac drive to lower the frequency, the starting torque is made to track the starting current with about 150% torque being produced by 150% starting current. This marked reduction in starting current (150% versus 600%) is due to the reduced frequency and voltage at the time of starting. Because the ac drive system provides starting at near zero frequency, a very small starting voltage is present, resulting in a reduced starting current. The lowered voltage provides adequate flux with reduced current, while the reduced frequency limits slip by lowering the synchronous speed of the motor—each aids in the reduction of the large inrush of starting current.

Types of Adjustable Frequency AC Drives

Three types of constant volts per hertz (V/Hz), solid-state, squirrel-cage induction motor drives are in common use today. The first two, the *rectifier/inverter drive system* and the *static frequency converter drive system*, are based on thyristor (SCR) technology while the third type, the *pulse-width modulation (PWM) drive system*, uses fast switching insulated gate bipolar transistor (IGBT) technology. Each system has many desirable features for driving squirrel-cage induction motors. Included in these features are soft start (no jerk), smooth speed control, S-curve START and STOP (constant acceleration/deceleration), speed and torque regulation, reversibility, dynamic or regenerative braking, and reduced starting current.

Rectifier/Inverter AC Motor Drive   The rectifier/inverter ac motor drive, as illustrated in the system block diagram of Figure 15.11(a), rectifies the incoming three-phase, 60 Hz line current using silicon-controlled rectifiers (SCRs) producing dc current that is then fed to the SCR inverter that produces an adjustable frequency, ac, rectangular pulse of current (for each phase) to drive the motor. This entire process may be under the control of a microprocessor-based control circuit that oversees the timing of the ON and OFF triggering of the SCR's gate circuits. The current inverter section of the drive produces current with frequencies ranging from near zero to twice the incoming line frequency (<10 Hz to 120 Hz). At start-up ($f <$ 10 Hz), the voltage to the motor is reduced by an increase in the firing angle of the SCR's gates in the controlled rectifier circuit.

Static Frequency Converter Drive System   The static frequency converter drive system characterized by the *cycloconverter* ac motor drive, as portrayed in the system block diagram of Figure 15.11(b), provides a direct ac-to-ac frequency conversion from the incoming three-phase line frequency. This type of motor drive incorporates a converter with a variable frequency output. In the cycloconverter, each phase of the three-phase induction motor is driven by its own dual converter consisting of 12 SCRs per phase. The adjustable output frequency of the cycloconverter is variable below the fixed line frequency of the input current. Both the induction motor's speed and voltage are varied by controlling the duration and phase of the gate pulses applied to the SCRs in the converter circuit.

These two first-generation drives have limitations in their control circuitry that result in poor low-speed performance (<10 Hz), restricting their application to the speed control of conveyors, pumps, and fans.

Pulse-Width Modulation (PWM) Drive System   The pulse-width modulation drive system, as depicted in the system block diagram of Figure 15.11(c), is a versatile, user friendly industrial drive system that can be programmed for volts per hertz (V/Hz) or *vector control* for improved low-speed operation, enabling the drive to control a wide range of applications from fans and conveyors to extruders and machine tools. With the introduction of the high-frequency, high-power, insulated-gate bipolar transistor (IGBT) for high-speed switching of power inverters, the performance of PWM drive systems has been improved and the use of PWM drives has become widespread, replacing dc motors and their drive systems.

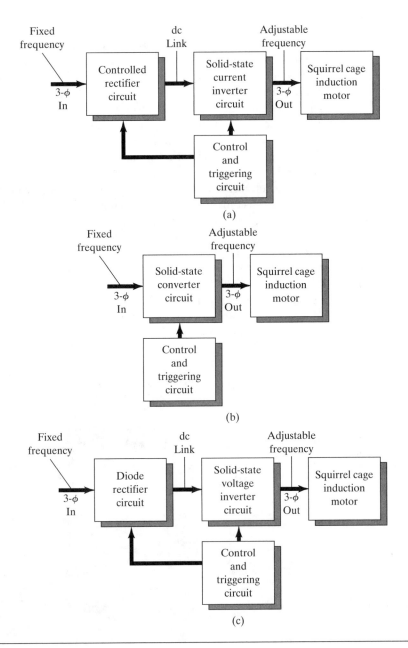

FIGURE 15.11

Block diagrams of adjustable-frequency ac drive systems. (a) Rectifier/inverter ac motor drive; (b) Static frequency converter drive system; (c) Pulse-width modulation (PWM) drive system.

Unlike the drives previously discussed, commercially available (off-the-shelf) pulse-width modulation drive systems have a great deal of sophistication and flexibility in their control of the inductions motor's speed, torque, and position for the most demanding drive application. This refinement is due to the inclusion of a microprocessor-based control system that incorporates solid state memory (RAM and ROM) for the storage of preprogrammed starting routines (*algorithms*) as well as an operator control station for selecting stored programs and programming the processor for custom applications. The modern PWM drive system, as pictured in Figure 15.12, is a powerful combination of software, hardware, and networking, enabling the ac induction motor to become an integral part of a **distributed control system** wherein the operation of local *work-cells* (including the induction motors) are under the control of a programmable logic controller (PLC).

Because the dc voltage into the inverter circuit does not need to be variable, the pulse-width modulation drive incorporates a simple three-phase, full-wave bridge rectifier (using power diodes) to produce a constant dc voltage that is applied to the PWM inverter utilizing IGBTs. Under microprocessor control, the base (gate) of the IGBTs are switched ON and OFF for varying time durations, producing short pulses of constant amplitude and varying widths, as shown in Figure 15.13 (a). The spacing, polarity (+ and −), and widths of the pulses are arranged so that their weighted average produces a *virtual fundamental frequency* sine wave with a

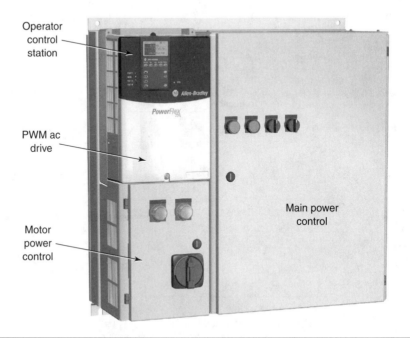

FIGURE 15.12

A commercial 0.37 to 15 kW (0.5 to 20 hp), panel mount, pulse-width modulation (PWM) drive system configured with a PWM ac squirrel-cage induction motor drive, an operator control station, communications link for networking, main power control, and motor start/stop—run/jog controls. Motor overload software is used to protect the motor from overload and overheating. *(Courtesy of Rockwell Automation, Inc.)*

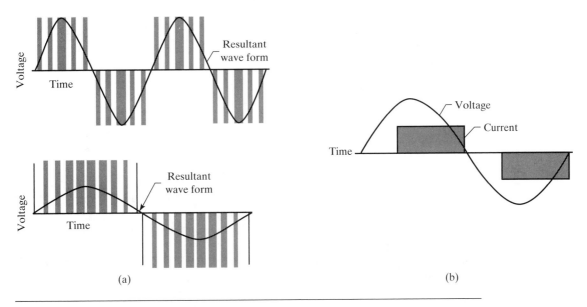

FIGURE 15.13
One phase of a three-phase output from: (a) A pulse-width modulated (PWM), insulated gate bipolar transistor (IGBT) switched inverter with the resultant voltage sine wave shown. The constant dc voltage into the inverter is chopped and sequenced into a series of constant amplitude pulses having variable spacing and width. As pictured, PWM varies both the frequency and amplitude of the resultant wave, keeping the V/Hz ratio constant. (b) A silicon-controlled rectifier (SCR) switched rectifier/inverter ac motor drive showing the rectangular current pulse. Since each of the three current pulses flows for 120°, their combination results in a revolving magnetic field that approximates a sine wave.

variable amplitude that is typically adjustable from near 0 to 400 Hz, producing a constant volts per hertz ratio.

Unlike the harmonic frequencies produced in conjunction with the rectangular wave of SCR-switched drive systems (Figure 15.13(b)), pulse-width modulation does not produce harmonics (5th and 7th) of the fundamental that, when present, are manifested in the drive torque as torque pulsations, called *cogging*. Cogging, an undersirable condition, occurs in SCR-switched drives when the drive is operated below 15 Hz.

Vector Control

From the discussion in Chapter 14 (pp. 476, 478) and from Figures 14.1 and 14.2, you know that current introduced into the stator coils causes a rotating magnetic field that links with the rotor conductors, thus inducing a current in them and causing a torque that spins the rotor.

The power producing torque ($P = \tau\omega$) that spins the motor shaft results from the interaction of two magnetic fields. The first is the rotating stator flux—the *motor flux*—that is developed from the *magnetizing current* component (I_{mag}) of the *stator current* (I_{stat}). The second is the flux resulting from the induced rotor current. This

FIGURE 15.14
The vector control algorithm, stored in the memory of the PWM adjustable drive system, provides a means of controlling the torque-producing current (I_{torq}). In *vector control*, the torque current tracks the torque demand of the load by providing full-load motor torque from zero to the maximum rated speed of the motor and its drive system.

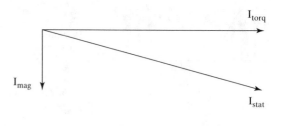

flux is termed the *torque flux* and the induced rotor current that produces this flux is the *torque current* (I_{torq}).

Like the currents in the transformer pictured in Figure 11.10 (p. 362), the magnetizing current (I_{mag}) is 90° out of phase with the component of stator current (I_{torq}) that provides torque to the motor shaft to produce power. The stator current (I_{stat}) is a composite (vector summation) of these two currents, as pictured in Figure 15.14.

To provide *vector control* of the squirrel-cage induction motor, the adjustable PWM drive system uses the vector control algorithm stored in the memory of the drive's microprocessor. This software routine is designed to provide a near constant magnetization current (I_{mag}) to the stator while at the same time proportioning the torque-producing current (I_{torq}) to provide rated starting torque without a large starting current. That is to say, the starting torque is made to track the starting current.

Because the torque current, I_{torq}, varies in direct proportion to the demand of the load attached to the motor's shaft over the entire operating range of the motor, the vector control feature of the PWM drive allows an off-the-shelf, ac squirrel-cage induction motor to be used in applications formerly reserved for high maintenance dc drive systems.

15.3   STEPPER MOTORS

Introduction

The typical *stepper motor* (also called step motor) is a reversible, brushless, four-pole dc motor. The motor shaft moves in *discrete angular increments* in step with pulsing currents in the stator windings. The stepper motor movement is directly controlled by digital signals, which drive *power solid-state devices* (transistors, MOS-FETs, etc.) ON and OFF, thereby switching current to the stator (field) windings in a sequential pattern. The continuous rotation of the stepper motor results from a continuous repetition of the pulse sequence developed by the digital circuitry of the electronic stepper motor controller.

Figure 15.15 pictures a simple block diagram of the major elements (pulse source, logic, and driver) of the stepper motor's control system. In its rudimentary form, the pulse source is a *pulse generator*, which generates square waves (pulses) with levels compatible with **TTL** (transistor-transistor logic; logic 1 = 5 V, logic 0 = 0 V). The *motor phase control logic* in its elementary form is made up of several **bistable**

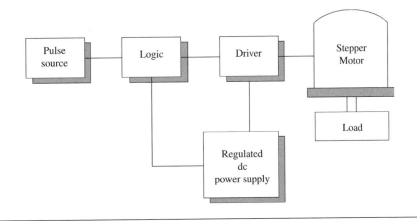

FIGURE 15.15
An elementary stepper motor control system.

multivibrators connected to generate four control signals, one for each phase of the stepper motor, in a repetitive time sequence, as pictured in Figure 15.16(a). As illustrated in Figure 15.16(b), the driver is basically a *solid-state switch* controlled by the logic, which directs current into the phase winding $(A, B, \overline{A}, \overline{B})$ of the stepper motor, causing the rotor to step. The solid driver is coupled to the phase control logic through an *optocoupler/isolator* to ensure that the digital control systems are separated from the electrical noise of the stepper motor.

Although the stepper motor has been available since 1935, it has become widely used only in recent years. The interest in stepper motors has come about because of readily available, low-cost, digital integrated solid-state control circuits as well as reliable microprocessor-based controllers, which allow the motor's rotor shaft and the attached load to be precisely and repeatedly moved without the use of a *feedback device* (**open loop**).

Types of Stepper Motors

There are three types of stepper motors available for industrial application: the variable-reluctance, the permanent magnet, and the permanent-magnet hybrid (simply called the hybrid) stepper motor. With the advent of modern solid-state motion controllers, the hybrid stepper motor is becoming widely applied; for many industrial applications, it is the motor of choice. The structure of each of the three types is briefly discussed.

The *variable-reluctance (VR) stepper motor*, illustrated in Figure 15.17, has a soft iron multipole rotor and a laminated core in the wound stator. When a set of stator coils is energized, a magnetic flux is created in the magnetic path between the stator and the soft iron rotor. Torque is developed to turn the rotor when the *reluctance* (magnetic resistance) in the magnetic path between the stator and rotor teeth is minimized. Initially, the teeth on the rotor are 15° out of alignment with the stator poles. The teeth on the rotor are attracted to the poles of the magnetically active

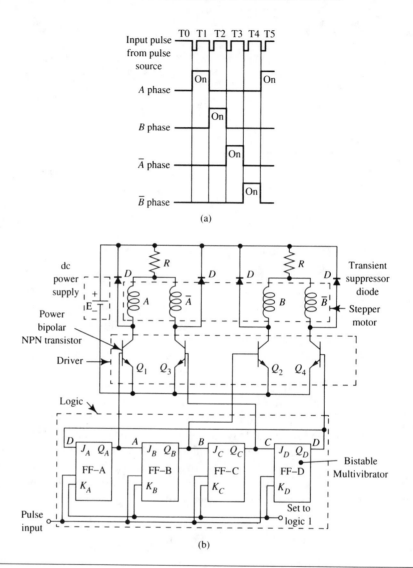

FIGURE 15.16

(a) Waveforms of one-ON, full-step switching (one phase energized during each period of time) developed in the logic section of the control circuit. (b) Logic and driver schematic for one-ON, full-step switching of the phase windings in the stepper motor.

stator, causing the rotor to align magnetically (rotate one 15° step) with the stator. In doing so, the reluctance of the magnetic path is minimized. Since the reluctance between the stator teeth and the rotor teeth changes with the angular position of the rotor, the motor is called a variable-reluctance stepper motor.

With each pulse from the pulse source, the motor phase control logic directs a different set of *stator coils* to be energized, causing the rotor to step one step. Thus, the stator field is rotated by switching from one set of stator poles to the next. As

FIGURE 15.17
Variable-reluctance stepper motor
(15° step angle) with the *A* phase
energized. The three coils are wound
on the stator using a four-pole
configuration.

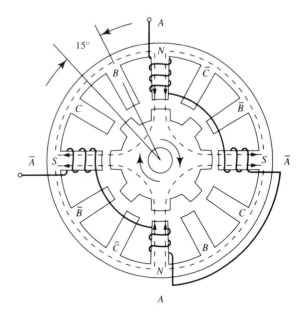

indicated by the letters used to designate the phases in Figure 15.17, the three-phase, four-pole stator is energized one phase after the other (A to B to C to A, etc.), causing the rotor to step.

Because the *detent torque* is zero, the load attached to the rotor shaft is free to turn once the stator windings are de-energized. Variable-reluctance motors are rarely used in industrial applications because of the lack of detent torque.

The *permanent-magnet (PM) stepper motor*, illustrated in Figure 15.18, has a permanent-magnet rotor that is polarized perpendicularly to the axis of the shaft (*radially magnetized rotor*), which provides detent torque (also called residual torque) to hold the load when the motor is not energized. When the four phases are energized in sequence, the rotor is stepped in 90° increments. Depending on the type of material used to form the magnetic rotor, the rotor can have more than two magnetic poles and is able to step in smaller angular intervals. The PM stepper motors have typical step angles of 90°, 45°, 15°, and 7.5°.

The *hybrid stepper motor*, illustrated in Figure 15.19(a), has a permanent magnet rotor that is *axially magnetized* (magnetized in the direction of the shaft). As shown in Figure 15.19(b), the cylindrically shaped permanent magnet is pressed onto the shaft and a *bonded* stack of soft steel-laminated, toothed rings is placed over each end of the magnet. The teeth of the two iron sections are displaced from one another by one-half tooth. As pictured in Figure 15.19(a), the stator is wound with **bifilar windings,** so that a single power supply can be used to power the four-pole, four-phase stepper motor. The power supply is attached to the center tap of the windings, as pictured in Figures 15.19(c) and 15.16(b).

The rotor of the 1.8° hybrid stepper motor typically has 50 teeth around its perimeter, whereas the stator has 48 teeth. The ratio of $\frac{48}{50}$ provides smoother operation with less jerk and less low-speed instability. Like the PM stepper motor, the

FIGURE 15.18
Permanent-magnet-rotor stepper
motor.

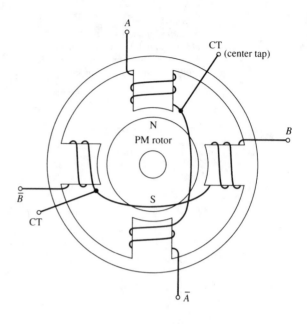

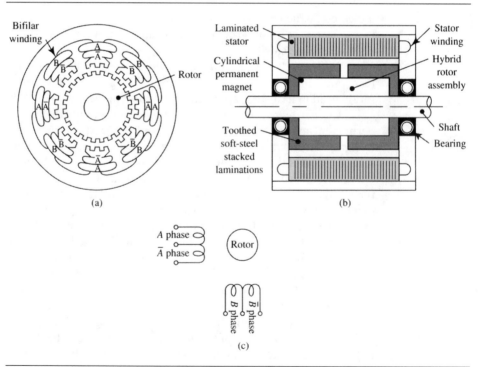

FIGURE 15.19
Hybrid-rotor stepper motor with (a) The bifilar stator windings and the toothed rotor shown; (b) The
makeup of the single stack hybrid rotor shown; (c) The schematic representation of the four-phrase,
six-wire stator winding shown.

hybrid motor has high detent torque to hold the load when the motor is not energized. The permanent-magnet hybrid stepper motor is produced in 7.2°, 3.6°, and 1.8° step angles, with the 1.8° hybrid motor being the motor of choice for precision motion control.

Operating Modes of the Logic Element

In modern stepper motor control systems (as illustrated in Figure 15.20), a **translator** is used to produce various operating modes to drive the stepper motor. The translator provides both the logic and the drive function for the stepper motor by accepting pulses from the pulse source and "translating" them into an appropriate sequence to control the stepping of the motor.

The translator can provide *one-ON full-step operation* (Figure 15.21(a)), *two-ON full-step operation* (Figure 15.21(b)), *one-ON-two-ON half-step operation* (Figure 15.21(c)), and *microstepping operation*. When a stepper motor is used with one-ON operation, one pulse into the translator provides current to one phase of the motor, causing one full step of motion. Two-ON operation provides one full step of motion for each pulse; however, two phases are driven at the same time, creating more torque at the shaft of the motor. When the motor is driven alternately one-ON and then two-ON, it half-steps for each pulse into the translator. Table 15.3 summarizes the sequence of phase excitation for four-phase stepper motor operation.

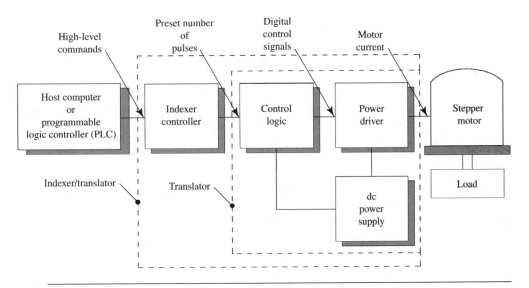

FIGURE 15.20
The block diagram of a typical stepper motor control system, where the blocks may represent either *discrete components* that are integrated to form the control system or *combined components* in a dedicated piece of hardware as with a translator or an indexer/translator (microprocessor-based programmable motion controller, which combines pulse source, control logic, and driver into a single package).

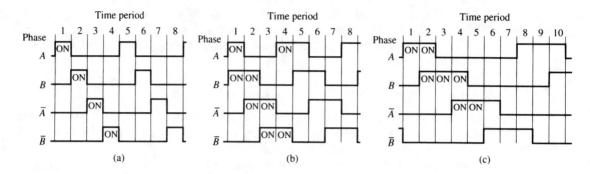

FIGURE 15.21
Stepper motor stator phase control signals from the translator part of the control system for (a) One-ON, full-step operation (one phase is ON during each time period); (b) Two-ON, full-step operation (two phases are ON during each time period); (c) one-ON-two-ON, half-step operation, which alternates between one phase ON per time period and two phases ON per time period.

Stepper Motor Resolution

The resolution of a stepper motor depends on the type of stepper motor and the method used to energize the phase windings of the stator. Resolution, measured in steps per revolution, can be determined from the motor's step angle specification by applying the following equation:

$$\text{Res} = \frac{360}{\beta} \tag{15.3}$$

where

Res = resolution of the stepper motor, in steps per revolution (steps/rev)
β = step angle, the angular distance the rotor advances per step (pulse), in °
360 = conversion factor, 360°/rev

Example 15.3

Determine the resolution of a four-phase hybrid stepper motor with a 1.8° step angle when it is
(a) Operated one-ON (one phase on at a time).
(b) Operated half-step (alternately one-ON, then two-ON)

SOLUTION

(a) The step angle for one-ON operation of the motor is the specified 1.8°/step. Find the resolution when the step angle is 1.8°.

Given

$$\text{Res} = \frac{360}{\beta}$$

Evaluate

$$\beta = 1.8°$$

TABLE 15.3 Truth Table for Four-Phase Stepper Motor Operation

Input Pulse	One-ON						Two-ON						Half-Step[a]					
	A	B	$\overline{A}$	$\overline{B}$	4-Bit Binary	Decimal Equiv.[b]	A	B	$\overline{A}$	$\overline{B}$	4-Bit Binary	Decimal Equiv.[b]	A	B	$\overline{A}$	$\overline{B}$	4-Bit Binary	Decimal Equiv.[b]
1	ON				1000[c]	8	ON	ON			1100	12	ON				1000	8
2		ON			0100	4		ON	ON		0110	6	ON	ON			1100	12
3			ON		0010	2			ON	ON	0011	3		ON			0100	4
4				ON	0001	1	ON			ON	1001	9		ON	ON		0110	6
5	ON				1000	8	ON	ON			1100	12			ON		0010	2
6		ON			0100	4		ON	ON		0110	6			ON	ON	0011	3
7			ON		0010	2			ON	ON	0011	3				ON	0001	1
8				ON	0001	1	ON			ON	1001	9	ON			ON	1001	9
9	ON				1000	8	ON	ON			1100	12	ON				1000	8

[a] Alternately one-ON and then two-ON.
[b] Decimal equivalent of the 4-bit binary number.
[c] 1 = ON, 0 = OFF.

Substitute $\text{Res} = \dfrac{360}{1.8}$

Solve $\text{Res} = 200$ steps/rev

(b) The step angle for half-step operation of the motor is one-half of the specified 1.8°/step. Find the resolution when the step angle is 0.90°.

Given $\text{Res} = \dfrac{360}{\beta}$

Evaluate $\beta = 0.90°$

Substitute $\text{Res} = \dfrac{360}{0.90}$

Solve $\text{Res} = 400$ steps/rev

To improve on the resolution and the operating characteristics associated with full-step operation of the 1.8° hybrid motor, a system employing *microstepping* is utilized. Microstepping uses two stator phases at a time that are simultaneously energized with currents that vary with time. Each ensuing pulse strengthens the current in one phase while lessening the current in the other. This technique creates very small, incremental changes in the position of the rotor. The resulting step angles are uniform and very small—in essence, a microstep. The two common microstep sizes are $\frac{1}{10}$ and $\frac{1}{125}$ of a full step, which is equivalent to a resolution of 2000 steps/rev and 25,000 steps/rev, respectively, when the full step is 1.8°.

At between 50 and 200 pulses per second (pps) (15 to 60 rev/min for a step angle of 1.8°), the stepper motor experiences rough operation and *misstepping* due to the natural resonance of the motor. By microstepping the motor, the resonance is damped (reduced in amplitude), allowing the motor to direct drive mechanisms without first going through a gear train or cogged belt drive system. The sophistication of the modern microstepping drive system provides for both constant acceleration (for starting and stopping with minimal jerk) and constant velocity (for smooth travel). As pictured in Figure 15.22, the pulse frequency is first varied to furnish constant rotor acceleration without a *misstep*; then the pulse frequency is fixed to provide constant velocity for rotating the load at a uniform full-load speed.

The information in Table 15.4 contrasts the fine *positioning resolution* (available with microstep operation) to the good positioning resolution available with full- and half-step operation of the 1.8° stepper motor. The table indicates that the positioning resolution of a five-thread-per-inch (.200-in pitch) zero backlash lead screw, directly coupled to the stepper motor, is .001 in per each full step, .0005 in per each half-step, .0001 in per each $\frac{1}{10}$ microstep, and .000008 in per each $\frac{1}{125}$ microstep.

Dynamic Operating Characteristics and Parameters

The operation of a given stepper motor depends on the specified parameters inherent in the design of that motor. In addition to the type of stepper motor, some or all

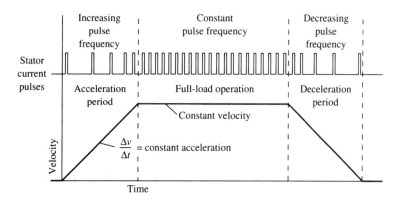

FIGURE 15.22
Typical stepper motor velocity profile with the regions of constant acceleration and constant velocity noted.

of the following parameters may be specified, and each should be considered when selecting a replacement or choosing a stepper motor for a new application. Table 15.5 lists the typical specifications for a stepper motor.

Step angle The step angle is the nominal angle (named angle 1.8°, 7.5°, etc.) through which the shaft of a stepper motor turns in response to an input of one pulse. The step angle determines the resolution of the motor, and it depends on the structure of the motor and the mode of operation.

Step angle accuracy The step angle accuracy is expressed as a plus or minus percent of the specified step angle, with ±5% being a typical accuracy. Because the error is noncumulative, the amount of error remains constant whether one step or 200 steps have been taken. For example, a step angle accuracy of 5% is 0.09° for a 1.8° step angle, which means that the deviation from true position would be 0.09°, whether the motor is rotated 1.8° or 360°.

TABLE 15.4 1.8° Stepper Motor Positioning Resolution for a Five-Thread-per-Inch Lead Screw

Operation Mode	Step Angle (°)	Resolution (steps/rev)	Lead Screw Travel per Step Angle (in)
Full-step, one-ON	1.8	200	.001
Full-step, two-ON	1.8	200	.001
Half-step, alternating	0.9	400	.0005
Microstep, $\frac{1}{10}$ full-step	0.18	2,000	.0001
Microstep, $\frac{1}{125}$ full-step	0.0144	25,000	.000008

TABLE 15.5 Specifications for a Four-Phase Hybrid 1.8° Stepper Motor

Parameter	Value
Operating voltage (dc)	6.0 V
Holding torque	50 oz-in
Step angle	1.8°
Current per phase	1.0 A
Resistance per phase	6.0 Ω
Inductance per phase	12 mH
Step angle accuracy (noncumulative)	±5%
Ambient operating temperature	−20°C to 60°C
Detent torque	5.0 oz-in
Shaft diameter	.250 in
Weight	1.2 lb

Holding torque Also called *static torque*, holding torque is the maximum external torque that can be applied to a stopped, energized stepper motor operating at its rated voltage without causing the rotor shaft to rotate continuously. Like all torque developed in a stepper motor, it depends on the current in the stator.

Detent torque Also called *residual torque*, detent torque is the amount of torque present at the shaft when the motor is not energized (unenergized motor). Detent torque is developed only in PM and hybrid stepper motors; when compared to the motor's holding torque, it is small (10% or less than the value of the holding torque).

Resistance per phase The phase resistance sets the current in the winding. Since torque depends on the available current in the winding, low resistance results in high torque. The current is calculated by applying the formula $I = E/R$ to the known winding resistance and the specified voltage per phase.

Inductance per phase The inductance determines how fast the current in the winding comes up to its full value once the winding is energized. High values of inductance slow the development of current in the coil. When high-speed operation is needed, motors with low values of inductance must be selected.

Rotor inertia Rotor inertia (moment of inertia), measured in lb-in^2 or kg·cm^2, is used to calculate the motor's acceleration.

Perhaps the most important of the parameters are those based on the speed-torque characteristic of the stepper motor. As pictured in Figure 15.23, the speed-torque characteristic of a typical stepper motor has several specified dynamic parameters:

Pull-in torque The maximum torque (combined inertial and frictional load) at which an energized stepper motor will be self-starting and run in **synchronism** (not skip a step while starting). The motor will start and run without skipped steps in the *self-starting and stopping* region of the speed-torque curve.

Pull-out torque The pull-out torque is the maximum torque (load) that can be applied to the *rotating* motor shaft without causing the motor to go out of synchronism (stopping). As illustrated in Figure 15.23, there are two operating ranges, self-starting and stopping, and the slew or pull-out range. Once started in the start-stop range, the motor can be accelerated beyond the pull-in torque into the slew range of operation without the loss of synchronism. The motor cannot be started in the slew range.

Maximum starting frequency The maximum starting frequency is the maximum frequency at which an unloaded stepper motor will be self-starting and also remain in synchronism with the pulse source.

As shown by the speed-torque curve of Figure 15.23, the pull-in torque decreases as the stepping frequency (f_s) increases. The speed of rotation in revolutions per minute can be determined from the stepping frequency in pulses per second by applying the following equation:

$$\omega = \frac{(f_s\beta) \times 60}{360} = \frac{f_s\beta}{6} \qquad \textbf{(15.4)}$$

FIGURE 15.23
Speed-torque characteristic of a four-phase hybrid 1.8° step angle stepper motor.

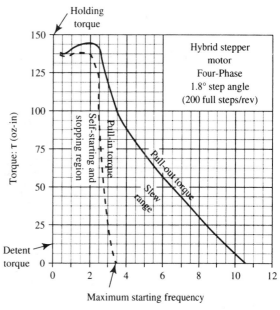

where

ω = angular velocity (speed) of the rotor shaft, in rev/min
f_s = stepping frequency (pulse rate) of the stator current, in pps
β = step angle, the angular distance the rotor advances per step (pulse), in °
360 = conversion factor, 360°/rev
60 = conversion factor, 60 s/min

Example 15.4

Determine the speed of rotation in revolutions per minute when the four-phase hybrid motor of Figure 15.23 is operating at 5000 pps with a torque of 50 oz-in at a full step angle of 1.8°.

SOLUTION   Find the speed in revolutions per minute.

Given $\omega = \dfrac{f_s \beta}{6}$

Evaluate f_s = 5000 pps

$\beta = 1.8°$

Substitute $\omega = \dfrac{5000 \times 1.8}{6}$

Solve $\omega = 1500$ rev/min

The angular distance moved by the stepper motor shaft can be expressed in revolutions when the step angle and the number of pulses into the stepper motor are known. The following equation is used to determine the angular displacement of the stepper motor's rotor shaft:

$$\theta = \frac{N \times \beta}{360} \qquad \textbf{(15.5)}$$

where

θ = angular displacement of the rotor shaft, in rev
N = number of steps (pulses) fed into the stepper motor
β = step angle, the angular distance the rotor advances per step (pulse), in °
360 = conversion factor, 360°/rev

Example 15.5

A permanent-magnet stepper motor with a 15° step angle is used to direct drive a ten-thread per inch (.100-in) lead screw. Determine
    (a) The resolution of the stepper motor in steps/revolution.
    (b) The distance the lead screw travels (in inches) for each 15° step of the stepper motor.

(c) The number of full 15° steps required to move the lead screw and the stepper-motor shaft through 17.5 revolutions.

(d) The shaft speed (in rev/min) when the stepping frequency is 220 pps.

SOLUTION   (a) Find the resolution of the 15° stepper motor.

Given $\text{Res} = \dfrac{360}{\beta}$

Evaluate $\beta = 15°$

Substitute $\text{Res} = \dfrac{360}{15}$

Solve $\text{Res} = 24 \text{ steps/rev}$

(b) Find the travel (in inches) of the lead screw when it is turned 15°.

Given   The lead screw advances .100 in for each revolution of the stepper motor shaft.

Observation Set up a proportion using the fact that a 360° rotation of the lead screw results in .100 in of travel.

$$\frac{15°}{360°} = \frac{s}{.100}$$

Solve for s:

$$s = \frac{15 \times .100}{360}$$

Solve $s = .00417 \Rightarrow .0042 \text{ in}$

(c) Find the number of steps needed to rotate the shaft through 17.5 rev.

Given $\theta = \dfrac{N \times \beta}{360}$; solving for N results in

$$N = \frac{360 \times \theta}{\beta}$$

Evaluate $\theta = 17.5 \text{ rev}$

$\beta = 15°$

Substitute $N = \dfrac{360 \times 17.5}{15}$

Solve $N = 420 \text{ steps}$

Observation An *indexer*, a microprocessor-based, programmable motion controller, is used to count out a preset number of pulses; it applies them to the translator section of the controller for the precise control of the stepper motor.

(d) Find the shaft speed (rev/min) of the motor when the stepping frequency is 220 pps.

Given	$\omega = \dfrac{f_s \beta}{6}$
Evaluate	$f_s = 220$ pps
	$\beta = 15°$
Substitute	$\omega = \dfrac{220 \times 15}{6}$
Solve	$\omega = 550$ rev/min

In conclusion, the stepper motor is driven by a digital control system (rather than continuous voltage), which counts the number of pulses applied to the motor, thereby producing a precise amount of shaft rotation. A properly designed and specified stepper motor system will repeatedly follow the digital control signals without the aid of a feedback device. Because the stepper motor is electronically controlled, it is well suited for use in computer-controlled automation systems.

CHAPTER SUMMARY

- DC motors are used in industrial applications where control of speed and torque is necessary.
- DC motors are commonly used with commercial solid-state controllers.
- DC motors are classified by the method used to produce the field (stator) flux.
- Interpoles, wound with compensating windings, prevent the neutral plane in dc motors from shifting.
- Starter circuits are used to limit current in dc motors.
- Speed regulation, expressed as a percent, is used to evaluate a dc motor's response to a change in its load.
- PM and shunt dc motors are constant-speed motors.
- Series, compound, and universal motors are variable-speed motors.
- Pulse-width modulation (PWM) adjustable frequency ac drive systems can be configured for either vector control or volts per hertz operation.
- Adjustable frequency ac squirrel-cage induction motor drives are used in applications formerly reserved for dc drive systems.
- Stepper motors are used in industrial automation systems, where they are usually operated without feedback (open loop).
- Permanent-magnet, variable-reluctance, and hybrid are three types of stepper motors.
- Translators take in digital signals (pulses) from the digital source and then translate them into control signals to drive the stepper motor.
- An indexer counts the number of pulses applied to the translator, providing precise positioning of the stepper motor shaft as well as the load.
- Full-step, half-step, and microstep are modes of stepper motor operation.
- The resolution of the stepper motor is improved when the step angle is made smaller, as with microstepping.
- The dynamic operating characteristics and parameters of the stepper motor are provided through the manufacturer's design specifications in conjunction with the motor's speed-torque curve.

SELECTED TECHNICAL TERMS

The following technical terms, abbreviations, and acronyms are defined in the glossary located after Chapter 16. You are encouraged to use the glossary to aid your understanding and to test your knowledge of these important terms.

armature	open loop
bifilar winding	runaway
bistable multivibrator	shunt field winding
commutation	step angle
commutator	synchronism
distributed control system	translator
interpoles	TTL

END-OF-CHAPTER QUESTIONS

Write T if the statement is true and F if the statement is false.

1. DC motors can be stopped and reversed quickly.
2. The PM dc motor has high starting torque but poor speed regulation.
3. Torque in dc motors is directly related to the current in the armature of the motor.
4. Once the armature starts to rotate, the armature current begins to decrease.
5. A dc motor starter is rarely used to start a dc motor.
6. Variation in operating speed of the shunt dc motor usually exceeds 20%.
7. When operated below 15 Hz, PWM drives do not undergo cogging.
8. The resolution of a stepper motor depends on the type of motor as well as the mode of operation.
9. Microstepping a stepper motor reduces low-speed resonance.
10. Holding torque is developed only in PM and hybrid stepper motors but not in variable-reluctance motors.

In the following, select the word or phrase that makes the statement true.

11. Stepper motors are usually operated (open loop, with feedback, from analog signals).
12. The dc motor having the highest starting torque is the (compound, universal, series) motor.
13. When the stepper motor is driven one-ON-two-ON, it will (full-step, half-step, microstep).
14. The universal motor can be operated as (either a series or shunt motor, a continuous-duty motor, either an ac or a dc current motor).
15. When de-energized, the rotor of the stepper motor is held by its (holding, detent, pull-in) torque.

Answer each of the following questions with a short answer in the form of a complete sentence. Include a restatement of the question in your answer.

16. What is the purpose of interpoles in dc motors?
17. When voltage speed control is used with the permanent-magnet motor, which parameters remain constant and which parameters are varied?

18. What is the principal function of an indexer in the control of a stepper motor?
19. Why does the speed of the shunt-wound motor increase when an external resistance is added to the field winding circuit?
20. What role does a translator play in the control of a stepper motor?

END-OF-CHAPTER PROBLEMS

Solve the following problems. Make sketches to aid in solving the problems; structure your work so it follows in an orderly progression and can easily be checked. Table 15.6 summarizes the formulas used in Chapter 15.

1. Determine the starting torque produced by a dc PM motor when the armature voltage is 180 V and the resistance of the armature winding is 15 Ω. Use Equation (15.1) and let $k\phi = 6.85$ so that the torque results in units of lb-in when the armature current is expressed in amperes.

2. A 90-V dc shunt motor has a full-load speed of 1725 rev/min and a full-load torque of 27.4 lb-in when the full-load current (combined armature and field) is 8.4 A. Determine
   (a) The full-load horsepower of the motor.
   (b) The initial starting current when a dc motor starter is used that limits the initial starting current to three times the full-load current.

3. The speed regulation of a compound motor is specified to be 22%. Determine the no-load speed when the base speed (full-load speed) is measured with a tachometer and found to be 2213 rev/min.

4. Determine the resolution of a four-phase hybrid stepper motor with a 3.6° step angle when it is
   (a) Operated two-ON.
   (b) Operated half-step.

5. Determine the amount of time (in seconds) it will take for a stepper motor with a 15.0° step angle, operating one-ON, to rotate through 28.0 rev when the pulse rate is 180.0 pps. *Note*: $t = \theta/\omega$ (Equation (6.13)).

TABLE 15.6 Summary of Formulas
Used in Chapter 15

Equation Number	Equation
15.1	$\tau = kI_A\phi$
15.2	Speed regulation $= \dfrac{n_{nl} - n_{fl}}{n_{fl}} \times 100\%$
15.3	$\text{Res} = \dfrac{360}{\beta}$
15.4	$\omega = \dfrac{f_s\beta}{6}$
15.5	$\theta = \dfrac{N \times \beta}{360}$

6. A 1.80° hybrid stepper motor operating half-step is directly coupled to an eight-thread-per-inch lead screw. Determine the distance traveled by the lead screw when a preset indexer sends 4355 pulses into the translator attached to the motor.

7. Determine the volts per hertz ratio for a 200-V, 60.0-Hz squirrel-cage induction motor.

8. A 230-V, 60-Hz, three-phase, two-pole squirrel-cage induction motor has a shaft speed of 3460 rev/min at full-load torque. If the motor is attached to an adjustable frequency ac drive that delivers a constant volts per hertz ratio, determine

   (a) The synchronous speed of the motor when the drive output frequency is set to 5.0 Hz and then 72.0 Hz.

   (b) The slip speed (n_{sl}) at full-load torque for each of the synchronous speeds.

   (c) The operating speed of the motor shaft (n_r) at full-load torque for each of the synchronous speeds.

   (d) The percent slip at full-load torque for each of the synchronous speeds.

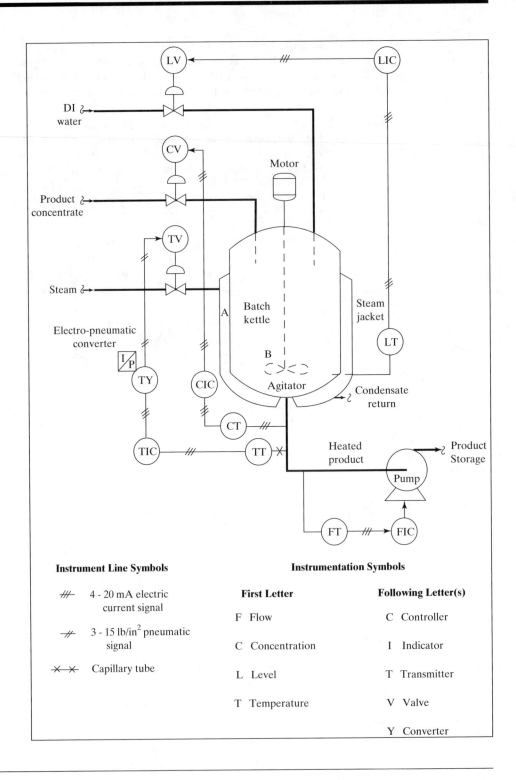

Instrument Line Symbols

///— 4 - 20 mA electric current signal

—//— 3 - 15 lb/in² pneumatic signal

—×—×— Capillary tube

Instrumentation Symbols

First Letter

F Flow

C Concentration

L Level

T Temperature

Following Letter(s)

C Controller

I Indicator

T Transmitter

V Valve

Y Converter

Chapter 16

Concepts of Automatic Control Systems

About the Illustration. A conceptual instrumentation diagram *is used to aid devel-opment of control systems without concern for specific hardware. This type of dia-gram employs* process equipment symbols *along with* instrumentation symbols *(circular **bubble***) to illustrate the principal points of measurement and control. As illustrated, this blending and heating process system is used to produce a sports drink from deionized water and product concentrate. Heating of the product is pro-vided by steam passing through a steam jacket (A), while mixing is accomplished by way of a motor-driven agitator (B). Heavy* process lines *indicate the flow of the prod-uct through the system, while lighter* signal lines *indicate the flow of electric current from the* transmitters *(CT, FT, LT, TT) to the indicating* controllers *(CIC, FIC, LIC, TIC) that develop the control signals (electric, pneumatic, etc.) for the actuators on the* control valves *(CV, LV, TV).* ■

INTRODUCTION

To improve productivity, reduce cost, heighten workplace safety, and enhance prod-uct quality, processes have been automated; that is to say, they are managed by an automatic control system.

An automatic control system normally incorporates a self-contained analog or digital control **instrument** that employs electronic, pneumatic, hydraulic, mechani-cal, or any combination of these technologies for the control of processes or machine tools. Today's predominant control technology is electronic, and many of today's digital electronic controllers are microprocessor based programmable logic controllers (PLCs) utilized to control batch and continuous processes.

This chapter introduces automatic process and servo control systems. Also studied are the elements making up the control system, the variables, and signals, as well as the principles, terms, and symbols common to automatic control systems.

549

CHAPTER CONTENTS

PERFORMANCE OBJECTIVES

Once you have read and studied each section; worked through each example with pencil, paper, and calculator; worked through the end-of-chapter problems; and answered the end-of-chapter questions, then you will be able to

- Comprehend basic control system terminology.
- Differentiate between closed-loop and open-loop control systems.
- Discern between process control and servo control systems.
- Identify and understand the function of the group of elements that make up a control system.
- Identify and classify the dynamic behavior of control elements.
- Understand the operation of continuous and discontinuous control modes.

16.1 CONTROL CONCEPTS

Terminology

To grasp the general principles of automatic control, it is necessary to first understand the terminology used to define a control system. Several important general terms associated with process control systems are pictured in the block diagram of Figure 16.1 and defined in Table 16.1.

In process control, it is the **controlled variable** that is maintained at a desired value called the **set point.** The controlled variable in a process might include one or more of the **process variables** such as the level of liquid in a tank, the temperature of an oven, or the amount of product flowing in a pipe.

For each controlled variable in a controlled process, there is a corresponding **manipulated variable** associated with it. In the case of product flowing in a pipe, the flow rate is manipulated by a **control valve.** It is the change in the manipulated variable that forces a change in the controlled variable, causing it to track the set point.

In any process, there are unwanted input variables, called **disturbance variables,** that influence the outcome of the process. For example, an open outside door would be a disturbance to the temperature in an air conditioned room. Disturbances have an adverse influence on the controlled variable by driving it away from the desired value—the set point.

FIGURE 16.1
The process block with the manipulated, controlled, and disturbance variables shown.

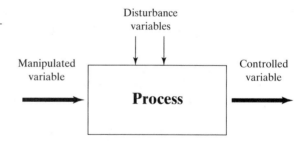

The function of the automatic control system is to adjust the manipulated variable so the desired value (set point) of the controlled variable is maintained even when influenced by the disturbance variables. Additionally, when the set point value is altered, the automatic control system responds by modifying the manipulated variable, causing the controlled variable to change to the new set point value.

Development of a Control System

In this section, three scenarios are provided to illustrate the level of sophistication in a heating system used to control the temperature within apartments in a multiple-story apartment house. The source of energy is from a hot water radiator system supplied by a central boiler.

No Control   The radiators in the apartment complex have no control valves with which to adjust the flow of hot water through the radiator. In this case, when the weather turns cold, the superintendent of the building fires up the boiler and turns on the circulation pumps to move hot water through the radiators. Since the

TABLE 16.1 Definitions of Terms and Variables Used to Define
            an Automatic Control System

Term/Variable	Definition
Control valve	A device that directly modifies the rate of flow of a fluid stream by adjusting the size of the flow passage.
Controlled variable	The process variable whose value is controlled by the control system.
Disturbance variable	An undesired input process variable that affects the controlled variable.
Manipulated variable	The process variable that is acted on by the controller so as to change its value.
Process variable	Any variable property of a process.
Set point	An input variable that sets the desired value of the controlled variable. It is expressed in the same units as the controlled variable; e.g., if the controlled variable is °C, then the set point will also be °C.

hot water continuously moves through the radiators, there is no direct control of the room temperature. This lack of control results in a very simple, low cost (no sensors, controller, etc.), inefficient, and impractical heating system.

Open-Loop Control   In this improved system, the radiators in the apartment house have been equipped with manually adjustable control valves that incorporate a pointer and a scale with relative calibrations (warm to hot) to aid in setting the temperature (*set and forget*). Once the valve is set, and assuming no human intervention and no direct comparison between the desired temperature (set point) and the actual temperature (measured value of the controlled variable), the system is operating as an **open-loop** control system.

The principal advantage of an open-loop control system is cost. It is inexpensive to build and it is inexpensive to operate and maintain. Since the controlled variable (temperature in this case) is not measured and no set point is provided, there is no need for an expensive controller or an operator to take corrective action. The major disadvantage of open-loop control is that any disturbances placed into the system by unexpected events are not correctable.

Closed-Loop Control   The shortcoming of the previous systems can be overcome by adding a wall-mounted thermometer to report the room temperature and by having a tenant (a person) present to visually read the thermometer. The tenant, acting as an operator, can compare the temperature reading to the desired temperature (the mental set point) of the controlled variable. Then, by manipulating the manual control valve to change the hot water flow rate (manipulated variable), the temperature (controlled variable) can be made to move toward the desired temperature (set point value), resulting in the room temperature being regulated.

The manually controlled heating system with a visual temperature indicator is operating as a **closed-loop** manual control system in which the tenant serves as the **controller.** This system, pictured in Figure 16.2, is a vast improvement over the impractical, no control heating system and a marked improvement over the open-loop, manual, set and forget control system. Since the rate of hot water flow (gal/min) through the radiator is now adjustable, the closed-loop control system, with *feedback* from the operator acting as the controller, is able to correct for disturbances as well as maintain the desired apartment temperature.

Closed-Loop Principles

Because an open-loop control system cannot ensure the desired output when a process is experiencing a disturbance, a close-loop system with feedback is utilized. In a closed-loop control system, as diagrammed in Figure 16.3, the controlled variable is directly measured (sensed), resulting in the following sequence of events.

First, a feedback signal, originating from the sensor, is transmitted to the controller where it is compared to the desired value (set point). Then, an error signal is produced, creating the manipulating signal needed to direct the change in the **final control element** (often a control valve) which directly controls the manipulated variable. Next, the change in the manipulated variable compels the controlled variable

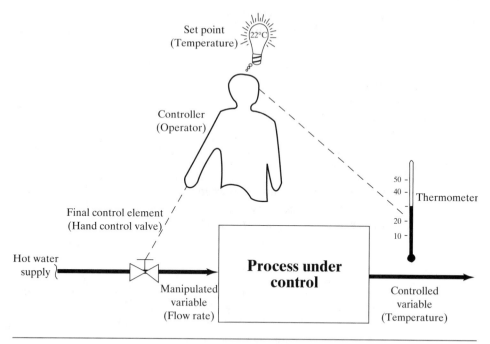

FIGURE 16.2
Closed-loop, manually operated temperature control system.

FIGURE 16.3
Conceptual diagram of a closed-loop
control system with feedback.

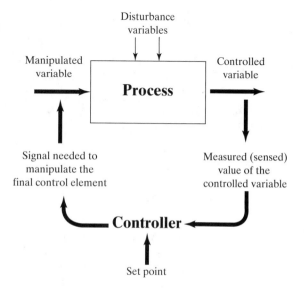

TABLE 16.2   Summary of the Functions Performed by
an Automatic Control System with Feedback

Function	Detail
Measure	Measure the value of the controlled variable.
Compare	Compute the error by algebraically adding the desired value and the measured value and then use the result to form a control action.
Manipulate	Using the control action, vary the manipulated variable to reduce the error between the desired value (set point) and the controlled variable.

to track the set point, finally eliminating the **error** from the control system. This
sequence of steps, as outlined here and summarized in Table 16.2, is the essence of
the operation of an automatic control system.

Example
16.1

An electric frying pan with a cover and an ON/OFF temperature controller is used
to cook a dinner entree. Name the variable and identify the unit of measurement for
   (a) The controlled variable and the set point.
   (b) The manipulated variable.
   (c) The disturbance variables.

SOLUTION

(a) Since the set point of the ON/OFF controller is expressed in units of temperature (°F or °C), the frying pan temperature (the controlled variable) is also
expressed as temperature in °F or °C.

Observation

From the definition of set point in Table 16.1, the set point is given in the same unit
of measure as the controlled variable.

(b) Since the flow of electric current is switched ON and OFF by the controller, the
electric current is the manipulated variable expressed in amperes (A).

Observation

The rate of heat (J/s) produced by the current passing through the resistance in the
heating element ($P = I^2R$) elevates the temperature of the frying pan. Since the resistance of the heating element is nearly constant, it is the current that is varied by
the controller; thus, the current is the manipulated variable.

(c) The disturbance variables include the ambient temperature of the room, the initial temperature of the entree in the frying pan, the thermal conductivity of the
metal in the pan and lid and the resulting rate of heat flow in and out of the pan
and lid, the setting of the vent opening in the cover, and the opening or closing
of the lid.

Since it provides a means of eliminating error, the use of feedback in an automatic control system simplifies the design of the control system by freeing the designer from having to know ahead of time the nature of the disturbance variables
and their effect on the controlled variables. In most cases, this results in a standard
approach to the design of automatic control systems in which the arrangement of

the **instrumentation** follows the same general pattern for each feedback loop regardless of the unique nature of the process or the number of controlled variables involved. Because of this redundancy in design, the process technician can easily comprehend the operation of all of the process *loops* in a complex automatic control system once one of the control loops is understood. Closed-loop control with feedback is the dominant model for use in automatic control systems.

Negative Feedback

Figure 16.3 pictures a closed-loop control system with feedback. The figure shows an arrow, coming from the controlled variable on the right to the controller, indicating the movement of the signal of the measured value of the controlled variable. This signal is a feedback signal and because its polarity is negative, it is called *negative feedback*.

Negative feedback is the cornerstone of all automatic control systems because it enables an automatic control system to be operated in a stable manner. With negative feedback, a control system will have an increase in stability within the system, a slower response to changes in the system, a wider bandwidth, and a reduction in the system gain. Although positive feedback is possible, it is not used in automatic control systems because it leads to destructive, unstable system operation.

16.2 BLOCK DIAGRAMS

Introduction

A block diagram is made up of *blocks*, representing each element in the automatic control system, and *lines*, connecting the blocks and representing the pathways through which signals flow. The intent of the block diagram is to convey the essential details of the structure of the elements making up the system and to provide an understanding of the movement of the signals in and out of the elements. Additionally, the block diagram provides a visual perception of the interdependence of the elements making up the system and their functional operation.

Basic Symbols

Only a few basic symbols are needed to portray an automatic control system in a simple, concise manner. Included in these symbols are the summing junction (a small circle), the block (a rectangle), the flow arrow (an arrow tip), and the node (a filled dot).

Summing Junction The summing junction, as pictured in Figure 16.4(a), represents the point where signals (represented by signal lines) are added to produce a resultant.

Block The block, as noted in Figure 16.4(b), represents an element in the block diagram in which one signal enters the element and one signal leaves the element. These two signals, one input and one output, form the **transfer function** (a mathematical expression) for that element. The transfer function defines the characteristic of an element in the control system.

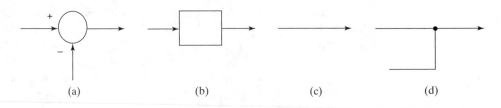

(a)   (b)   (c)   (d)

FIGURE 16.4
Symbols used in a control system block diagram: (a) Summing junction; (b) Block; (c) Flow arrow; (d) Node.

Flow Arrow The flow arrow, as shown in Figure 16.4(c), points the direction of signal flow through the signal line. A symbol (+ or −) may be added to the tip of the flow arrow to indicate the polarity of the signal in that line. With no symbol present, the polarity is assumed to be positive.

Node The node, as shown in Figure 16.4(d), represents a junction point in the block diagram where a signal branches.

Elements of an Automatic Control System

The block diagram pictured in Figure 16.5 is that of a general closed-loop process control system depicting the elements, variables, and control signals. From a structural point of view, the block diagram, as a whole, is perceived as a single functional unit controlling a process. In operation, it is seen as four *interdependent* parts, each carrying out a specific function.

Input The input portion of the control system consists of the *set point* (SP), and the *reference input element* that is used to convert the set point para-

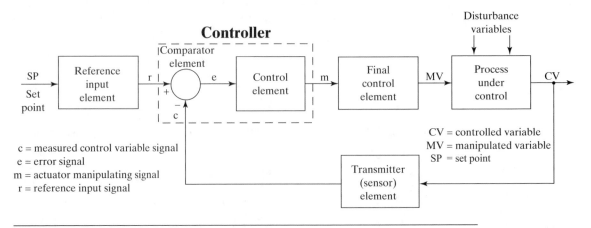

c = measured control variable signal
e = error signal
m = actuator manipulating signal
r = reference input signal

CV = controlled variable
MV = manipulated variable
SP = set point

FIGURE 16.5
Block diagram of a closed-loop process control system.

meter (temperature, flow rate, level, etc.) into a standard control signal format (r) representing the desired system output. The input portion, which provides the reference signal, initiates and governs the response of the system.

Output The output portion of the control system is the *controlled variable* (CV), one of the process variables, whose value is controlled by the control system. The controlled variable is the measurable result of the process under control and it is the value of the controlled variable that is sensed to originate the feedback signal. The set point and the controlled variable quantities are expressed in the same units, e.g., °C, gal/min, m^3, etc.

Feedback Path The feedback path of the control system incorporates the **transmitter** element, a device that includes a **sensor** to sense the value of the controlled variable and a **converter** to convert the sensed parameter to a standard control signal format (c) that represents the measured value of the controlled variable. This signal (c) must be of the same form (current, voltage, air pressure, etc.), type (analog or digital), and scale as the signal from the reference element (r) since the difference between them will result in the error signal (e). Table 16.3 lists several of the analog control signal formats.

Forward Path The forward path portion of the control system consists of four elements (comparator, control, final control, and process), two of which are integrated into the controller. The *comparator element* is the error-measuring element (error detector) that employs a summing junction to compare values of the reference input signal (r) and the measured value of the controlled variable (the output signal from the transmitter element (c)) to produce the error signal (e). The error signal is sent to the input of the *control element* for formatting into one of several *control modes* that serve to convert the error signal into the controller output signal, the manipulating signal (m). The manipulating signal is used to signal the actuator component (pneumatic diaphragm, electric solenoid, motor, etc.) located on the *final control element* (control valve, heating element, damper, etc.) to take action and change the value of the manipulated variable (MV). This action causes the controlled variable (CV) of the *process under control* to move toward the set point value (SP), thereby reducing the system error and bringing the process into control.

TABLE 16.3 Analog Control Signal Formats

Form	Range	Unit
Pneumatic (air)	3–15	lb/in^2
Electric current	4–20	mA (dc)
Electric voltage	±5	V (dc)
	±10	V (dc)

16.3 CONTROL SYSTEMS

Introduction

As previously noted, control systems may be classified as open-loop or closed-loop, depending on whether or not feedback is incorporated into the system. In practice, control systems may be classified in a number of different ways, including the nature of the signal—analog or digital—or the location of the controller—distributed or centralized, that is, near the final control element or in a remote central control room. Of the several ways of classifying a control system, the one most often used divides control systems into two groups, servo control (servomechanism) and process control.

Servo Control

A *servomechanism* utilizes electromechanical elements (gears, motors, links, etc.) to direct the positioning (movement) of physical objects such as an *x-y* table, a robot arm, or a solar array of photovoltaic cells.

In servo control, the position, velocity (speed), and/or acceleration of the load (an object) is made to follow the reference signal from the set point. The response of a servo system to a change in the reference signal value (set point) is very fast, taking less than a second. Because most servomechanisms are employed to keep the output position in direct correspondence with the input reference signal, they are referred to as *follow-up systems*. Follow-up systems are characterized by a control system that undergoes frequent changes in the reference signal.

Servo systems are easily identified by their final control element (called an actuator). Common components of the actuator include electric motors, hydraulic/pneumatic cylinders, and gear trains. The feedback element (sensor) in a servo system may employ a precision potentiometer, an optical encoder, a variable reluctance **transducer,** a tachometer-generator, etc.

A Servo Control System

A position control servo system is shown in the pictorial diagram of Figure 16.6(a). The servomechanism pictured uses a permanent-magnet dc motor as the *servomotor*, powering a synchronous-belt system and a rack and pinion drive, to position the load, θ_o, the controlled variable shown in Figure 16.6(b).

The servomotor positions the load through the synchronous-belt drive by rotating the pinion in the rack and pinion drive. The position of the load (θ_o) attached to the rack is sensed by the *feedback potentiometer*, which produces an output voltage proportional to the load's position. By using a two-step timing-belt drive (Figure 16.6(a)) to greatly reduce the angular velocity, the shaft of the feedback potentiometer is rotated through an arc (θ_f) of 340°, (0°–340°). The rotation of 340° represents a linear movement of the load through its entire range of travel from the left stop to the right stop.

The output voltage from the feedback potentiometer depends on the division of resistance resulting from the position of the potentiometer's wiper as it rides

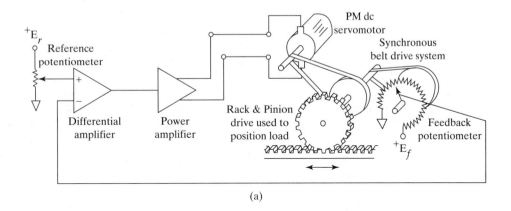

(a)

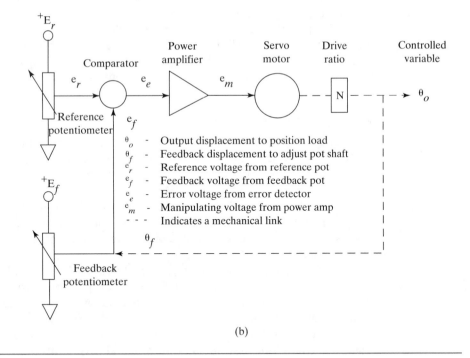

(b)

FIGURE 16.6

A closed-loop position control servo system: (a) Pictorial diagram (*From J. Michael Jacob,* Industrial Control Electronics *(Prentice Hall, Inc., 1988)*); (b) Block diagram.

along the resistive wire within the potentiometer. The output voltage from the potentiometer is the feedback signal (e_f) that is sent to the differential amplifier (comparator element) where it is subtracted from the reference signal voltage, e_r. The difference, e_e (error signal), from the differential amplifier, is sent to the power amplifier where it is amplified. The output voltage from the amplifier, e_m (manipulating signal), is used to energize the motor, causing the load to move.

Once the voltage signal from the position feedback potentiometer is equal to the voltage signal from the reference potentiometer ($e_f = e_r$), then the system is at rest. Because there is no error signal ($e_e = 0$ V), there is no output from the differential amplifier, resulting in no signal from the power amplifier ($e_m = 0$ V) to drive the motor. Thus, when the reference signal is changed, signaling the load to move to a new position, the servo system automatically changes the voltage to the servomotor (changing the manipulated variable) so that the new load position (the controlled variable) can be established.

From Figure 16.6, the variables (controlled, reference input, and manipulated) of a position servomechanism are the shaft angles (θ) of the pinion gear, reference input (command angle), and motor, respectively. The reference input element and feedback element (potentiometers in both of these cases) convert the mechanical displacement angles into electrical signals.

The positioning of a load utilizing a closed-loop servomechanism results in high accuracy with very little deviation from the desired position, a fast response by the servo system to changes in the reference input ($\ll 1.0$ s), and excellent stability while the dynamic response is being carried out by the system.

Servo Applications

Industrial Robots   Industrial robots represent one of the more common uses of servomechanism control. Their speed, precision, and flexibility lower the cost of manufactured goods by lowering production costs while improving the quality of the product.

An industrial robot, as defined by the Robot Institute of America (RIA), is a programmable multifunction manipulator designed to move material, parts, tools, or specialized devices through variable programmed motions for the performance of a variety of tasks.

The manipulator (the robot's arm and wrist) is built from links and joints, allowing flexibility in movement. The flexibility of the robot, provided by the several joints that are controlled by a servo system, allows the manipulator to move the end-of-arm tooling into any position within the robot's work envelope. The closed-loop servo control systems, Figure 16.7(b), use position and velocity feedback signals to control the movement of the arm and wrist joints of the manipulator, providing rapid and smooth movement of the tooling being manipulated.

Figure 16.7(a) pictures a Selective Compliance Assembly Robot Arm (SCARA robot) having one vertical (linear) joint and two revolute (rotating) joints. This electrically driven robot has been designed for fast, precise assembly work with repeatability of ± 0.05 mm ($\pm.002$ in). In addition to general light duty assembly of parts up to 5.0 kg (11 lb), the SCARA robot finds application in clean rooms.

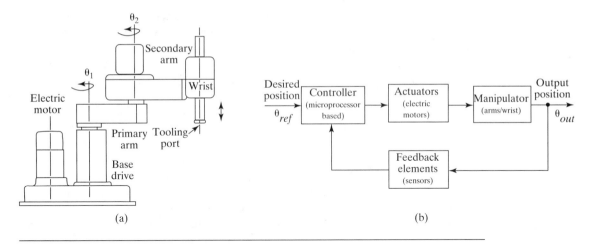

FIGURE 16.7
Multi-axis robot: (a) Basic configuration of a Selective Compliance Assembly Robot Arm (SCARA robot) with two revolute (revolving) and one linear movable joints; (b) The block diagram of a feedback control system for a multi-axis manipulator (robot arm and wrist).

The typical general end-of-arm tooling for all types of robots includes spray guns for painting, welding torches for metal joining, spotwelding guns for auto body assembly, grippers for part placement in units being assembled, pneumatic screwdrivers and wrenches for installing fasteners, and deburring tools for burr removal.

Machining Centers Machining centers employ numerical control systems that incorporate closed-loop servo systems to position the tooling and to control the motion of the tooling. The typical machining center is equipped with a programmable automatic tool changer, an arm designed to take preset tools from a magazine and insert them into a spindle, thereby enabling the machining center to perform several different operations on the workpiece.

Predetermined instructions are used to control a machining center and to specify position, direction of travel, speed of travel, cutting speed, and tool sequence. Numerically controlled machining centers are able to carry out various machining processes including drilling, turning, and milling.

Process Control

The other major class of control systems is process control. In process control, the variables that are controlled and manipulated are those commonly used in conjunction with the production of products such as food, dairy, beverages, pharmaceuticals, gasoline, microchips, and electric energy. The most commonly controlled and manipulated variables in a process are the temperature, flow rate, level, and pressure. Other variables include color, force, composition, humidity, viscosity, pH, density, conductivity, and proximity.

Unlike servo controlled systems where rapid changes in the reference input are typical, process control experiences only occasional moderate set point changes of 5 to 10% of the range of the set point value. With the set point virtually constant, the design of process control systems centers around how well the output responds to changes in the load (changes in the disturbance and controlled variables). Thus, the process control system has a relatively slow response rate, on the order of seconds or minutes, or, depending on the product being produced, even hours. Because most process control systems are employed to regulate the controlled variables at constant set point values and maintain constant process conditions, they are referred to as *regulator systems*.

Process Applications

Closed-loop process control systems are subdivided into two categories, continuous and batch processes. In a **continuous process,** material is continuously entering and leaving the process. As the material passes through the various stations in the process, one or more process variables of the material being processed are manipulated.

Extruding polymer (plastic—polypropylene, polyethylene, etc.) into sheets is an example of a continuous process. Figure 16.8 pictures the compounded resin, in the form of solid beads, being fed into the hopper, where it is passed into the screw and barrel of the extruder, passing through the thermocouple controlled heating zone (170–250°C/340–480°F).

The polymer beads enter the extruder's feed section (hopper end) as a solid and as the material is heated and conveyed by the tapered screw, it is compressed, causing the material to become plasticized. Once the material is forced (14–28 MPa/2000–4000 lb/in^2) past the choke plate and metered into the die, it is in a liquid state. The liquid product (polymer) is extruded through the pre-shaped die, setting the thickness of the sheet. Once extruded, the sheet passes through water-cooled cooling rollers and then onto the shear, where it is cut into sheets of the desired length.

In a **batch process,** the process is carried out in a series of sequenced, timed operations performed on the materials used to form the finished product. Some

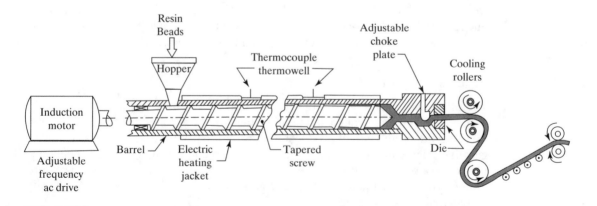

FIGURE 16.8

An extrusion process being used in a *continuous process* to produce plastic sheet material.

FIGURE 16.9

A *chip* (a device formed on a wafer) is pictured with an etched pocket into which a dopant has been thermally diffused to create either an N-type or P-type material. Thermal diffusion employs a chemical *batch process*.

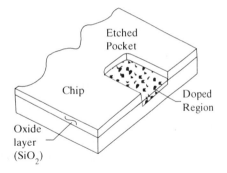

batch processes have several operations and each operation may have multiple steps. This is the case for processing silicon into semiconductors. This process has four basic operations, with each operation having several steps.

In semiconductor wafer fabrication (a batch process), silicon wafers undergo thermal diffusion (a chemical process) to carry out the *doping* operation—one of the basic operations used in the manufacture of semiconductors. In this operation, a boat (a quartz wafer holder) filled with silicon wafers is heated to around 1000°C while being exposed for a prescribed amount of time to a controlled atmosphere containing vapors of an appropriate chemical dopant. The doping operation creates either N-type or P-type material in previously etched pockets in the wafer's surface (Figure 16.9). These pockets form the N or P junctions for the manufacture of transistors, resistors, diodes, or capacitors. Once the doping operation is complete, the wafers are unloaded into a teflon boat for transfer to the next tool for further processing, and the sequence of doping is started over again on the next batch of silicon wafers.

16.4 DYNAMIC BEHAVIOR

Transfer Function

As was previously noted in Figure 16.5, most blocks in the block diagram of a closed-loop process control system have two signals associated with them. These two signals, one input and one output, form the *transfer function* (a mathematical expression) for each block in the control system, defining the characteristic of the element represented by the block. Thus, each element has a transfer function (K) that mathematically defines its characteristics. When the various elements are configured into a closed-loop control system, then the system also has a transfer function that defines its characteristics.

The transfer function (K) is made up of two parts, the magnitude and the phase. The magnitude is the size relationship between the input and the output signals, while the phase is the timing relationship (response time) between the input and output. For example, the magnitude may be such that the output signal is three times as large as the input signal, and the phase may be such that the output lags the input by 5 s.

> ➢ **As a Rule** The transfer function (K) of an element describes the magnitude (size) and phase (response time) relationship between the output signal and the input signal of that element.

Gain Transfer Function

The several elements represented by the blocks making up the control system, as well as the system itself, each have different dynamic behaviors, that is, different time dependent behaviors. Most of the elements have no lag in their response time between the input and output signals, as illustrated in Figure 16.10, and are classed as *non-dynamic*. This type of behavior is not dynamic at all since there is no phase difference (no time delay) between the input and the output signals. From a mathematical perspective, the response is algebraic (not time dependent) in nature. The output signal responds in step with the change in the input signal, and this response is immediate.

In effect, the magnitude and phase of the output signal are directly proportional to that of the input signal; the proportionality constant, K, is the function that relates the output to the input (e.g., output = $K \times$ input).

The transfer function of an element in a control system is given by the following expression.

$$\text{transfer function} = \text{output/input} \tag{16.1}$$

where

K = transfer function in mixed units or unitless
input = the input signal into an element represented by a block
output = the output signal from an element represented by a block

The transfer function provides a simple, concise, and complete way of describing an element's performance. When the element has its input and output signals expressed in the same unit, such as the amplifier in a servomechanism (Figure 16.6), then the transfer function is simply a unitless number. Whether mixed units or unitless, the transfer function is referred to as *gain* when there is no phase difference (no time delay) between the input and the output signals. As stated in the following equation, the output, c, is a function of the gain, K.

FIGURE 16.10
An example of an individual control element with no time dependence. A step input is used to study the response curve (output curve). Since the input is in a steady state before its change, the resulting output curve can be inspected as it transitions from one steady state to a new steady state.

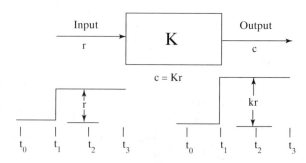

$$c = Kr \qquad\qquad \textbf{(16.2)}$$

where

    K = the gain transfer function (non-dynamic) in mixed units or unitless
    r = the input signal into an element represented by a block
    c = the output signal from an element represented by a block

> ➤ **As a Rule** The gain transfer function (K) of an element has the dimension of the output unit over the input unit—resulting in either a mixed unit or no unit at all.

Example 16.2

The pneumatic actuator attached to the control valve, pictured in Figure 16.11(a,b), is non-dynamic as it undergoes little or no time delay when operating. Also, the valve response is fast enough to be non-dynamic. The input/output characteristic of the actuator/control valve (a final control element) is described by the graph of Figure 16.11(c). Determine

    (a) The actuator/control valve gain (K).
    (b) The equation for the output flow rate (c) in terms of the input pneumatic pressure (r).
    (c) The output flow rate of the control valve when the input pneumatic pressure is 9.0 lb/in^2.
    (d) The input pressure to the actuator when the output flow rate is 175 gal/min.

SOLUTION

    (a) The gain, K, of the actuator/control valve is equal to the slope (rise over run) of the curve pictured in Figure 16.11(c).

Given

$$K = \frac{\Delta c}{\Delta r}$$

Evaluate

$$\Delta c = 250 - 0 = 250 \text{ gal/min}$$

$$\Delta r = 15 - 3 = 12 \text{ lb/in}^2$$

Substitute

$$K = 250/12$$

Solve

$$K = 20.83 \, \frac{\text{gal/min}}{\text{lb/in}^2}$$

    (b) Using the general equation for a straight line, $y = mx + b$, write the equation for the output flow rate (c) in terms of the input pneumatic pressure (r).

Given

$y = mx + b$. Restate the equation in terms of c, K, and r where $c = y$, $K = m$, $r = x$.

$$c = Kr + b$$

Solve for b.

$$b = c - Kr$$

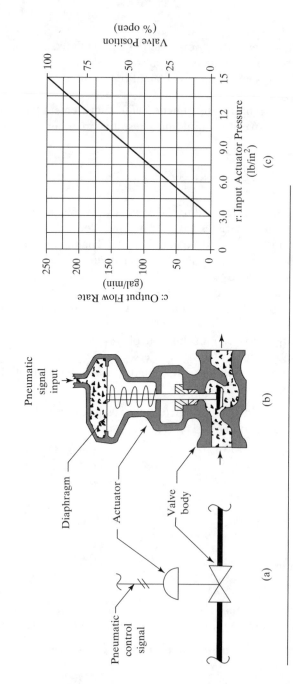

FIGURE 16.11

Control valve: (a) General instrumentation symbol shown with pneumatic diaphragm pilot assembly (actuator); (b) Variable-flow, pneumatically operated control valve used as the manipulating element (final control element) to vary the flow rate; (c) Control valve input/output graph, showing the 3.0–15 lb/in² air pressure that proportionally moves the valve from open to closed (0–250 gal/min).

Evaluate	$c = 250$ gal/min (Figure 16.11(c) when $r = 15$ lb/in^2)
	$K = 20.83 \dfrac{\text{gal/min}}{\text{lb/in}^2}$
	$r = 15$ lb/in^2
Substitute	$b = 250 - 20.83(15)$
Solve	$b = -62.45$ gal/min

State the equation $c = Kr + b$ in terms of r.

$c = 20.83r - 62.45$ gal/min

(c) Using the equation $c = 20.83r - 62.45$ gal/min, solve for the output flow rate (c) of the control valve when the input pneumatic pressure (r) is 9.0 lb/in^2.

Given	$c = 20.83r - 62.45$ gal/min
Evaluate	$r = 9.0$ lb/in^2
Substitute	$c = 20.83(9.0) - 62.45$
Solve	$c = 125$ gal/min
Observation	Verify the answer of 125 gal/min by entering the curve of Figure 16.11(c) at 9.0 lb/in^2 on the horizontal axis, projecting vertically up, striking the curve, and then projecting horizontally to the left, reading 125 gal/min from the vertical axis.

(d) Using the equation $c = 20.83r - 62.45$ gal/min, solve for the input pressure to the actuator (r) when the output flow rate (c) is 175 gal/min.

Given	$c = 20.83r - 62.45$ gal/min

Solve for r.

$r = (c + 62.45)/(20.83)$ lb/in^2

Evaluate	$c = 175$ gal/min
Substitute	$r = (175 + 62.45)/(20.83)$
Solve	$r = 11.40$ lb/in^2
Observation	Verify the answer of 11.40 lb/in^2 by entering the curve of Figure 16.11(c) at 175 gal/min on the vertical axis, projecting horizontally across to the right, striking the curve, and then projecting vertically down, reading 11.4 lb/in^2 from the horizontal axis.

Example 16.3

The feedback potentiometer of Figure 16.6(a), the sensor in the feedback path, has a gain transfer function as pictured in Figure 16.12. Determine

(a) The potentiometer's gain (K).

(b) The equation for the output voltage (c) in terms of the input angular displacement (r).

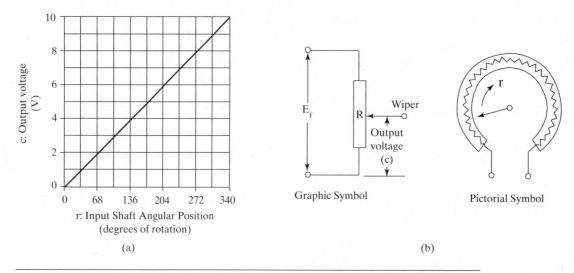

FIGURE 16.12
The potentiometer transfer function is volts per degree of rotation, a mixed unit: (a) An input/output graph of the feedback potentiometer of Figure 16.6(a); (b) The potentiometer, an angular position sensor, is used to change its shaft rotation, r (measured in degrees of rotation), an indicator of the load's position, into a voltage feedback signal (c).

(c) The output voltage of the potentiometer when the input angular displacement is 170° of rotation.

(d) The input angular displacement to the potentiometer when the output voltage is 8.0 V.

SOLUTION	**(a)** The gain, K, of the potentiometer (sensor) is equal to the slope (rise over run) of the curve pictured in Figure 16.12.
Given	$K = \dfrac{\Delta V}{\Delta^\circ} = \dfrac{\Delta c}{\Delta r}$
Evaluate	$\Delta c = 10 - 0 = 10\text{ V}$
	$\Delta r = 340 - 0 = 340^\circ$
Substitute	$K = 10/340$
Solve	$K = 0.0294 = 29.4 \text{ mV/degree of rotation}$
	(b) Using the general equation for a straight line, $y = mx + b$, write the equation for the output voltage (c) in terms of the angular displacement (r).
Given	$y = mx + b$. Restate the equation in terms of c, K, and r where $c = y$, $K = m$, $r = x$.
	$c = Kr + b$

Solve for b.

$$b = c - Kr$$

Evaluate $c = 10$ V (Figure 16.12(a) when $r = 340°$)

$K = 29.4$ mV/degree of rotation

$r = 340°$

Substitute $b = 10 - (29.4 \times 10^{-3})(340)$

Solve $b = 0.0$ V

Form the equation $c = Kr + b$ in terms of r.

$$c = (29.4 \times 10^{-3})r + 0 = (29.4 \times 10^{-3})r$$

(c) Using the equation $c = (29.4 \times 10^{-3})r$, solve for the output voltage (c) of the potentiometer when the input angular displacement (r) is 170° of rotation.

Given $c = (29.4 \times 10^{-3})r$

Evaluate $r = 170°$

Substitute $c = (29.4 \times 10^{-3})170$

Solve $c = 5.0$ V

Observation   Verify the answer of 5.0 V by entering the curve of Figure 16.12(a) at 170° on the horizontal axis, projecting vertically up, striking the curve, and then projecting horizontally to the left, reading 5.0 V from the vertical axis.

(d) Using the equation $c = (29.4 \times 10^{-3})r$, solve for the input angular displacement to the potentioner (r) when the output voltage (c) is 8.0 V.

Given $c = (29.4 \times 10^{-3})r$; solve for r.

$r = (c)/(29.4 \times 10^{-3})$ degrees of rotation

Evaluate $c = 8.0$ V

Substitute $r = 8.0/(29.4 \times 10^{-3})$

Solve $r = 272°$

Observation   Verify the answer of 272° by entering the curve of Figure 16.12(a) at 8.0 V on the vertical axis, projecting horizontally across to the right, striking the curve, and then projecting vertically down, reading 272° from the horizontal axis.

Gain is the simplest of the transfer functions. It contains a single number representing the magnitude and has no phase (time component) since it is non-dynamic. The output tracks the input well. In electronics, it is usual to think of gain as being unitless. But as you know, actuators and sensors input one type of quantity and output an entirely different quantity. So the gain transfer function (K) is expressed with mixed units and, in some cases (such as for amplifiers and attenuators), with no units at all.

Exponential Transfer Lag

Exponential transfer lag, commonly called **first-order lag,** is a time delay that depends on the time constant (τ) of an element in a process control system. When the element is subjected to a step input signal, the output is not instantaneous (as with the gain transfer function of non-dynamic elements); instead, the output responds by rising exponentially from its initial level to a new stable final level. Its output is not a step mirroring the input step but instead follows a response curve like that pictured in Figure 16.13(a).

In Figure 16. 13(a), it is seen that some time must pass ($t \geq 5\tau$), after the application of the input signal, before the output comes to a new steady state. The increase in the output from its former state to its new state is *transient* behavior and the graph of the output curve (represented by Figure 16.13(a)) is not linear, but is an *exponential function*. During the transient period of time between the initial and final state, the instantaneous value of the output is determined by applying Equation 16.3 for an increasing output level (a rise) or Equation 16.4 for a decreasing output level (a fall).

$$c(t) = Kr(1 - e^{-t/\tau}) \tag{16.3}$$

$$c(t) = Kr(e^{-t/\tau}) \tag{16.4}$$

where

K = the non-dynamic, steady state gain expressed in mixed units or unitless
r = the step change of the physical quantity into the element
e = the base of the natural logarithm; equal to 2.718 282
t = the time in seconds after the application of the input step change
τ = the time constant in seconds; the time required to reach 63.2% of an instantaneous step change

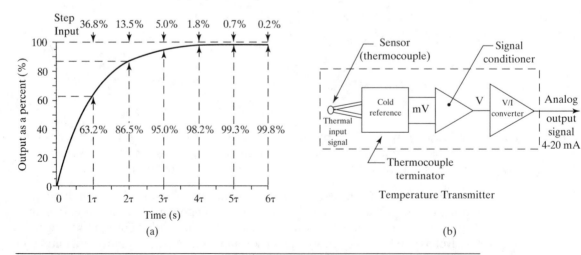

FIGURE 16.13
First-order lag: (a) Response curve to a step input; (b) Temperature transmitter element with beaded-type thermocouple sensor and voltage to current (V/I) signal converter.

c = the physical quantity out of the element before or after the transition period

$c(t)$ = the physical quantity out of the element as a function of time; i.e., when time (t) is between 0 and 5τ seconds, the output signal is in transition; when time (t) is greater than or equal to 5τ seconds ($t \geq 5\tau$), the output is considered to have reached its new steady level.

➤ **As a Rule** After responding to a step input, and once a first-order element reaches its steady output state ($t \geq 5\tau$), its new amplitude will simply be the steady state gain of the element (K) times the height of the input step (r). That is to say, after some transition time ($t \geq 5\tau$), a first-order element behaves like a non-dynamic element by assuming a stable, steady output.

Example 16.4

Assuming the 20.0°C to 300.0°C temperature transmitter of Figure 16.13(b) tracks the dynamic performance of the exponential response curve of Figure 16.13(a), and that the current out of the V/I converter is 4.00 mA at 20.0°C, and the steady state gain (K) is 0.0571 mA/°C, determine

    (a) The instantaneous amplitude of the output quantity, $c(t)$, at times equal to one, two, three, four, five, and six time constants ($t = 1\tau, 2\tau, 3\tau, 4\tau, 5\tau, 6\tau$) for a temperature step change, at the input to the thermocouple sensor, of 105.0°C (above the initial 20.0°C).

    (b) The time for the transition (in seconds) of the output signal from its initial state to its final stable state if the *beaded-type* thermocouple sensor of Figure 16.13(b) has a time constant of 2.0 s.

    (c) The time dependent transfer function for a first-order lag such as that found in a thermocouple sensor in a temperature transmitter element.

SOLUTION   **(a)** Using Equation 16.3, find the instantaneous amplitude of the output, $c(t)$, at $t = 1\tau, 2\tau, 3\tau, 4\tau, 5\tau, 6\tau$ s when the change in temperature is 105.0°C above the starting temperature of 20.0°C and the steady state gain (K) is 0.0571 mA/°C.

Given

$$c(t) = Kr(1 - e^{-t/\tau})$$

Evaluate

$$r = 105.0°C$$

$$K = 0.0571 \text{ mA/°C}$$

$$Kr = 105.0(0.0571 \times 10^{-3}) = 0.00600 \text{ A} = 6.00 \text{ mA}$$

$$t = 1\tau, 2\tau, 3\tau, 4\tau, 5\tau, 6\tau \text{ s}$$

Substitute

for 1τ   **for 2τ**

$c(t) = Kr(1 - e^{-t/\tau})$ $c(t) = Kr(1 - e^{-t/\tau})$

$c(t) = 0.00600(1 - e^{-1\tau/\tau})$ $c(t) = 0.00600(1 - e^{-2\tau/\tau})$

$c(t) = 0.00600(1 - e^{-1})$ $c(t) = 0.00600(1 - e^{-2})$

$c(t) = 0.00600(0.6321)$ $c(t) = 0.00600(0.8647)$

Solve $c(t) = 3.79 \text{ mA}$ $c(t) = 5.19 \text{ mA}$

Substitute	**for 3τ**	**for 4τ**
	$c(t) = Kr(1 - e^{-t/\tau})$	$c(t) = Kr(1 - e^{-t/\tau})$
	$c(t) = 0.00600(1 - e^{-3\tau/\tau})$	$c(t) = 0.00600(1 - e^{-4\tau/\tau})$
	$c(t) = 0.00600(1 - e^{-3})$	$c(t) = 0.00600(1 - e^{-4})$
	$c(t) = 0.00600(0.9502)$	$c(t) = 0.00600(0.9817)$
Solve	$c(t) = 5.70 \text{ mA}$	$c(t) = 5.89 \text{ mA}$
Substitute	**for 5τ**	**for 6τ**
	$c(t) = Kr(1 - e^{-t/\tau})$	$c(t) = Kr(1 - e^{-t/\tau})$
	$c(t) = 0.00600(1 - e^{-5\tau/\tau})$	$c(t) = 0.00600(1 - e^{-6\tau/\tau})$
	$c(t) = 0.00600(1 - e^{-5})$	$c(t) = 0.00600(1 - e^{-6})$
	$c(t) = 0.00600(0.9933)$	$c(t) = 0.00600(0.9975)$
Solve	$c(t) = 5.96 \text{ mA}$	$c(t) = 5.99 \text{ mA}$

Observation The step change in temperature at the input to the temperature transmitter causes the converter output current to transition from 4.00 mA at 20.0°C to a final stable output signal of approximately 10.0 mA (4.00 + 5.99 = 9.99 mA) at 125.0°C (20.0 + 105.0 = 125.0°C) in six time constants ($t = 6\tau$ s). At the time equal to 3τ, the output current (as pictured in Figure 16.14(a)) is within 5.0% of its final value; at 4τ, the current is within 2.0%; and at 5τ, the response is considered to be complete because only 0.7% remains.

(b) Find the transition time of the transmitter's output when the thermocouple's time constant (τ) is 2.0 s.

Observation The time constant (τ) of 2.0 s for this thermocouple is the response time required for the output to reach 63.2% of the instantaneous step change in the input temperature.

Given As a rule of thumb, the transition time for an element with transient behavior is approximately five time constants.

$t = 5\tau \text{ (s)}$

Evaluate $\tau = 2.0 \text{ s}$

Substitute $t = 5 \times 2.0$

Solve $t = 10 \text{ s}$

Observation What sets a first-order element apart from a steady state (non-dynamic) element in a process control system is the length of time it takes for the element to respond to a change in the input. When the time constant is very short (≈ 0 s), the transition time is nearly instantaneous and the element is considered to be non-dynamic.

Solve **(c)** Figure 16.14(b) illustrates the time dependent transfer function for a first-order lag.

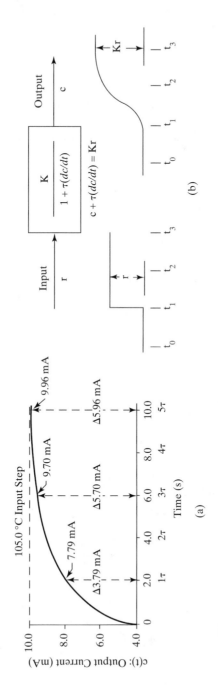

FIGURE 16.14

The response of a first-order element to a step at the input: (a) The output signal response, of the temperature transmitter of Figure 16.13(b), to a 105.0 °C step change in temperature at the input to the thermocouple sensor where $c(t) = 6.00(1 - e^{-t/2.0}) + 4.00$ mA; (b) An individual control element with the time dependent, dynamic, first-order lag transfer function noted.

From this example, the quickness of an element's response to a change at its input (its dynamic performance) is determined by the dynamic characteristics of the first-order lag of that particular element within the process control system. First-order lags are usually found in the process control system itself and, to a lesser degree, in control valves and *primary elements* (sensors). Primary elements, with first-order lag, include temperature, level, flow, and pressure transmitters.

For first-order lag elements, one of the important parameters making up an element's dynamic characteristic is its time constant. From the response curve of Figure 16.14(a), the maximum rate of change in the response, to the input step change occurs at the origin of the response curve from 0.0 to 2.0 s, the time equal to one time constant; i.e., for the first time constant the current moved from 4.0 mA to 7.78 mA (63.2%) in the first 2.0 s.

Since a process control system is a regulator system, the presence of a first-order lag in an element of a few seconds or more is acceptable as most process control systems respond slowly to change in the process variables. In addition to the time constant, other first-order lag parameters include rise time, dead time, and frequency response.

In a process control system, five time constants (5τ) is a reasonable approximation for the time needed by a first-order element to transition from a stable condition to a new steady state. A time of five time constants will place the new output value within 1% (0.7%) of the nominal set point value, which is well within the tolerance limit of most process control systems.

Dead Time

In response to change in the controlled variable or movement in the set point value, some process control loops exhibit a delay in time before any measurable dynamic response at the manipulated variable of the final control element is noticed. During the time delay, the control system appears dead as there is no observable response and no control signal is available to initiate corrective action.

This loss of control is due to **dead time** that has somehow found its way into the dynamics of the process control system, possibly due to a design flaw, improper sensor placement, or poor dynamic properties of the final control element. Figure 16.15 pictures several output response curves, each with dead time in their output.

Because dead time causes the overall loop control to deteriorate, it is sometimes difficult to establish good, stable process control without a microprocessor based controller. With traditional controllers, there is no ready way to compensate for systemic dead time, but with a microcomputer, an algorithm (a mathematical model) of the process can be created to include compensation for the dead time.

In the blending and heating instrumentation diagram, pictured on the first page of this chapter, a steam jacket is used to heat the product and a temperature transmitter (TT) is used to sense the temperature. The temperature transmitter signals the temperature indicating controller (TIC) to open or close the control valve to increase or decrease the rate of flow of steam (the manipulated variable) controlling the temperature of the heated product.

FIGURE 16.15

Response of system elements, with dead time, to a step input: (a) Non-dynamic element: once the dead time expires, the output tracks the input with its gain; (b) First-order element: once the dead time expires, the output rises exponentially for five time constants, completing its response by returning to its steady state gain; (c) Second-order element: once the dead time expires, the output passes through its damped response, returning to its steady state gain.

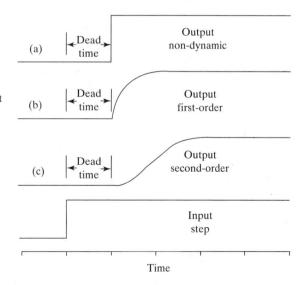

Transportation lag (a type of dead time) may occur in this process control loop if the temperature transmitter (TT) is located too far downstream from the outlet of the batch kettle. As the thermal sensor is moved farther and farther down the outlet pipe, more and more dead time is created in the feedback loop. Since velocity is equal to displacement divided by time ($v = s/t$), then the dead time is equal to the distance from the bottom of the batch kettle to the location of the *thermowell* (point where the thermal sensor is installed), divided by the velocity of the product inside the pipe.

Good design, and proper selection of sensor element parameters, minimal backlash, and free travel in mechanical final control devices, and careful positioning of sensors, can help to eliminate dead time in process control loops and increase the stability of the entire system.

Higher-Order Lags

When several first-order lag elements with differing time constants are cascaded in series within a control loop or when energy storage elements such as inductors, capacitors, or springs; or energy converting elements such as mechanical dampers; or electrical resistance are present, then the dynamics of the system is described with a second-order transfer function having a characteristic S-shaped response curve. Several second-order response curves are pictured in Figure 16.16.

The dynamics of a loop element with a second-order transfer function is characterized by its damping coefficient and its resonant frequency. The shape of the second-order response curve, resulting from a step input, is determined by the value of the *damping coefficient,* while the response time is dependent on the steady state *gain* and the value of the natural *resonance frequency* of the element.

The damping coefficient (ζ) ranges from a decimal fraction less than one, for example, 0.1, 0.2, etc., representing an underdamped condition to an integer value,

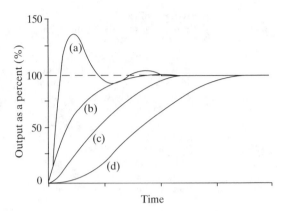

FIGURE 16.16
Response of a second-order element to a step input. A first-order element output response curve is included for reference: (a) Underdamped response curve; (b) First-order response curve; (c) Critically damped response curve; (d) Overdamped response curve.

such as 1, 2, etc., representing a critically damped condition when the coefficient is equal to 1 and an overdamped condition when the coefficient is greater than 1. When a loop element is overdamped ($\zeta > 1$), the system response to a step input is subdued, behaving in a sluggish manner. On the other hand, with an element underdamped ($\zeta < 1$), the system response is fast acting, resulting in the output overshooting and oscillating before coming to rest at a new level. When the system is set up to be critically damped ($\zeta = 1$), it transitions smoothly and rapidly to a new level without overshoot or oscillation.

Unlike the exponential response curve of first-order elements, the second-order output response curve does not have its maximum rate of change at the origin; it occurs at a later period of time in the curve. Mathematically, a second-order response curve is very complex, requiring higher-order differential equations for its description.

16.5 CONTROL MODES

Comparator and Control Element

As was pictured in Figure 16.5, the comparator element (the error detector unit) and the control element (the unit that implements the control modes) make up the main components of the controller. The controller provides automatic regulation to the controlled variable (CV) by computing an error signal and using it to produce the control action.

In operation, as the automatic regulator of the process loop, the process controller begins its operation by creating an error signal at the comparator's summing junction. The summing junction compares values of the set point's reference input signal to that of the feedback element's signal (the measured value of the controlled process variable), producing the error signal. As previously shown in Figure 16.5, the summing junction of the comparator adds the positive reference signal (r) to the negative feedback signal (c) to create the error signal (e) as noted in the following equation.

$$e = r - c \tag{16.5}$$

where

> e = the error signal—result may be positive or negative
> r = the reference signal derived from the set point
> c = the signal from the transmitter, obtained by measuring the value of the controlled variable
> *Note*: All signals are expressed in the same units.

The error signal, the output from the comparator, is sent on to the control element for formatting into one of several control modes that serve to change the error signal into the manipulating signal (m), the controller's output signal. The manipulating signal is used to signal the final control element to take action by changing the value of the manipulated variable (MV), thereby forcing the controlled variable (CV) to move toward the set point value, reducing the error (e) in the system and bringing the process into control.

ON-OFF Control

The simplest controllers use an ON-OFF control device that has a two-position control action—full ON or full OFF. ON-OFF control is classed as *discontinuous* control, because it is ON part of the time and OFF part of the time. Although inaccurate, this method of control is satisfactory for systems that respond slowly and have extended cycle times, as in a residential heating and cooling system.

ON-OFF control is the most common type of controller. It is employed in residential heating and cooling systems and in domestic hot water heaters. In heating and cooling systems, the ON-OFF controller is called a *thermostat*, a two-position *automatic switch* with a bimetal sensing element that controls a set of contacts or a mercury tilt switch. The thermostat's switch contacts are used to turn the system on or off by signaling the contactor in a two-wire control system.

In addition to the thermostat (a type of temperature switch), other types of two-position automatic switches are used in ON-OFF control systems. These include float switches, pressure switches, and limit switches.

In a home heating system, the furnace will come on when the temperature falls below the set point temperature and will go off after the temperature rises above the set point temperature. As seen in Figure 16.17, this cycling action causes the desired temperature to *oscillate* around the set point in a narrow region called the **dead band**. Since no control takes place within the dead band, an error is present in the control system.

The error represented by the dead band is usually not noticeable, as the interval of temperature represented by the dead band is small, 1 to 4 percent of the full range. Since no control action takes place during the time the temperature is within the dead band, the system does not undergo prolonged cycling in an attempt to keep the temperature at the set point. The reduction in cycling results in a decrease in the stress on the system's mechanical components, that is, the motor, relay contacts, drive belt, control valve, etc.

The rate at which the system cycles ON and OFF in an oscillatory pattern depends on the thermal inertia of the space to be heated. Because a residence has a

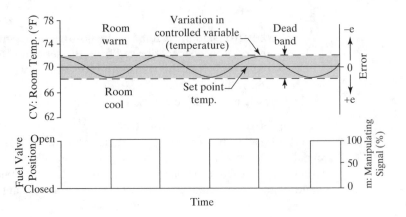

FIGURE 16.17

The thermostat in a home heating system controls the opening and closing of the furnace fuel valve. The control system adds energy when it cycles ON at the bottom of the 4°F dead band, and it cycles OFF when the temperature reaches the top of the dead band range.

sizable thermal inertia, the temperature within a room doesn't rapidly change, resulting in extended periods of time when the furnace is OFF, providing satisfactory control of the temperature within the residence.

ON-OFF controllers are simple, reliable, fast acting, and inexpensive. They are available in a variety of forms including mechanical, pneumatic, electrical, hydraulic, and digital.

Proportional Mode (P)

In proportional mode control, the dead band of the discontinuous ON-OFF control is replaced by a continuous *proportional band* that provides smooth, responsive process control. Proportional control, the primary continuous control mode, produces a continuous controlling action that is proportional to the error signal. The control action, provided to the final control element by the control signal from the proportional controller, is a modulating action that produces a proportional adjustment in the manipulated variable (MV), altering the controlled variable (CV).

When the error signal is small, the proportional control element outputs a small manipulating signal, producing a moderate controlling action (movement, heating, etc.) in the final control element. Conversely, when the error is large, the controller outputs a large manipulating signal, resulting in a sizeable controlling action in the final control element. The *proportional sensitivity* of the controller is the gain (K_c) produced by the control element. As noted in the following equation (Equation 16.6), the proportional sensitivity of the control element (K_c) is equal to the change in the manipulating signal (Δm) per unit change in the error signal (Δe).

$$K_c = \frac{\Delta m}{\Delta e}$$ **16.6**

where

K_c = the proportional sensitivity (the gain) of the controller, a unitless quantity

Δm = the change in the output manipulating signal

Δe = the change in the input error signal

Note: Both Δm and Δe are expressed in the same units.

The gain, or as it is commonly called, the proportional sensitivity, is the parameter adjusted by a technician to obtain the desired proportional band for the process under control. The proportional band is the key parameter in a proportional controller. The size of the proportional band (PB) is equivalent to the change in the error signal, producing sufficient change in the manipulating signal to cause the final control element to go from full OFF to full ON. The proportional band is usually expressed as a graph of its transfer function.

➢ **As a Rule** The proportional band of a controller is defined as the span of the input error signal (Δe), expressed as a percent, that results in a full change in the output manipulating signal, expressed as a percent.

The size of the proportional band (PB), expressed as a percent, is inversely related to the proportional sensitivity (K_c) of the controller, as noted in the following equation.

$$\text{PB} = \frac{1}{K_c} \times 100\% \qquad \qquad \textbf{(16.7)}$$

where

PB = the proportional band as a percent

K_c = the proportional sensitivity, a unitless quantity

Example 16.5

Construct a proportional control transfer curve for each of two proportional bands. One band is to be 20% while the other is to be 80%. Assume that the final control element is a pneumatic-actuated variable-flow valve (Figure 16.11) controlling the temperature of a batch kettle by varying the flow rate of steam entering the steam jacket surrounding the kettle. The system is set up so the control valve is half open (50%) when the system is operating normally near the set point with near zero error. From the 80% transfer curve, determine

(a) Both the position of the valve and the % value of the manipulating signal (m) when the error signal changes from near zero to +20%.

(b) Both the position of the valve and the % value of the manipulating signal (m) when the error signal changes from near zero to –30%.

(c) The proportional sensitivity (gain), K_c, of the control element for each of the transfer curves. Use Equation 16.6.

SOLUTION

Figure 16.18 pictures the proportional control transfer curves for the 20% and the 80% proportional bands. Note that the width of the proportional band is equal to

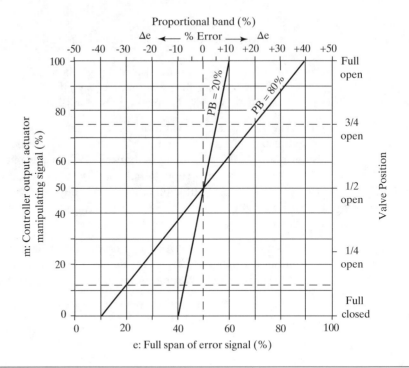

FIGURE 16.18
The 20% and 80% proportional band transfer curves for Example 16.5.

the interval of error signal (Δe) that causes the manipulating variable (m) to change from 0 to 100%; that is, a 20% proportional band equals the error interval from −10% to +10%.

(a) From the graph of Figure 16.18, enter the curve from the top at the +20% point. Project down, strike the 80% transfer curve, and project horizontally to the left and right as pictured. Read 75% for the manipulating variable, m, and $\frac{3}{4}$ open for the valve position.

(b) Enter the curve from the top at the −30% point. Project down, strike the 80% transfer curve, and project horizontally to the left and right as pictured in Figure 16.18. Read 12% for the manipulating variable, m, and $\frac{1}{8}$ open for the valve position.

Observation   When the temperature is above the set point, the error signal is negative ($|c| > |r|$ and $e = r - c$), resulting in the control valve shutting down to lower the steam flow and reduce the temperature in the batch kettle.

(c) Use Equation 16.6 to determine the proportional sensitivity, K_c, of the curves.

Given $$K_c = \frac{\Delta m}{\Delta e} \qquad\qquad K_c = \frac{\Delta m}{\Delta e}$$

Evaluate	$\Delta m = 100\%$	$\Delta m = 100\%$
	$\Delta e = 80\%$	$\Delta e = 20\%$
Substitute	$K_c = 100\%/80\%$	$K_c = 100\%/20\%$
Solve	$K_c = 1.25 = 1.3$	$K_c = 5.0$

Based on the transfer curves of the proportional bands (Figure 16.18), the curve with less slope (wide band) has a larger PB percentage ($>50\%$) and is less responsive ($K_c = 1.3$) to changes in the controlled variable than the curve with more slope (narrow band), that has a smaller PB percentage ($\leq20\%$), $K_c = 5.0$. The response to change is dependent on the gain, K_c. The proportional controller is *tuned* to control the process by adjusting the size of the proportional band. This is done by changing the proportional sensitivity (K_c) control. When in operation, and properly tuned, the proportional controller is stable and responsive to changes in the controlled variable.

Proportional-Integral Mode (PI)

Integral mode control is added to proportional control to compensate for the one shortcoming in the proportional controller. When operating at steady state, the controlled variable (under proportional control) is never quite equal to the set point value; that is, there is always a difference between these values. This difference, called an *offset*, is seen as an ongoing persistent error between the desired value of the set point and the actual value of the controlled variable.

With the addition of an integral mode to the proportional mode, the offset is eliminated. Integration, the mathematical operation carried out on the error signal by the integral mode, changes the value of the manipulating variable, m (the output of the PI controller), at a rate proportional to the size of the error. With a large error signal, the output signal from the integral control element changes quickly. When the error is small, as with the offset, the integral mode slowly increases the correction action, thus avoiding overcorrection.

As an example, under integral control, when the error signal is at 2% and moves to 6%, the control valve (the final control element) moves three times as fast at 6% as it would at 2% to correct for the error. As long as an error signal (e) is present at the input to the PI controller, its output signal (m) will continue to change until the error is equal to zero. With the error at zero, the change in error is also zero and the control system is operating at an error free steady state.

The PI controller combines the error eliminating characteristic of the integral control element with the stable response to step change of the proportional control element, bringing the desired value of the set point equal to the actual value of the controlled variable.

Proportional-Integral-Derivative Mode (PID)

The addition of the derivative mode to the proportional mode adds lead time to the process controller to counteract the accumulation of lag from around the loop, thereby providing a much faster response to step change in the error signal than that provided by the proportional controller alone. With the addition of a derivative mode to the proportional mode, lag is reduced.

Differentiation, the mathematical operation carried out on the error signal by the derivative control element, changes the value of the manipulating variable, m, at a rate proportional to the rate of change of the error. Thus, fast moving changes in the error signal, such as a sudden change in the controlled temperature resulting from a large addition of liquid to a batch kettle, produce a very large, fast-rising output signal from the derivative controller element.

In general, the quicker the error changes, the greater the output from the derivative control element; conversely, the slower the error changes, the smaller the output. Steady state error, large or small, will produce zero output from the derivative control element. The rate control, provided by the derivative control element, damps the response to load disturbances, providing a stabilizing effect on the overall process control system.

The proportional-integral-derivative controller (PID controller) is a continuous controller used to control processes that experience sudden, large changes in load or set point. Due to its sophistication and complex tuning, it is not intended for all applications. However, when called for, it is the most effective continuous controller. The PID controller combines the rapid response of the proportional control element, the automatic elimination of residual error of the integral control element, and the lag reduction and stabilization of the derivative control element.

CHAPTER SUMMARY

- The function of the automatic control system is to adjust the manipulated variable so the desired value of the controlled variable is maintained even when influenced by disturbance variables or changes in the set point.
- Automatic control systems incorporate analog or digital controllers employing electronic, pneumatic, hydraulic, or mechanical technologies for process or servo control.
- Control systems are divided into two groups, process control and servo control (servomechanism).
- A servo control system is a follow-up system, responding to changes in position very quickly and precisely, while a process control system is a regulator system, maintaining constant process conditions.
- Each element in a control system has a transfer function that mathematically defines its dynamic characteristics as non-dynamic, first-order lag, second-order lag, etc.
- Common control modes include: ON-OFF, proportional (P), proportional-integral (PI), and proportional-integral-derivative (PID).

SELECTED TECHNICAL TERMS

The following technical terms, abbreviations, and acronyms are defined in the glossary located after this chapter. You are encouraged to use the glossary to aid your understanding and to test your knowledge of these important terms.

bubble	first-order lag
closed loop	instrument
control valve	instrumentation
controlled variable	manipulated variable
controller	open loop
converter	process variable
dead band	sensor
dead time	set point
disturbance variable	transducer
error	transfer function
final control element	transmitter

END-OF-CHAPTER QUESTIONS

Write T if the statement is true and F if the statement is false.

1. In a batch process, material is continuously entering and leaving the process.
2. Positive feedback enables an automatic control system to be operated in a stable manner.
3. The transfer function defines the characteristics of the control system's elements.
4. A sensor is found in the feedback path of a process control system.
5. Control elements having no lag in their response time are classed as dynamic.
6. An automobile's power steering system is a regulator system and not a follow-up system.
7. Disturbance variables are needed because they influence the outcome of the process.
8. A primary element may be an integral part of a transmitter.
9. Most process control systems respond quickly to change in their process variables.
10. Transportation lag, a dead time lag, may be improved by positioning the loop sensor closer to the process.

In the following, select the word or phrase that makes the statement true.

11. The batch process is carried out in a series of (random, partial, sequenced), timed operations performed on the materials used to form the finished product.
12. The length of time delay in a first-order lag element depends on the time (variable interval, constant) of the element.
13. The gain transfer function of an element may have a (partial, mixed, constant) unit of measure.
14. As a rule, the transition time for an element with first-order lag is about (five, four, three) time constants.
15. An overdamped, second-order damping coefficient is represented by a value of (0.5, 1, 3).

Answer each of the following questions with a short answer in the form of a complete sentence. Include a restatement of the question in your answer.

16. Identify the controlled variable, manipulated variable, and the disturbance variables associated with a natural gas oven operating in a residential kitchen.

17. List the primary difference between a process control system and a servo control system.

18. Describe the difference in ride quality of an automotive suspension system that is underdamped (Figure 16.16(a)) to one that is critically damped (Figure 16.16(c)).

19. Using a ladder diagram format, sketch a two-wire control system to control the operation of a heat pump motor. An ON-OFF thermostat is employed to signal the contactor to start and stop the heat pump motor. Provide a 15 s time delay before turning ON the air handler motor contactor once the heat pump is activated.

20. List the definitive characteristics for each of these continuous control modes: proportional (P), integral (I), and derivative (D).

END-OF-CHAPTER PROBLEMS

Solve the following problems. Make sketches to aid in solving the problems; structure your work so it follows in an orderly progression and can easily be checked. Table 16.4 summarizes the formulas used in Chapter 16.

1. Determine the value of the error signal (e) from the summing junction of an electronic controller when
   (a) The reference signal (r) is 11.0 mA and the measured process variable signal (c) from a flow transmitter in the feedback path is 13.8 mA.
   (b) The reference signal is 10.0 mA and the signal from an ultrasonic level transmitter in the feedback path is 8.4 mA.

2. The reference input element of a pneumatic controller has a set point range of 0.0 to 250.0 gal/min and a corresponding pneumatic output signal of 3.0 to 15.0 lb/in^2. Determine

   (a) The gain transfer function for the reference input element in units of lb/in^2 per gal/min. Use $K = \dfrac{\Delta c}{\Delta r}$.
   (b) The value of the pneumatic output signal for a set point of 175.0 gal/min.

TABLE 16.4 Summary of Formulas
Used in Chapter 16

Equation Number	Equation
16.1	transfer function = output/input
16.2	$c = Kr$
16.3	$c(t) = Kr(1 - e^{-t/\tau})$
16.4	$c(t) = Kr(e^{-t/\tau})$
16.5	$e = r - c$
16.6	$K_c = \dfrac{\Delta m}{\Delta e}$
16.7	$\text{PB} = \dfrac{1}{K_c} \times 100\%$

3. A ceramic RTD (resistance temperature device) sensor with a first-order lag is used to measure the air temperature in a forced air drying system. Assuming a step change in air temperature and a time constant of 6.70 s, determine the time (in seconds) for the RTD sensor to transition from its initial state to its final stable state.

4. From the PB = 20% curve of Figure 16.18, determine the position of the valve and the percent value of the manipulating signal (*m*) when the error signal changes from near zero to –5.0%.

5. A permanent-magnet dc motor has a no-load shaft speed of 250 rev/min with the application of 30.0 V and 1250 rev/min with the application of 150 V. Determine
   (a) The gain transfer function (*K*) for the unloaded PM motor in units of rev/min per V. Use $K = \dfrac{\Delta c}{\Delta r}$.
   (b) The value of the applied voltage for a no-load shaft speed of 330 rev/min.

6. Determine the proportional sensitivity, K_c, of a proportional controller when
   (a) The proportional band is 140%.
   (b) The proportional band is 8.00%.

7. A certain temperature transmitter (sensor/converter) produces an output signal that varies from 4.0 mA at the minimum temperature of the controlled variable to 20 mA at the maximum temperature. Using $E = IR$, determine the resistance value (Ω) of the input resistance, *R*, of an *I/E* (current to voltage) converter used to change the temperature transmitter current signal from 4.0–20.0 mA to a voltage signal of 2.0–10.0 V.

8. A digital thermocouple thermometer, used to measure the temperature of a product being processed, indicates a steady state temperature of 255.0°C. Determine the temperature seen on the readout 12.0 s after the thermocouple probe is removed from the product and returned to the cooler 28.0°C temperature of the manufacturing plant. Assume that $K = 1$ and that the probe's time constant is 8.70 s. (*Hint:* Temp = (c(t) + 28.0) °C

9. Using the general equation for a straight line, $y = mx + b$, write the equation for the 20% proportional band of Figure 16.18. Let $y = m$ (manipulating signal as a decimal fraction), $m = K_c$ (the slope of the transfer curve), $x = e$ (error signal as a decimal fraction from the bottom scale); determine *b* (the *y*-intercept). Using the equation, solve for the percent value of *m* and determine the valve position when $e = 57.5\%$ (bottom scale).

10. A flow transmitter, with a linear input/output graph, has an input range of 0.0–150.0 gal/min and an output signal range of 4.0–20.0 mA.
   (a) Construct the input/output graph with the output signal (*c*) on the vertical axis and the input flow rate (*r*) on the horizontal axis.
   (b) From the slope of the plotted curve determine the gain transfer function,
   $K = \dfrac{\Delta c}{\Delta r}$, for the flow transmitter.
   (c) Using the general equation for a straight line, write the equation for the output signal (*c*) in terms of the input flow rate (*r*); i.e., $c = Kr + b$.
   (d) Using the equation from part **(c)** determine the value of the output signal for a flow rate of 52.5 gal/min.

Glossary of Selected Technical Terms, Abbreviations, and Acronyms

abrasive wear   The loss of material due to a rubbing action between objects of different hardness, or when abrasive particles contaminate the area between moving surfaces.

acceleration   The rate of change in the velocity of a moving body. Acceleration is expressed in meters per second per second (SI) or feet per second (BES).

accuracy   The degree to which a measurement or a calculation conforms to a recognized standard or a specified value.

addition and subtraction rule   Round the answer (sum or difference) to the same precision as the least precise number used in the calculation.

adhesive wear   The loss of material when two surfaces that have adhered to each other are moved apart.

AGMA   American Gear Manufacturers' Association.

AMA   Actual mechanical advantage; the mechanical advantage.

ambient temperature   The temperature of the cooling medium, usually air, immediately surrounding an electrical device.

ampacity   Current-carrying capacity expressed in amperes.

ampere (A)   The SI unit of electrical current, the rate of flow of electrons. One volt across 1 Ω of resistance causes a current flow of 1 A. A flow of 1 C/s equals 1 A.

angular acceleration   The rate of change of angular velocity.

angular displacement   The amount of rotation of a body about an axis.

angular velocity   The rate at which a body revolves around a fixed axis.

arcing   A prolonged electrical discharge or series of prolonged discharges between two electrodes. (Both produce a bright-colored flame.)

armature   The moving element in an electromechanical device, such as the rotating part of a generator or motor or the movable part of a relay.

asperities   Local microscopic points of roughness, or unevenness, on a surface.

AWG   American Wire Gauge.

axial   Of, pertaining to, or like an axis; moving about an axis.

backlash   The shortest distance between nondriving tooth surfaces of adjacent teeth in mating gears or the amount of clearance between the teeth of mating gears; usually measured at the common pitch circle.

bearings   Any part of a machine in or on which another part revolves, slides, etc.

BES   British Engineering System.

bifilar winding   A method of winding in which the wire is folded back on itself and then wound double.

bifurcated   Divided into two branches. A bifurcated contact is a movable contact that is forked to provide two contact-mating surfaces in parallel for a more reliable contact.

bimetal   A strip formed of two dissimilar metals welded together.

bistable multivibrator (flip-flop)   A circuit having two stable states; it will stay in either one indefinitely until triggered, after which it immediately switches to the other state.

Boolean algebra   A system of mathematical logic dealing with classes, propositions, ON-OFF circuit elements, etc., associated by operators such as AND, OR, and NOT, thereby permitting computations and demonstration as in any mathematical system. Boolean algebra is named after George Boole, English mathematician and logician, who introduced it in 1849.

breakdown torque   The maximum torque a motor will develop, without an abrupt drop in speed, as the rated voltage is applied at the rated frequency.

bubble   The circular symbol used to denote and identify the purpose of an instrument or function. It may contain a tag number. *Bubble* is a synonym for "balloon."

cam   The projecting part of a wheel or curved plate, etc., that is shaped in such a way as to cause an eccentric or alternating motion of any required velocity or direction in another piece engaging or meeting it.

capacitance   In a capacitor or a system of conductors and dielectrics, that property permitting the storage of electrically separated charges when potential differences exist between the conductors.

CEMF   Counter electromotive force; the back-induced voltage in an ac or dc motor or generator that opposes the applied voltage.

chordal effect   The ripple present in the speed of a chain due to the pitch radius of the chain changing size as each link engages a tooth on the sprocket and is advanced.

closed loop   The type of control system that uses feedback; an arrangement where a sensor measures the value of the controlled variable and transmits the information to the controller.

clutch   An apparatus that can temporarily connect or disconnect power to a drive shaft, gear train, etc.

coefficient   In mathematics, a number or algebraic symbol prefixed as a multiplier to a variable or unknown quantity; in $2y$ and ay, 2 and a are the coefficients of y.

coefficient of friction   The ratio between the friction force (the force opposing the motion on an object) and the normal force (the force pressing the surfaces together). It varies with the type of materials in contact.

commutation   The switching sequence of drive voltage into motor phase windings necessary to ensure continuous motor rotation.

commutator   The part of the armature to which the coils of a motor are connected. The motor brushes ride on the outer edges of the commutator bars.

compound machine   A combination of two or more simple machines.

constant-velocity joint   Used for the smooth transmission of torque through large angles that are constantly changing.

control relay   An electromechanical device in which contacts are opened and/or closed by variations in the conditions of one electric circuit, thereby affecting the operation of other devices in the same or other electric circuits.

control valve   A final controlling element; a device that directly modifies the rate of flow of one or more fluid process streams by adjusting the size of the flow passage; not a hand-actuated ON-OFF valve or a self-actuated check valve.

controlled variable   The process variable whose value is controlled by the control system and whose value is directly measured (sensed) to originate a feedback signal.

controller   A device that operates automatically to regulate a controlled variable by computing the error signal and using it to produce the control action.

converter   A device that receives an input signal in one form and transmits an output signal in another form, e.g., voltage-current converter, electro-pneumatic converter, etc.

coulomb (C)   The quantity of electricity that passes any point in an electric circuit in 1 s when the current is maintained constant at 1 A.

coupling   Any mechanical device serving to unite two or more parts or things.

dashpot   An apparatus for deadening the blow of any falling weight and preventing any jar in the machinery.

dead band   The range of values through which an input can be varied without initiating an observable response; the controlled variable range for which an ON-OFF controller remains in one of its two stable states; the range of values, on either side of the set point of the controlled variable, where the error cannot be corrected by the controller.

dead time   The interval of time between the initiation of an input change and the start of the resulting observable response.

delta connection   A three-phase circuit in which the windings of the system are connected in the form of the Greek letter delta; a Δ-shaped connection.

digital caliper   A measuring tool used to make outside, inside, and depth measurements with discrimination to .001 in.

displacement   The vector quantity representing change of position.

distributed control system   A functionally integrated system consisting of subsystems that are physically separate and remotely located from one another.

disturbance variable   An undesired process input variable that affects the controlled variable but is not controlled by the system.

ductility   The quality of a metal whereby it can be stretched, drawn, or hammered thin without breaking.

dwell   A momentary stopping of motion in some part of a machine to allow time for the completion of a function.

eddy current   Electrical current circulating in the core of a transformer as a result of induction. These currents produce unwanted heat in the core and windings of the transformer.

efficiency   The ratio of the useful output of a physical quantity that can be stored, transferred, or transformed by a device to the total input of the device.

EHL   Elastohydrodynamic lubrication.

EMF   Electromotive force; the force resulting from differences in potential that causes an electric current.

EMI Electromagnetic interference.

engineering notation Numbers written in powers-of-ten notation having exponents that are multiples of 3 (e.g., 83.7×10^{-3}).

EP Extreme pressure; additional SAE rating.

equilibrium A state of balance; a condition in which opposing forces exactly balance or equal each other.

error The signal in a controller that is obtained by subtracting the measured value of the controlled variable from the set point value.

eutectic alloy An alloy or mixture whose sharply defined melting point is lower than that of any other alloy or mixture composed of the same ingredients, and has no plastic range.

eutectic point The temperature at which a eutectic alloy melts.

exponent A number written above and to the right of a symbol or quantity to show how many times the symbol or quantity is to be used as a factor; for example, in 2^3 and a^3, 3 is the exponent.

fault current The current that can flow in any part of a circuit under abnormal conditions.

ferromagnetic The property of a material to become highly magnetic in a relatively weak magnetic field.

final control element The device that directly controls the value of the manipulated variable of a control loop. Often the final control element is a control valve.

first-order lag The most common type of dynamic component encountered in process control. First-order lag derives its name from the first-order differential equation that describes its behavior.

FLC Full-load current.

flux See *magnetic flux*.

flux density The number of flux lines per unit area.

flux lines Any of the imaginary lines of a magnetic field that curve from the north pole to the south pole of a magnet or electromagnet.

force Any influence or agency that causes a body to move or accelerate; a vector quantity that tends to produce a change in the motion of objects.

friction Resistance to motion of surfaces that touch; resistance of a body in motion to the air, water, etc., through which it travels or to the surface on which it travels.

fulcrum The support on which a lever turns or rests in moving or lifting something.

full-load slip The difference between the synchronous and full-load speeds of a motor.

gear A wheel having teeth that fit into the teeth of another wheel. If the wheels are of different sizes, they will turn at different speeds.

gear train A combination of two or more gears.

gearmotor A train of gears and a motor used for reducing or increasing the speed of the driven object.

GFI Ground fault interrupter; a device that, unlike a fuse, opens a circuit when it detects a small, set amount of ground current (often 5 mA). It must be manually reset to restore circuit voltage.

helix   Anything having a spiral, coiled form, such as a screw thread, a watch spring, or a snail shell.

hydrodynamic lubrication   A film that completely separates two sliding surfaces.

hysteresis   A lagging of an effect behind its cause, as when a magnetic body is subjected to a varying force (the property of a magnetic substance that causes the magnetization to lag behind the magnetizing force).

ideal machine   A machine that has no loss in energy. The work into the machine is exactly equal to the work out of the machine.

idler gear   A gear that does not drive a shaft but is inserted between other gears to alter the direction of rotation of those gears.

IMA   Ideal mechanical advantage; the velocity ratio.

impedance   The apparent resistance in an alternating-current circuit, which is made up of two components, reactance and true or ohmic resistance.

inclined plane   A plank or other plane surface set at an acute angle to a horizontal surface. It is a simple machine.

inductance   The property of an electric circuit by which an electromotive force is induced in it or in a nearby circuit by a change of current in either circuit; the property that opposes any change in the existing current. Inductance is present only when the current is changing.

inertia   A measure of an object's resistance to a change in velocity. The larger an object's inertia, the greater the force needed to accelerate or decelerate it.

inertial load   A load, such as a flywheel or other heavy rotating object, that tends to oppose acceleration.

instrument   A device used directly or indirectly to measure and/or control a variable; includes computing devices, primary elements, final control elements, and electrical devices.

instrumentation   A collection of instruments or their application for the purpose of control, observation, measurement, or any combination of these.

interpoles   Small auxiliary poles placed between the main poles of a direct-current generator or motor to reduce sparking at the commutator.

ionize   To charge an atom positively or negatively by the removal or addition of an electron.

ISO   International Organization for Standardization.

joule (J)   SI unit for energy and work; 1 J of work is done when an applied force of 1 N moves an object 1 m.

kilogram (kg)   The SI unit of mass, a fundamental unit equal to 1000 g or the mass of 1 L of water at a temperature of 4°C.

kinematic   Motion in the abstract, without reference to force or mass.

kinetic energy   The energy a body has because it is in motion. It is equal to one-half the product of the mass of a body and the square of its velocity.

kinetic force of friction   The friction force opposing movement.

law of acceleration   (Newton's second law) When a net outside force F acts on an object of mass m and causes it to accelerate, the acceleration can be computed by the formula $a = F/m$, and the acceleration is in the direction of the net outside force.

law of action and reaction   (Newton's third law) Forces always occur in pairs; that is, for every action, there is an equal and opposite reaction.

law of inertia   (Newton's first law) An object at rest will remain at rest and an object in motion will remain in motion at the same speed and direction unless it is acted upon by an outside force.

lever   A bar that turns on a fixed support called a fulcrum and is used to transmit effort and motion. It is a simple machine.

lope   A sinusoidal pattern of acceleration and deceleration in an output shaft connected at some angle to an input shaft with a universal joint.

lubrication   The application of a lubricant to machinery, etc., in order to reduce friction in operation.

magnetic contactor   A contactor that is actuated electromagnetically.

magnetic flux   The total of the magnetic flux lines in any given field.

manipulated variable   A quantity that is varied as a function of the error signal so as to change the value of the controlled variable.

mass   A measure of the quantity of matter a body contains; the property of a physical body that gives the body inertia. Mass is a property not dependent on gravity.

meter (m)   The SI unit of length, equal to the distance traveled by light in a vacuum in 1/299 792 458 s.

micrometer   An instrument for measuring distances with very small resolution (.0001 in). Available as an outside, inside, or depth micrometer.

mmf (magnetomotive force)   The force producing a magnetic flux.

moment of inertia   A measure of the resistance of a body to angular acceleration.

MOV   Metal-oxide varistor.

multiplication and division rule   Determine (by inspection) which of the quantities (numbers) has the fewest number of significant figures. The answer (product or quotient) is then rounded to the same number of significant figures as in this quantity.

NARM   National Association of Relay Manufacturers.

NEC   National electric code; a construction code for electrical equipment and distribution system installation.

NEMA   National Electrical Manufacturers' Association.

newton (N)   The SI unit of force. It is the force required to give an acceleration of 1 m/s^2 to a mass of 1 kg.

NLGI   National Lubricating Grease Institute.

open loop   The type of control system that does not use feedback; a process in which the output is not monitored and there is no feedback to modify the input to correct output inaccuracies.

optocoupler A solid-state device that provides electrical isolation between two circuits by transmitting signals optically with a light source (light-emitting diode) and receiving them with a light sensor (phototransistor).

overcurrent In a circuit, the current that will cause an excessive or even dangerous rise in temperature in the conductor or its insulation.

overload A load greater than that which a system is designed to carry. It is characterized by waveform distortion or overheating.

pawl A mechanical device allowing rotation in only one direction.

permeability The measure of how much better a given material is than air as a path of magnetic flux lines.

pitting The deformation of surface material due to heavy loads transmitted between two surfaces passing over one another, resulting in small particles falling out of the surface.

PLC Programmable logic controller; a self-contained microprocessor-based controller with multiple inputs and outputs that activates relays and other input/output devices from an alterable stored program.

pole The movable contact in a switch or relay.

potential energy The energy a body has because of its position or structure rather than as a result of its motion. A coiled spring or a raised weight has potential energy.

precision In measurement, the degree that individual measurements agree with each other, i.e., repeatability; in common use, it implies exactness.

primary winding A transformer winding that carries current and normally sets up a current in one or more secondary windings. The input winding of a transformer.

prime mover A machine that supplies input power to a drive. Common prime movers include electric motors, internal combustion engines, and hydraulic and air motors.

process variable Any variable property of a process; the actual value of a variable, as reported by the sensor. The directly controlled variable is frequently referred to as the process variable.

PTFE Polytetrafluoroethylene; a solid lubricant.

pulley A wheel with a grooved rim in which a rope, cable, belt, etc., can run to change the direction of the pull and so lift a load. It is a simple machine.

radial Pertaining to or placed like a radius (i.e., extending or moving outward from a central point).

radian In a circle, the angle included within an arc equal to the radius of the circle. Numerically, it is approximately equal to $57.2958°$. A complete circle contains 2π radians.

rated torque The torque produced when a motor is continuously producing the rated output power while operating at the rated terminal voltage and rated line frequency.

RCC Rated continuous current.

reactance Opposition to the flow of alternating current. Capacitive reactance is the opposition offered by capacitors, and inductive reactance is the opposition offered by a coil or other inductance. Both reactances are measured in ohms.

rectilinear motion Movement in a straight line.

resistance The opposition offered by a body or substance to the passage of an electric current through it, resulting in a change of electrical energy into heat or other forms of energy.

rheostat   A variable resistor that has one fixed terminal and a movable contact.

RMA   Rubber Manufacturers' Association.

rms   Root mean square; the effective value of an alternating current or voltage.

roots and powers rule   The root or power of a number should have the same number of significant figures as the base number.

rotor   The rotating member of an electrical machine.

runaway   Any additive condition to which continued exposure will eventually destroy a device.

SAE   Society of Automotive Engineers.

scientific notation   Numbers written in powers-of-ten notation having the decimal point placed after the leftmost nonzero digit.

screw   A cylinder with an inclined plane wound around it and fitting into or making a threaded cylindrical hole. It is a simple machine, used to raise a load over the threads by applying a small force.

secondary winding   The winding on the output side of a transformer.

sensor   That part of a loop or instrument that first senses the value of a process variable, and that assumes a corresponding, predetermined, and intelligible output; the part of the control system that monitors the system's output; also known as a primary element.

sequential   Occurring one after another, or in sequence.

set point   An input variable that sets the desired value of the controlled variable; expressed in the same units as the controlled variable.

SF   Service factor; the measure of the reserve margin built into a motor or drive element.

sheaves   A grooved wheel in a pulley block on which a rope works.

SHM   Simple harmonic motion; back and forth motion, such as that of a pendulum, in which the distance on one side of equilibrium always equals the distance on the other side; the acceleration is toward the point of equilibrium and directly proportional to the distance from it.

short circuit   An abnormal connection of relatively low resistance between two points of a circuit.

shunt field winding   A type of field coil used in a dc motor that is connected in parallel (shunt) with the armature.

SI   Système International d'Unités (International System of Units); the metric system.

single-phase   A power system having only two wires with one voltage.

sinusoidal   Of or having to do with a sine wave.

skewed   Turned to one side; twisted.

slip   The difference between the synchronous speed of a motor and the speed at which it operates.

solenoid   A current-carrying coil surrounding a movable iron core, especially a spiral or cylindrical coil of wire that acts like a magnet when a current passes through it. A solenoid is used to convert electrical energy into mechanical work (force and displacement).

specific heat   The quantity of heat needed to raise a unit mass of material 1°C.

sprocket   Any of a number of toothlike projections, as on the rim of a wheel, to engage with the links of a chain.

starting torque   Also called pull-in torque; the maximum load torque with which motors can start and come to synchronous speed.

stator   The nonrotating part in a motor.

step angle   Degrees of angular motion for a single step of a stepper motor.

synchronism   The state in which a stepper motor is rotating at a speed corresponding correctly to the applied step-pulse frequency. Load torques in excess of the motor's rated torque will cause a loss of synchronism. The condition is not damaging to a stepper motor.

three-phase   A combination of circuits energized by alternating electromotive forces that differ in phase by one-third of a cycle, or 120 electrical degrees.

throw   The stationary contact(s) in a switch or relay.

tolerance   A permissible deviation from a specified value.

torque   A force that produces a rotating or twisting action.

transducer   A general term for a device that receives information in the form of one quantity, converts it to information in the form of the same or another quantity, and produces a resultant output signal; it may be a converter, sensor, transmitter, relay, or other device.

transfer function   A mathematical relationship between an input signal and a corresponding output signal; may be given in terms of LaPlace transforms and as the ratio of output to input.

transformer   A device for changing an alternating electric current into one of higher or lower voltage by electromagnetic induction.

translator   A device that transforms signals from the form in which they were generated into a form that can be used by another device.

transmitter   A device that senses a process variable through the medium of a sensor and has the output whose steady-state value varies only as a predetermined function of the process variable. The sensor may or may not be integral with the transmitter.

TTL   Transistor-transistor logic; a common family of digital logic integrated circuits (IC) that are commonly used in digital electronic systems. TTL signals have two distinct states described with voltage levels: logic 1 is 2.4 to 5.5 V, and logic 0 is 0 to 0.8 V.

universal joint   A joint or coupling that permits a swing at any angle within certain limits, especially one used to transmit rotary motion from one shaft to another not in line with it.

VAR   Volt-amps-reactive; the unit of reactive power.

velocity   A vector quantity that includes both magnitude (speed) and direction in relation to a given frame of reference.

viscosimeter   A device for measuring the degree to which a liquid resists a change in shape.

viscosity   The frictional resistance offered by one part or layer of a liquid as it moves past an adjacent part or layer of the same liquid. A resistance to flow due to internal friction of a liquid's molecules.

watt (W) The SI unit of power, equal to 1 J/s. In the case of electrical power, the number of watts is equal to the current in amperes multiplied by the electrical potential in volts.

wheel and axle An axle on which a wheel is fastened. As a simple machine, one of its uses is to lift weights by winding a rope or chain onto the axle as the wheel is turned.

windlass An apparatus operated by hand or machine, for hauling or hoisting, consisting of a drum or cylinder upon which is wound the rope, cable, or chain that is attached to the object to be lifted.

wye connection A three-phase circuit in which the windings of the system are connected in the form of the letter Y; also called a star connection.

Appendixes Contents

597

Appendix *A*

Selected Reference Tables

TABLE A1   Selected Conversion Factors for SI and BES*

Displacement (length)
1 m = 100 cm = 1000 mm = 3.281 ft = 39.37 in
1 km = .6214 mi = 3281 ft
1 in = 2.54 cm = 25.4 mm
1 mi = 5280 ft
1 yd = 3 ft = 36 in
1 ft = 12 in
1 revolution = 360° = 2π rad = 6.2832 rad

Area
$\overline{1 \ m^2}$ = 10.76 ft² = 1550 in²   **1 yd² = 9 ft²**
1 cm² = .1550 in²   **1 ft² = 144 in²**
1 m² = 10 000 cm²
1 cm² = 100 mm²

Time
$\overline{1 \ h}$ **= 60 min = 3600 s**
1 min = 60 s

Force
$\overline{1 \ N}$ = .2248 lb   **1 lb = 16 oz**
1 lb = 4.448 N   **1 ton = 2000 lb**

Mass
$\overline{1 \ kg}$ = .0685 slug   1 kg = 1000 g
1 slug = 14.6 kg   1 g = 1000 mg

Velocity (speed)
1 m/s = 3.60 km/h = 2.24 mi/h = 3.28 ft/s
60 mi/h = 88 ft/s
1 ft/s = 0.3048 m/s
1 rev/min = 0.1047 rad/s
60 rev/min = 1 cycle/s = 1 Hz

Work (energy, torque)
$\overline{1 \ J}$ = .738 ft-lb   1 ft-lb = 1.36 J
1 kWh = 3.6 MJ   1 N·m = .738 lb-ft
1 Btu = 1.06 kJ = 778 ft-lb

Power
$\overline{\textbf{1 hp = 746 W = 550 ft-lb/s}}$
1 hp = 33,000 ft-lb/min

1 W = .738 ft-lb/s = 3.412 Btu/h

1 kW = 1.34 hp

General Constants
$\overline{\text{Acceleration}}$ due to gravity (BES) = 32.2 ft/s²
Acceleration due to gravity (SI) = 9.81 m/s²
π = 3.1416

***Boldface** physical quantities are exact.

TABLE A2   Temperature Conversion

To use the table, locate the known temperature in the °F–°C column. Locate the converted temperature in the °F or °C column.*

°F–°C	°F	°C	°F–°C	°F	°C
−20	−4.0	−29	55	131	12.8
−15	5.0	−26	60	140	15.6
−10	14	−23	65	149	18.3
−5	23	−21	70	158	21.1
0	32	−18	75	167	23.9
5	41	−15	80	176	26.7
10	50	−12	85	185	29.4
15	59	−9.4	90	194	32.2
20	68	−6.7	95	203	35.0
25	77	−3.9	100	212	37.8
30	86	−1.1	105	221	40.6
35	95	1.7	110	230	43.3
40	104	4.4	115	239	46.1
45	113	7.2	120	248	48.9
50	122	10.0	125	257	51.7

*For temperatures not listed, use the following formulas.

$$F = (1.8 \times C) + 32$$
$$C = 0.5556 (F - 32)$$

where F = degrees Fahrenheit and C = degrees Celsius.

TABLE A3   Fraction-to-Decimal-to-Millimeter Conversion

Fraction	Decimal Inch		Millimeter (mm)
	.01	.001	
$\frac{1}{32}$	.03	.031	0.8
$\frac{1}{16}$	.06	.062	1.6
$\frac{3}{32}$	.09	.094	2.4
$\frac{1}{8}$	.12	.125	3.2
$\frac{5}{32}$	.16	.156	4.0
$\frac{3}{16}$	.19	.188	4.8
$\frac{7}{32}$	.22	.219	5.6
$\frac{1}{4}$	.25	.250	6.4
$\frac{9}{32}$	.28	.281	7.1
$\frac{5}{16}$	.31	.312	7.9
$\frac{11}{32}$	.34	.344	8.7
$\frac{3}{8}$	.38	.375	9.5
$\frac{13}{32}$	.41	.406	10.3
$\frac{7}{16}$	.44	.438	11.1
$\frac{15}{32}$	.47	.469	11.9
$\frac{1}{2}$	.50	.500	12.7
$\frac{17}{32}$	.53	.531	13.5
$\frac{9}{16}$	.56	.562	14.3
$\frac{19}{32}$	.59	.594	15.1
$\frac{5}{8}$	.62	.625	15.9
$\frac{21}{32}$	.66	.656	16.7
$\frac{11}{16}$	.69	.688	17.5
$\frac{23}{32}$	.72	.719	18.3
$\frac{3}{4}$	.75	.750	19.1
$\frac{25}{32}$	.78	.781	19.9
$\frac{13}{16}$	.81	.812	20.6
$\frac{27}{32}$	.84	.844	21.4
$\frac{7}{8}$	.88	.875	22.2
$\frac{29}{32}$	.91	.906	23.0
$\frac{15}{16}$	.94	.938	23.8
$\frac{31}{32}$	.97	.969	24.6
$\frac{1}{1}$	1.00	1.000	25.4

TABLE A4   SAE Bolt Grade Markings and Torque Specifications for Steel Bolts and Screws

Grade Marking	Specification	Tensile Strength (lb/in²)	Torque (dry) UNC Thread (lb-ft)					
			$\frac{1}{4}$	$\frac{3}{8}$	$\frac{1}{2}$	$\frac{5}{8}$	$\frac{3}{4}$	1
	SAE Grade 1 SAE Grade 2	74,000	6	20	50	100	155	310
	SAE Grade 3	100,000	9	30	70	145	235	550
	SAE Grade 5	120,000	10	35	80	155	260	590
	SAE Grade 7	133,000	13	45	110	215	360	840
	SAE Grade 8	150,000	14	48	120	230	380	900

TABLE A5 Unified American Standard and Metric Threads

	Unified American Standard			Metric		
Number or Size	Diameter (in)	UNC Threads per Inch	UNF Threads per Inch	Preferred Nominal Size (mm)	Coarse-Thread Pitch	Fine-Thread Pitch
0	.060	—	80	1.6	0.35	—
1	.073	64	72	2.0	0.40	—
2	.086	56	64	2.5	0.45	—
3	.099	48	56			
4	.112	40	48	3.0	0.50	—
5	.125	40	44			
6	.138	32	40			
8	.164	32	36	4.0	0.70	—
10	.190	24	32	5.0	0.80	—
12	.216	24	28			
$\frac{1}{4}$	.250	20	28			
$\frac{5}{16}$	.312	18	24	8.0	1.25	1.00
$\frac{3}{8}$	.375	16	24	10.0	1.50	1.25
$\frac{7}{16}$	.438	14	20			
$\frac{1}{2}$	.500	13	20	12.0	1.75	1.25
$\frac{9}{16}$	.563	12	18			
$\frac{5}{8}$	.625	11	18	16.0	2.00	1.50
$\frac{3}{4}$	.750	10	16	20.0	2.50	1.50
$\frac{7}{8}$	.875	9	14			
1	1.000	8	12	24.0	3.00	2.00

TABLE A6   Relay and Lever Switch Contact Forms

Form	Term	Symbol	Form	Term	Symbol
A	Make		J	Make make-before break	
B	Break				
C	Break make (transfer)		K	Center off	
			L	Break make make	
D	Make-before break (continuity transfer)				
			U	Double make (contact on arm)	
E	Break make-before break		V	Double break (contact on arm)	
F	Make make		W	Double break double make (contact on arm)	
G	Break break				
			X	Double make	
H	Break break make		Y	Double break	
I	Make break-before make		Z	Double break double make	

TABLE A7   Ampacities of Two or Three Insulated Conductors*

Size	Temperature Rating of Conductor						Size
	60°C (140°F)	75°C (167°F)	90°C (194°F)	60°C (140°F)	75°C (167°F)	90°C (194°F)	
AWG kcmil	TYPES TW, UF	TYPES RH, RHW, THHW, THW, THWN, XHHW, ZW	TYPES THHN, THHW, THW-2, THWN-2, RHH, RWH-2, USE-2, XHHW, XHHW-2, ZW-2	TYPE TW	TYPES RH, RHW, THHW, THW, THWN, XHHW	TYPES THHN, THHW, THW-2, THWN-2, RHH, RWH-2, USE-2, XHHW, XHHW-2, ZW-2	AWG kcmil
	COPPER			ALUMINUM OR COPPER-CLAD ALUMINUM			
14	16†	18†	21†	—	—	—	14
12	20†	24†	27†	16†	18†	21†	12
10	27†	33†	36†	21†	25†	28†	10
8	36	43	48	28	33	37	8
6	48	58	65	38	45	51	6
4	66	79	89	51	61	69	4
3	76	90	102	59	70	79	3
2	88	105	119	69	83	93	2
1	102	121	137	80	95	106	1
1/0	121	145	163	94	113	127	1/0
2/0	138	166	186	108	129	146	2/0
3/0	158	189	214	124	147	167	3/0
4/0	187	223	253	147	176	197	4/0
250	205	245	276	160	192	217	250
300	234	281	317	185	221	250	300
350	255	305	345	202	242	273	350
400	274	328	371	218	261	295	400
500	315	378	427	254	303	342	500
600	343	413	468	279	335	378	600
700	376	452	514	310	371	420	700
750	387	466	529	321	384	435	750
800	397	479	543	331	397	450	800
900	415	500	570	350	421	477	900
1000	448	542	617	382	460	521	1000
Ambient Temp. °C	For ambient temperatures other than 30°C (86°F), multiply the ampacities shown above by the appropriate factor shown below.						Ambient Temp. °F
21–25	1.08	1.05	1.04	1.08	1.05	1.04	70–77
26–30	1.00	1.00	1.00	1.00	1.00	1.00	79–86
31–35	.91	.94	.96	.91	.94	.96	88–95
36–40	.82	.88	.91	.82	.88	.91	97–104
41–45	.71	.82	.87	.71	.82	.87	106–113
46–50	.58	.75	.82	.58	.75	.82	115–122
51–55	.41	.67	.76	.41	.67	.76	124–131
56–60	—	.58	.71	—	.58	.71	133–140
61–70	—	.33	.58	—	.33	.58	142–158
71–80	—	—	.41	—	—	.41	160–176

*Rated 0 through 2000 V, within an overall covering (multiconductor cable), in raceway in free air based on ambient temperature of 30°C (86°F).
†Unless otherwise specifically permitted elsewhere in this Code, the overcurrent protection for conductor types marked with a dagger (†) shall not exceed 15 amperes for No. 14, 20 amperes for No. 12, and 30 amperes for No. 10 copper; or 15 amperes for No. 12 and 25 amperes for No. 10 aluminum and copper-clad aluminum.

TABLE A8   Capacitor Multipliers for Correcting the Load Power Factor

To determine the value of the power factor capacitor required to correct the load power factor from the original to the desired value, first select the power factor multiplier (pf_{mult}) from the following table and then use the following equation to compute the required capacity of the power factor capacitor. Select a commercially available value from the lower table of selected values for 240-, 480-, and 600-V, three-phase, 60-Hz power capacitor.

$$C_{kvar} = P_{load} \times pf_{mult}$$

where

C_{kvar} = capacitor, rated in kvar
P_{load} = power in load, in kW
pf_{mult} = power factor multiplier from following table

Original Power Factor (%)	Desired Power Factor (%)			
	100*	95	90	85
60	1.33	1.00	0.85	0.71
64	1.20	0.87	0.72	0.58
68	1.08	0.75	0.59	0.46
72	0.96	0.64	0.48	0.34
76	0.86	0.53	0.37	0.24
80	0.75	0.42	0.27	0.13
84	0.65	0.32	0.16	0.03
88	0.54	0.21	0.06	
92	0.43	0.10		
96	0.29			

*Motor power factors must not be overcorrected (>100%) as this may cause high transient voltages, currents, and torques, resulting in possible damage to the motor and/or injury to personnel.

Selected Commercial Values of 240-, 480-, and 600-V, Three-Phase, 60-Hz Power Capacitors

Capacitor Ratings (kvar)			
6	25	60	120
8	30	70	150
10	35	80	200
15	40	90	250
20	50	100	300

Example: Correct 120 kW, 480-V load from 72% to 95% power factor. From table, pf_{mult} = 0.64; C_{kvar} = 120 × 0.64 = 76.8 kvar. Select an 80-kvar, 480-V capacitor.

TABLE A9   Horsepower Ratings for NEMA Starters Used in Full-Voltage Starting

NEMA Size	Polyphase Motors			Single-Phase Motors	
	200 V	230 V	460 V	115 V	230 V
00	1.5	1.5	2	$\frac{1}{3}$	1
0	3	3	5	1	2
1	7.5	7.5	10	2	3
2	10	15	25	—	7.5
3	25	30	50	—	15
4	40	50	100		
5	75	100	200		
6	150	200	400		
7	—	300	600		
8	—	450	900		
9	—	800	1600		

TABLE A10   Ohm's Law and Power Law Formulas for DC Circuits

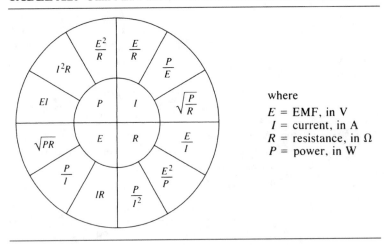

where

E = EMF, in V
I = current, in A
R = resistance, in Ω
P = power, in W

TABLE A11   Formulas for Circumference and Area of a Circle

$$C = 2\pi r = \pi d \qquad A = \pi r^2 = \frac{(\pi d^2)}{4}$$

$$d = 2r$$

where

C = circumference (in, ft, cm, m, or any linear unit)
A = area (in^2, ft^2, cm^2, m^2, or corresponding square unit)
r = radius (in, ft, cm, m, or any linear unit)
d = diameter (in, ft, cm, m, or any linear unit)

TABLE B1 Mathematical Symbols

Symbol		Symbol	
$\times, \cdot$	Multiply by, AND	$\geq$	Greater than or equal to
$\div$	Divide by	$\leq$	Less than or equal to
$+$	Postive, add, OR	$\therefore$	Therefore
$-$	Negative, substract	$\angle$	Angle
$\pm$	Plus or minus	$|a|$	Absolute value of a
$=$	Equals	Δ	Interval
$\equiv$	Identically equal	$\%$	Percent
$\approx$	Approximately equal	$\sqrt{\ }$	Radical sign
$\neq$	Not equal	$\llcorner$	Right angle
$>$	Greater than	∞	Undefined, infinity
$\gg$	Much greater than	$\Rightarrow$	Yields
$<$	Less than	$\propto$	Proportional to
$\ll$	Much less than	$\parallel$	Parallel to

TABLE B2 Greek Alphabet

Name	Capital	Lowercase	Name	Capital	Lowercase
Alpha	A	α	Nu	N	ν
Beta	B	β	Xi	Ξ	ξ
Gamma	Γ	γ	Omicron	O	o
Delta	Δ	δ	Pi	Π	π
Epsilon	E	ϵ	Rho	P	ρ
Zeta	Z	ζ	Sigma	Σ	σ
Eta	H	η	Tau	T	τ
Theta	Θ	θ	Upsilon	Υ	υ
Iota	I	ι	Phi	Φ	ϕ
Kappa	K	κ	Chi	X	χ
Lambda	Λ	λ	Psi	Ψ	ψ
Mu	M	μ	Omega	Ω	ω

TABLE B3 SI Prefixes

Symbol	Engineering Notation	Prefix Name
E	10^{18}	exa
P	10^{15}	peta
T	10^{12}	tera
G	10^{9}	giga
M	10^{6}	mega
k	10^{3}	kilo
m	10^{-3}	milli
μ	10^{-6}	micro
n	10^{-9}	nano
p	10^{-12}	pico
f	10^{-15}	femto
a	10^{-18}	atto

TABLE B4 Electrical Graphic Symbols (Standard Elementary Diagram Symbols)

The diagram symbols shown below have been adopted by the Square D Company and conform where applicable to standards established by the National Electrical Manufacturers Association (NEMA).

SWITCHES								
DISCONNECT	CIRCUIT INTERRUPTER	CIRCUIT BREAKER W/THERMAL O L	CIRCUIT BREAKER W/MAGNETIC O L	CIRCUIT BREAKER W/THERMAL AND MAGNETIC O L	LIMIT SWITCHES		FOOT SWITCHES	
					NORMALLY OPEN	NORMALLY CLOSED	N.O.	N C
					HELD CLOSED	HELD OPEN		

PRESSURE & VACUUM SWITCHES		LIQUID LEVEL SWITCH		TEMPERATURE ACTUATED SWITCH		FLOW SWITCH (AIR, WATER, ETC.)	
N.O.	N.C.	N.O.	N.C.	N.O.	N.C.	N.O.	N.C.

SPEED (PLUGGING)	ANTI-PLUG	SELECTOR			
F F	F	2 POSITION	3 POSITION	2 POS SEL PUSH BUTTON	
R	R	I-CONTACT CLOSED	I-CONTACT CLOSED	I-CONTACT CLOSED	

2 POS SEL PUSH BUTTON table:

CONTACTS	SELECTOR POSITION	
	A	B
	BUTTON	BUTTON
	FREE	DEPRES'D	FREE	DEPRES'D
1 - 2	I			
3 - 4		I	I	I

PUSH BUTTONS								PILOT LIGHTS	
MOMENTARY CONTACT						MAINTAINED CONTACT	ILLUMINATED	INDICATE COLOR BY LETTER	
SINGLE CIRCUIT		DOUBLE CIRCUIT	MUSHROOM HEAD	WOBBLE STICK	TWO SINGLE CKT	ONE DOUBLE CKT		NON PUSH-TO-TEST	PUSH-TO-TEST
N.O.	N.C.	N.O. & N.C.							

CONTACTS						COILS		OVERLOAD RELAYS		INDUCTORS
INSTANT OPERATING				TIMED CONTACTS - CONTACT ACTION RETARDED AFTER COIL IS		SHUNT	SERIES	THERMAL	MAGNETIC	IRON CORE
WITH BLOWOUT		WITHOUT BLOWOUT		ENERGIZED	DE-ENERGIZED					
N.O.	N.C.	N.O.	N.C.	N.O.T.C.	N.C.T.O.	N.O.T.O.	N.C.T.C.			AIR CORE

TRANSFORMERS					AC MOTORS				DC MOTORS			
AUTO	IRON CORE	AIR CORE	CURRENT	DUAL VOLTAGE	SINGLE PHASE	3 PHASE SQUIRREL CAGE	2 PHASE 4 WIRE	WOUND ROTOR	ARMATURE	SHUNT FIELD	SERIES FIELD	COMM. OR COMPENS FIELD
										(SHOW 4 LOOPS)	(SHOW 3 LOOPS)	(SHOW 2 LOOPS)

(continued)

WIRING					CONNECTIONS		RESISTORS			CAPACITORS	
NOT CONNECTED	CONNECTED	POWER	CONTROL	WIRING TERMINAL	MECHANICAL	FIXED	ADJ BY FIXED TAPS	RHEOSTAT, POT OR ADJ TAP	FIXED	ADJ	

Under WIRING TERMINAL: o / GROUND
Under MECHANICAL: GROUND / MECHANICAL INTERLOCK
Under FIXED (resistors): RES / H / HEATING ELEMENT
Under ADJ BY FIXED TAPS: RES
Under RHEOSTAT, POT OR ADJ TAP: RH

ANNUNCIATOR	BELL	BUZZER	HORN SIREN. ETC	METER	METER SHUNT	HALF WAVE RECTIFIER	FULL WAVE RECTIFIER	BATTERY	FUSE	THERMO-COUPLE

METER: INDICATE TYPE BY LETTER — VM, AM
FULL WAVE RECTIFIER: + AC DC DC AC +
FUSE: POWER OR CONTROL

IGNITRON TUBE	SEMICONDUCTORS										
	DIODE	TUNNEL DIODE	UNIDIRECTIONAL BREAKDOWN (ZENER) DIODE	BIDIRECTIONAL BREAKDOWN DIODE	PHOTOSENSITIVE CELL	TRIAC (BIDIRECTIONAL TRIODE THYRISTOR)	SILICON CONTROLLED SCR	PROGRAMMABLE UNIT-JUNCTION TRANSISTOR (PUT)	TRANSISTOR	UNIJUNCTION TRANSISTOR	GATE TURN OFF THYRISTOR

TRANSISTOR: PNP TYPE, NPN TYPE (B, C, E)
UNIJUNCTION TRANSISTOR: P BASE, N BASE (E, B2, B1)
GATE TURN OFF THYRISTOR: G, A, K
IGNITRON TUBE: DOT IN ANY TUBE DENOTES GAS

SUPPLEMENTARY CONTACT SYMBOLS

SPST N O		SPST N C		SPDT		TERMS
SINGLE BREAK	DOUBLE BREAK	SINGLE BREAK	DOUBLE BREAK	SINGLE BREAK	DOUBLE BREAK	SPST - SINGLE POLE SINGLE THROW
DPST, 2 N O		DPST, 2 N C		DPDT		SPDT - SINGLE POLE DOUBLE THROW
SINGLE BREAK	DOUBLE BREAK	SINGLE BREAK	DOUBLE BREAK	SINGLE BREAK	DOUBLE BREAK	DPST - DOUBLE POLE SINGLE THROW

DPDT - DOUBLE POLE DOUBLE THROW
N O - NORMALLY OPEN
N C - NORMALLY CLOSED

SYMBOLS FOR STATIC SWITCHING CONTROL DEVICES

STATIC SWITCHING CONTROL IS A METHOD OF SWITCHING ELECTRICAL CIRCUITS WITHOUT THE USE OF CONTACTS, PRIMARILY BY SOLID STATE DEVICES. USE THE SYMBOLS SHOWN IN TABLE ABOVE EXCEPT ENCLOSED IN A DIAMOND:

EXAMPLES: INPUT "COIL", OUTPUT N O., LIMIT SWITCH N O.

CONTROL AND POWER CONNECTIONS - 600 VOLTS OR LESS - ACROSS-THE-LINE STARTERS
(From NEMA Standard ICS 2-321A.60)

		1 PHASE	2 PHASE 4 WIRE	3 PHASE
LINE MARKINGS		L1, L2	L1,L3 - PHASE 1 / L2,L4 - PHASE 2	L1, L2, L3
GROUND WHEN USED		L1 IS ALWAYS UNGROUNDED	—	L2
MOTOR RUNNING OVERCURRENT UNITS IN	1 ELEMENT	L1	—	—
	2 ELEMENT	—	L1, L4	—
	3 ELEMENT	—	—	L1, L2, L3
CONTROL CIRCUIT CONNECTED TO		L1, L2	L1, L3	L1, L2
FOR REVERSING INTERCHANGE LINES		—	L1, L3	L1, L3

Maintenance, Installation, and Adjustment of Mechanical and Electrical Machine Elements

As a general rule, only qualified and trained personnel who are authorized to do so should carry out the maintenance, installation, and adjustment of mechanical and electrical machine elements. It is the responsibility of these individuals to take unfailing care when performing maintenance on equipment and to follow and comply with all national and local safety and electrical codes, including the *National Electrical Code* (NEC) and the Occupational Safety and Health Act (OSHA). Standard maintenance work procedures and practices must be followed. Included in these are lockout/tagout procedures, blocking of equipment, recordkeeping, etc.

C1   BELT DRIVES

Installation of V-Belts

A. Loosen mounting bolts and move the motor so that the belts can be slipped over the pulleys. Never *run* the belt onto the pulleys or use a screwdriver to pry the belt onto the pulleys, as this will break the cords and ruin the belt.

B. Slack belts on the top side of drive.

C. Initially, tighten belts to the approximate specified tension.

D. Seat the belts in the pulley groove by running the drive.

E. Once the belts are seated (after a few minutes), stop and retighten the belts to the final tension.

F. Follow up in a day or two to verify that satisfactory operation and tension exist.

V-Belt Tension

A. Tension classical multiple (type A, B, etc.) and light-duty V-belts (type 2L, 3L, etc.) so that they stretch in length by 2%. This is done by measuring the distance between two marks on the belt while tightening the belts. Never exceed a 5% increase (stretch) in the belt length.

B. Narrow groove belts (types 3V, 5V, 8V) must be tensioned with a tensioning tool (gauge), since they require very high tension for correct operation and cannot be set by feel.

C. Maintaining specified belt tension will ensure long belt life.

D. Belts that are tightened below the specified tension will slip excessively, causing heat, wear, and noise (squeal).

E. Never use belt dressing on V-belts to correct slippage, as this will only shorten the belt life; instead, correctly tension the belt(s) (if they are not stretched beyond their specified length); if worn or stretched excessively, replace with new belt(s) correctly tensioned.

F. When changing the belts of multiple belt drives, change all belts using matched sets of selected belts from the same manufacturer. Don't mix belts from different manufacturers, because their wear characteristics and their nominal dimensions may vary.

V-Belt Drive System Maintenance

A. Standard V-belts should be operated in an oil-free environment where the surrounding temperature (ambient temperature) is below 140°F. The pulley groove and belt must be dry and smooth to ensure sufficient friction to move the load without excessive slippage. Overheating of the belt is indicated by cracking on the bottom and sides of the belt, signaling that the belt may need replacement. *Premium-quality* belts designed for high-temperature operation (greater than 140°F) are available for applications requiring a high ambient temperature.

B. Pulleys will wear with use and they must be replaced or rebuilt when:
  • a new, correctly tensioned belt *rides* .06 to .08 in below the top of the pulley.
  • the pulley side wall is *cupped* more than .03 to .04 in.

C. The drive and the driven pulleys must be maintained in alignment with one another to prevent wear in both the pulley side wall and the faces of the belt.

D. Pulleys must run concentric to (with the same center as) the shaft to prevent the belt from whipping and wearing rapidly. Bent, dented, chipped, or wobbling pulleys must not be allowed to remain in service, because serious damage (heating and wear) to the belts may occur and excessive stress (loading) will be placed on the bearings.

E. Pulleys must run smoothly and in balance so that vibration (and noise) is not present, which will cause uneven delivery of power to the load and excessive loading of the bearings. Computer-based *machinery analyzers* are available for detecting vibration and pinpointing its cause (unbalance, misalignment, bent shaft, mechanical looseness, etc.).

Codes for V-Belts

A. *Classical multiple belts (types A, B, etc.)*   Sizes range from 26 to 660 in. For example, a *size/length code* of C90 is interpreted as a C ($\frac{7}{8}$-in) belt that is 90 in long. When replacing multiple belts, each belt should be from the same manufacturer and each should have the same *size/length code* and *match code*. These codes are printed on the belt as a size/length code followed by the manufacturer's name and then the match code, as in "B42 manufacturer's name 50."

B. *Light-duty (FHP) belts (types 2L, 3L, etc.)*   Sizes range from 10 to 100 in. For example, a size/length code of 2L200 is interpreted as a 2L ($\frac{1}{4}$-in) belt that is 20.0 in long.

C. *Narrow groove belts (types 3V, 5V, 8V)*   Sizes range from 25 to 500 in. For example, a size/length code of 5V1320 is interpreted as a 5V ($\frac{5}{8}$-in) belt that is 132 in long.

Timing Belt Installation

A. When installing timing belts, remove the tension from the belt by backing off the tensioner. The new belt is installed loosely on the pulleys, ensuring that the belt teeth engage the pulley teeth. Once in place, the belt is tensioned to remove the slack. Because of the slip-proof nature of the belt, it requires only small amounts of tension, just enough to provide a tight and secure fit, preventing the belt from *skipping a tooth* when operating under load. *Note*: Under no circumstance should the belt be forced onto the pulleys, because such action could damage the belt.

B. The input and output shafts must be checked for alignment so that they are parallel, ensuring that the belt will run true.

C. The pulleys must run concentric to the shaft with no wobble or vibration.

C2   ROLLER CHAIN DRIVES

Installation of Roller Chain

A. Initially inspect the sprocket for wear and check the shafts for alignment (when horizontal, both shafts must be level and parallel). Clean the sprocket of dirt, grease, and grime.

B. When the center distance between the shafts is adjustable, reduce it before installing the chain. The chain being installed must be free from dirt, and it must be lubricated so that the pins and bushings are completely coated with oil.

C. Wrap the chain around both sprockets and install the *connecting pin link* through the ends of the chain. Snap the *free link plate* into place and secure it with the spring clip or cotter pins.

D. Adjust the *sag* in the chain by pulling the chain taut across the bottom, causing the excess chain to collect at the top. Measure the sag while adjusting the center distance (if adjustable) so that the sag is equal to $\frac{1}{8}$ in per each 6 in of center distance (i.e., 30-in center distance would have $\frac{5}{8}$-in sag), or about 2% of the center distance. When the center distance is not adjustable, the chain must be shortened by removing links from the end of the chain until the correct sag is measured.

E. Check for lubrication and, when indicated, apply lubrication to the chain.

F. Operate the drive system for a few minutes and then recheck the sag of the chain. Recheck the chain once again after 24 to 48 h of operation.

Lubrication of the Roller Chain

A. To reduce wear and to ensure long life and trouble-free operation, the roller chain must be lubricated at regular intervals during operation.

B. The joints of the chain must be lubricated and the chain must be kept clean and free of grit and contaminants.

Inspection and Adjustment of the Roller Chain Drive System

A. To ensure trouble-free operation, frequent inspection of the chain and sprocket for wear and alignment must be carried out to detect developing trouble.

B. Check for wear on the inside of the link plates and on the sides of the sprocket; if wear is present, the drive system is out of alignment.

C. Inspect the sprockets for excessive tooth wear; a hooked appearance of the teeth is a signal that the sprocket needs replacement or reversal.

D. From time to time, disassemble the chain and measure it to determine the amount of elongation. Wear resulting in elongation is a normal occurrence, and over time the chain will come to a point where the pitch of each link elongates too much (greater than 3% of its original length), causing the chain to go out of tolerance. When this occurs, the chain must be replaced, since it has reached the end of its life; it is riding higher on the sprocket teeth and will eventually jump off the sprocket teeth, causing damage to the drive system. Never install a new chain on badly worn sprockets, as this will result in instant damage to the chain.

E. As the chain ages, it lengthens (elongates) due to wear between the pins and rollers. Remove links from the end of the chain to maintain the specified sag in the chain.

C3   FLEXIBLE COUPLINGS

Flexible-Coupling Alignment

A. Connecting two axial shafts (end to end) with a flexible coupling requires that the center lines of the two shafts be closely aligned ($\pm.005$ in) along the same axis.

B. Carry out the alignment by mechanically moving the two machine elements (by shimming or turning the base adjusters) to bring the two shafts into vertical and horizontal alignment. The direction and amount of movement of the two independently mounted shafts is determined by measuring the amount of misalignment with a straightedge and taper and feeler gauges and/or a dial indicator. Recently, laser alignment systems using computer interpretation are being used to align the center lines of the two axial shafts to within $\pm.0005$ in.

Inspection and Maintenance of Flexible Couplers

A. Check the tightness of the setscrews or bolts (using a torque wrench), and inspect to see that the keys are correctly positioned in the keyway of the couplers.

B. Regularly check the level and condition of the lubricant in the flexible couplers that require lubrication.

C. Inspect flexible couplers that use rubber, neoprene, polyurethane, etc., for wear (cracks). When wear is found, check the alignment of the shafts and the tightness of the equipment base bolts. If the wear is severe, then replace the coupler.

C4   BEARINGS

Motor Bearing Lubrication

A. When fractional-horsepower motors with sleeve bearings are lubricated with oil, the oil should be light machine oil (SAE 10 nondetergent) applied to the oil port of the bearing on a regular, periodic basis, as recommended by the motor manufacturer. Only a few drops of oil are necessary at each oiling, since excess oiling must be avoided. Do not attempt to lubricate small motors that do not have an oil port, because these motors are lubricated for life at the time of manufacture.

B. When grease is used for lubrication of sleeve bearings in larger-horsepower motors, NLGI grade 2 grease is commonly used. Before lubricating, wipe the grease fitting with a clean cloth, removing dust and grit from the grease fitting; also remove the grease plug below the bearing and then apply just enough grease from the grease gun to displace the old grease from the bearing. Operate the motor for 5 min before reinstalling the grease plug. Don't overfill the bearing, because excess heating will result from the *churning* (high-speed stirring) of the lubricant.

Ball and Roller Bearing Maintenance

A. When removing antifriction bearings for maintenance or when reinstalling antifriction bearings after maintenance, always clean the shaft, housing, and keyways of dirt and grit. When handling a bearing, never allow your skin to touch the bearing surfaces, because the natural acid from your skin will corrode the bearing metal.

B. Wash bearings in clean solvent and dry them with a lint-free cloth. When using compressed air to dry a bearing, always hold the bearing so it cannot spin. Never spin a bearing with compressed air, as this will damage the bearing and is a very dangerous practice.

C. When repacking an antifriction bearing, always use clean grease. Pack the bearing only half full so that excessive heat is not produced by the churning of the lubricant. Bearing operating temperatures greater than 40°F above ambient temperature should be avoided. Number 2 grease will perform well under normal operating conditions.

D. Always cover exposed bearings to prevent dust and grit from entering the bearing, since any form of contaminant will shorten the life of the bearing.

E. When pressing a bearing on or off the shaft, always support the bearing by the inner ring next to the shaft and never apply pressure to the outer ring, because this will lead to bearing failure.

F. Bearings greater than 4 inches in diameter must be expanded by heating to 200°F in a temperature-controlled oven or in an oil bath before installing onto the shaft. The shaft may also be cooled with dry ice to shrink it in size.

G. The motor case and the bearings supporting the armature in the motor must be grounded when they are driving belt-driven equipment to prevent the buildup and discharge of static electricity through the bearings, which will lead to bearing failure.

C5 SOLENOIDS

General Maintenance

A. The plunger of the linear solenoid may become sluggish from the buildup of dirt and grime. When this occurs, remove the plunger and clean it in a gum-reducing solvent. Dry it thoroughly before reinstalling it into the solenoid.

B. When lubrication is called for, never use petroleum-based oils, because this will lead to sludge and gum buildup. Instead, use a silicon lubricant.

C. A pronounced hum (greater than its normal characteristic hum) in an ac solenoid may be due to excessive wear between the plunger and the pole. Replacement may be the only solution to quieter operation.

D. When inspecting an ac solenoid, check to see that the plunger fully seats against the mating pole piece. If it doesn't, the coil will overheat. Remove any buildup of dirt that might be present and adjust the linkage associated with the solenoid to ensure proper seating.

C6   MAGNETIC CLUTCHES

General Maintenance

A. Once installed and aligned, a new clutch is *burnished* (worn in) to ensure full torque capacity and long life. Burnishing of the armature plate and the rotor surfaces is carried out by operating the clutch under load at a reduced current (30% to 50% of normal rated voltage) and low speed (150 rev/min) for 2 to 3 min, causing the surfaces to slip continuously and wear into each other.

B. Never lubricate the friction surfaces.

C. Under normal operation, do not allow clutches to slip for extended periods of time, because this will shorten the clutch life due to overheating.

D. Keep the clutch properly ventilated and free from oil and grime.

C7   CONTROL RELAYS

Contact Maintenance

A. The industrial class of control relays has provisions for replacing worn, burnt, or welded contacts with new sets of contacts in the form of a replaceable modular unit. Additionally, the coil of the industrial control relay is also a replaceable modular unit.

B. General-purpose control relays are designed to plug into a socket. Since many of these are sealed, they are simply unplugged and replaced with a new relay when they malfunction.

C8   MOTOR CONTACTORS/STARTERS

General Maintenance

A. In order to prevent the contactor contacts from becoming welded together, refrain from excessive numbers of starts of large motors or repeated attempts to start a motor after the overload relay is reset. When welded contacts are found, they may be separated and burnished and put back into service. However, excessively burned, oxidized, or pitted contacts must be replaced with sets of replacement contacts supplied by the manufacturer of the motor contactor/starter.

B. If the motor is not overloaded and the overload relays operate frequently, then replace the heaters in the thermal overload relay portion of the motor starter.

C9   1-ϕ, SQUIRREL-CAGE INDUCTION MOTORS

General Maintenance

A. Motor types using a starting capacitor (capacitor-start and capacitor motors) are predisposed to several areas of failure requiring routine replacement and/or maintenance. The two areas of weakness in these motors are the starting capacitor and the centrifugal switch or current relay used to disconnect the start winding and the ac electrolytic capacitor (starting capacitor) from the line once the motor is started. An open starting capacitor will also open the start winding, preventing the motor from starting. When replacing the ac electrolytic capacitor (starting capacitor), the same capacitance (μF) and working voltage must be used.

B. Since split-phase, capacitor-start, and capacitor single-phase induction motors use a centrifugal switch in the starting circuit, each type may experience *failure to start*, caused by the switch not closing, or *failure to run*, due to the switch not opening once the motor has started. When either condition is found, the centrifugal switch mechanism or the contacts may be the cause. Welded contacts (caused by repeated starting and stopping of the motor) prevent the switch from opening, whereas a dirty and gummy mechanism may prevent the switch contacts from either closing or opening. Open contacts may result from badly oxidized contacts. Each of these conditions may be remedied by cleaning the mechanism and/or cleaning and burnishing the contacts. If these measures fail to solve the starting problem, then replacement of the centrifugal switch may be necessary. On rare occasions, the problem may be an open or shorted start winding, which may be determined by measuring the resistance of the winding.

C. Because of the low starting torque of the shaded-pole motor, it may experience slow or sluggish starting when the bearing lubrication becomes gummy due to dust in the oil or when there is insufficient lubrication and the bearings are running dry. Disassembly, cleaning, and relubrication of the bearings will solve this problem.

D. When a motor that is protected by an internal thermal protector fails to reset after sufficient cool down, the open thermal protector must be replaced. Before removing the protector, check the thermal switch for continuity using an ohmmeter. Always replace with the same type (i.e., automatic or manual) of thermal protector.

C10   3-φ, SQUIRREL-CAGE INDUCTION MOTORS

General Maintenance

A. Because of the robust nature of the three-phase, squirrel-cage induction motor, it normally requires very little maintenance; since it has no brushes, commutator, slip rings, or capacitors. However, when the induction motor is torn down for maintenance, examine the rotor bars for tightness and any evidence of scraping, which is an indication that the rotor is out of round due to bearing wear or that foreign material has been ingested into the air gap between the rotor and stator.

B. Check the fan blades for dirt buildup and cracked, chipped, or broken blades. Remove dirt and replace the fan if it is found to be defective.

C. Inspect the inside of the motor and remove any foreign material that has been drawn into the motor.

C11   WOUND-ROTOR INDUCTION MOTORS

Brush Replacement and Slip Ring and Rheostat Maintenance

A. Inspect the slip rings for dirt, roughness, burn spots, or lack of concentricity (roundness). Clean the slip rings with a burnishing stick. Clean dust and carbon particles from around the collector rings and the wiring. When the motor is operating properly, the surface of the rings should have a soft tarnish-brown patina (copper oxide color).

B. Check the condition of the brushes and replace burned (overheated) or worn brushes. When installing new brushes, contour them to the arc of the slip rings by seating them using sandpaper.

C. When it is adjustable, set the brush pressure by tensioning the brush springs to the manufacturer's specifications with the aid of a tension gauge.

D. The rotor rheostat must change values smoothly from a high value to zero. When dead spots or erratic operation are noted, the rheostat must be replaced. The rheostat may be tested by measuring its operation with an ohmmeter.

C12   DC MOTORS

Brush Replacement and Commutator Maintenance

A. In commutator motors, brush replacement is a common maintenance procedure. The brushes are changed by removing and disconnecting them from the motor. New brushes are installed in their holders, reconnected to the motor, and then *seated* on (contoured to) the commutator by slipping a piece of sandpaper between the brushes and the commutator. By manually rotating the armature shaft, the brushes are sanded to the shape of the arc of the commutator. Fine sandpaper is used for the final seating of the brushes.

B. When adjustable, set the brush pressure by tensioning the brush springs to the manufacturer's specifications with the aid of a tension gauge.

C. Inspect the surface of the commutator for excess wear and buildup of contamination. A properly operating commutator will have a smooth coating of copper oxide called *patina*, which appears as soft, light tarnish brown. If a hard dark glaze is present, it is the result of oil or grease getting on the commutator and collecting dust and carbon particles. Remove this glaze, along with any burned spots or roughness, using a commutator burnishing stick.

D. Inspect the mica insulation between each commutator bar. The mica must be below the top surface of the bars. As the copper wears, the mica is brought up to the top level of the bars. When this happens, the mica must be *undercut* using an undercutting tool designed for this task.

E. Some sparking between the brushes and commutator is a common occurrence. However, when the sparking is extreme, it must be corrected, and it may be due to high mica, worn brushes, a dirty commutator, improper brush pressure, a worn or faulty brush holder, or a raised commutator bar.

Answers to Chapter Exercises

CHAPTER 1
Exercise 1.1
1. 863.375×10^3
2. 55.1×10^{-5}
3. 9.35×10^{-3}
4. 18.45×10^2
5. 3.25×10^2
6. 0.78×10^{-7}
7. 52.45×10^{-2}
8. 2.875×10^3
9. 3.95×10^{-3}
10. 62.885×10^3
11. $1.952\,47 \times 10^3$
12. 3168×10^{-5}
13. 0.2903×10^4
14. 5180.7×10^{-6}
15. 1490.2×10^{-2}

Exercise 1.2
1. 0.0235
2. 2 058 000
3. 0.000 2828
4. 12 420
5. 0.006 935
6. 601 500
7. 8.62
8. 0.000 3486
9. 982.2
10. 5850
11. 0.000 1486
12. 88.31
13. 0.0252
14. 362
15. 0.001 486

Exercise 1.3
1. 5.328×10^3
2. 165.875×10^3
3. 75×10^{-6}
4. 58.0×10^{-3}
5. $2.500\,875 \times 10^6$
6. 8.2×10^{-3}
7. 6×10^{-3}
8. 27.125×10^3
9. 437×10^{-6}
10. 0.8435×10^3
11. $1.326\,28 \times 10^3$
12. $6.90\,10^{-3}$
13. 300×10^{-3}
14. $0.902\,00 \times 10^3$
15. 63.0800×10^3

Exercise 1.4
1. 3.205×10^3
2. $1.658\,36 \times 10^5$
3. 7.5×10^{-5}
4. 5.80×10^{-3}
5. $2.506\,258 \times 10^6$
6. 5.550×10^3
7. 6.83×10^{-2}
8. 2.7130×10^4
9. 4.37×10^{-4}
10. $5.780\,22 \times 10^5$
11. 8.29×10^{-3}
12. 8.38×10^{-1}
13. 1.328×10^2
14. 4.078×10^2
15. 1.302×10^{-2}

Exercise 1.5
1. 3.95 min; approximate
2. 3 wheels; exact
3. 100 cm; exact
4. 2 ends; exact
5. 545 mi/h; approximate
6. 74.3°F; approximate
7. 24 teeth; exact
8. 1.125-in dia.; approximate

Exercise 1.6

		Precise to:
1.	.345 ft	Three significant digits
2.	.0589 in	Three significant digits
3.	3.005 cm	Four significant digits
4.	$5\bar{2}00$ lb	Three significant digits
5.	40 600 m	Five significant digits
6.	.00032 in	Two significant digits
7.	24.008 kg	Five significant digits
8.	9024.06 mi	Six significant digits
9.	2000 lb	Four significant digits
10.	0.0075 mm	Two significant digits

	Precise to:
11. 1.30×10^{-3} m	Three significant digits
12. 400×10^{-2} kg	Three significant digits

Exercise 1.7

1. 455
2. 88.0
3. 45.0
4. 0.005 21 or 5.21×10^{-3}
5. 56.6
6. 1.99×10^3
7. 0.309
8. $20\overline{0}0$ or 2.00×10^3
9. 0.0392 or 39.2×10^{-3}
10. 7.00
11. $48\overline{0}\,000$ or 480×10^3
12. 0.0700 or 7.00×10^{-2}
13. 5.20×10^2 or 520
14. 3.00×10^3 or $30\overline{0}0$
15. 8.00×10^{-3}
16. 5.05×10^3
17. 1.89×10^{-3}
18. 3.00×10^6

Exercise 1.8

1. 520; unit
2. 0.32; hundredths
3. 2.725; thousandths
4. $928\,5\overline{0}0$; tens
5. 88; units
6. 0.0442; ten thousandths
7. 45.68
8. 60 900
9. 0.9
10. 0.32
11. 176.27
12. 70
13. $1\overline{0}$
14. 15.20
15. 34×10^6
16. -150×10^3

Exercise 1.9

1. 3.1×10^2
2. 75.83
3. 62.8

4. 5.000
5. 0.24
6. 2200
7. 0.911
8. 75
9. 37×10^{-3}
10. 2 120 000

Exercise 1.10

1. 4.5
2. 20
3. 2.7
4. 6593
5. 0.048
6. 11.18
7. 7.7×10^2
8. 4.4×10^2
9. 86

Exercise 1.11

1. 0.003 125 N
2. 47 W
3. 12 000 s
4. 0.014 m
5. 234 000 J
6. 0.525 A
7. 32 m/s
8. 200 kg

Exercise 1.12

1. 5.820 kJ
2. 0.75 Ms
3. 92 mA
4. 3.325 kW
5. 450 μs
6. 2.5 mm
7. 4.527 kN
8. 750 nA
9. 0.52 m
10. 182 A
11. 335 000 N
12. 0.0025 g
13. 0.000 518 s
14. 83 200 000 W
15. 0.30 mA
16. 4.00 kW
17. 0.16 Mm
18. -155 μs
19. 0.24 g
20. 0.61 m

Exercise 1.13

1. 126 in
2. 3.39 lb
3. 37 ft/s
4. .56 yd^2
5. 2.273 hp
6. 719 mm
7. 5.06 kWh
8. 0.850 g
9. 7.75 m/s
10. $28\,\overline{0}00$ cm^2
11. 93.52 m
12. 2.81 lb
13. 82.9 J
14. 157 mi/h
15. 9.53 mm

CHAPTER 2

Exercise 2.1

1. $F = 9.5$ kN
2. $m = 45$ slug
3. $a = 2.8$ m/s^2
4. $F = .56$ lb
5. $F = 15$ N
6. $m = 45.9$ kg
7. $a = 3.0$ m/s^2
8. $F = 3.6 \times 10^2$ lb

Exercise 2.2

1. **(a)** $v = 18.3$ ft/s
     (b) $v = 12.5$ mi/h
2. **(a)** $a = 104$ km/h/s
     (b) $a = 28.9$ m/s^2
3. **(a)** $v = 66.0$ ft/s
     (b) $a = 3.00$ mi/h/s
     (c) $s = 660$ ft
4. **(a)** $s = .32$ ft
     (b) $t = 0.39$ s
5. **(a)** $t = -4.35$
     (b) $s = 205$ ft
     (c) $v_f = 25.2$ ft/s
6. $a = -4.00$ ft/s^2
7. **(a)** $a = 0.35$ m/s^2
     (b) $F = 11$ N
     (c) $F = 16$ N
8. **(a)** $t = 0.400$ s; $s = 0.100$ m
     (b) $t = 0.400$ s; $s = 0.100$ m
     (c) $t = 3.60$ s; $s = 1.80$ m
     (d) $t_{\text{total}} = 4.40$ s

CHAPTER 3
Exercise 3.1
1. 173 ft-lb
2. 3.81 kJ
3. $\eta = -0.817$
4. $W_{out} = 8.4 \times 10^2$ ft-lb
5. $W_{lost} = 13$ J
6. AMA = 8.6
7. $\eta = 0.65$
8. IMA = 1.9×10^2
9. (a) IMA = 3.0
   (b) AMA = 1.80
   (c) $\eta = 0.60$
10. (a) IMA = 1.3
    (b) AMA = 1.1
    (c) $\eta = .83$

Exercise 3.2
1. IMA = 2.6
2. $F_E = 1.1$ kN
3. (a) class 3
   (b) class 2
   (c) class 3
4. $F_R = .60$ lb
5. $F_E = 66$ mN
6. $F_E = .30$ lb
7. (a) class 2
   (b) class 3
   (c) class 2
8. $F_E = 90$ lb
9. (a) $F_E = 0.51$ kN
   (b) 194 N · m = 194 N · m
10. $F_R = 54$ lb

CHAPTER 4
Exercise 4.1
1.
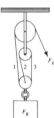

2.
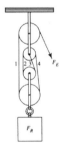

3. (a) IMA = 1
   (b) AMA = 0.95
4. (a) $n = 2$ and IMA = 2
   (b) AMA = 1.63
   (c) $\eta = 0.815$
5. (a) $F_E = 83$ lb
   (b) $s_E = 15$ ft
   (c) AMA = 1.79
   (d) $\eta = 89.5\%$
6. (a) $s_R = 2.50$ ft
   (b) $\eta = 0.65$
   (c) $F_E = 2\overline{0}0$ lb

Exercise 4.2
1. IMA = 29
2. (a) IMA = 2.4
   (b) AMA = 2.16
   (c) $\eta = 0.90$
3. (a) $r_w = 60$ cm
   (b) $F_E = 30$ N
4. (a) IMA = 7.3
   (b) $\eta = 0.64$
5. (a) AMA = 5.3
   (b) $d_a = .88$ in
   (c) IMA = 7.4
6. (a) IMA = 66
   (b) $s_E = 99$ ft
   (c) $F_E = 68$ lb

Exercise 4.3
1. (a) IMA = 7.8
   (b) $F_E = 35$ lb
2. $F_E = 34$ lb; 32 lb is not enough to move the crate.
3. (a) IMA = 2.5
   (b) $F_R = 11$ kN
4. (a) IMA = 2010
   (b) $F_E = 3.5$ lb
5. (a) IMA = 940
   (b) FE = 25 N
   (c) $\eta = 5.1\%$
6. $r = 0.15$ m

Exercise 4.4
1. $\eta_{total} = 0.36$
2. (a) $n = 3$ and IMA = 3
   (b) IMA = 11
   (c) IMA = 3.0
   (d) $F_E = 7.1 \times 10^2$ lb
3. (a) $\eta_{total} = 0.48$
   (b) $F_E = 1.5 \times 10^3$ lb
4. (a) $IMA_{total} = 1.5$
   (b) $F_E = 6.7$ N
5. (a) $IMA_{w\&a} = 11$
   (b) $IMA_{cam} = 6.3$
   (c) $IMA_{total} = 69$
6. (a) $IMA_{lever} = 4.0$
   (b) $IMA_{inc\ pl} = 6.0$
   (c) $IMA_{total} = 24$

CHAPTER 6
Exercise 6.1
1. $W = 1200$ ft-lb
2. (a) $F_E = F_f = 44$ N
   (b) $W = 530$ J
   (c) $W_{effort} = W_{frict} = 530$ J
   (d) Zero, since the crate was not lifted.
3. $F = 6.6$ lb
4. KE = 4.43 kJ
5. (a) $W = 110$ ft-lb
   (b) $c_{rr} = .12$ in
6. (a) $W = 180$ ft-lb
   (b) $s_E = 27$ in
   (c) $F_E = 80$ lb

Exercise 6.2
1. 9.0 kW
2. 8.95 hp
3. $P = 1900$ ft-lb/s
4. $P = 18$ kW
5. $\eta = 0.86$
6. 1.3×10^5 gal/h

Exercise 6.3
1. (a) $\tau = 31$ lb-ft
   (b) $\tau = 42$ N·m
2. $\tau = 38$ lb-ft
3. $s = 0.31$ m
4. $F = 310$ N
5. $F = 3500$ lb
6. (a) $F = 800$ lb
   (b) $\tau = 740$ lb-ft

Exercise 6.4

1. **(a)** $\theta = 4.0$ rad
   (b) $\theta = 230$
   (c) $\theta = 0.64$ rev
2. 38 rad/s
3. **(a)** $\omega = 11.0$ rev/min
   (b) $\omega = 1.15$ rad/s
4. **(a)** $v = 40$ ft
   (b) $v = 12$ m/s
5. **(a)** $\alpha = 76$ rad/s^2
   (b) $\theta = 220$ rad
6. **(a)** $a = .543$ ft/s^2
   (b) $\omega = 190$ rev/min
   (c) $\alpha = 1.1$ rad/s^2

Exercise 6.5

1. $P = 98$ hp
2. $P = 4100$ W
3. $\tau = 1.50$ lb-ft
4. $\tau = 660$ lb-ft
5. $\omega = 80$ rev/min
6. **(a)** $P_{out} = 23$ hp
   (b) $\eta = 0.72$

CHAPTER 7
Exercise 7.1

1. IMA = 0.20 or 0.20:1.0
2. AMA = 0.17
3. $\tau_d = 2.5$ lb-ft
4. $P_{in} = 91.2$ ft-lb/s
5. $\omega_D = 2880$ rev/min
6. $\tau_d = 3.5$ lb-ft
7. Gear A: $N = 26$ teeth,
   CD = .6875 in;
   Gear B: $PD = .5625$ in,
   DP = 32, ccw,
   $\omega = 118$ rev/min;
   Gear C: $N = 29$ teeth,
   DP = 48, ccw,
   $\omega = 118$ rev/min;
   Gear D: $PD = .3750$,
   CD = .4896 in, cw,
   $\omega = 190$ rev/min
8. Gear A: $N = 36$ teeth,
   CD = .5625 in;
   Gear B: $PD = .3750$ in, cw,
   $\omega = 360.0$ rev/min,

DP = 48;
Gear C: $N = 50$ teeth, cw,
   $\omega = 360.0$ rev/min,
   CD = .6600;
Gear D: DP = 50, ccw,
   $\omega = 1125$ rev/min;
Gear E: $N = 28$ teeth, ccw,
   $\omega = 1125$ rev/min,
   CD = 1.063;
Gear F: $PD = 1.250$ in, cw,
   $\omega = 787.5$ rev/min,
   DP = 32

Exercise 7.2

1. $PD = 1.67$ in
2. IMA = 2.7; Pulley ratio, 2.7:1
3. $l = 300$ mm
4. AC = 174°; correction factor
   = 0.99; corrected power
   rating = 1.70 hp
5. **(a)** SF = 160%
   (b) drive system power rating =
   .53 hp
6. **(a)** $\omega_d = 345$ rev/min
   (b) $\omega_D = 1035$ rev/min
   (c) IMA = 10; the pulley ratio
   is 10:1.
   (d) direction = ccw

Exercise 7.3

1. heavy, double strand, general
   duty, 1.750-in pitch
2. The number 25 single-strand
   roller chain is rollerless.
3. IMA = 0.25; sprocket ratio =
   1:4
4. $\omega_d = 3100$ rev/min
5. $PD = 2.15$ in
6. IMA = 2.7; sprocket ratio =
   2.7:1

CHAPTER 9
Exercise 9.1

1. EMF = 120 V
2. $Q = 22.5$ kC

3. $T = 16.7$ ms
4. $\omega_D = 3600$ rev/min
5. $W = 108$ kJ

Exercise 9.2

1. $P = 4.8$ W
2. $R = 0.720$ Ω
3. **(a)** $R_{AWG} = 12.9$ Ω
   (b) $R = 15.6$ Ω
4. $P = 400$ W
5. $T = 80.3$°C
6. ampacity = 33 A;
   overcurrent protection = 30 A

Exercise 9.3

1. $X_L = 45$ Ω
2. $X_C = 3.9$ kΩ
3. **(a)** $E_L = 268$ V
   (b) $E_s = 276$ V
4. $Z = 18.2$ Ω
5. $I = 1.0$ A

Exercise 9.4

1. **(a)** $\cos \theta = 0.95$
   (b) lagging
2. $P = 3.16$ kW
3. $P = 2.0$ kW
4. $P_{lost} = 900$ W
5. **(a)** 11.2 kW
   (b) $P_{in} = 13.2$ kW
   (c) $I = 31$ A

CHAPTER 10
Exercise 10.1

1. $\phi = 0.13$ μWb
2. $B = 0.67$ T
3. mmf = 74.4 A (amp-turns)
4. $N = 250$ turns
5. $\mu = 0.5$ mWb/A·m

Index